INSECTS AND OTHER TERRESTRIAL ARTHROPODS: BIOLOGY, CHEMISTRY AND BEHAVIOR

THE ENTOMOLOGICAL GUIDE TO *RHIPICEPHALUS*

INSECTS AND OTHER TERRESTRIAL ARTHROPODS: BIOLOGY, CHEMISTRY AND BEHAVIOR

Additional books and e-books in this series can be found on Nova's website under the Series tab.

INSECTS AND OTHER TERRESTRIAL ARTHROPODS:
BIOLOGY, CHEMISTRY AND BEHAVIOR

THE ENTOMOLOGICAL GUIDE TO *RHIPICEPHALUS*

SACHIN KUMAR
RAQUEL COSSIO BAYUGAR
ANIL KUMAR SHARMA
ESTEFHAN MIRANDA MIRANDA
AND
ASHOK KUMAR CHAUBEY
EDITORS

nova
science publishers
New York

Library of Congress Cataloging-in-Publication Data

ISBN: 978-1-53619-619-1
Names: Kumar, Sachin (Veterinary research), editor.
Title: The entomological guide to *rhipicephalus* / Sachin Kumar, Raquel Cossio Bayugar, Anil Kumar Sharma, Estefhan Miranda Miranda, Ashok Kumar Chaubey, editors.
Description: New York : Nova Science Publishers, [2021] | Series: Insects and other terrestrial arthropods: biology, chemistry and behavior | Includes bibliographical references and index. |
Identifiers: LCCN 2021019806 (print) | LCCN 2021019807 (ebook) | ISBN 9781536196191 (hardcover) | ISBN 9781536196351 (adobe pdf)
Subjects: LCSH: *Rhipicephalus*. | Tick-borne diseases in animals. | Ticks--Control.
Classification: LCC QL458.2.I9 E58 2021 (print) | LCC QL458.2.I9 (ebook) | DDC 595.4/29--dc23
LC record available at https://lccn.loc.gov/2021019806
LC ebook record available at https://lccn.loc.gov/2021019807

Published by Nova Science Publishers, Inc. † New York

CONTENTS

PREFACE

Ticks are among the most competent and versatile vectors of pathogens and are second to mosquitoes as vectors of a number of human pathogens. They are the most important vector of pathogens affecting cattle worldwide. Problems with tick-borne diseases were related to the introduction of improved breeds of cattle into tick-infested areas because of their greater productivity compared to well-adapted indigenous breeds. The global loss due to ticks and tick borne diseases (TTBDs) was estimated to be between \$13.9 and \$18.7 billion annually while in India the cost of controlling TTBDs has been estimated at \$498.7 million/annum. Also, cattle infested with ticks and infected with tick-borne disease agents were moved into areas where these tick species had not previously existed. This book is written by an international collection of tick experts of prestigious organizations and covers in-depth information on different aspects of ticks i.e., biology, acaricide resistance, tick-borne diseases, tick management strategies etc. It is a valuable resource for students, academic researchers and professionals because it covers the whole range of ticks and tick-borne diseases. This handbook was assembled through the efforts of five editors and the book chapters' authors, each of whom contributed to different components of the handbook.

Chapter 1 - In Brazil, the ixodidic fauna to date is composed of 75 species, distributed in nine genera, being: *Antricola* Cooley and Kohls (3 species), *Argas* Latreille (1 species), *Nothoaspis* Keirans and Clifford (1 species), *Ornithodoros* Koch (19 species), *Amblyomma* Koch (33 species), *Dermacentor* Koch (1 species), *Haemaphysalis* Koch (3 species), *Ixodes* Latreille (12 species), *Rhipicephalus* Koch (2 species). The species that parasitize domestic animals are the subject of numerous researches in the country, mainly those of the genus *Rhipicephalus*, which until now are: *Rhipicephalus sanguineus* s. l and *Rhipicephalus microplus*. *R. sanguineus*, known as the red dog tick, is a trioxene tick, found in dogs, and can parasitize other mammals. It is a vector of pathogens such as *Babesia canis*, *Ehrlichia canis*, *Hepatozoon canis*, *Rickettsia rickettsii* among others. Due to the proximity of dogs to humans, this tick represents a high risk to public health. *Rhipicephalus microplus* is a monoxene tick and has a preference in parasitizing cattle. It is widely distributed in the Brazilian territory, being the species of economic interest for causing great damages in the cattle industry, besides being transmitter of infectious agents responsible for the cattle tick fever.

Chapter 2 - Tick-borne diseases (TBDs) have a global economic impact and researchers are particularly concerned about their control and prevention. The aim of this chapter is to provide a brief introduction to some of the common tick-borne pathogens, vectors responsible for their transmission, geographical distribution of these TBDs and the risk factors that are associated with these diseases especially in developing countries. The main focus of this chapter is on theileriosis, anaplasmosis, babesiosis, ehrlichiosis and heartwater as these diseases affect the milk, meat and hide production causing significant losses to the livestock sector.

Chapter 3 - Ticks and tick-borne diseases (TBDs) are responsible for financial constrains for livestock sector in tropical and subtropical regions of the world and ultimately affects the livelihoods of poor farming communities as they are exclusively dependent on their animals. Livestock sector has become an integral part of Pakistan's economy and large number and variety of ruminants are reared, by the owners, in order to meet an increasing demand of milk, meat and hide products. It is timely to review current status of TBDs in Pakistan with a note on their vector ticks in order to identify the gaps in the knowledge of TBDs and their control. Anaplasmosis, babesiosis and theileriosis are the three major TBDs that are reported from various parts in Pakistan. This article will provide a brief background on key TBDs and ticks and reviews the current status of ticks and TBDs in Pakistan. This extensive review of literature will add to the understanding of these diseases and will propose areas for future research and will draw attention to the need for improved tools for the diagnosis and control of TBDs in Pakistan.

Chapter 4 - Ticks and tick borne diseases are a major threat to the livestock and economy of developing countries including India. Ticks, as vectors of several zoonotic diseases, are second only to mosquitoes as vectors. The diseases spread by ticks are a major constraint to animal productivity while causing morbidity and mortality in both animals and humans. *Rhipicephalus (Boophilus) microplus* and *Hyalomma anatolicum* have been recognized since long as vectors of importance for transmitting diseases like babesiosis, theilieriosis and CCHF. There is need to recognize the importance of ticks as vectors in India and devise better methods for their control. This review is focused on the major tick borne animal and human diseases in India and the current situation in India regarding control of ticks including the development of resistance, progress in tick vaccine and the development of potential phytoacaricides as alternatives to chemical acaricides.

Chapter 5 - Ticks are the global problems, most important vectors of Rickettsia and protozoa (*Anaplasma*, *Babesia* and *Theileria*), rickettsiae (*Ehrlichia, Yersinia pestis, Francisella tularensis, Coxiella burnetii)*, viruses (Tick-borne encephalitis, Nairobi sheep disease virus, louping ill virus, Bhanja viruses, west Nile virus, African swine fever virus) and spirochaetes (*Borrelia burgdogferi* and *B. hermsii*). Several efforts have been made to eradicate ticks and tick-borne diseases by using various chemical control strategies in many developing countries. However, problems associated with the continuous usage of these acaricides (Organochlorides, macrocyclic lactones, pyrethroides, carbamates, organophosphates, formamidines) Integrated tick management is recommended. Various plants e.g., neem tree (*Azadiratcha indica*), African spider (*Gynandropsis gynandra*), tamarind tree (*Tamarindus Indica*), Jamaica pepper (*Pimenta dioica*), and wormwood (*Artmisia brevifolia*) have been evaluated for their anti-tick properties. So due to having folded benefits there are more opportunities for testing and applications of these plants as anti-tick. Futuristic approach involves the development of combination of chemical and vaccine strategies, directed towards the both tick eradication and transmission. New emerging technologies should be used for gene discovery, cocktail vaccines formulation and delivery. Usage of endosymbiotic approach, role played by *Buphagus erythrorhynchus* for tick control and fungal biocontrol agents will be used in near future research for the control of tick vectors.

Chapter 6 - Ticks exert a huge effect on animal health through blood-sucking and also, they are capable of transmitting a wide range of pathogens. The study revealed that about 80% of the world cattle population is at risk of ticks associated diseases. *Rhipicephalus microplus* is the most prevalent tick in most of the Indian states. Losses caused by bovine tick burdens in tropical countries have a tremendous economic impact on production systems. The control of *R. microplus* is mostly based on the use of conventional chemical acaricides. However, such a control measure has been found to be ineffective and also involving large costs. Therefore, need for the development of alternative approaches, management of animal husbandry practices, manual removal of ticks, selection for host resistance, nutritional management, selection of cattle to increase the frequency of alleles associated with tick resistance, differences in immune response profile, and associated genes, etc. have come to center stage. The phenotype(s) could also be used to improve tick control by using cattle management. Differences in the abilityof cattle to become resistant to ticks are well documented and known by the livestock keepers. Indigenous cattle those are native to India are known tick resistant individuals. Resistance to ticks is a heritable character and several workers have estimated the heritability for resistance to ticks.

Individual selection for tick resistance can be done in three possible ways; selection under normal environmental conditions, selection under challenging environment, and an indirect approach with the use of indicators of resistance can be done. Resistant animals can then be multiplied through natural breeding as well as the use of reproductive bio-techniques. After multiplying for a number of generations a resistant population can be obtained. These measures can be adopted as they are cost-effective and reliably measures the cattle tick resistance in thousands of individual animals in commercial and smallholder farmer herds in tropical and subtropical areas globally. The selection of animals for tick resistance has gained momentum as studies have indicated a good heritability estimate for tick burden in cattle, which may result in the success of the breeding program for tick resistance. This measure will yield cattle that are permanently resistant to ticks and also it may facilitate select animals at a younger age, to reduce generation interval and increase genetic gain, etc.

Chapter 7 - Currently, chemical acaricide use remains the most common method of *Rhipicephalus microplus* control; however, its excessive use has led to the appearance of tick strains resistant to the main classes of conventional acaricides. When there is a lack of response to tickticide treatment, the farmers' reaction is to increase the frequency of treatments and the dose of acaricides, which increases the cost and selective pressure for resistance. Due to the above, there has been a need to implement strategies to intelligently use acaricides and thus increase their useful life and delay the appearance of resistance. This chapter describes the main classes of conventional acaricides, growth inhibitors, and macrocyclic lactones used in the control of *R. microplus* ticks. Additionally, the most common strategies that involve the use of chemical acaricides for tick control are described, such as the use of synergized acaricides, acaricide mixture formulations, use of growth inhibitors and development of new acaricides with mechanisms of action that are different from conventional ones. It is clear that the use of chemical acaricides as the only form of tick control is a mistake. Therefore, the authors must consider that tick control should include the use of chemical acaricides together with a systematized combination of strategies and technologies for integrated tick control.

Chapter 8 - The cattle tick infestations and the bovine infectious diseases transmitted by them, are controlled mainly by the massive use of acaricides on the livestock, however, this strategy has led to the emergence of tick resistance to most acaricide formulations, and the lack of acaricide effectiveness has become a severe threat to the strategies to control of ticks and tick borne diseases, consequently, it is a necessity to monitor the spread of acaricide resistancein order to design preventive and alternative measures of tick control in endemic areas, for a long time this has been done by toxicological bioassays using living field-isolated ticks of various development stages exposed to the different acaricide formulations, these procedures are time-consuming and expensive because they require abundant specialized labor and facilities, therefore, the use of fast molecular test is required to overcome the disadvantages of outdated bioassays. The design of a molecular test requires in the first place, the identification of the molecular mechanism of acaricide resistance in the cattle tick to the different acaricide chemical formulations, acaricide resistance is a genetically inherited phenomenon that involves the expression products of one or several genes. Two major genetic-driven mechanisms are recognized nowadays, namely: a) enhanced metabolic detoxification involving an activity increase of one or several xenobiotic metabolizing enzymes, b) altered target sites involving DNA mutations of the genes that code for the proteins that are targeted by the acaricide, this chapter describes themolecular and biochemical mechanisms that cause resistance to the most common acaricides used for tick control.

Chapter 9 – The authors describe the selection, culture, maintenance and rearing of the cattle tick *Rhipicephalus* (*Boophilus) microplus* at CENAPA-SENASICA-SADER, Mexico, where reference susceptible and acaricide-resistant tick strains are reared and toxicologically tested. The purpose of maintaining these tick strains is to evaluate the efficacy of diverse acaricides prior to requesting commercial permits to the Mexican Agriculture Ministry. Ticks are also used for assessments of acaricide resistance and compared to field-isolated ticks in acaricide bioassays, such as the larval package test and engorged female test. Five reference cattle tick strains are maintained for these purposes, and all of them are constantly toxicologically characterized. A complete description of their susceptibility and resistance to acaricide formulations, such as organophosphates, pyrethroids, amitraz, fipronil and ivermectin, is provided along with detailed procedures of acaricide bioassays and strain provenance.

Chapter 10 - In India, *Rhipicephalus microplus* is the most economically important tick species, widely prevalent and infesting animals. The infestation causes many problems in animals including skin damage, blood loss, fatal diseases like babesiosis and anaplasmosis, reduced growth and milk production. Frequent application of chemical acaricides on hosts is the most common tick management method adopted by the animal owners. Indiscriminate use of different generations of chemical acaricides during the last several decades has led to the development of many problem linked with environment, human, animal health and development of resistant tick populations. During the last ten years, the resistance to many commonly used chemical acaricides in tick populations collected from different states of India has been reported. Along with organophosphates, synthetic pyrethroids and amidines, the tick population in India is developing resistance to ivermectin. The rapid and sensitive molecular and biochemical techniques have paved ways for early detection and understanding the possible mechanisms of resistance. The possible mechanisms involving metabolic detoxification mediated by esterases, glutathione S-transferase and monooxygenase as well as point mutations at target sites were reported in Indian ticks. The chapter will provide the status of acaricide resistance in ticks, the possible mechanisms of resistance operating in the tick population, factors contributing to the development of resistance and the managemental strategies.

Chapter 11 - The application of ixodicides has long been the alternative of choice for tick control, however, the use of ixodicides has currently limited efficacy in reducing tick infestations, due to the selection of resistant individuals within tick populations. Acaricide resistance is a growing problem that needs to be addressed, because it is currently affecting the competitiveness of livestock and the economy of thousands of producers in Mexico. The solution to this problem requires more research focused on the use of 'omics' and reverse genetic approaches in order to identify new target genes for diagnostics or immunoprophylactic purpouses, since the improvement of diagnostic tests and a better vaccine, will thus contribute to design better strategies and better tick control programs, in order to better management of acaricide resistance. The aim of this review is to compile different current topics related to the evolution of acaricide resistance and reverse genetics research, in order to make clear how identification of new target genes or genetic traits, can lead to improve Molecular diagnostics, vaccine development and therefore the methods currently used for tick control.

Chapter 12 - Ticks are highly specific obligate haematophagous ectoparasites of mammals, birds and reptiles, distributed globally and are of enormous medical and veterinary relevance owing to the direct damage they cause to their hosts. Moreover, they act as vectors of a large variety of human and animal pathogens. Today, most emerging infectious diseases arise from zoonotic pathogens, and many of them are transmitted by tick vectors. The cattle tick, *Rhipicephalus microplus* is one of the most damaging tick attacking bovines in the tropical and subtropical regions of the world. In this book chapter, the authors would like to depict the prevalence of ticks in the western India and the factors responsible for their predominance in this region. The tick control programme is focused mainly on repeated use of organophosphates, pyrethroids, formamidines, phenylpyrazoles and macrocyclic lactones. However, the application of the acaricides is poorly supervised, resulting in economic problems produced by phenomenon such as acaricide resistance. Single host life cycle with short life cycle and high fecundity, the *R. microplus* has been subjected to maximum exposure of acaricides and thus developed a significant ability to become resistant to the toxic effect of the chemicals used to manage them. Control strategies against ticks rely mainly on judicious use of acaricides. Non-chemical natural product-based tick management protocol has also shown significant success in recent years. The indigenous ethno-veterinary and remedial knowledge of Indian farmers propose a range of herbs having insecticidal and acaricidal properties. The integrated approaches including pasture management with limited applications of effective chemical acaricides, strategic use of phytoformulation, housing management, strengthening of dairy cattle management with alternative feeding technologies and immunization against ticks and ticks borne diseases may be the most viable option for management of acaricide failure. Integrated tick management will certainly reduce the over reliance on acaricides and lessen selection pressure on ticks thereby preserving the chemicals. Theavailabilityofdifferent control measures against ticks, their advantages and disadvantages, and their economic impact will be discussed in this chapter with the hope that the coordinated activities will significantly improve the health status of animals and the environment by minimizing the use of non-effective acaricides which will reduce the economic burden to the farmers.

Chapter 13 - Ticks and tick-borne diseases (TTBDs) of cattle pose a serious threat to the growth of thedairy industry and cause a significant reduction in profit by severe loss in lactation. The problem of TTBDs has been ranked high in terms of their adverse impact on the livelihood of poor farming communities in developing countries including India. The tick, *Rhipicephalus (Boophilus) microplus* is the most widely distributed ticks and serves as the vector of bacterial and protozoan diseases causing heavy economic losses directly and indirectly in animal husbandry. At present, tick control in India is based on large scale repeated use of synthetic acaricides belonging to three major classes, namely organophosphates, carbamates, and pyrethroids. Besides various indirect effects like environmental pollution due tothe large-scale use of synthetic acaricides, there is an emerging problem of thedevelopment of resistance in the ticks to these acaricides. Emphasis should be made on a reliable compound against the ticks which are effective against all the stages of the ticks and does not leave any residues in the environment. Compounds of plant origin constitute an important segment of such effort and have been used in therecent past against all the stages of economically important tick species with encouraging results. This chapter will focus on the different herbs and their active principles which are used to control ticks in India. Several medicinal plants have been screened for acaricidal activity against the female tick, *Rhipicephalus microplus* throughout the world. There are different herbs like *Allium sativum, Carica papaya, Datura stramonium, Azadirachta indica, Calotropis procera, Annona squamosa*, *Cymbopogon winterianus* and many more are found to have anti-tick activity against *R. microplus.* The established plants extracts with acaricidal activity will be eco-friendly and will be beneficial for controlling chemical-resistant tick infestation in animals.

Chapter 14 - It has been well known that some hosts are resistant to the cattle tick *Rhipicephalus microplus*, which occurs due to immunological, physical and mechanical attributes of the skin of resistant animals. On the other hand, the development of resistance in dogs to the tick *Rhipicephalus sanguineus* sensu lato (s.l.) was refuted until some years ago. However, evidences accumulated over recent years indicate that some breeds, such as beagles and probably miniature pinschers, develop resistance against *R. sanguineus*s.l. Ticks that feed on resistant hosts have impairment in their biological parameters, such as reduced numbers of ticks feeding to engorgement, reduced egg production and egg viability. Despite the fact that hosts resistant to ticks have been known since the beginning of the last century, it was recently understood that ticks can distinguish resistant from susceptible hosts using their sense organs. In this chapter the authors will briefly describe immunological features and address the behavioural, electrophysiological and chemical features that are involved in how *R. sanguineus*s.l. and *R. microplus* perceive resistant dogs and bovines, respectively. The use of repellent chemical compounds (allomones) identified on resistant animals to control *R. sanguineus*s.l. and *R. microplus* on susceptible animals will be discussed.

Chapter 15 - Ticks are obligate hematophagous ectoparasites of humans, domestic, and wild animals, and have been recognized as one of the most important vectors of diseases in terms of diversity of pathogen transmission. The organisms known to be transmitted by ticks include viruses, bacteria, fungi, and protozoans. The Ixodidae family (hard ticks) comprises fourteen genera, with *Rhipicephalus* representing the most widespread genus of ticks affecting animals. In this chapter the authors review the medical and veterinary importance of the most important *Rhipicephalus* ticks such as the ones included in the complex *R.*

(Boophilus) microplus; R. appendiculatus, and *R. sanguineus.* The cattle tick *R. microplus*, and the brown ear tick *R. appendiculatus* are the vectors of babesiosis, anaplasmosis, and theileriosis; the most devastating tick-borne diseases that constrain the development of livestock in tropical and subtropical regions of the world. The brown dog tick, *R. sanguineus* is the most widely distributed species of ticks worldwide and is involved in the transmission of several diseases to humans and dogs.

Chapter 16 - Vaccines are the preparation of microorganisms (live, attenuated, or inactivated) or their components, such as proteins, carbohydrates, or nucleic acids, to be administered into an organism to induce immunity. The goal of vaccinations is to protect an individual against future pathogen invasion. The *Rhipicephalus microplus* tick is considered one of the most harmful parasites to cattle in tropical and sub-tropical regions, infecting approximately 80% of livestock worldwide. There is an urgent need to develop control strategies that consider the safety of the environment and animal and human health. Vaccination against *R. microplus* is developed from tick antigens and is regarded as an economical and safe alternative to acaricides, is environmentally friendly, and safe for administration. Monoantigenic vaccines are commercially available, containing the *R. microplus* Bm86 antigen. However, these vaccines' efficacy has been highly variable due to allelic variations in the *Bm86* gene, supporting the need to continue vaccination research.

Several antigens have been evaluated as vaccine candidates, including Bm86, subolesin, Ferritin-2, aquaporins, serpins, vitellin, and VDAC. This chapter will describe the key aspects of the immune response required from vaccine antigens and revise the findings reported from the results of tick antigen vaccines. The authors will address the importance of using reverse vaccinology and immuno-bioinformatics to identify new *R. microplus* antigens isolated from the intestine, salivary gland, and ovary to develop a multiantigenic vaccine against *R. microplus*.

ABOUT THE EDITORS

Dr. Sachin Kumar (Ph.D, NPDF)

Senior Researcher

Indian Veterinary Research Institute, Izatnagar, Bareilly (U.P.), India

I, Dr Sachin Kumar, am working as a Senior Researcher in Division of Parasitology, Indian Veterinary Research Institute, Izatnagar, Bareilly, Uttar Pradesh, India. My PhD work was also focused on cattle ticks and I have been completed Post Doctoral from IVRI, Izatnagar, Bareilly. I am exclusively working on chemical acaricide resistance in cattle ticks and their mechanism. The last 12-13 years, I have developed acaricide resistance data of a number of tick populations collected from different states of India. I have also screened many plant extracts and identified two plants having acaricidal activity. I have developed one natural formulation were developed for the control of tick infestation and validated on resistant tick in field condition. I have published 02 patents and more than 45 international research papers in reputed journal. I have been completed more than 03 projects as a Principal Investigator.

Dr. Raquel Cossio-Bayugar (Ph.D)

Senior Researcher

National Center for Disciplinary Research in Animal Health and Safety

Dr. Raquel Cossio-Bayugar is a senior scientific researcher at the National Center for Disciplinary Research in Animal Health and Safety (CENID-SAI), which is one of 77 research centers at the National Institute for Agricultural, Livestock and Forestry Research (INIFAP) distributed along the geography of Mexico, Dr. Cossio-Bayugar's main area of scientific research is Cattle tick acaricide-resistance, in vitro-tick cell culture, Tick-physiology, tick-immunology and tick response to pathogens, on which she has publish 40 research papers. Cossio-Bayugar obtained a degree in Biomedical research from the National

Autonomous University of Mexico and a Ph.D. degree from Texas A&M University, since then she has been working as a researcher.

Dr. Anil Kumar Sharma (Ph.D)

Senior Researcher
Indian Veterinary Research Institute, Izatnagar, Bareilly (U.P.), India

I, Dr Anil Kumar Sharma, am working as a Senior Researcher in Entomology Laboratory, Division of Parasitology, ICAR-Indian Veterinary Research Institute, Izatnagar, Bareilly (U.P.), India. I am exclusively working on acaricide resistance in ticks and development of herbal acaricides for the management of resistant tick infestations in animals for the last 12 years. I have developed acaricide resistance data of a number of tick populations collected from different states of India. I have also screened many plant extracts and identified two plants having acaricidal activity. Two herbal technologies were developed and validated in field condition. One antitick herbal technology is commercialized. More than 35 international and 02 patents have been published.

Dr. Estefan Miranda-Miranda (Ph.D)

Senior Scientific Researcher
National Center for Disciplinary Research in Animal Health and Safety (CENID-SAI), Maxico

Dr. Estefan Miranda-Miranda has been a senior scientific researcher for 35 years at the National Center for Disciplinary Research in Animal Health and Safety (CENID-SAI), which is one of 77 research centers at the National Institute for Agricultural, Livestock and Forestry Research (INIFAP) distributed along the geography of Mexico, Dr. Miranda-Miranda main area of scientific research is parasite antiparasitic-resistance, Molecular Biology of parasites, Parasite-physiology,. Miranda-Miranda obtained a degree in Parasitology from the Morelos Autonomous University of Mexico and a Ph.D. degree from Texas A&M University, since then he has been working full time as a researcher and published numerous scientific papers on the subject.

Dr. Ashok Kumar Chaubey (Ph.D, FSPPS, FAIR)
Professor
Ch. Charan Singh University, Meerut-250001, Uttar Pradesh, India

Dr. Ashok Kumar Chaubey; born: 1 January 1962, Ballia (U.P.); Graduation, Post-Graduation and Doctoral Degree from Allahabad University, Allahabad. He has published 96 full length papers of National and International Journals of repute; 03 edited books and 237 short communications; guided 11 Ph. D and 52 M. Phil. Students; completed 06 research projects. He has attended 71 National and International conferences, acted as Chairperson and delivered Invited/Keynote lectures. He is recipient of 23 awards and honours notably Dr. J.C. Edward Gold Medal; Plant Pathology Leadership Award, Plant Nematology Leadership Award etc., Fellows of various science societies/organizations viz. Society of Plant Protection Sciences; Nematological Society of India; Academy of Environmental Biology; Zoological Society of India; Applied Zoologists Research Association; Helminthological Society etc. He is also Member, Research Board of Advisers, ABI (USA) and Editorial Board of various Journals. He has visited Australia, South Africa, Netherland, Czech Republic; Dubai and South Korea.

In: The Entomological Guide to *Rhipicephalus* ISBN: 978-1-53619-619-1
Editors: Sachin Kumar, Raquel Cossio Bayugar et al.

Chapter 1

BIOLOGY, ECOLOGY AND IMPORTANCE OF *RHIPICEPHALUS* TICKS IN BRAZIL

Renato Andreotti[1,*], Marcos Valério Garcia[2], Leandro de Oliveira Souza Higa[2] and Francisco Tobias Barradas Piña[3]

[1]Empresa Brasileira de Pesquisa Agropecuária, Embrapa Gado de Corte, Campo Grande, MS, Brazil
[2]Laboratório de Biologia do Carrapato, Embrapa Gado de Corte, Bolsista Fundapam, Campo Grande, MS, Brazil
[3]Instituto Nacional de Investigaciones Forestales Agrícolas y Pecuarias, Campo Experimental La Posta, Mexico City, Mexico

ABSTRACT

In Brazil, the ixodidic fauna to date is composed of 75 species, distributed in nine genera, being: *Antricola* Cooley and Kohls (3 species), *Argas* Latreille (1 species), *Nothoaspis* Keirans and Clifford (1 species), *Ornithodoros* Koch (19 species), *Amblyomma* Koch (33 species), *Dermacentor* Koch (1 species), *Haemaphysalis* Koch (3 species), *Ixodes* Latreille (12 species), *Rhipicephalus* Koch (2 species). The species that parasitize domestic animals are the subject of numerous researches in the country, mainly those of the genus *Rhipicephalus*, which until now are: *Rhipicephalus sanguineus* s. l and *Rhipicephalus microplus*. *R. sanguineus*, known as the red dog tick, is a trioxene tick, found in dogs, and can parasitize other mammals. It is a vector of pathogens such as *Babesia canis*, *Ehrlichia canis*, *Hepatozoon canis*, *Rickettsia rickettsii* among others. Due to the proximity of dogs to humans, this tick represents a high risk to public health. *Rhipicephalus microplus* is a monoxene tick and has a preference in parasitizing cattle. It is widely distributed in the Brazilian territory, being the species of economic interest for causing great damages in the cattle industry, besides being transmitter of infectious agents responsible for the cattle tick fever.

Keywords: cattle tick, brown dog tick, ectoparasite, bovine, livestock

---

* Corresponding Author's Email: renato.andreotti@embrapa.br.

INTRODUCTION

Every year new records of tick species are reported around the world; to date more than 956 species are described, all of which belong to suborder Ixodida, and these are divided into three families, two of which are the most important and abundant.

The family Ixodidae popularly known as "hard ticks" is represented by more than 736 species, divided into 14 genera, commonly known as "hard" ticks, the second family with the largest number of individuals is the Argasidae with more than 218 species, represented with five genera and are known as "soft" ticks. Finally, the Nuttalliellidae family, with only one species (Guglielmone et al., 2010; Nava et al., 2017).

In Brazil, the tick fauna to date is composed of 75 species belonging to two families: Ixodidae, the most abundant with 51 species, and Argasidae, with 24 species. Ixodids are divided into five genera: *Amblyomma*, *Ixodes*, *Haemaphysalis*, *Rhipicephalus* and *Dermacentor*. The *Amblyomma* genus is the most abundant with (33) and is important in public health, while the genus *Rhipicephalus* is of greater economic importance.

Argasids are composed of four genera: *Antricola, Argas, Nothoaspis and Ornithodoros*, the latter with the largest number of species. (Martins et al., 2019; Labruna et al., 2020; Muñoz-Leal et al., 2020; Onofrio et al., 2020).

In this context, it is worth mentioning that among the five genera of ixodids found in the country, the genus *Rhipicephalus* comprises only two species out of a total of 85 registered in the world (Guglielmone and Nava et al., 2014). However, these two species of ticks of the genus *Rhipicephalus* are widely studied because they arouse a great interest in the scientific community as well as regarding the importance in animal and public health and for being responsible or for the notable economic impact that in the case of the tick *Rhipicephalus microplus* causes in cattle industry (Andreotti et al., 2019), and no less important but also a cause of great concern and damage, *R. sanguineus*, one of the main ectoparasites of domestic canids.

It is worth mentioning that, to date, the most used tool in the control of these ticks (*R. microplus* and *R. sanguineus*) is still the use of chemotherapeutic drugs, which has been responsible for major environmental problems, public health and economic losses. Not to mention the emergence of resistant tick populations of the most chemical class varieties.

In view of the present context, this chapter will briefly discuss these two tick species of importance in Brazil: *R. microplus* and *R. sanguineus*.

RHIPICEPHALUS MICROPLUS (CANESTRINI, 1888)

The taxonomic classification of this tick species according to the *National Center for Biotechnology Information* (NCBI-ID: 6941), of the United States of America.

- Kingdom - Metazoa
- Phylum - Arthropoda
- Class - Arachnida
- Subclass - Acari
- Superorder - Parasitiformes

- Order - Ixodida
- Superfamily - Ixodoidea
- Family - Ixodidae
- Subfamily- *Rhipicephalinae*
- Genus - *Rhipicephalus*
- Subgenus - *Boophilus*
- Species - *Rhipicephalus (Boophilus) microplus.*

Rhipicephalus microplus popularly known as the cattle tick has a wide geographical distribution throughout the world, and especially in Latin America (Figure 1).

Until the year 2000, this species was called *Boophilus microplus*, however a study of molecular analyzes carried out by Murrell and Barker (2003) reclassified it in the genus *Rhipicephalus,* changing its name to *Rhipicephalus* (*Boophilus*).

Since then, the genus *Boophilus* in this species has been maintained as a subgenus, facilitating the recovery of documents written prior to the studies carried out. (Figure 2)

With regard to its importance, this tick causes great losses in the world livestock, and only in Brazil does it cause an estimated annual loss of USD $3.24 billion (Grisi et al., 2014), in addition to being a transmitter of the etiological agents of cattle tick fever, a disease caused by bacteria of the genus *Anaplasma* and protozoa of the genus *Babesia* (Guglielmone et al., 2006) and which causes weakness in animals.

One of the important bottlenecks to be faced by ranchers in the creation and maintenance of both beef cattle and milk in Brazil is the presence of the *R. microplus* tick and the negative economic impact caused by this ectoparasite (Andreotti et al., 2019).

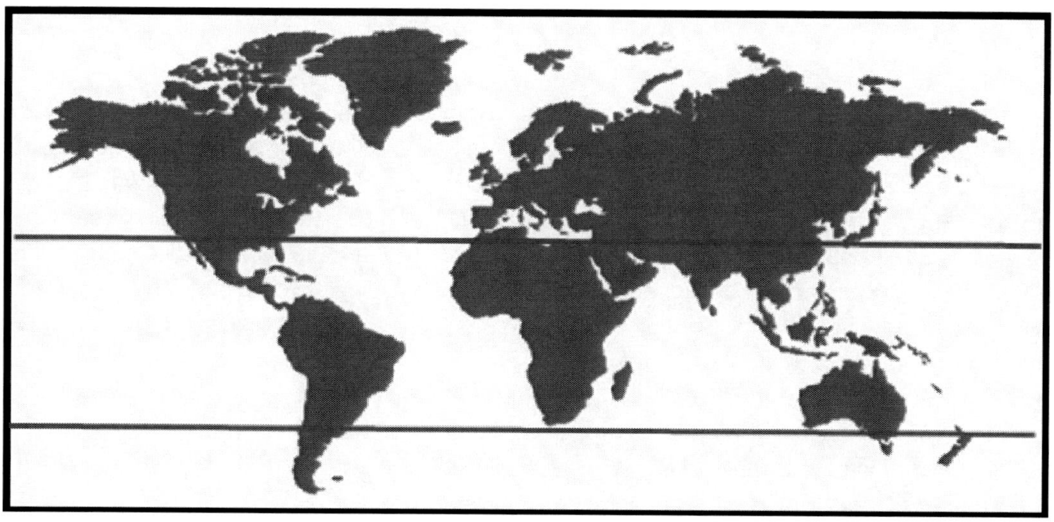

Figure 1. Worldwide geographical distribution of *Rhipicephalus microplus*, between the parallels 32° N and 32° S in 2002 - areas marked in black.

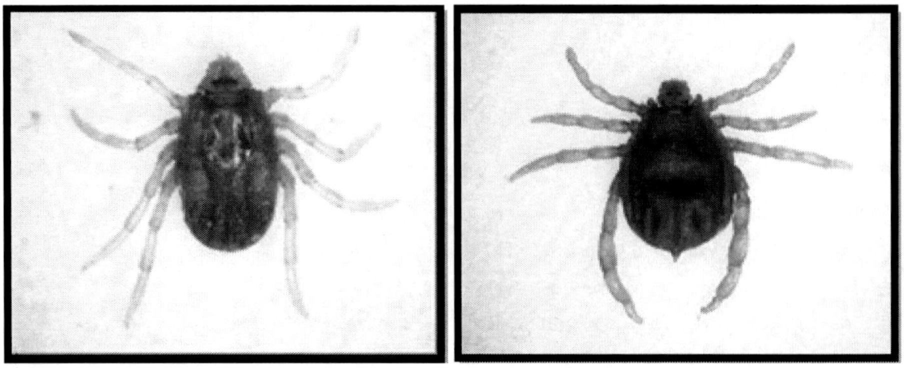

(a) (b)

Figure 2. *Rhipicephalus microplus*. a: dorsal female. b: dorsal male.

Main Morphological Characteristics

Rhipicephalus microplus presents the base of the hexagonal gnatosome with rostrum, that is, hypostomium and chelicerae with short palps, with the third segment of the palp making a flap at the junction with the second segment, it has its shield without ornamentation, the eyes are present and the festoons absent, in addition to round or oval stigmas. Males have two pairs of well-developed adanal plates ventrally, usually have a caudal extension and may be smaller than females (Monteiro, 2011).

Geographic Distribution

Probably the *R. microplus* tick is originally from India and Java Island, Asia. According to historical records, exploratory expeditions that also transported goods and possibly parasitic animals, made it possible for them to spread, including ticks, to other places, such as in tropical and subtropical regions. Subsequently, it came to be established between the 32° North and 32° South parallels, with some foci in the 35 ° South parallel (Nuñes et al., 1982).

In Brazil, everything indicates that the entry of this tick occurred around the beginning of the 18th century in the same way as previously mentioned, through expeditions. Since today there are populations in all states regardless of climatic differences, of course, in some regions they exist in a greater number of annual generations, such as the cerrado regions.

It is a monoxene tick and has cattle as the main host that according to Verissimo et al., (1997) the animals of taurine breeds (*Bos taurus*) and their crossings are the most affected by this species of tick. But it is worth remembering that they can parasitize other animals from amphibians, horses, pigs, sheep and also wild animals such as deer for example or any other that come to live in the same environment as cattle (Nava et al., 2017).

In general, due to the favorable climatic conditions in Brazil, this tick has adapted in all regions of the country. However, climatic factors directly influence the annual generations of this tick. For example, in the South of the country, 3 generations can be observed throughout the year, while in the Southeast and Midwest, 4 to 5 generations can be observed (Pereira et al., 2008; Cruz, 2017). On the other hand, in the Caatinga biome, the survival of larvae of this tick species in the dry season is not favorable and that the generations are directly influenced by the rains (De Barros et al., 2017).

Regarding the parasitic phase, the parasitism period, which ranges from the fixation of the larva to the detachment of the engorged female, lasts an average of 21 days.

BIOLOGICAL CYCLE

Rhipicephalus microplus has a monoxenic life cycle, and has a predilection for parasitizing cattle, with preference for *Bos taurus* and their crosses over *Bos indicus* (Gonzales, 1975).

It is important to remember that the life cycle of this tick species can be divided into two stages, the parasitic phase and the free life phase. The parasitic phase begins with the fixation of the larva in a sensitive host until it reaches the adult stage, which in the case of engorged females detach themselves from the host and fall to the ground, from this moment onwards the free life phase begins, which after falling to the soil the engorged female searches for a suitable location and begins oviposition with subsequent incubation of eggs and subsequent hatching of the larvae (Figure 3).

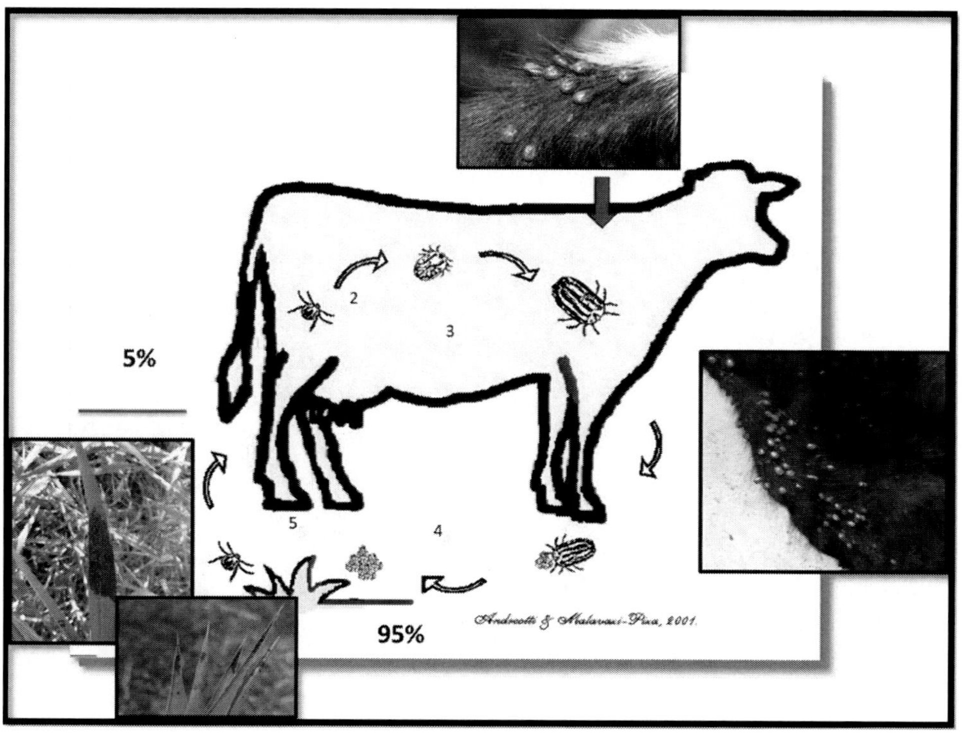

Figure 3. Biological cycle of *Rhipicephalus microplus*, illustrating the phase of free and parasitic life.

Non-Parasitic Phase

It starts when the engorged female detaches from the animal and falls to the ground, usually this phenomenon occurs in the early morning and late afternoon, periods with the most favorable climatic conditions. After the detachment, the engorged female looks for a place on the ground that is safe and protected from light, intense heat and natural enemies (Hitchcock, 1955).

In an interval no longer than 5 days after the engorged female detachment, in favorable climatic conditions, there is a pre-laying period, which is necessary for the ovary to mature, produce and mature the eggs. Always remembering that all these events are directly influenced by climatic conditions (Legg, 1930). After this period, oviposition begins (Figure 4).

After oviposition, the female dies, ending her life cycle and leaving the egg mass for incubation. Normally an engorged female has the capacity to convert 50% of its body weight into eggs and this can reach approximately 3,000. After the necessary incubation time of the eggs has elapsed, which can also vary according to climatic conditions (for example low temperatures can prolong the incubation period), the hexapod larvae hatch. It's almost translucent color is modified after exposure and contact with air and thus, chitin starts to acquire a reddish hue. After a short period of quiescence the larvae rise in groups to the tips of the leaves of the grass, where they remain grouped waiting for the host.

As previously mentioned, climatic conditions directly influence the duration of the non-parasitic phase. In spring and summer, the time from the release of the engorged female to the appearance of its larvae in the pasture is reduced, and that during the autumn and winter seasons this time period is longer, making the non-parasitic phase longer as well.

Figure 4. *Rhipicephalus microplus* engorged female performing oviposition.

Figure 5. *Rhipicephalus microplus* larvae on pasture waiting for a host.

It is important to remember that 95% of the ticks are found in the pasture, either in the egg, larvae and/or engorged female stages and only 5% of the tick population are parasitizing the cattle (Pereira et al., 2008). This becomes a major problem with regard to control, given that the actions to combat *R. microplus* are intended only for ticks in the parasitic phase that represents the minority of the population (Pereira et al., 2008).

Parasitic Phase

It starts with the larva's fixation in the host, some regions of the animal's body are preferred (dewlap, between legs, udder, posterior and perineum), be it because of the temperature, skin thickness as well as to protect itself from self-cleaning performed by the hosts (Pereira et al., 2008).

In the interval from 4 to 7 days after the larva fixation, the stage changes, changing to a nymph that after a period of 9 to 16 days undergoes ecdysis transforming into adults. In turn, adults copulate and engorged females will detach from the host between 18 and 35 days, and males remain in the host in search of new copulation females (Gonzales, 1974). Despite this length of fixation time reported in the literature, the parasitic phase lasts an average of 21 days.

The parasitic phase does not suffer as much from climatic conditions as it is fixed to the host that maintains a constant body temperature, unlike the free-living phase, in which climatic conditions are exposed (Pereira et al., 2008). In addition the fixed tick controls its feeding and hydration as well. (Figure 6).

However, it is important to know the biology of this tick as well as the population dynamics in the pasture environment. With this information, control can be conducted more safely and, consequently, be successful in combating this ectoparasite.

Figure 6. Angus calf with high tick infestation and with cattle tick fever.

In Brazil, strategic control is based on the use of acaricides as a tool, always assessing previously which chemical basis the tick population in question is sensitive to. It is recommended to use the control on all animals in a given pasture to control the population with 5 applications at intervals of a maximum of 21 days at the end of the unfavorable period for the population of larvae in the field, that is, late winter or early spring. when found in the smallest population of larvae in the environment.

Rhipicephalus Sanguineus (Latreille) *Sensu Lato*

The taxonomic classification of this tick species according to *the National Center for Biotechnology Information* (NCBI-ID: 34632), of the United States of America.

- Kingdom - Metazoa
- Phylum - Arthropoda
- Class - Arachnida
- Subclass - Acari
- Superorder - Parasitiformes
- Order - Ixodida
- Superfamily - Ixodoidea
- Family - Ixodidae
- Subfamily - *Rhipicephalinae*
- Genus - *Rhipicephalus*
- Species - *Rhipicephalus sanguíneus.*

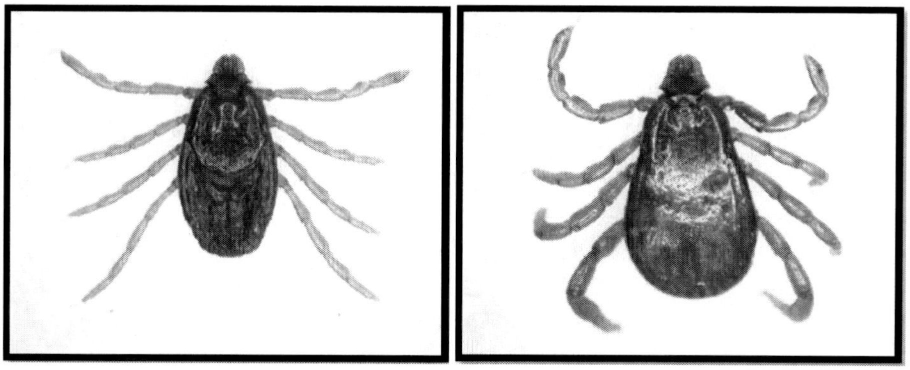

(a) (b)

Figure 7. *Rhipicephalus sanguineus* s. l. a: dorsal female. b: dorsal male.

Rhipicephalus sanguineus s. l Popularly known as the brown dog tick is an ectoparasite, it differs from *R. microplus* in that it has a trioxene life cycle (Figure 6). It has a wide geographical distribution, present in all continents, is considered a cosmopolitan tick, has as primary host domestic dogs (Labruna 2004), but can also occasionally parasitize other species of mammals even humans (Estrada-Peña and Jongejan, 1999; Walker et al., 2005; Nava et al., 2017).

With regard to the importance of *R. sanguineus* l. s is that currently in Brazil it can be incriminated as one of the main ectoparasites (of pet animals), together with fleas, as a result of this position, it has deserved great interest and attention from the pharmaceutical industries linked to veterinary lines (Labruna, 2004).

It is worth remembering that *R. sanguineus* s l. (Figure 7) are potential vectors of various infectious agents for humans and dogs, such as *Ehrlichia canis, Anaplasma platys, Babesia canis vogeli, B. gibsoni, Hepatozoon canis, Rickettsia rickettsii, R. conorii and R. massiliae* (Walker et al., 2000; Parola et al., 2005; Cicuttin et al., 2015), in Brazil is the main vector of *H. canis* (O'Dwyer and Massard, 2001).

MAIN MORPHOLOGICAL CHARACTERISTICS

Rhipicephalus sanguineus s. l usually presents a red or brown color, its main characteristic is the stigma in the form of a comma, observed mainly in the male and less accentuated in the female, eyes and festoons are present. Its size can vary in engorged females that reach a length of 12 mm. The palps are short and the base of the chapter is dorsally hexagonal, the thigh of the first pair of legs has two spines and the legs may become slightly larger from the anterior pair to the posterior pair. The anal sulcus surrounds the posterior half of the anus and extends to the median sulcus. Male ticks have adanal plaques and accessory shields. The larva has three pairs of small, light brown colored legs while nymphs have four pairs of legs with a reddish brown color (Taylor et al., 2017).

Geographic Distribution

Rhipicephalus sanguineus s. l is present on all continents, as previously mentioned, has a predilection to parasitize the domestic dog and is a cosmopolitan tick. Although it originated in the Afrotropical region, its distribution is due to human migrations around the world, taking the domestic dog with it (Walker et al., 2000). The introduction of *R. sanguineus* s. l in the Americas may have occurred during European colonization in the late 15th century, or even earlier, as there are fossil reports of dogs from Central and South America dating from before the 15th century (Leonard et al., 2002).

Biological Cycle

Rhipicephalus sanguineus sl has a more complex cycle than *R. microplus*, as it is a species that needs three hosts to complete its life cycle so it is called a trioxene tick, comprising the parasitic phase or cycle and the non-parasitic phase in which it is in the environment.

Non-Parasitic Phase

During this phase the instars: engorged female, eggs, larvae, engorged larvae, nymphs, engorged nymphs and non-fed adults remain in the place close to the environment where the host usually rests (dog house, kennel, cracks and cracks in the walls), this behavior is called nidicula.

Parasitic Phase

Like other ticks, this phase begins with the encounter of any instar (larvae, nymphs and adults) with the host. In the case of tick larvae when they find the host, the feeding process (engorgement) lasts an average of approximately 2 to 4 days, then the larva, already full (engorged), detaches from the animal and seeks safe shelter to undergo ecdysis (changes), this process can take up to 8 days, depending on weather conditions.

Then, this engorged larva of the origin of the nymph, which again seeks the host, and for a time of more or less from 5 to 6 days, when it is already fed, detaches from the host and again seeks shelter so that it can suffer ecdysis, this period can vary from 14 to 16 days, and turn into adult males or females. Once the adult ticks in the host, the male copulates the female, and after a period of approximately seven to nine days already completely engorged, it detaches itself and goes looking for a safe place and close to the host's habitat. Under conditions of ideal temperature and humidity, the female starts laying eggs around the third day, a period called pre-laying. After oviposition, the eggs are incubated, which also needs good climatic conditions, and finally around 30 days the larvae hatch.

References

Andreotti, R., Garcia, M. V., Koller, W. W. (2019) Carrapatos na cadeia produtiva de bovinos In: Controle estratégico dos carrapatos nos bovinos. Brasilia, DF. *Embrapa,* 9: 125.

Cicuttin, G. L., Tarragona, E. L., De Salvo, M. N., Mangold, A. J., Nava, S. (2015). Infection with *Ehrlichia canis* and *Anaplasma platys* (Rickettsiales: Anaplasmataceae) in two lineages of *Rhipicephalus sanguineus* sensu lato (Acari: Ixodidae) from Argentina. *Ticks and Tick Borne Disease,* 6: 724-9.

Cruz, B. C. (2017). Aspectos ecológicos, biológicos e de resistência de *Rhipicephalus* (*Boophilus*) *microplus* (Acari: Ixodidae) na região de Jaboticabal, São Paulo, Brasil [Ecological, biological and resistance aspects of *Rhipicephalus* (*Boophilus*) *microplus* (Acari: Ixodidae) in the Jaboticabal region, São Paulo, Brazil] (Tese doutorado) - Universidade Estadual Paulista, Faculdade de Ciências Agrárias e Veterinárias, *Universidade Estadual Paulista "Julio de Mesquita Filho,"* Jaboticabal.

De Barros, M. N. D. L., Riet-Correa, F., Azevedo, S. S., Labruna, M. B. (2017). Off-host development and survival of *Rhipicephalus* (*Boophilus*) *microplus* in the Brazilian semiarid. *Veterinary Parasitology: Regional Studies and Reports,* 9: 17-24.

Estrada-Peña, A. and Jongejan, F. (1999) Ticks feeding on humans: a review of records on human-biting Ixodoidea with special reference to pathogen transmission. *Experimental and Applied Acarology,* 23: 685-715. https://doi.org/10.1023/A:1006241108739.

Gonzales, J. C. (1975). *O controle dos carrapatos dos bovinos* [*Control of bovine ticks*]. Porto Alegre: Sulina, 104 p.

Gonzales, J. C. (1995). *O controle do carrapato do boi* [*The control of the ox tick*]. Porto Alegre: Edição do Autor. 2. Ed, 235 p.

Gonzales, J. C., Silva, N. R., Wagner, E. M. (1974). O ciclo parasitário do *Boophilus microplus* em bovinos estabulados [The parasitic cycle of Boophilus microplus in cows]. *Arquivo da Faculdade de Veteterinaria,* UFRGS, 2: 25-34.

Grisi, L., Leite, R. C., Martins, J. R. S., Barros, A. T. M., Andreotti, R., Cançado, P. H. D., Leon, A. A. P., Pereira, J. B., Villela, H. S. (2014). Reassessment of the potential economic impact of cattle parasites in Brazil. *Brazilian Journal of Veterinary Parasitology,* 23 (2): 150-156.

Guglielmone, A. A., Beati, L., Barros-Battesti, D. M. (2006). Ticks (Ixodidae) on humans in South America. *Experimental and Applied Acarology,* 40 (2): 83-100.

Guglielmone, A. A., Robbins, G. R., Apanaskevich, A. D., Petney, N. T., Estrada-Peña, A., Horak, G. I., Shao, R., Barker, C. S. (2010) The Argasidae, Ixodidae and Nuttalliellidae (Acari: Ixodida) of the world: a list of valid species names. *Zootaxa,* 2528: 1-28.

Hitchcock, L. F. (1955) Studies on the parasitic stages of the cattle fever tick, *Boophilus microplus* (Canestrini) (Acarina: Ixodidae). *Australian Journal Zoologycal*, 3: 145-155.

Labruna, M. B. (2004). Carta Acarológica [Acarological letter]. *Revista Brasileira de Parasitologia Veterinária,* 23(1): 199-202.

Labruna, M. B., Onofrio, V. C., Barros-Battesti, D. M., Gianizella, S. L., Venzal, J. M., Guglielmone, A. A. (2020). Synonymy of *Ixodes aragaoi* with *Ixodes fuscipes*, and reinstatement of *Ixodes spinosus* (Acari: Ixodidae). *Ticks and Tick Borne Diseases,* 11 (2): 101349.

Legg, J. (1930). Some observations on the life history of the cattle tick (*Boophilus australis*). *Proceedings of the Royal Society of Queensland,* 1930 - https://cabdirect.org, 41: 121-132.

Leonard, J. A., Wayne, R. K., Wheeler, J., Valadez, R., Guillen, S., Vila, C. (2002). Ancient DNA evidence for Old World origin of New World dogs. *Science,* Washington, 298: 1613-1616.

Martins, T. F., Luz, H. R., Munõz-Leal, S., Ramirez, D. G., Milanelo, L., Marques, S., Sanches, T. C., Onofrio, V. C., Acosta, I. C. L., BenattI, H. R., Maturano, R., Oliveira, P. B., Albuquerque, G. R., Marcili, A., Flausino, W., Silveira, L. F., Mcintosh, D., Faccini, J. L. H., Labruna, M. B. (2019). A new species of *Amblyomma* (Acari: Ixodidae) associated with monkeys and passerines of the Atlantic rainforest biome, Southeastern Brazil. *Ticks and Tick-Borne Disease,* 10 (6): 101259. https://doi.org/10.1016/j.ttbdis. 2019.07.003.

Monteiro, S. G. (2011). *Parasitologia na medicina veterinária* [*Parasitology in veterinary medicine*] (Vol. 1). São Paulo: Roca.

Muñoz-Leal, S., Martins, M. M., Nava, S., Gabriel, A., Landulfo, G. A., Simons, S. M., Rodrigues, V. S., Ramos, V. N., Suzin, A., Szabó, M. P. J., Labruna, M. B. (2020). *Ornithodoros cerradoensis* n. sp. (Acari: Argasidae), a member of the *Ornithodoros talaje* (Guérin-Méneville, 1849) group, parasite of rodents in the Brazilian Savannah. *Ticks and Tick borne Disease,* 11 (5): 101497. https://doi.org/10.1016/j.ttbdis.2020. 101497.

Murrell, A., Barker, S. C. (2003). Synonymy of *Boophilus* Curtice, 1891 with *Rhipicephalus* Koch, 1844 (Acari: Ixodidae). *Systematic Parasitology,* 56 (1): 169-172.

Nava, S., Venzal, J. M., Gonzalez-Acuña, D., Martins, T. F., Guglielmone, A. A. (2017). *Ticks of the Southern Cone of América: Diagnosis, Distribution, and Hosts with Taxonomy, Ecology and Sanitary Importance.* Academic Press, 339 p.

Nuñes, J. L., Muñoz Cobenas, M. E., Moltedo, H. L. (1982) *Boophilus microplus*, la garrapata comun del ganado vacuno. Buenos Aires: *Hemisfério Sur,* 19 p.

O'dwyer, L. H. O., Massard, C. L. (2001). Aspectos gerais da hepatozoonose canina. *Clínica Veterinária,* 6 (31): 34-39.

Onofrio, V. C., Guglielmone, A. A., Barros-Battesti, D. M., Gianizellae, S. L., Marcili, A., Quadros, R. M., Marques, S., Labruna, M. B. (2020). Description of a new species of *Ixodes* (Acari: Ixodidae) and first report of *Ixodes lasallei* and *Ixodes bocatorensis* in Brazil. *Ticks and Tick borne Disease.* 11: 101423. https://doi.org/10.1016/j.ttbdis.2020. 101423.

Parola, P., Paddock, C. D., Raoult, D. (2005). Tick-borne Rickettsioses around the world: Emerging diseases challenging old concepts. *Clinical Microbiology Reviews,* 18 (4): 719-756.

Pereira, M. C., Labruna, M. B., Szabó, M. P. J., Klafke, G. M. (2008). *Rhipicephalus* (*Boophilus*) *microplus*: biologia, controle e resistência. São Paulo: *MedVet,* p. 21-25, 65-90.

Taylor, M. A., Coop, R. L., Wall, R. L. (2017). *Parasitologia Veterinária* [*Veterinary Parasitology*]. Rio de Janeiro, Brasil: Guanabara Koogan.

Veríssimo, C. J., Silva, R. G., Oliveira, A. A. D., Ribeiro, W. R., Rocha, U. F. (1997). Contagens de ínstares do carrapato *Boophilus microplus* em bovinos mestiços [Intestation counts of the Boophilus microplus tick in crossbred cattle]. *Boletim de Indústria Animal,* 54 (2): 21-26.

Walker, J. B., Keirans, J. E., Horak, I. G. (Ed.) (2005). *The genus Rhipicephalus (Acari: Ixodidae): a guide to the brown ticks of the world.* Cambridge: Cambridge University Press: 643 p.

In: The Entomological Guide to *Rhipicephalus* ISBN: 978-1-53619-619-1
Editors: Sachin Kumar, Raquel Cossio Bayugar et al.

Chapter 2

EPIDEMIOLOGY OF TICK-BORNE DISEASES

Adil Khan[1], Ijaz Ali[1], Zia Ur Rahman[1],
Furhan Iqbal[2], Nasreen Nasreen[1,*]
and Sadaf Niaz[1]
[1]Department of Zoology, Abdul Wali Khan University Mardan, Pakistan
[2]Institute of Pure and Applied Biology, Zoology Division,
Bahauddin Zakariya Univrsity Multan, Pakistan

ABSTRACT

Tick-borne diseases (TBDs) have a global economic impact and researchers are particularly concerned about their control and prevention. The aim of this chapter is to provide a brief introduction to some of the common tick-borne pathogens, vectors responsible for their transmission, geographical distribution of these TBDs and the risk factors that are associated with these diseases especially in developing countries. The main focus of this chapter is on theileriosis, anaplasmosis, babesiosis, ehrlichiosis and heartwater as these diseases affect the milk, meat and hide production causing significant losses to the livestock sector.

Keywords: anaplasmosis, babesiosis, ehrlichiosis, heartwater disease, theileriosis

ABBREVATIONS

Tick-borne diseases	TBDs
Red blood cells	RBCs
Species	Spp.
Babesia	B.

* Corresponding Author's Email: nasreen@awkum.edu.pk.

Human granulocytic ehrlichiosis agent	HE
Hyalomma	*Hy.*
Anaplasma	A.
Ehrlichia	*E.*
Tick infestation prevelance	TIP
Amblyomma	*A.*
Dogs per household ratio	DHR
Micrometer	μm
Millimeter	mm
Milligram	Mg
Kilogram	Kg
Alpha	α-
Major surface protein 4	MSP4
Indirect flouresence Antibidy	IFA
complement-enzyme linked immuno sorbent assay	cElisa
Canine Monocytic Ehrlichiosis	CME

INTRODUCTION

Tick-borne diseases (TBDs) are among the major constraints to the livestock industry in developing countries as they are adversely affecting the economic performance. Ticks are responsible for the spread of several pathogens to livestock that can cause paralysis or toxicosis and physical damage to them and in extreme cases may lead to the death of infected animals (Durrani et al., 2008). Among the most commonly reported TBDs, theileriosis, anaplasmosis, babesiosis, ehrlichiosis and heartwater disease are among the mostly commonly reported diseases from the livestock sector of developing countries (Mukhebi et al., 1992; De Vos, 1991; Allsopp, 2015). Theileriosis is a TBD caused by the intraerythrocytic protozoan parasite of the genus *Theileria* which is responsible for large economic loss to livestock in tropical and sub-tropical parts of the world (Gul et al., 2015). Babesiosis, most commonly reported TBDs in animals worldwide, is also caused by intraerythrocytic protozoan parasites of the genus *Babesia*. Ehrlichiosis is an emerging zoonotic disease caused by obligate gram-negative bacteria of the genera *Ehrlichia* within the family Rickettsiacea and order Rickettsiales (Goldstein et al., 2001). Heartwater or cowdriosis is an economically important TBD disease of livestock that is caused by the rickettsia *Ehrlichia ruminantium* (Uilenberg, 1983). Anaplasmosis is a non-contagious disease caused by several *Anaplasma* species and it is considered as one of the top 10 economically important diseases causing a huge economic loss in domestic animals (Belal et al., 2014; Aubry and Geale, 2011). In this chapter, we are going to acoount these TBDs with an emphasis on the epidemiological factors that are associated with the transmission of these TBDs in livestock especially that of developing countries where livestock is the major source of income of population.

THEILERIOSIS

Theileriosis is a tick born disease caused by intraerythrocytic protozoan parasite of the genus *Theileria* and it is responsible for considerable economic losses to livestock in tropical and sub-tropical parts of the world (Gul et al., 2015). It was first reported by Arnald Theiler in Egypt and now it is considered as one the most important tick-borne pathogen (Levine, 1985).

Morphology

During screening of Giemsa-stained blood smear, *Theileria* spp. appear like pear spherical, ovoid or irregular with an apical complex consist of rhoptries (Hadi et al., 2012; Yin et al. 2007).

Taxonomy

The phylum Apicomplexa consist of large group of eukaryotic organisms and its further divided into four principal groups: the Gregarinasina (gregarines), Coccidia, Haemospororida (haemosporidians) and the Piroplasmorida (piroplasmids). The Piroplasmorida comprises two main genera (*Babesia* and *Theileria*) responsible for the economic important diseases of domestic and wild animals (Adl et al., 2012).

Theileria Species

Seven species of *Theileria* including *Theileria ovis, Theileria separata, Theileria uilenbergi, Theileria recondita, Theileria lestoquardi, Theileria luwenshuni* and *Theileria annulata*as been reported from various parts of the world causing ovine theileriosis (Schnittger et al., 2003; Li et al., 2009; Zhang et al., 2014; Khan et al., 2020). In cattle, five species of *Theileria* are documented to causes bovine theileriosis. Among them *Theileria parva* and *Theileria annulata* are most pathogenic causing East Coast fever and tropical theileriosis, respectively. While *Theileria sergenti/buffeli/orientalis* causes mild infection in cattle and the disease caused by them is called bovine benign theileriosis (Soosaraei et al., 2018). Two species *T. mutans* and *T. velifera* also causes mild infection in buffalo and cattle (Norval et al., 1992).

Clinical Signs of Theileriosis

The clinical signs of infection includes fever, nasal discharge, jaundice, lacrimation, enlargement of superficial lymph nodes, rapid weight loss, anorexia, reduced appetite, tachycardia, tachypnea, anemia and hepatosplenomegaly. Rarely, hemoglobinurea is also observed (Hussein et al., 2007; Naz et al., 2012; El-Ashker, et al., 2015; Shah et al., 2017).

Life Cycle of *Theileria*

Theileria passes through three phases during its reproduction: schizogony, gemogony and sporogony (Levine, 1985). In the schizogony phase sporozoites enter into vertebrate host via the tick bite during blood feeding. In the blood stream sporozoites penetrate into lymphocytes and form schizonts (Kock's blue bodies). Synchronized with host cell division the schizonts produce merozoites. Large numbers of merozoites formed inside the lymphocytes by division and is eventually released. Now merozoites gain entry into erythrocytes and form comma shaped or spherical ovoid structures known as piroplasms (Mehlhorn and Schein, 1994). The gamogony phase occur inside the gut cells of tick after ingestions of the piroplasms by the tick. Micro- and macro-gamonts form and then fuse in the gut to form a zygote which converts into motile kinete (Levine, 1985). Kinetes that infect the tick gut epithelial cells and migrate to the haemolymph and subsequently infect the salivary glands. Inside the salivary gland kinetes undergo sporogony forming numerous sporozoites. These sporoziotes are then injected into the host blood stream during tick feeding. It's estimated that in one salivary gland cell can contain approximately 50,000 sporozoites (Mehlhorn and Schein, 1994).

Transmission

Ixodid ticks of the genera *Rhipicephalus, Hyalomma, Haemaphysalis* and *Amblyomma* are responsible for transmit the different *Theileria* spp. to their mammalian hosts (Spitalska et al., 2005). Two tick species namely *Hyalomma anatolicum anatolicum* and *Haemaphysalis qinghaiensis* are known to transmit *Theileria* spp. to small ruminants (Bakheit et al., 2006; Liu et al., 2008). In cattle, *T. annulata* is transmitted by *Hyalomma detritum detritum, Hyalomma anatolicum, Hyalomma lusitanicum* and *Hyalomma dromedarii* (Aktas et al., 2004). While *T. parva* is reported to be transmitted by *Rhipicephalus appendiculatus* and *Rhipicephalus zambeziensis* ((Lawrence et al., 1994)). *Theileria orientalis* is transmitted by *Haemaphysalis longicornis* (Yin et al., 2004). *Theileria mutans* and *Theileria velifera* are transmitted by *Amblyomma spp,* while *Theileria taurotragi* is transmitted *by R. appendiculatus* (Uilenberg et al., 1982; Norval et al., 1992; Jabbar et al., 2015).

Geographical Distribution

Theileriosis is mainly reported from northern Asia, Africa, Middle East and southern Europe however it is primarily restricted to tropical and sub-tropical regions of the world (Yang et al., 2014). Table 1 lists the countries from where theileriosis has been reported. Prevalence of *Theileria* spp. infection varies between 5.8 to 73% (Table 1).

Table 1. Prevalence of theileriosis as reported in literature from various countries

Localization	Prevalence (%)	References
Australia	5%	(Eamens et al., 2013)
Bangladesh	5.8%	(Al Mahmud et al., 2015)
China	26%	(Ge et al., 2012).
Cuba	73%	(Díaz-Sánchez et al., 2018)
Ethiopia	44.8%	(Gebrekidan et al., 2014)
Greece	24.6%	(Papadopoulos et al., 1996)
Hungary	67.9%	(Farkas et al., 2013)
India	29.5	(Tuli et al., 2015)
Iran	60%	(Heidarpour Bami et al., 2010)
Iran	18.6%	(Razmi and Yaghfoori 2013)
Iraq	26.5%	(Al-Fetly, 2012)
Israel	26.4%	(Steinman et al., 2012)
New Zealand	27.1%	(McFadden et al., 2016)
Oman	28%	(Al-Fahdi et al., 2017)
Pakistan	35%	(Durrani et al., 2011)
Palestine	10.8%	(Azmi et al., 2019)
South Africa	16.5%	(Mbizeni et al., 2013)
Spain	7.7%	(Garcia Sanmartin et al., 2008)
Sudan	61.2%	(Taha et al., 2013).
Tunisia	5.8%	(Rjeib e al., 2014)
Turkey	18.3%	(Altay et al., 2004)
U.S.A	12.7%	(Thompson et al., 2020)

Risk Factors Associated with Theileriosis

A number of epidemiological factors are reported to be associated with the prevalence of theileriosis. Ullah et al. (2018) had reported higher infection of in animals during summer season as compared to winter. Differences in prevalences has also been reported between stall fed (24.91%) and nomadic animals (15.43%) as well as between male (22.92%) and female (16.78%) animals. Ullah et al. (2018) had concluded that high prevalence of the infection was probably due to the overcrowded and filthy stall condition which facilitate optimum environment for the disease occurrence. However, age, breed and herd sizes were not found associated with disease (Ullah et al., 2018). Tuli et al. (2015) had reported that age, sex, location and management of farms were significantly associated with the prevalence of infection. Animals older than one year showed higher rate of infection. Unlikely above this study reported a signicant bred wise prevalence. Higher prevalence of infection was reported in crossbred cattle followed by indigenous cattle and buffaloes.

Durrani et al. (2012) had analyzed the association of various epidemiological factors with the prevalence of *Theileria* in small ruminants and found that the presence of tick in small ruminants was significantly associated with the prevalence of theileriosis. They also observed that herd composition was important for the spread of theileriosis as higher prevalence of *Theileria* in herds composed of sheep only compared to herds consisting of only goat or having both sheep and goats kept together.

In a similar study in cattle, conducted by Kispotta et al. (2016), it was found that sex, large herd size (>10), temperature (≥ 103ºF), presence of ticks on animals and no ivermectin treatment were associated with higher prevalence of bovine theileriosis.

In a study of bovine theileriosis in Sudan, Salih et al. 2007 found that sampling location, management system and age are risk factors associated with the prevalence of *T. annulata* infection while *T. parva* infection was associated with sampling location and season. The prevalence of theileriosis was highest in Central Sudan followed by Northern Sudan, White Nile, Blue Nile, and Western Sudan respectively. Animals older than four years animals and held under a semi-extensive management system had significant higher prevalence of *T. annulata. Theileria parva* had the highest prevalence in April (p = <0.001) and was while lowestin July (p = <0.05)

Guidi et al. (2015) had documented that sex, age and keeping horses with or near cattle were significantly associated with *Theileria* infection in horses. García-Bocanegra et al. (2013) also reported that increased age, presence of ticks and vaccination against other diseases had significant association with theileriosis in equids. Steinman et al. (2012) conducted a study to analyze various risk factors associated with theileriosis in horses and they found that relative humidity and minimum land surface temperature were the environmental characters that showed negative correlation with prevalence of theileriosis.

Treatment

Hooshmand-Rad (1989) has tested the chemotherapeutic efficacy of various chemical compounds including drugs such as parvaquone (Clexon) and buparvaquone (Butalex) against theileriosis. However, some of these compounds are effective (El Hussein et al., 1993; Hashemi-Fesharki, 1997) but they are not easily and quickly eliminated from the animal and therefore may constitute a public health hazard (McHardy, et al., 1985). Recently, the therapeutic effect of the alkaloids extracted from a medical plant (*Peganum harmala*) has been investigated (Mirzaiedehaghi, 2006; Derakhshanfar and Mirzaei, 2008).

BABESIOSIS

Babesiosis is caused by intraerythrocytic protozoan parasites of the genus *Babesia*. It is among the most common TBDs in animals worldwide. In the 19th century, Victor Babes discovered a small microorganism in red blood cells (RBCs) of cattle in Romania and reported it as the cause of red water fever, also known as bovine haemoglobinuria, in cattle. He also found similar microorganisms in sheep blood and in a subsequent taxonomic revision the genus *Babesia* was established (Babes, 1888). Recently, *Babesia* infection has been identified in humans and is increasingly considered an important zoonotic infection (Jabbar et al., 2015).

Morphology and Size Description

Size can be used as a diagnostic character to divide *Babesia* spp. into two groups; larger *Babesia* spp. including *Babesia caballi*, *Babesia bigemina*, *Babesia canis* with size ranging from 2.5 to 5.0 µm and small *Babesia* spp. ranging in size from 1.0 to 2.5 µm which includes *Babesia gibsoni*, *Babesia bovis*, *Babesia microti* and *Babesia rodhaini* (Laha et al., 2015). The orientation of the parasite in RBCs is correlated size because large pyriform parasites meet at their pointed ends at an acute angle to each other and small forms make an obtuse angle to each other (Ruprah, 1985). *Babesia* spp. can be pear shaped, round, oval, leaf shaped, amoeboid or irregularly shaped (Laha et al., 2015).

Taxonomy

The genus *Babesia* is placed in the phylum Apicomplexa, class Aconoidasida (Piroplasmea), order Piroplasmida and family Babesiidae (Abdela and Jilo, 2016).

Babesia Species

The genus *Babesia* contains around one-hundred species that cause babesiosis in a variety of mammals and birds. Bovine babesiosis is caused by *Babesia bovis, Babesia divergens, Babesia bigemina*, and *Babesia major*. Among these, *Babesia bovis* and *Babesia bigemina* have a considerable economic effect due to their impact on cattle health and productivity (Uilenberg, 1995). Ovine babesiosis is caused by *Babesia ovis, Babesia motasi* and *Babesia crassa* (Uilenberg 2001). The disease caused by *Babesia motassi* may be acute or chronic (Morel, 1989), whereas *Babesia crassa* appears to have little or no pathogenicity (Friedhoff, 1997). *Babesia ovis* is reported to be highly pathogenic in sheep (Hashemi, 1997). *Babesia microti* is known to infect humans (Kim, 2007). Canine babesiosis is caused by two species of *Babesia*: *Babesia Canis* and *Babesia gibsoni* while *Babesia caballi* and *Babesia equi* causes equine babesiosis. *Babesia trautmanni* causes disease in swine while *Babesia cati* and *Babesia felis* cause babesiosis in feline (Laha et al., 2015; Sabu et al., 2013).

Clinical Signs of Babesiosis

Babesiosis is characterized by high fever and intravascular haemolysis (anaemia), icterus, haemoglobinurea and ultimately death of infected animals (Zintl et al., 2003).

Life Cycle of *Babesia*

The life cycle of babesia is similar to *Theileria* spp. but there are two primary differences between them: 1) A pre-erythrocytic phase during schizogony is absent for the genus *Babesia* but occurrs in the genus *Theileria*, and 2). In *Babesia* spp. the zygote divides and forms vermicules which infect various organs of ticks including the ovaries, such that the infection transmitted vertically to the next generation of ticks through the egg, whereas transovarial transmission is absent in *Theileria* (Homer, et al, 2000).

Transmission of Babesiosis

Ticks are the major vector of of *Babesia* species. A number of tick genera are reported as vectors for *Babesia* species.

Table 2. A list of *Babesia* species, the genus of their vector ticks, and host animals as reported in literature

Species	Host	Tick vector (Genus)	Reference
B. beliceri	Cattle	*Hyalomma*	Luo et al., 2002
B. bigemina	Cattle, buffalo	*Boophilus, Rhipicephalus*	Habela et al., 1995
B. bovis	Cattle, buffalo	*Boophilus, Rhipicephalus*	Habela et al., 1995
B. divergens	Cattle	*Ixodes*	Donnelly and Peirce, 1975
B. jakimovi	Cattle, reindeer	*Ixodes*	Uilenberg, 2006
B. major	Cattle	*Haemaphysalis*	Uilenberg, 2006
B. occultans	Cattle	*Hyalomma*	Luo et al., 2002
B. ovate	Cattle	*Haemaphysalis*	Sivakuma et al., 2014
B. orientalis	Buffalo	*Rhipicephalus*	He, L et al., 2017
B. crassa	Sheep, goats	Unknown	Bai et al., 2002
B. motasi	Sheep, goats	*Haemaphysalis*	Bai et al., 2002
B. ovis	Sheep, goats	*Rhipicephalus*	Altay et al., 2008
B. caballi	Horse, donkey, mule	*Dermacentor, Hyalomma, Rhipicephalus*	De Waal et al., 1987; Uilenberg, 2006
B. perroncitoi	Pig	Unknown	Uilenberg, 2006
B. trautmanni	Pig	*Rhipicephalus*	De Waal et al., 1992
B. canis	Dog, cat	*Dermacentor*	Zahler et al., 1998
B. rossi	Dog	*Haemaphysalis*	Penzhorn et al., 2020
B. vogeli	Dog	*Rhipicephalus*	Rojas et al., 2014
Babesia sp.	Dog	Unknown	Uilenberg, 2006
B. gibsoni	Dog	*Haemaphysalis, Rhipicephalus*	Yamane et al., 1993
B. canis presentii	Cat	Unknown	Uilenberg, 2006
B. felis	Cat	Unknown	Uilenberg, 2006

Geographical Distribution

Geographically, babesiosis has been reported from Asia, Africa, Central and South America, Southern Europe and Australia. It is more commonly reported from tropical and sub-tropical regions of the world (Spickler and Roth, 2008). A number of studies have been conducted around the world and have reported numerous *Babesia* species in a variety of host animals. However, the prevalence of *Babesia* species varies from county to country. For example, the prevalence of *Babesia* spp. was 78% in Mozambique (Tembue et al., 2011), 70.6% in Venezuela (Mujica et al., 2011), 69% in Portugal (Cardoso et al., 2013), 34.9% in

Tanzania (Swai et al., 2007), 34% in Pakistan (Iqbal et al., 2011), 31.86% in India (Das et al., 2015), 31.57% in China (Niu et al., 2015), 17% in the US (Westmoreland et al., 2019), 17.41 in Pakistan (Ijaz et al., 2013), 13.2% in Spain (García-Bocanegra et al., 2013), 11.7% in Ethiopia (Abdela et al., 2018), 11.33% in Africa (Ibrahim et al., 3013), 10.7% in Europe (Michel et al., 2014), 10.3% in Italy (Del Pino et al., 2016), 8.25% in Turkey, 5.92% in Sudan (Awad et al., 2011), 5.73% in Germany (Hunfeld et al., 2002), 1.5% in Bangladesh (Rahman et al., 2015), 0.6% in Poland (Siński et al., 2006) and 0.6% in Uganda (Muhanguzi et al., 2010).

Risk Factors Associated with Babesiosis

Guidi et al. (2015) conducted a study to determine risk factors associated with babesiosis in horses. They analyzed breed, gender, age, control measures against mosquitos and against helminths, primary activity (leisure, sport, livestock cutting), housing during day, cohabitation with cows, and land cover in the enclosure (wetland, grassland, forest, tree, crops, building). Among all these factors those most significantly associated with higher risk of infection were enclosures in wetlands, cohabitation with cows, García-Bocanegra et al. (2013) also reported that factors such as presence of ticks on animals was significantly associated with babesiosis in equids. Tefera et al. (2011) analyzed various risk factors and concluded that body condition and presence of ticks on animals were significantly associated with babesiosis in donkeys.

Michel et al. (2014) conducted a study in wild ruminants and reported that age, host species, and altitude were significant risk factors that were associated with the prevalence of babesiosis. Iqbal et al. (2011) reported that gender, age, herd size and presence of herding dogs were associated with various *Babesia* spp. in small ruminants from Pakistan.

Abdela et al. (2018) studied different risk factors associated with bovine babesiosis included age, sex, body condition, management system, and tick infestation and found that age, body condition, and tick infestation were significantly associated with the presence of *Babesia* spp. On the other hand, sex and management system were found not to be associated with *Babesia* infection. Recently, Zhou et al. (2019) conducted a study in cattle and reported that age and altitude are significantly associated with *Babesia* infection while gender and presence of cats had no effect on babesiosis in cattle.

Sing et al. (2014) documented that the prevalence of canine babesiosis was higher in summer as compared to winter and was higher in young dogs, while breed and sex of the host were not significantly associated with the *Babesia* infection.

EHRLICHIOSIS

Morphology and Size Descriptions

Electron microscopy reveals that *Ehrlichia* have two forms: reticulate cells (0.4–0.6 mm X 0.7–1.9 mm) and the dense-cored cells (0.4–0.6 mm) present within the intracellular vacuoles (morulae) in mammalian host cells. Dense-cored cells have a dense nucleo while reticulate cells have uniformly dispersed nucleoid filaments and ribosomes, sometimes forming long projections of the cell wall, protrusions of the cytoplasmic membrane into the periplasmic space or budding protoplast fragments into the periplasmic space (Popov et al., 1995). Sells et al. (1976) described the shape of *Ehrlichia* spp. as ovoid, round or rod-shape.

Taxonomy

Ehrlichiosis is an emerging tick-borne zoonotic disease caused by obligate intracellular gram-negative bacteria of the genera *Ehrlichia* within the family Rickettsiacea and order Rickettsiales (Santino et al., 2002; Ismail et al. 2010).

Ehrlichia Species

On the basis of *16S rRNA* sequence homology and antigenic relationship, three genogroups of *Ehrlichia* have been described: (i) *Ehrlichia senetsu* genogroup (*Ehrlichia resticii, Ehrlichia sennetsu* and *Ehrlichia bovis*) (ii) *Ehrlichia phagocytophila* genogroup (*Ehrlichia equi, Ehrlichia phagocytophila,* human granulocytic ehrlichiosis agent (HGE) and *Anaplasma platys*); and (iii) *Ehrlichia canis* genogroup (*Ehrlichia chaffeensis, Ehrlichia canis* and *Ehrlichia ewingii*) (Anderson et al., 1991; Rikihisa, 1991; Hardalo et al., 1995; Wen et al., 1995; Kruss et al., 2003). *Ehrlichia* species infect various hosts including, humans, cattle, ruminants, dogs, horses, cats, and mice (Wen et al., 1995; Kruss *et al.*, 2003).

Clinical Signs

Clinical signs related to these tick-borne pathogens are vague and may include fever and shifting limb lameness (Littman et al., 2006). However, severe clinicopathologic changes and disease can occur, such as Lyme nephritis (Littman et al., 2018), thrombocytopenia or death in extreme cases (Neer et al., 2002). In humans, the symptoms include fever, rash, myalgia, vomiting, diarrhea and headache (Dumler and Bakken, 1995; Schutze and Jacobs, 1997).

Life Cycle

Ehrlichia spp. replicate in tick and trematodes. Generally, the life cycle of *Ehrlichia* spp. goes through three phases of development; elementary bodies, initial bodies and morulae (Forbes et al., 1998). The small elementary bodies (0.2 to 0.5 µm) inside the host cell undergo division and form initial bodies (1.0 to 1.5 µm) and eventually morulae (2 to 5 µm) that consist of approximately 100 elementary bodies (Kruss et al., 2003) that rupture and relase infect other cells.

Ehrlichia spp. Transmission

A number of tick vectors are involved in the transmission of different *Ehrlichia* species (Table 3).

Table 3. A list of *Ehrlichia* species, their vector ticks, and host animals as reported in literature

Ehrlichia species	Tick vector	Hosts	References
E. chaffeensis	*Amblyomma americanum, Dermacentor variabilis*	Human and various Mammals	Shaw et al., (2001) and Parola et al., (2005)
E. ewingii	*Amblyomma americanum*	Human and dog	Shaw et al., (2001) and Parola and Raoult, (2001)
E. ruminantium	*Amblyomma hebraeum, A. astrion, A. cohaerens, A. gemma, A. marmoreum, A. lepidum, A. pomposum, A. variegatum A. americanum*8	Mainly cattle	Jongejan and Uilenberg, (2004) Walker and Olwage (1987)
E. canis	*Rhipicephalus sanguineus*	Dog	Shaw et al., (2001)
E. equi	*Ixodes pacificus*	Horse	Richter et al. (1996)
Human Granulocytic Ehrlichiosis	*Ixodes dammin*	Human	Pancholi et al., (1995)

Geographical Distributions

Ehrlichia species infect animals in most parts of the world but mainly in tropical and temperate climates (Neer, 1998). Prevalence rates of different *Ehrlichia* spp. have been reported around the world; 47.4% in Korea (Schulze et al., 2011), 47.35% in Cuba (Navarrete et al., 2018), 46.87% in South Africa (Mdladla et al., 2016), 37.40% in Southeastern Europe (Christova et al., 2003), 33% in Malaysia (Konto et al., 2017), 31.21% in Brazil (Aguiar et al., 2007), 30.8% in Mexico (Movilla et al, 2016), 28% in Pakistan (Malik et al., 2018), 24.7% in Grenada (Yabsley et al., 2008), 20.93% in southeastern Virginia (Wright et al., 2014), 12% in Italy (Santino et al., 2002), 7.1% in New Jersey (Schulze et al., 2011), 4.7% in Japan (Inokuma et al., 1999), 3.2% in Costa Rica (Barrantes-González et al., 2016), 2.60% in India (Bhattacharjee et al., 2013), 2.53% in UK (Ogden et al., 1998), 2.1% in Romania (Mircean et al., 2012), 1.18% in Switzerland (Wicki et al., 2000), 1.5% in Minnesota (Beall et al., 2008), (Bowman et al., 2009) and 0.5% in Canada (Schulze et al., 2011).

Risk Factors

Barrantes-González et al., (2016) analyzed various risk factors including age, breed, area, number of dogs, dogs per household ratio (DHR) and number of dogs infested with ticks [tick infestation prevalence (TIP)]. All of these factors were found significantly associated with the prevalence of canine ehrlichiosis.

Mdladla et al., (2016) conducted a study to identify various epidemiological factors associated with ehrlichiosis in small ruminants. They consider factors like sampling areas, status of region (endemic or non-endemic), production system (commercial, research and small hold), sex, age, breed and tick infestation with respect to the prevalence of *Ehrlichia* spp. They had reported that none of the studied parameters were associated with *Ehrlichia* spp. except tick infestation ($P< 0.05$).

Carvalho et al. (2008) had documented that sex, age, breed have no significant association with the prevalence of *Ehrlichia* spp. in dog while presence of ticks on doge and location had impact on the prevalence of *Ehrlichia* spp. in dogs from Brazil. Mircean et al. (2012) had analyzed various risk factors i.e., age, gender, breed, dog category (Guard, pets, shelter, stay and hunting), lifestyle (indoor or outdoor) and prophylactic treatment in dogs from Rominia. Among all these studied parameters only prophylactic treatment was found associated with prevalence of *Ehrlichia* spp. Malik et al. (2018) studied risk factor such as sex, age, ticks on dog, body temperature, mucus membrane, haematuria and vomiting in dogs from Punjab (Pakistan) and correlated them with prevalence with *Ehrlichia canis* and reported that none of the studied parameter was associated with ehrlichiosis in dogs from Pakistan.

Treatment

Doxycycline, a semi-synthetic tetracycline, has been the first-line drug in the treatment of CME, while in the acute CME doxycycline has been shown to be very effective in eliminating the infection (Gal et al., 2007). Dosing recommendation for doxycycline in CME is 5 mg/Kg, orally, twice daily, for at least 28 days (Neer et al., 2002). Recently report, rifampin at 15 mg/kg, orally, twice daily, for 7 days, was as effective as doxycycline in eliminating experimental subclinical *E. canis* infection in two dogs (Schaefer et al., 2008).

Heartwater Disease

Morphology

Erlichia ruminantium (formely *Cowdria ruminantium* (Cowdry 1925)) ranges in shape from small pleomorphic to ellipsoidal to coccoid and sometimes they are seen as rod shaped organisms inside the vacuoles in the cytoplasm of vascular endothelial cells of ruminants (Ristic et al., 1984).

Taxonomy

Heartwater or cowdriosis is an economically important tick-transmitted disease of livestock that is caused by the rickettsia *Cowdria ruminantium* (now *Erlichia ruminantium*) (Uilenberg, 1983). The pathogen *Erlichia rurninantium* is an obligately intracellular Gram negative coccal bacterium that is placed in order Rickettsiales, family Anaplasmataceae, and genus *Erlichia* (Dumler et.al. 2001).

Symptoms

The infection is characterized by fever, nervous signs, hydrothorax, and hydropericardium and is often fatal in susceptible animals (Van de Pypekamp and Prozesky 1987).

Vector

There are 12 species of *Amblyomma* which are involve in the transmission of *Erlichia ruminantium.* Namely *Amblyomma yariegatum, Amblyomma pomposum, Amblyomma lepidum, Amblyomma astrion, Amblyomma hebraeum, Amblyomma cohaerens, Amblyomma tholloni, Amblyomma maculatum, Amblyomma gemma, Amblyomma sparsum, Amblyomma cajennense* and *Amblyomma marmoreum*. These vectors are found many parts of the world but mainly they are African in origin (Walker and Olwage, 1987).

Susceptible Hosts

Earlier in 1902 it was considered that the heart water disease is only of goat and sheep (Lounsbury, 1902). But later it has been also reported from cattle, domestic (Asian) buffalo *(Bubalis bulbalis)* (Yunker, 1996). The disease is not only limited to domesticated ruminants it has been also reported from some wild animals like African buffalo (*Synceros caffer)* (Andrew and Norval 1989) as well as from reptile (leopard tortoises) and birds (guinea fowl) (Bezuidenhout et al., 1985).

Geographical Localization

The infection is occurring in each country of Africa where the tick *Amblyomma*, are present but mainly it's reported from sub-Saharan Africa (Yunker, 1996). However, it has been spread to a number of islands in the Caribbean, Atlantic Oceans, Indian and poses a threat to the American (Norval et al., 1992).

ANAPLASMOSIS

Livestock industries face a major problem from tick-borne diseases (TBDs) in developing countries by affecting its economy, mainly by the spread of severe pathogens, causing paralysis or toxicosis and physical damage to animals. (Durrani et al., 2008). Anaplasmosis is one of the tick-borne infectious but noncontagious disease caused by several anaplasma species (Razzaq et al., 2015). This is considered as one of the top 10 economically important diseases causing a huge economic loss in domestic animals formally called as gall sickness (Naimat et al., 2019).

Cellular Morphology

Anaplasma genus is often coccoid to ellipsoidal and pleomorphic residing in vacuoles of erythrocytes of mammalian hosts (Dumler et al., 2001). The organisms develop independently, regularly in firm packed formations of cells termed morulae. These inclusions are seen in the monocytes, granulocytes and more impotently the neutrophils (Rikihisa, 2011). *A. centrale* is found in the center, *A. marginale* is usually observed at the margin (De Waal, 2000) and *A. ovis* can be seen as deep purple and almost spherical granules inside the erythrocytes on microscopy (Stoltsz, 2004). The inclusion bodies of *A. ovis* are indistinguishable from those of *A. marginale* (Metinou, 2017). *A. ovis* inclusion bodies are located in the marginal of the erythrocytes in 60 to 65% of the cases (Jansson Lagerkvist, 2017), while *A. marginale* is found marginal in 90% of the cases (Metinou, 2017). The rest have been found either submarginal or central location of the cell. Additionally, there is a high similarity between the *Anaplasma* inclusions and other intra-erythrocyte structures like Heinz bodies, Howell-Jolly bodies, basophilic stippling in reticulocytes or staining artifacts (Noaman, 2013)

Taxonomic Classification

Anaplasma were first identified by Sir Arnold Theiler, who first recognized as a “marginal point” were found responsible to cause some specific disease (Theiler, 1910). Belong to one of the three officially recognized family anaplasmataceae of the order Rickettsiales of α-Proteobacteria (Dumler et al., 2001). Which are small, Gram-negative, obligate intracellular bacteria that reside within the endothelial cells of blood vessels or the cytoplasm of blood cells such as neutrophils, monocytes, macrophages and erythrocytes (Lee et al., 2015). Which are known as important pathogens of medical and veterinary importance (Dumler et al., 2005).

Species

The genus *Anaplasma* is composed of 6 species *A. marginale, A. centrale, A. ovis, A. bovis, A. phagocytophilum* and *A. platys* that different from each other’s in theirs host and cell tropism. *A. marginale, A. centrale,* and *A. ovis* infect the RBCs of ruminants (Renneker et al., 2013). *A. bovis* infects the monocytes of ruminants and small mammals. *A. phagocytophilum* is the causative agent of human granulocytic anaplasmosis and infects the neutrophils of humans, ruminants, dogs, and horses (Zhan et al., 2010). Finally, *A. platys* is a platelet pathogen that infects dogs, and can cause canine cyclic thrombocytopenia. Of these *Anaplasma* species, *A. marginale, A. centrale,* and *A. ovis* are prominent pathogens in ruminants worldwide (Liu et al., 2012). In particular, *A. marginale* and *A. ovis* have considerable economic importance. *A. marginale* is highly pathogenic in cattle, while *A. centrale* is less pathogenic. *A. ovis* is moderately pathogenic in sheep, goats, and wild ruminants and causes acute disease in animals exposed to stress or other predisposing factors (Seong et al., 2015).

Clinical Signs

Anaplasmosis is characterized by fever, weight loss, decreased milk production, pale mucous membranes, severe anaemia, jaundice, hyper-excitability, abortion and mortality without hemoglobinemia and hemoglobinuria during acute phase of the infection (Atif et al. 2013). The most marked clinical signs of Anaplasmosis are anemia and jaundice, the latter occurring late in the disease. Affected animals may die suddenly from hypoxia if they are handled roughly (Quinn et al., 2002). Animals that survive the acute phase of the disease develop persistent infection (Kocan et al., 2003).

Life Cycles

In the life cycles of *Anaplasma* spp. there is a reproduction stage taking place both in ticks and vertebrate animals (Rikihisa, 1991). Which act as a reservoir of infection and causes a persistent infection in the vertebrate hosts. Ticks can't transmit *Anaplasma* spp. transovarially (from eggs to new generation of ticks) but they transmit it transstadially (from larvae to nymphs and from nymphs to adults). In the next stage when they feed on infected animals uninfected ticks become infected and can transmit the infectious agent to other mammals. When ticks take blood from infected hosts, pathogen first enter the midgut epithelium of tick, where its primary replication takes place (Ueti et al., 2009). Then the pathogens move to the tick salivary glands to invade the epithelial cells. Where their second replication cycles occurred and enter the salivary gland secretion when the tick feeds on the next vertebrate host (Rar and Golovljova., 2011). After biting anaplasma enters in erythrocyte and start replication. As a result, busting of RBCs occur known as hemolysis (Jongejan et al., 2004).

Ticks Vector

Anaplasmosis are biologically transmitted by about 20 types of ticks including *Rhipicephalus* spp., *Hyalomma* spp., *Boophilus, Demacentor* spp., *Ixodes* spp., among these *Boophilus microplus* is the major transmitting tick agent (Kocan et al., 2003). *A. ovis* is transmitted by *Dermacentor andersoni* in the New World and by *Rhipicephalus bursa* and other ticks in the Old World in small ruminants. (Friedhoff, 1997). Mechanical transmission has been demonstrated with a number of biting flies of the family Tabanidae (horseflies), and with mosquitoes of the genus *Psorophora* (Rajput et al., 2005) or through fomites by infected hypodermic needle, castrating, and dehorning instruments, blood transfusion embryo transplant (Demessie and Derso 2015).

Tick species that are identified in Pakistan from cows and buffaloes were *Hyalomma (Hy.) aegyptium, Hy. anatolicum anatolicum Hy, marginatum, Boophilus microplus, Rhipicephalus sanguineus* (Ramzan et al. 2008). In many countries like India, South Africa and Brazil the main carrier of *Anaplasma spp* is *Rhipicephalus (Boophilus) microplus* (Gosh 2007). In Pakistan the major tick vectors of bovine Anaplasmosis are not known. Mechanical transmission occurs through various contaminated instruments and Biting flies (Aubry and Geale 2011).

Geographical Distribution

Anaplasma is endemic in all six populated continents of the world; mostly in tropical and subtropical region (Sajid et al., 2014). Including much of Africa and Australia, but it is also frequently documented in temperate areas of the world like the United States and Europe. However, due to global change driving tick populations into new environments the pattern of disease distribution constantly changing (Khan et al., 2019). The infection is higher in female than male animals due to hormonal disturbances, due to its use in milk production, draught power and breeding system which pose it to weakened immune system (Demessie & Derso 2015).

The infection rate of anaplasmosis is different from one region to another region e.g., United States (4.6-17.7%), Costa Rica (60-91%), Columbia (63-100%), Jamaica (71%) (Kocan et al. 2003), South Africa (72%) (Molad et al. 2006), Mozambique (33-87%) (Alfredo et al. 2005), Turkey (27-29%) (Keles et al. 2001), Iran (19.35%) (Razmi et al. 2006), India (0.68%) (Shaw 1989), Bangladesh (71%) (Chawdhury et al. 2006), Sri Lanka (17-98%) (Jorgensen et al. 1992). The pattern of disease distribution is varying constantly because of global warming and driving of tick from one place to another (Jonsson and Reid, 2000).

Anaplasmosis is worldwide in distribution and has been reported in both cattle and buffaloes in different countries such as Italy (50%), China (32.35%), Costa Rica (59.6%). Morocco (21.9%) and Iran (50%). In Pakistan limited information's are available from few areas such as Islamabad (66.66%), Khushab (31.43%), Sargodha (37.14%), Attock (75.92%) and Rawalpindi (24.57%) (Khan et al., 2019).

Pakistan and India have 8.33% and 26.15% prevalence in small ruminants by microscopy (Nasreen et al., 2016; Rajasokkappan and Selvaraju, 2016). Renneker et al., 2013 study samples using major surface protein 4 (MSP-4) for PCR. Samples were collected from Turkey, Iraq, Sudan and Portugal of which 31.4%, 66.6% 41.6% and 82.5%, respectively, were positive. Shabana et al., 2018 reported *A. ovis* 25.3% sheep and 15.5% goats, and *A. phagocytophilum* 38.1% sheep and 20.8% goats from Saudi Arabia.

Han et al. 2017 investigated 14.3% occurrence of *A. ovis* strains in small ruminant from 12 provinces of China. In Europe Hungary show 99.4% seropositivity by cELISA and Norway have 55% prevalence of *A. phagocytophilum* in lambs by IFA and in France 52.0%

A. ovis prevalence were reported (Sandor Hornok et al., 2007; Lise Grova et al., 2011; Cabezas-Cruz et al. 2019).

Risk Factors

Severity of the disease depends on the age of animals, with older animals tending to have more severe clinical signs than younger animals (Kocan et al., 2003). Hot, rainy and humid weather linked with abundance of ticks are favorable conditions for its occurrence (El-Metenawy., 2000). The animals recovered from anaplasmosis become carriers. Splenectomy and intercurrent infections lower the resistance of the animals and render them more susceptible to the disease. Malnutrition and pregnancy are the others stress factors which increase the susceptibility of animals to anaplasmosis (Yasini et al., 2012). Spatio-temporal conditions such as vector habitat, bacterial populations, animal grazing systems, hygiene, and management practices affect the intensity of *Anaplasma* infection (Kocan et al., 2012).

Treatment

Use of drugs and vaccines have been attempted to prevent anaplasmosis. The predominant antibiotics used in the control of anaplasmosis are the tetracyclines. Other antibiotics which have also shown good results are enrofloxacin (Facury-Filho et al., 2012) and immidocarb (Stuen et al., 2013) although these are not frequently used (Coetzee et al., 2006). It has reported that during the earlier stages of the disease tetracycline is most effective, but it could be quite a problem to diagnose the disease during early stage of the disease (Kocan et al., 2010).

Conclusion

Ticks are associated with weight loss and reduced milk production in animals. Tick-borne diseases are a major economic concern to the global livestock industry, especially in developing nations. Absence of rural poultry, not performing acaricide treatments, traditional rural housing systems and grazing were important risk factors associated with higher tick prevalence in livestock farms. Age, gender, breed and animal species significantly affected the intensity of tick infestation. The outcomes of this study will be useful in the planning of integrated control strategies for ticks and tick-borne diseases.

References

Abdela, N. and Jilo, K. (2016). Bovine Babesiosis and its current status in Ethiopia: A Systemic Review. *Advances in Biological Research*, 10(3): 138-146.

Abdela, N., Ibrahim, N. and Begna, F. (2018). Prevalence, risk factors and vectors identification of bovine anaplasmosis and babesiosis in and around Jimma town, Southwestern Ethiopia. *Acta tropica*, 177: 9-18.

Adl, S.M., Simpson, A.G., Lane, C.E., Lukeš, J., Bass, D. and Bowser, S.S. (2012). The revised classification of eukaryotes. *Journal of Eukaryotic Microbiology*, 59: 429–493.

Aguiar, D.M., Cavalcante, G.T., Pinter, A., Gennari, S.M., Camargo, L.M.A. and Labruna, M.B. (2007). Prevalence of *Ehrlichia canis* (Rickettsiales: Anaplasmataceae) in dogs and *Rhipicephalus sanguineus* (Acari: Ixodidae) ticks from Brazil. *Journal of medical entomology*, 44(1): 126-132.

Ahmadi-Hamedani, M., Khaki, Z., Rahbari, S., Kazemi, B. and Bandehpour, M. (2009). Molecular identification of anaplasmosis in goats using a new PCR-RFLP method. *Iran Journal of Veterinary Research*, 10(4): 367-372.

Aktas, M., Altay, K. and Dumanli, N. (2007). Determination of prevalence and risk factors for infection with *Babesia ovis* in small ruminants from Turkey by polymerase chain reaction. *Parasitology Research*, 100(4): 797-802.

Aktas, M., Dumanli, N. and Angin, M. (2004). Cattle infestation by *Hyalomma* ticks and prevalence of *Theileria* in *Hyalomma* species in the east of Turkey. *Veterinary Parasitology*, 119(1), pp.1-8.

Al Mahmud, M.A., Belal, S.S.H. and Hossain, M.A. (2015). Prevalence of theileriosis and babesiosis in cattle in Sirajganj district of Bangladesh. *Research in Agriculture Livestock and Fisheries*, 2(1): 79-86.

Al-Fahdi, A., Alqamashoui, B., Al-Hamidhi, S., Kose, O., Tageldin, M.H., Bobade, P., Johnson, E.H., Hussain, A.R., Karagenc, T., Tait, A. and Shiels, B. (2017). Molecular surveillance of *Theileria* parasites of livestock in Oman. *Ticks and tick-borne diseases*, 8(5): 741-748.

Al-Fetly, D.R.H. (2012). Detection of *Theileria* spp. in blood samples and estimation of haematological and biochemical changes in sheep in Al-Diwaniya province. *Kufa Journal for Veterinary Medical Sciences*, 3(2): 45-53.

Alfredo, A.A.N., Jonsson, N.N., Finch, T.M., Neves, L., Mollov, J.B. and Jorgensen, W.K. (2005). Serological survey of *Babesia bovis* and *Anaplasma marginale* in cattle in Tete Province, Mozambique. *Tropical Animal Health Production*, 37(2): 121-31.

Allsopp, B.A. (2015). Heartwater-*Ehrlichia ruminantium* infection. *Scientific and tehchical Review of the Office International des Epizooties*, 34 (2), 557-568.

Altay, K., Aktaş, M. and Dumanli, N. (2004). *Theileria* infections in small ruminants in the east and southeast Anatolia. *Blood*, 2005.

Altay, K., Aktas, M. and Dumanli, N. (2008). Detection of *Babesia ovis* by PCR in *Rhipicephalus bursa* collected from naturally infested sheep and goats. *Research in Veterinary Science*, 85(1): 116-119.

Anderson, B.E., Dawson, J.E., Jones, D.C. and Wilson, K.H. (1991). *Ehrlichia chaffeensis*, a new species associated with human ehrlichiosis. *Journal of clinical microbiology*, 29(12): 2838-2842.

Andrew, H.R. and Norval, R.A.I. (1989). The carrier status of sheep, cattle and African buffalo recovered from heartwater. *Veterinary Parasitology*, 34(3): 261-266.

Atif, F.A., Khan, M.S., Roheen, T., Muhammad, F., Younus, M., Avais, M. and Ullah, S. (2013). Seroprevalence of *Anaplasma marginale* infection among cattle from three districts of the Northern Punjab, Pakistan. *The Journal of Animal & Plant Sciences*, 23(4): 995-998.

Aubry, P. and Geale, D.W. (2011). A review of bovine anaplasmosis. *Transboundary and Emerging Diseases,* 58(1): 1-30.

Awad, H., Antunes, S., Galindo, R.C., do Rosário, V.E., De la Fuente, J., Domingos, A. and El Hussein, A.M. (2011). Prevalence and genetic diversity of *Babesia* and *Anaplasma* species in cattle in Sudan. *Veterinary Parasitology*, 181(2-4):146-152.

Azmi, K., Al-Jawabreh, A. and Abdeen, Z. (2019). Molecular detection of *Theileria ovis* and *Theleiria equi* in livestock from Palestine. *Scientific Reports,* 9: 11557.

Babes, V. (1888). Sur l'hémoglobinurie bactérienne du bœuf [On bacterial hemoglobinuria in beef]. *Comptes rendus de l'Académie des Sciences*, 107 : 692-694.

Bai, Q., Liu, G., Liu, D., Ren, J. and Li, X. (2002). Isolation and preliminary characterization of a large *Babesia* sp. from sheep and goats in the eastern part of Gansu Province, China. *Parasitology Research*, 88(1): S16-S21.

Bakheit, M.A., Seitzer, U. and Ahmed, J.S. (2006). A new recombinant protein-based ELISA for the diagnosis of malignant theileriosis of sheep and goats. *Parasitology Research*, 98(2): 145-149.

Bakken, J.S., Haller, I., Riddell, D., Walls, J.J. and Dumler, J.S. (2002). The serological response of patients infected with the agent of human granulocytic ehrlichiosis. *Clinical Infectious Diseases*, 34(1): 22-27.

Barrantes-González, A.V., Jiménez-Rocha, A.E., Romero-Zuñiga, J.J. and Dolz, G. (2016). Serology, molecular detection and risk factors of *Ehrlichia canis* infection in dogs in Costa Rica. *Ticks and Tick-Borne Diseases*, 7(6): 1245-1251.

Beall, M.J., Chandrashekar, R., Eberts, M.D., Cyr, K.E., Diniz, P.P.V., Mainville, C., Hegarty, B.C., Crawford, J.M. and Breitschwerdt, E.B. (2008). Serological and molecular prevalence of *Borrelia burgdorferi*, *Anaplasma phagocytophilum*, and *Ehrlichia* species in dogs from Minnesota. *Vector-Borne and Zoonotic Diseases*, 8(4): 455-464.

Belal, S.S.H., Al Mahmud, M.A. and Ferdous, M.J. (2014). Prevalence of anaplasmosis in cattle in Sirajganj district of Bangladesh. *Research in Agriculture Livestock and Fisheries*, 1(1), pp.97-103.

Belkahia, H., Said, M.B., El Hamdi, S., Yahiaoui, M., Gharbi, M., Daaloul-Jedidi, M., Mhadhbi, M., Jedidi, M., Darghouth, M.A., Klabi, I. and Zribi, L. (2014). First molecular identification and genetic characterization of *Anaplasma ovis* in sheep from Tunisia. *Small Ruminant Research*, 121(2-3): 404-410.

Berthelsson, J. (2017). *Anaplasma spp. infection in smallholder goat flocks around Gaborone, Botswana.* Degree Project 30 credits within the Veterinary Medicine Programme, Swedish University of Agricultural Sciences, Uppsala ISSN 1652-8697.

Bezuidenhout, J.D., Paterson, C.L. and Barnard, B.J. (1985). *In vitro* cultivation of *Cowdria ruminantium*. *The Onderstepoort journal of veterinary research*, 52(2): 113.

Bhattacharjee, K. and Sarmah, P.C. (2013). Prevalence of haemoparasites in pet, working and stray dogs of Assam and North-East India: A hospital-based study. *Veterinary World*, 6(11): 874.

Bowman, D., Little, S.E., Lorentzen, L., Shields, J., Sullivan, M.P. and Carlin, E.P. (2009). Prevalence and geographic distribution of *Dirofilaria immitis*, *Borrelia burgdorferi*, *Ehrlichia canis*, and *Anaplasma phagocytophilum* in dogs in the United States: results of a national clinic-based serologic survey. *Veterinary Parasitology*, 160(1-2): 138-148.

Bradway, D.S., de Echaide, S.T., Knowles, D.P., Hennager, S.G. and McElwain, T.F. (2001). Sensitivity and specificity of the complement fixation test for detection of cattle persistently infected with *Anaplasma marginale. Journal of Veterinary Diagnostic Investigation*, 13(1): 79-81.

Cabezas-Cruz, A., Gallois, M., Fontugne, M., Allain, E., Denoual, M., Moutailler, S., Devillers, E., Zientara, S., Memmi, M., Chauvin, A. and Agoulon, A. (2019). Epidemiology and genetic diversity of *Anaplasma ovis* in goats in Corsica, France. *Parasites & Vectors*, 12(1): 3.

Cardoso, L., Cortes, H.C.E., Reis, A., Rodrigues, P., Simões, M., Lopes, A.P., Vila-Viçosa, M.J., Talmi-Frank, D., Eyal, O., Solano-Gallego, L. and Baneth, G. (2013). Prevalence of *Babesia microti*-like infection in red foxes (Vulpes vulpes) from Portugal. *Veterinary Parasitology*, 196(1-2): 90-95.

Carvalho, F.S., Wenceslau, A.A., Carlos, R.S.A. and Albuquerque, G.R. (2008). Epidemiological and molecular study of *Ehrlichia canis* in dogs in Bahia, Brazil. *Genetics and Molecular Research,* 7(3): 657-662.

Chowdhury, S., Hossain, M.A., Barua, S.R. and Islam, S. (2006). Occurrence of common blood parasites of cattle in Sirajgonj Sadar area of Bangladesh, *Bangladesh Journal of Veterinary Medicine*, 4(2): 143-145.

Christova, I., Van De Pol, J., Yazar, S., Velo, E. and Schouls, L. (2003). Identification of *Borrelia burgdorferi sensu lato*, *Anaplasma* and *Ehrlichia* species, and spotted fever group Rickettsiae in ticks from Southeastern Europe. *European Journal of Clinical Microbiology and Infectious Diseases*, 22(9): 535-542.

Coetzee, J.F., Apley, M.D. and Kocan, K.M. (2006). Comparison of the efficacy of enrofloxacin, imidocarb, and oxytetracycline for clearance of persistent *Anaplasma marginale* infections in cattle. *Veterinary Therapeutics*, 7(4): 347.

Das, M.K., Baidya, S., Mahato, A., Pandit, S., Ghosh, J.D., Chaudhuri, S. and Das, M. (2015). Incidence of canine babesiosis in and around Kolkata, West Bengal, India. *Exploratory Animal and Medical Research*, 5(1), pp.102-107.

De Vos, A.J. (1991). Distribution, economic importance and control measures for *Babesia* and *Anaplasma. Recent developments in the control of anaplasmosis, babesiosis and cowdriosis*. Proceedings of, pp.3-12.

De Waal, D.T. (2000). Anaplasmosis control and diagnosis in South Africa. *Annals of the NewYork Academy of Sciences*, 916(1): 474-483.

De Waal, D.T. and Potgieter, F.T. (1987). The transstadial transmission of *Babesia caballi* by *Rhipicephalus evertsi evertsi. Onderstepoort Journal of Veterinary Research,* 54: 655-656.

De Waal, D.T., Lopez Rebollar, L.M. and Potgieter, F.T. (1992). The transovarial transmission of *Babesia trautmanni* by *Rhipicephalus simus* to domestic pigs. *Onderstepoort Journal of Veterinary Research,* 59: 219-221.

Del Pino, L.E.B., Roberto, N., Vincenzo, V., Francesca, I., Antonella, C., Luca, A.G., Francesco, B. and Teresa, S.M. (2016). *Babesia caballi* and *Theileria equi* infections in horses in Central-Southern Italy: Sero-molecular survey and associated risk factors. *Ticks and Tick-Borne Diseases*, 7(3): 462-469.

Demessie, Y. and Derso, S. (2015). Tick borne hemoparasitic diseases of ruminants: A review. *Advances in Biological Research*, 9(4): 210-224.

Derakhshanfar, A. and Mirzaei, M. (2008). Effect of *Peganum harmala* (wild rue) extract on experimental ovine malignant theileriosis: pathological and parasitological findings. *Onderstepoort Journal of Veterinary Research*, 75(1): 67-72.

Díaz-Sánchez, A.A., Pires, M.S., Estrada, C.Y., Cañizares, E.V., del Castillo Domínguez, S.L., Cabezas-Cruz, A., Rivero, E.L., Da Fonseca, A.H., Massard, C.L. and Corona-González, B. (2018). First molecular evidence of *Babesia caballi* and *Theileria equi* infections in horses in Cuba. *Parasitology Research*, 117(10): 3109-3118.

Donnelly, J. and Peirce, M.A. (1975). Experiments on the transmission of *Babesia divergens* to cattle by the tick Ixodes ricinus. *International Journal for Parasitology*, 5(3): 363-367.

Dumler, J.S. and Bakken, J.S. (1995). Ehrlichial diseases of humans: emerging tick-borne infections. *Clinical Infectious Diseases*, 20(5): 1102-1110.

Dumler, J.S., Barbet, A.F., Bekker, C.P., Dasch, G.A., Palmer, G.H., Ray, S.C., Rikihisa, Y. and Rurangirwa, F.R. (2001). Reorganization of genera in the families Rickettsiaceae and Anaplasmataceae in the order Rickettsiales: unification of some species of *Ehrlichia* with *Anaplasma, Cowdria* with *Ehrlichia* and *Ehrlichia* with *Neorickettsia,* descriptions of six new species combinations and designation of *Ehrlichia equi* and 'HGE agent' as subjective synonyms of *Ehrlichia phagocytophila. International Journal of Systematic and Evolutionary Microbiology*, 51(6): 2145-2165.

Dumler, J.S., Choi, K.S., Garcia-Garcia, J.C., Barat, N.S., Scorpio, D.G., Garyu, J.W., Grab, D.J. and Bakken, J.S. (2005). Human granulocytic anaplasmosis and *Anaplasma phagocytophilum. Emerging Infectious Diseases*, 11(12): 1828.

Durrani, A.Z., Shakoori, A.R. and Kamal, N. (2008). Bionomics of *Hyalomma* ticks in three districts of Punjab, Pakistan. *Animal & Poultry Science*, 18: 121-124.

Durrani, A.Z., Younus, M., Kamal, N., Mehmood, N. and Shakoori, A.R. (2011). Prevalence of ovine *Theileria* species in District Lahore, Pakistan. *Pakistan Journal of Zoology*, 43(1): 57-60.

Durrani, S., Khan, Z., Khattak, R.M., Andleeb, M., Ali, M., Hameed, H., Taqddas, A., Faryal, M., Kiran, S., Anwar, H. and Riaz, M. (2012). A comparison of the presence of *Theileria ovis* by PCR amplification of their SSU rRNA gene in small ruminants from two provinces of Pakistan. *Asian Pacific Journal of Tropical Disease*, 2(1): 43-47.

Eamens, G.J., Bailey, G., Gonsalves, J.R. and Jenkins, C. (2013). Distribution and temporal prevalence of *T heileria orientalis* major piroplasm surface protein types in eastern Australian cattle herds. *Australian Veterinary Journal,* 91(8), pp.332-340.

El Hussein, A.M., El Ghali, A.A. and Mohammed, S.A. (1993). Efficacy of buparvaquone in the treatment of malignant theileriosis of sheep in Ed–Damer Province Nile State Sudan: a field trial. *Sudan Journal of Veterinary Research*, 12: 51–57.

El-Ashker, M., Hotzel, H., Gwida, M., El-Beskawy, M., Silaghi, C. and Tomaso, H. (2015). Molecular biological identification of *Babesia, Theileria*, and *Anaplasma* species in cattle in Egypt using PCR assays, gene sequence analysis and a novel DNA microarray. *Veterinary Parasitology*, 207(3-4): 329-334.

El-Metenawy, T.M. (2000). Prevalence of blood parasites among cattle at the central area of Saudi Arabia. *Veterinary Parasitology*, 87(2-3): 231-236.

Evason, M., Stull, J.W., Pearl, D.L., Peregrine, A.S., Jardine, C., Buch, J.S., Lailer, Z., O'Connor, T., Chandrashekar, R. and Weese, J.S. (2019). Prevalence of *Borrelia burgdorferi, Anaplasma* spp., *Ehrlichia* spp. and *Dirofilaria immitis* in Canadian dogs, 2008 to 2015: a repeat cross-sectional study. *Parasites & Vectors*, 12(1): 64.

Facury-Filho, E.J., Carvalho, A.Ú.D., Ferreira, P.M., Moura, M.F., Apolinário, B.C., Santos, L.D.P.H. and Ribeiro, M.F.B. (2012). Effectiveness of enrofloxacin for the treatment of experimentally induced bovine anaplasmosis. *Revista Brasileira de Parasitologia Veterinária*, 21(1): 32-36.

Farkas, R., Tánczos, B., Gyurkovszky, M., Földvári, G., Solymosi, N., Edelhofer, R. and Hornok, S. (2013). Serological and molecular detection of *Theileria equi* infection in horses in Hungary. *Veterinary parasitology*, 192(1-3), pp.143-148.

Forbes, B.A., Sahm, D.F. and Weissfeld, A.S. (1998). Obligate intracellular and non culturable bacterial agents. In: *Bailey & Scott's diagnostic microbiology.* 10th ed. Miss. Mosby. St. Louis., Pp.751-765

Friedhoff, K.T. (1997). Tick-borne diseases of sheep and goats caused by *Babesia, Theileria* or *Anaplama* spp. *Parassitologia*, 39(2): 99-109.

Gal, A., Loeb, E., Yisaschar-Mekuzas, Y. and Baneth, G. (2008). Detection of *Ehrlichia canis* by PCR in different tissues obtained during necropsy from dogs surveyed for naturally occurring canine monocytic ehrlichiosis. *The Veterinary Journal,* 175(2): 212-217.

García-Bocanegra, I., Arenas-Montes, A., Hernández, E., Adaszek, Ł., Carbonero, A., Almería, S., Jaén-Téllez, J.A., Gutiérrez-Palomino, P. and Arenas, A. (2013). Seroprevalence and risk factors associated with *Babesia caballi* and *Theileria equi* infection in equids. *The Veterinary Journal*, 195(2): 172-178.

García-Bocanegra, I., Arenas-Montes, A., Hernández, E., Adaszek, Ł., Carbonero, A., Almería, S., Jaén-Téllez, J.A., Gutiérrez-Palomino, P. and Arenas, A. (2013). Seroprevalence and risk factors associated with *Babesia caballi* and *Theileria equi* infection in equids. *The Veterinary Journal*, 195(2): 172-178.

Garcia-Sanmartin, J., Barandika, J.F., Juste, R.A., Garcia-Perez, A.L. and Hurtado, A., (2008). Distribution and molecular detection of *Theileria* and *Babesia* in questing ticks from northern Spain. *Medical and Veterinary Entomology*, 22(4): 318-325.

Ge, Y., Pan, W. and Yin, H. (2012). Prevalence of *Theileria* infections in goats and sheep in southeastern China. *Veterinary Parasitology*, 186(3-4): 466-469.

Gebrekidan, H., Hailu, A., Kassahun, A., Rohoušová, I., Maia, C., Talmi-Frank, D., Warburg, A. and Baneth, G. (2014). *Theileria* infection in domestic ruminants in northern Ethiopia. *Veterinary Parasitology*, 200(1-2): 31-38.

Ghosh, S., Azhahianambia, P. and Yadav, M.P. (2007). Upcoming and future strategies of tick control: a review. *Journal of Vector Borne Diseases*, 44(2): 79-89.

Goff, W.L., Stiller, D., Roeder, R.A., Johnson, L.W., Falk, D., Gorham, J.R. and McGuire, T.C., (1990). Comparison of a DNA probe, complement-fixation and indirect immunofluorescence tests for diagnosing *Anaplasma marginale* in suspected carrier cattle. *Veterinary Microbiology*, 24(3-4): 381-390.

Goldstein, E.J., Thompson, C., Spielman, A. and Krause, P.J. (2001). Coinfecting deer-associated zoonoses: Lyme disease, babesiosis, and ehrlichiosis. *Clinical Infectious Diseases*, 33(5): 676-685.

Gubbels, M.J., Hong, Y., Weide, M., Qi, B., Nijman, I.J., Guangyuan, L. and Jongejan, F. (2000). Molecular characterization of the *Theileria buffeli/orientalis* group. *International Journal of Parasitology*, 30: 943-952.

Guidi, E., Pradier, S., Lebert, I. and Leblond, A. (2015). Piroplasmosis in an endemic area: analysis of the risk factors and their implications in the control of Theileriosis and Babesiosis in horses. *Parasitology Research*, 114(1): 71-83.

Gul, N., Ayaz, S., Gul, I., Adnan, M., Shams, S. and Akbar, N. (2015). Tropical theileriosis and east coast fever in cattle: present, past and future perspective. *International Journal of Current Microbiology and Applied Sciences*, 4(8):1000-18.

Habela, M., Solano, A., Anton, J.M., Boticario, D., Moreno, F. and Rol, J.A. (1995). Aislamiento, identification y transmision experimental de *Babesia bigemina* en Extremadura (España) [Isolation, identification and experimental transmission of Babesia bigemina in Extremadura (Spain)]. *IV Congreso Ibericode Parasitologia, Libro de Resúmenes, Comm. N° 121-P,* pp. 104-105.

Hadi, A.M. and Al-Amery, A.M.A. (2012). Isolation of *Theileria* and *Babesia* from gut and overy of hard ticks: *Hyalomma a. anatolicum* in Baghdad. *Diyala Agricultural Sciences Journal*, 4(2): 1-8.

Han, R., Yang, J., Liu, Z., Gao, S., Niu, Q., Hassan, M.A., Luo, J. and Yin, H. (2017). Characterization of *Anaplasma ovis* strains using the major surface protein 1a repeat sequences. *Parasites & Vectors*, 10(1): 447.

Hardalo, C.J., Quagliarello, V. and Dumler, J.S. (1995). Human granulocytic ehrlichiosis in Connecticut: report of a fatal case. *Clinical Infectious Diseases*, 21(4): 910-914.

Hashemi-Fesharki, R. (1997). Tick-borne diseases of sheep and goats and their related vectors in Iran. *Parassitologia*, 39(2): 115-117.

He, L., Liu, Q., Yao, B., Zhou, Y., Hu, M., Fang, R. and Zhao, J. (2017). A historical overview of research on *Babesia orientalis*, a protozoan parasite infecting water buffalo. *Frontiers in Microbiology*, 8:1323.

Heidarpour Bami, M., Khazraiinia, P., Haddadzadeh, H.R. and Kazemi, B. (2010). Identification of *Theileria* species in sheep in the eastern half of Iran using nested PCR-RFLP and microscopic techniques. *Iranian Journal of Veterinary Research*, 11(3): 262-266.

Homer, MJ., Aguilar-Delfin, I, Telford III, S.R., Krause, P.J. and Persing, D.H. (2000). Babesiosis. *Clinical Microbiology Reviews*, 13(3):451-469.

Hooshmand-Rad, P., (1989). Chemotherapy in ovine malignant theileriosis. *Archives of Razi Institute*, 40(1):.1-8.

Hunfeld, K.P., Lambert, A., Kampen, H., Albert, S., Epe, C., Brade, V. and Tenter, A.M. (2002). Seroprevalence of *Babesia* infections in humans exposed to ticks in midwestern Germany. *Journal of Clinical Microbiology*, 40(7): 2431-2436.

Hussein, A.H., Mohammed, N.A.E.S. and Mohammed, H.K. (2007). Theileriosis and babesiosis in cattle: haemogram and some biochemical parameters. In *Proceedings of XIII International Congress of International Society of Animal Hygiene,* pp. 143-150.

Ibrahim, H.M., Moumouni, P.F.A., Mohammed-Geba, K., Sheir, S.K., Hashem, I.S., Cao, S., Terkawi, M.A., Kamyingkird, K., Nishikawa, Y., Suzuki, H. and Xuan, X. (2013). Molecular and serological prevalence of *Babesia bigemina* and *Babesia bovis* in cattle and water buffalos under small-scale dairy farming in Beheira and Faiyum Provinces, Egypt. *Veterinary Parasitology*, 198(1-2): 187-192.

Ijaz, M., Rehman, A., Ali, M.M., Umair, M., Khalid, S., Mehmood, K. and Hanif, A. (2013). Clinico-epidemiology and therapeutical trials on babesiosis in sheep and goats in Lahore, Pakistan. *Journal of Animal and Plant Science*, 23(2): 666-669.

Inokuma, H., Ohno, K. and Yamamoto, S. (1999). Serosurvey of *Ehrlichia canis* and *Hepatozoon canis* infection in dogs in Yamaguchi Prefecture, Japan. *Journal of Veterinary Medical Science*, 61(10): 1153-1155.

Iqbal, F., Fatima, M., Shahnawaz, S., Naeem, M., Shaikh, R.S., Ali, M., Shaikh, A.S., Aktas, M. and Ali, M. (2011). A study on the determination of risk factors associated with babesiosis and prevalence of *Babesia* sp., by PCR amplification, in small ruminants from Southern Punjab (Pakistan). *Parasite: journal de la Société Française de Parasitologie*, 18(3): 229.

Ismail, N., Bloch, K.C. and McBride, J.W. (2010). Human ehrlichiosis and anaplasmosis. *Clinics in Laboratory Medicine*, 30(1): 261-292.

Jabbar, A., Abbas, T., Saddiqi, H.A., Qamar, M.F. and Gasser, R.B. (2015). Tick-borne diseases of bovines in Pakistan: major scope for future research and improved control. *Parasites & Vectors,* 8(1): 283.

Jansson Lagerkvist M. (2017*). Cross-sectional study of Anaplasma spp. in goats and sheep in Mongolia.* Degree Project 30 credits within the Veterinary Medicine Programme, Swedish University of Agricultural Sciences, Uppsala. ISSN: 1652-8697.

Jongejan, F. and Uilenberg, G. (2004). The global importance of ticks. *Parasitology*, 129(S1): S3-S14.

Jonson, N.N. and Reid, S.W.J. (2000). Guest editorial', *Veterinary Journal*, 160: 87-89.

Jonsson, N.N., Bock, R.E. and Jorgensen, W.K. (2008). Productivity and health effects of anaplasmosis and babesiosis on *Bos indicus* cattle and their crosses, and the effects of differing intensity of tick control in Australia. *Veterinary Parasitology*, 155(1-2): 1-9.

Jorgensen, W.K., Weilgama, D.J., Navaratne, M. and Dalgliesh, R.J. (1992). Prevalence of *Babesia bovis* and *Anaplasma marginale* at selected localities in Sri Lanka, *Tropical Animal Health Production,* 24: 9-14.

Keles, I., Deger, S., Altug, N., Karaea, M. and Akdenir, C. (2001). Tick-borne diseases in cattle, clinical and haematological findings, diagnosis, treatment, seasonal distribution, breed, sex and age factors and the transmitters of the disease', *Veteriner Fakultesi Dergisi*, 12(1-2): 26-32.

Khan, A., Mitchell III, R.D., Niaz, S., Ayaz, S., Khattak, I., Naeem, H., de León, A.A.P. and Zaman, M.A. (2019). Seroprevalence of *Anaplasma* spp. among sheep and goats in Charsadda District, Pakistan. *Small Ruminant Research*, 176:5-10.

Khan, N.U., Sarwar, M.S., Ayaz, S., Ali, H., Ali, A., Khan, I., Khan, M.A., Khan, A.U., Hussain, M., Ali, M. and Rashid, G. (2019). 1. Prevalence and risk factors analysis associated with anaplasmosis in symptomatic cattle under field conditions in southern Khyber Pakhtoonkhwa, Pakistan. *Pure and Applied Biology (PAB)*, 8(4): 2119-2127.

Kim, J.Y., Shin-Hyeong, C., Hyun-Na, J., Tsuji, M., Sung-Ran, C., Park, J., Gyung-Tae, C., Jung-Won, J., Cheun, H., Hyeong-Woo, L., Young-Hee, L. and Tong-Soo, K. (2007). First case of human Babesiosis in Korea: Detection and characterization of a novel type of *Babesia* sp. (KO1) similar to ovine *Babesia. Journal of Clinical Microbiology*, 45: 2084-2087.

Kispotta, S., Islam, M.F., Hoque, M.F., Rahman, M.S., Borman, A., Haque, M.A. and Rahman, M.R. (2016). Study of prevalence and associated risk factors of anaplasmosis and theileriasis in cattle. *Asian Journal of Medical and Biological Research*, 2(4): 567-576.

Kocan, K.M., Busby, A.T., Allison, R.W., Breshears, M.A., Coburn, L., Galindo, R.C., Ayllón, N., Blouin, E.F. and de la Fuente, J. (2012). Sheep experimentally infected with a human isolate of *Anaplasma phagocytophilum* serve as a host for infection of *Ixodes scapularis* ticks. *Ticks and Tick-Borne Diseases*, 3(3): 147-153.

Kocan, K.M., de la Fuente, J., Blouin, E.F., Coetzee, J.F. and Ewing, S.A. (2010). The natural history of *Anaplasma marginale*. *Veterinary Parasitology*, 167(2-4): 95-107.

Kocan, K.M., De la Fuente, J., Guglielmone, A.A. and Meléndez, R.D. (2003). Antigens and alternatives for control of *Anaplasma marginale* infection in cattle. *Clinical Microbiology Reviews*, 16(4): 698-712.

Konto, M., Tukur, S.M., Watanabe, M., Abd-Rani, P.A.M., Sharma, R.S.K., Fong, L.S. and Watanabe, M. (2017). Molecular and serological prevalence of *Anaplasma* and *Ehrlichia* sp. among stray dogs in East Malaysia. *Tropical Biomedicine*, 34: 570-575.

Krauss, H., Weber, A., Appel, M., Enders, B., Isenberg, H.D., Schiefer, H.G., Slenczka, W., von Graevenitz, A. and Zahner, H. (2003). *Zoonoses: infectious diseases transmissible from animals to humans*. 3rd edition Washington, DC: ASM press. p. 456.

Laha, R., Das, M. and Sen, A. (2015). Morphology, epidemiology, and phylogeny of Babesia: An overview. *Tropical Parasitology*, 5(2): 94-100.

Lawrence, J.A., De Vos, A.J. and Irvin, A. (1994). *East Coast fever In: Infectious Diseases of Livestock*. Coetzer J.A.W., Thompson G.R., Tustin R.C., editors.

Lee, S.H., Kim, K.T., Yun, S.H., Choi, E., Lee, G.H., Park, Y.S., Cho, K.H., Yi, S., Kwon, O.D., Kim, T.H. and Kwak, D. (2015). Serological and molecular detection of *Anaplasma phagocytophilum* in horses reared in Korea. *Veterinary Medicine*, 60(10): 533-538.

Lee, S.O., Na, D.K., Kim, C.M., Li, Y.H., Cho, Y.H., Park, J.H., Lee, J.H., Eo, S.K., Klein, T.A. and Chae, J.S. (2005). Identification and prevalence of *Ehrlichia chaffeensis* infection in *Haemaphysalis longicornis* ticks from Korea by PCR, sequencing and phylogenetic analysis based on 16S rRNA gene. *Journal of Veterinary Science*, 6(2):151-155.

Levine, N.D. (1971). Taxonomy of the piroplasms. *Transactions of the American Microscopical Society*, 90(1): 2-33.

Levine, N.D. (1985). *Veterinary protozoology* (No. 636.0896016 L665). Ames: Iowa State University Press.

Li, Y., Luo, J., Guan, G., Ma, M., Liu, A., Liu, J. Ren, Q., Niu, Q., Lu, B., Gao, J., Liu, Z., Dang, Z., Tian, Z., Zang, B., He, Z., Bai, Q. and Yin, H. (2009). Experimental transmission of *Theileria uilenbergi* infective for small ruminants by *Haemaphysalis longicornis* and *Haemaphysalis qinghaiensis*. *Parasitology Research*, 104: 1227–1231.

Littlewood, W. (1915). *Annual report for 1914.* Ministry of Agriculture Egypt, Veterinary Service.

Littman, M.P., Gerber, B., Goldstein, R.E., Labato, M.A., Lappin, M.R. and Moore, G.E. (2018). ACVIM consensus update on Lyme borreliosis in dogs and cats. *Journal of Veterinary Internal Medicine*, 32(3): 887-903.

Littman, M.P., Goldstein, R.E., Labato, M.A., Lappin, M.R. and Moore, G.E. (2006). ACVIM small animal consensus statement on Lyme disease in dogs: diagnosis, treatment, and prevention. *Journal of Veterinary Internal Medicine*, 20(2): 422-434.

Liu, Z., Hou, J., Bakheit, M.A., Salih, D.A., Luo, J., Yin, H., Ahmed, J.S. and Seitzer, U. (2008). Development of loop-mediated isothermal amplification (LAMP) assay for rapid diagnosis of ovine theileriosis in China. *Parasitology Research*, 103(6): 1407-1412.

Liu, Z., Ma, M., Wang, Z., Wang, J., Peng, Y., Li, Y., Guan, G., Luo, J. and Yin, H. (2012). Molecular survey and genetic identification of *Anaplasma* species in goats from central and southern China. *Appl. Environ. Microbiol*, 78(2): 464-470.

Lounsbury, C.P. (1902). Heartwater in calves. *Agricultural Journal of the Cape of Good Hope*, 21(1):65-169.

Luo, J., Yin, H., Guan, G., Zhang, Q. and Lu, W. (2002). Description of a new *Babesia* sp. infective for cattle in China. *Parasitology Research*, 88(1): S13-S15.

Mahoney, D.F. (1977). *Babesia* of domestic animals. In: *Parasitic protozoa*, Kreier J.P. (Ed). Academic Press, New York. p. 152.

Malik, M.I., Qamar, M., Ain, Q., Hussain, M.F., Dahmani, M., Ayaz, M., Mahmood, A.K., Davoust, B., Shaikh, R.S. and Iqbal, F. (2018). Molecular detection of *Ehrlichia canis* in dogs from three districts in Punjab (Pakistan). *Veterinary Medicine and Science*, 4(2): 126-132.

Mbizeni, S., Potgieter, F.T., Troskie, C., Mans, B.J., Penzhorn, B.L. and Latif, A.A. (2013). Field and laboratory studies on Corridor disease (*Theileria parva* infection) in cattle population at the livestock/game interface of uPhongolo-Mkuze area, South Africa. *Ticks and tick-borne diseases*, 4(3): 227-234.

McFadden, A.M.J., Gias, E., Heuer, C., Stevens McFadden, F.J. and Pulford, D.J., (2016). Prevalence and spatial distribution of cattle herds infected with *Theileria orientalis* in New Zealand between 2012 and 2013. *New Zealand Veterinary Journal*, 64(1): 55-59.

McHardy, N., Wekesa, L.S., Hudson, A.T. and Randall, A.W. (1985). Antitheilerial activity of BW720C (buparvaquone): A comparison with parvaquone. *Research in Veterinary Science,* 39: 29-33.

Mdladla, K., Dzomba, E.F. and Muchadeyi, F.C. (2016). Seroprevalence of *Ehrlichia ruminantium* antibodies and its associated risk factors in indigenous goats of South Africa. *Preventive Veterinary Medicine*, 125: 99-105.

Mehlhorn, H., Schein, E. and Ahmed, J.S. (1994). *Theileria.* In: Kreier JP (eds) *Parasitic Protozoa*, part 2, vol 7. Academic Press, San Diego, Calif., pp 217–304.

Metinou, K.S. (2017*). Molecular Identification and Characterization of Anaplasma Haemoparasites Isolated from Cattle and Sheep in Homabay County,* Kenya (Doctoral dissertation, University of Nairobi).

Michel, A.O., Mathis, A. and Ryser-Degiorgis, M.P. (2014). *Babesia* spp. in European wild ruminant species: parasite diversity and risk factors for infection. *Veterinary Research,* 45(1):65.

Mircean, V., Dumitrache, M.O., Györke, A., Pantchev, N., Jodies, R., Mihalca, A.D. and Cozma, V. (2012). Seroprevalence and geographic distribution of *Dirofilaria immitis* and tick-borne infections (*Anaplasma phagocytophilum, Borrelia burgdorferi sensu lato,* and *Ehrlichia canis*) in dogs from Romania. *Vector-Borne and Zoonotic Diseases*, 12(7): 595-604.

Mirzaiedehaghi, M. (2006). Treatment of natural ovine malignant theileriosis with a chloroform extract of the plant *Peganum harmala*: research communication. *Onderstepoort Journal of Veterinary Research*, 73(2): 153-155.

Molad, T., Mazuz, M.L., Fleiderovitz, L., Fish, L., Savitsky, I., Krigel, Y., Leibovitz, B., Moltoy, J., Jongejan, F., and Shkap, V. (2006). Molecular and serological detection of *A. centrale* and *A. marginale* infected cattle grazing within an endemic area. *Veterinary microbiology*, 113(1-2), 55-62.

Morel, P. (1989). *Tick-borne diseases of livestock in Africa. Manual of tropical veterinary parasitology.* Fischer M and Ralph S (eds), 301-391.

Movilla, R., García, C., Siebert, S., and Roura, X. (2016). Countrywide serological evaluation of canine prevalence for *Anaplasma* spp., *Borrelia burgdorferi* (*sensu lato*), *Dirofilaria immitis* and *Ehrlichia canis* in Mexico. *Parasites & Vectors*, 9(1): 421.

Muhanguzi, D., Matovu, E., and Waiswa, C. (2010). Prevalence and characterization of *Theileria* and *Babesia* species in cattle under different husbandry systems in western Uganda. *International Journal of Animal Veterinary Advances*, 2(2): 51-58.

Mujica, F.F., Perrone, T., Forlano, M., Coronado, A., Meléndez, R.D., Barrios, N., and Granda, F. (2011). Serological prevalence of Babesia caballi and Theileria equi in horses of Lara State, Venezuela. *Veterinary Parasitology*, 178(1-2): 180-183.

Mukhebi, A.W., Perry, B.D. and Kruska, R. (1992). Estimated economics of theileriosis control in Africa. *Preventive veterinary medicine,* 12(1-2): 73-85.

Nasreen, K.S., Khan, A., Niaz, S., and Akhtar, N. (2016). Serodiagnosis and haematological effect of anaplasmosis in goats and sheep of District Mardan, Khyber Pakhtunkhwa, Pakistan. *World Journal of Zoology*, 11(2): 67-80.

Nasreen, Khan, A., Niaz, S., Hassan, M., Khan, A., Ahmed, H., Khattak, I., Zeb, J., Naeem, H., Hassan, M.A., Ulucesme, M.C. and Ozubek, S. (2020). Molecular detection of small ruminant piroplasmosis and first report of *Theileria luwenshuni* (Apicomplexa: Theileridae) in small ruminants of Pakistan. *Experimental Parasitology*, 212: p107872.

Navarrete, M.G., Cordeiro, M.D., Silva, C.B., Massard, C.L., López, E.R., Rodríguez, J.C. A., ribeiro, C.C.D.U., Rodriguez, O.F. and Fonseca, A.H. (2018). Serological and molecular diagnosis of *Ehrlichia canis* and associated risk factors in dogs domiciled in western Cuba. *Veterinary Parasitology: Regional Studies and Reports*, 14: 170-175.

Naz, S., Maqbool, A., Ahmed, S., Ashraf, K., Ahmed, N., Saeed, K. and Nagra, I.A. (2012). Prevalence of theileriosis in small ruminants in Lahore-Pakistan. *Journal of Veterinary and Animal Science*, 2: 6-20.

Neer, T.M., Breitschwerdt, E.B., Greene, R.T., & Lappin, M.R. (2002). Consensus statement on ehrlichial disease of small animals from the infectious disease study group of the ACVIM. *Journal of Veterinary Internal Medicine*, 16(3): 309-315.

Neer, T.M., Breitschwerdt, E.B., Greene, R.T., and Lappin, M.R. (2002). Consensus statement on ehrlichial disease of small animals from the infectious disease study group of the ACVIM. *Journal of Veterinary Internal Medicine*, 16(3): 309-315.

Neer, T.M., (1998). Canine monocytic and granulocytic ehrlichiosis. *Infectious Diseases of the Dog and Cat*, 2nd ed. Philadelphia, PA: WB Saunders, pp.139-147.

Niu, Q., Liu, Z., Yu, P., Yang, J., Abdallah, M.O., Guan, G., and Yin, H. (2015). Genetic characterization and molecular survey of *Babesia bovis*, *Babesia bigemina* and *Babesia ovata* in cattle, dairy cattle and yaks in China. *Parasites & Vectors*, 8(1): 518.

Noaman, V. (2013). Discrimination between *Anaplasma marginale* and *Anaplasma ovis* by PCR-RFLP. *World Applied Science Journal*, 21(2): 190-95.

Noaman, V., and Bastani, D. (2016). Molecular study on infection rates of *Anaplasma ovis* and *Anaplasma marginale* in sheep and cattle in West-Azerbaijan province, Iran. In *Veterinary Research Forum*, 7(2):163. Faculty of Veterinary Medicine, Urmia University, Urmia, Iran.

Norval, R.A.I., Andrew, H.R., Yunker, C.E., and Burridge, M.J. (1992). Biological processes in the epidemiology of heartwater. In *Tick Vector Biology* (pp. 71-86). Springer, Berlin, Heidelberg.

Norval, R.A.I., Perry, B.D. and Young, A.S. (1992). *The epidemiology of theileriosis in Africa.* Academic Press. Orlando, FL.

Ogden, N.H., Bown, K., Horrocks, B.K., Woldehiwet, Z., and Bennett, M. (1998). Granulocytic *Ehrlichia* infection in ixodid ticks and mammals in woodlands and uplands of the UK. *Medical and Veterinary Entomology*, 12(4): 423-429.

Pancholi, P., Kolbert, C.P., Mitchell, P.D., Reed, K.D., Dumler, J.S., Bakken, J.S., Telford 3rd, S.R and Persing, D.H. (1995). *Ixodes dammini* as a potential vector of human granulocytic ehrlichiosis. *Journal of Infectious Diseases*, 172(4): 1007-1012.

Papadopoulos, B., Brossard, M. and Perié, N.M. (1996). Piroplasms of domestic animals in the Macedonia region of Greece 3. Piroplasms of small ruminants. *Veterinary parasitology,* 63(1-2): 67-74.

Parola, P., and Raoult, D. (2001). Ticks and tickborne bacterial diseases in humans: an emerging infectious threat. *Clinical Infectious Diseases*, 32(6): 897-928.

Parola, P., Davoust, B., and Raoult, D. (2005). Tick-and flea-borne rickettsial emerging zoonoses. *Veterinary Research*, 36(3): 469-492.

Penzhorn, B.L., Harrison-White, R.F., and Stoltsz, W.H. (2020). Completing the cycle: *Haemaphysalis elliptica*, the vector of *Babesia rossi*, is the most prevalent tick infesting black-backed jackals (*Canis mesomelas*), an indigenous reservoir host of *B. rossi* in South Africa. *Ticks and Tick-Borne Diseases*, 11(2): 101325.

Popov, V.L., Chen, S.M., Feng, H.M. and Walker, O.H. (1995). Ultrastructural variation of cultured Ehrlichia chaffeensis. *Journal of Medical Microbiology*, 43(6): 411-421.

Quinn, P.J., Markey, B.K., Carter, M.E., Donnelly, W.J.C., and Leonard, F.C. (2002). *Veterinary microbiology and microbial disease*. Blackwell science.

Rahman, A.S.M.S., Sumon, S.M.M.R., Khan, M.A.H.N.A., and Islam, M.T. (2015). Current status of subclinical form of babesiosis and anaplasmosis in cattle at Rangpur district in Bangladesh. *Progressive Agriculture*, 26(1): 51-59.

Rajasokkappan, S., and Selvaraju, G. (2016). Prevalence of anaplasmosis in goats in Ramanathapuram district of Tamil Nadu. *International Journal of Science, Environment and Technology*, 5(2): 511-514.

Rajput, Z.I., Hu, S.H., Arijo, A.G., Habib, M., and Khalid, M. (2005). Comparative study of *Anaplasma* parasites in tick carrying buffaloes and cattle. *Journal of Zhejiang University. Science. B*, 6(11): 1057.

Ramzan M, Khan, MS, Avais, M, Khan, JA, Pervez, K and Shahzad, W. (2008). Prevalence of ectoparasites and comparative efficacy of different drugs against tick infestation in cattle, *Journal of Animal & Plant Sciences*, 18(1): 17-19.

Rar, V. and Golovljova, I. (2011). *Anaplasma*, *Ehrlichia*, and "*Candidatus Neoehrlichia*" bacteria: pathogenicity, biodiversity, and molecular genetic characteristics, a review. *Infection, Genetics and Evolution*, 11(8): 1842-1861.

Razmi, G. and Yaghfoori, S. (2013). Molecular surveillance of *Theileria ovis*, *Theileria lestoquardi* and *Theileria annulata* infection in sheep and ixodid ticks in Iran. *Onderstepoort Journal of Veterinary Research,* 80(1): 01-05.

Razzaq, F., Khosa, T., Ahmad, S., Hussain, M., Saeed, Z., Khan, M.A. and Iqbal, F. (2015). Prevalence of *Anaplasma phagocytophilum* in horses from Southern Punjab (Pakistan). *Tropical Biomedicine*, 32(2): 233-239.

Renneker, S., Abdo, J., Salih, D.E. A., Karagenç, T., Bilgiç, H., Torina, A. and Seitzer, U. (2013). Can *Anaplasma ovis* in small ruminants be neglected any longer?. *Transboundary and emerging diseases*, 60, 105-112.

Richter Jr, P.J., Kimsey, R.B., Madigan, J.E., Barlough, J.E., Dumler, J.S., and Brooks, D.L. (1996). *Ixodes pacificus* (Acari: Ixodidae) as a vector of *Ehrlichia equi* (Rickettsiales: Ehrlichieae). *Journal of Medical Entomology*, 33(1): 1-5.

Rikihisa, Y. (1991). The tribe Ehrlichieae and ehrlichial diseases. *Clinical Microbiology Reviews*, 4(3): 286-308.

Rikihisa, Y. (2011). Mechanisms of obligatory intracellular infection with *Anaplasma phagocytophilum. Clinical Microbiology Reviews*, 24(3): 469-489.

Ristic, M., Huxsoll, D.L., and Holt, J.G. (1984). Tribe Ehrlichieae (Philip 1957). *Bergey's Manual of Systematic Bacteriology*, 1: 704-710.

Rjeibi, M.R., Gharbi, M., Mhadhbi, M., Mabrouk, W., Ayari, B., Nasfi, I., Jedidi, M., Sassi, L., Rekik, M. and Darghouth, M.A. (2014). Prevalence of piroplasms in small ruminants in North-West Tunisia and the first genetic characterisation of *Babesia ovis* in Africa. *Parasite*, 21.

Rojas, A., Rojas, D., Montenegro, V., Gutiérrez, R., Yasur-Landau, D. and Baneth, G. (2014). Vector-borne pathogens in dogs from Costa Rica: first molecular description of *Babesia vogeli* and *Hepatozoon canis* infections with a high prevalence of monocytic ehrlichiosis and the manifestations of co-infection. *Veterinary Parasitology*, 199(3-4): 121-128.

Ruprah, N.S. (1985). *A textbook of clinical protozoology*. Oxonian press.

Sabu, L., Sreekrishnan, R., Devada, K., Rejitha, T.S. and Lakshmanan, B. (2013). Babesia species in a cat-a case report. *Journal of Veterinary Parasitology*, 27(1): 68-69.

Sajid, M.S., Siddique, R.M., Khan, S.A., Zafar, I. and Khan, M.N. (2014). Prevalence and risk factors of anaplasmosis in cattle and buffalo populations of district Khanewal, Punjab, Pakistan. *Global Veterinaria*, 12(1): 146-153.

Saleem, S., Ijaz, M., Farooqi, S.H., Rashid, M.I., Khan, A., Masud, A., Mehmood, K. and Zhang, H. (2018). First molecular evidence of equine granulocytic anaplasmosis in Pakistan. *Acta Tropica*, 180: 18-25.

Salih, D.A., Hussein, A.E., Kyule, M.N., Zessin, K.H., Ahmed, J.S. and Seitzer, U. (2007). Determination of potential risk factors associated with *Theileria annulata* and *Theileria parva* infections of cattle in the Sudan. *Parasitology Research*, 101(5): 1285-1288.

Santino, I., Grillo, R., Nicoletti, M., Santapaola, D., Speziale, D., Sessa, R. and Del Piano, M. (2002). Prevalence of IgG antibodies against *Borrelia burgdorferi* sl and Ehrlichia phagocytophila in sera of patients presenting symptoms of Lyme disease in a central region of Italy. *International Journal of Immunopathology and Pharmacology*, 15(3): 245-248.

Schaefer, J.J., Kahn, J., Needham, G.R., Rikihisa, Y., Ewing, S.A. and Stich, R.W. (2008). Antibiotic clearance of *Ehrlichia canis* from dogs infected by intravenous inoculation of carrier blood. *Annals of the New York Academy of Sciences*, 1149: 263.

Schnittger, L., Yin, H., Gubbels, M.J., Beyer, D., Niemann, S., Jongejan, F. and Ahmed, J.S. (2003). Phylogeny of sheep and goat *Theileria* and *Babesia* parasites. *Parasitology Research*, 91(5): 398-406.

Schulze, T.L., Jordan, R.A., White, J.C., Roegner, V.E. and Healy, S.P. (2011). Geographical distribution and prevalence of selected *Borrelia, Ehrlichia*, and *Rickettsia* infections in

Amblyomma americanum (Acari: Ixodidae) in New Jersey. *Journal of the American Mosquito Control Association*, 27(3), 236-244.

Schutze, G.E. and Jacobs, R.F., (1997). Human monocytic ehrlichiosis in children. *Pediatrics*, 100(1): 10-e10.

Sells, D.M., Hildebrandt, P.K., Lewis, G.E., Nyindo, M.B. and Ristic, M. (1976). Ultrastructural observations on *Ehrlichia equi* organisms in equine granulocytes. *Infection and Immunity*, 13(1): 273-280.

Seong, G., Han, Y.J., Chae, J.B., Chae, J.S., Yu, D.H., Lee, Y.S., Park, J., Park, B.K., Yoo, J.G. and Choi, K.S. (2015). Detection of *Anaplasma* sp. in Korean native goats (Capra aegagrus hircus) on Jeju Island, Korea. *The Korean Journal of Parasitology*, 53(6): 765.

Shabana, I.I., Alhadlag, N.M. and Zaraket, H. (2018). Diagnostic tools of caprine and ovine anaplasmosis: a direct comparative study. *BMC Veterinary Research*, 14(1): 165.

Shah, S.S. A., Khan, M.I. and Rahman, H.U. (2017). Epidemiological and hematological investigations of tick-borne diseases in small ruminants in Peshawar and Khyber agency. Pakistan. *Journal of Advance Parasitology*, 4(2).

Sharma, A., Singla, L.D., Kaur, P. and Bal, M.S. (2015). PCR and ELISA vis-à-vis microscopy for detection of bovine anaplasmosis: a study on associated risk of an upcoming problem in North India. *The Scientific World Journal*, 2015.

Shaw, A.A. (1989). Investigation on some infections in the exotic, pure and cross bred cattle of Kashmir valley. *Indian Journal of Comparative Microbiology, Immunology and Infectious Diseases*, 10(1): 33-38.

Shaw, S.E., Day, M.J., Birtles, R.J. and Breitschwerdt, E.B. (2001). Tick-borne infectious diseases of dogs. *Trends in Parasitology*, 17(2): 74-80.

Singh, A., Singh, H., Singh, N.K., Singh, N.D. and Rath, S.S. (2014). Canine babesiosis in northwestern India: molecular detection and assessment of risk factors. *Biomed Research International*, 2014.

Siński, E., Bajer, A., Welc, R., Pawełczyk, A., Ogrzewalska, M. and Behnke, J.M. (2006). *Babesia microti*: prevalence in wild rodents and *Ixodes ricinus* ticks from the Mazury Lakes District of North-Eastern Poland. *International Journal of Medical Microbiology*, 296: 137-143.

Sivakumar, T., Tattiyapong, M., Okubo, K., Suganuma, K., Hayashida, K., Igarashi, I., Zakimi, S., Matsumoto, K., Inokuma, H. and Yokoyama, N. (2014). PCR detection of *Babesia ovata* from questing ticks in Japan. *Ticks and Tick-Borne Diseases*, 5(3): 305-310.

Soosaraei, M., Haghi, M.M., Etemadifar, F., Fakhar, M., Teshnizi, S.H., Hezarjaribi, H.Z. and Asfaram, S. (2018). Status of theileriosis among herbivores in Iran: A systematic review and meta-analysis, *Veterinary World*, 11(3): 332-341.

Spickler, A.R. and Roth, J.A. (2008). *Emerging and Exotic Diseases of Animal*, 3rd ed. CFSPH Iowa State University, pp. 132–135

Spitalska, E., Namavari, M.M., Hosseini, M.H., Shad-Del, F., Amrabadi, O.R. and Sparagano, O.A. (2005). Molecular surveillance of tick-borne diseases in Iranian small ruminants. *Small Ruminant Research*, 57(2-3): 245-248.

Steinman, A., Zimmerman, T., Klement, E., Lensky, I.M., Berlin, D., Gottlieb, Y. and Baneth, G. (2012). Demographic and environmental risk factors for infection by *Theileria equi* in 590 horses in Israel. *Veterinary Parasitology*, 187(3-4): 558-562.

Stoltsz, W.H. (2004). Ovine and caprine anaplasmosis. *Infectious Diseases of Livestock*, 2.

Stuen, S., Granquist, E.G. and Silaghi, C. (2013). *Anaplasma phagocytophilum*—a widespread multi-host pathogen with highly adaptive strategies. *Frontiers in Cellular and Infection Microbiology*, 3: 31.

Swai, E.S., Karimuribo, E.D., French, N.P., Fitzpatrick, J.L., Bryant, M.J., Kambarage, D.M. and Ogden, N.H. (2007). Seroprevalence of *Babesia bigeminai* in small holder dairy cattle in Tanzania and associated risk factors. *Journal of the South African Veterinary Association*, 78(1): 15-20.

Taha, K.M., Salih, D.A., Ali, A.M., Omer, R.A. and El Hussein, A.M. (2013). Naturally occurring infections of cattle with *Theileria lestoquardi* and sheep with *Theileria annulata* in the Sudan. *Veterinary Parasitology*, 191(1-2): 143-145.

Tefera, M., Worku, A., Tolosa, T. and Bitew, M. (2011). Prevalence and risk factors for donkey babesiosis in and around Debre Zeit, Central Ethiopia. *Veterinar. Researh*, 4:56-60.

Telford III, S.R., Gorenflot, A., Brasseur, P. and Spielman, A. (1993). Babesial infections in humans and wildlife. *Parasitic Protozoa*, 5: 1-47.

Tembue, A.A., Silva, F.J., Silva, J.B., Santos, T.M., Santos, H.A., Soares, C.O. and Fonseca, A.H. (2011). Risk factors associated with the frequency of antibodies against *Babesia bovis* and *Babesia bigemina* in cattle in southern Mozambique. *Pesquisa Veterinária Brasileira*, 31(8): 663-666.

Theiler, A. (1910). *Anaplasma marginale (gen. and spec. nov.): The marginal points in the blood of cattle suffering from a specific disease*. Pretoria: Government Printing and Stationary Office.

Thompson, A.T., White, S., Shaw, D., Egizi, A., Lahmers, K., Ruder, M.G. and Yabsley, M.J. (2020). *Theileria orientalis* Ikeda in host-seeking *Haemaphysalis longicornis* in Virginia, USA. *Ticks and Tick-borne Diseases*, 11(5):101450.

Tuli, A., Singla, L.D., Sharma, A., Bal, M.S., Filia, G. and Kaur, P. (2015). Molecular epidemiology, risk factors and hematochemical alterations induced by *Theileria annulata* in bovines of Punjab (India). *Acta Parasitologica*, 60(3): 378.

Ueti, M.W., Knowles, D.P., Davitt, C.M., Scoles, G.A., Baszler, T.V. and Palmer, G.H. (2009). Quantitative differences in salivary pathogen load during tick transmission underlie strain-specific variation in transmission efficiency of *Anaplasma marginale. Infection and Immunity*, 77(1): 70-75.

Uilenberg G. (2001) Babesiosis. In: Wallingford MW (ed) *Encyclopedia of artropod-transmitted infection of man and domesticated animals.* CABI Publishing, UK.

Uilenberg, G. (1981) Theilerial species of domestic livestock. In *Advances in the Control of Theileriosis* (pp. 4-37). Springer, Dordrecht.

Uilenberg, G. (1983). Heartwater (*Cowdria ruminantium* infection): current status. *Advances in Veterinary Science and Comparative Medicine*, 27: 427-480.

Uilenberg, G. (1995). International collaborative research: significance of tick-borne hemoparasitic diseases to world animal health. *Veterinary Parasitology*, 57(1-3): 19-41.

Uilenberg, G. (2006). Babesia - a historical overview. *Veterinary Parasitology*, 138(1-2): 3-10.

Uilenberg, G., Perie, N.M., Lawrence, J.A., De Vos, A.J., Paling, R.W. and Spanjer, A.A.M. (1982). Causal agents of bovine theileriosis in southern Africa. *Tropical animal health and production*, 14(3), pp.127-140.

Ullah, N., Durrani, A.Z., Avais, M., Ahmad, N., Ullah, S., Ullah, S., Khan, M.A., Ul-Haq, I. and Khan, N.U. (2018). A first report on prevalence of caprine theileriosis and its association with host biomarkers in Southern Khyber Pakhtunkhwa, Pakistan. *Small Ruminant Research*, 159*:* 56-61.

Van de Pypekamp, H.E. and Prozesky, L. (1987). *Heartwater. An overview of the clinical signs, susceptibility and differential diagnoses of the disease in domestic ruminants.*

Walker. J.B. and Olwage A. (1987): The tick vectors of *Cowdria ruminatium (*Ixodoidea: Ixodidae, genus *Amblyomma*) and their distribution. *Onderstepoort Journal of Veterinary Research,* 54: 353-379.

Wen, B., Rikihisa, Y., Mott, J., Fuerst, P.A., Kawahara, M. and Suto, C. (1995). *Ehrlichia muris* sp. nov., identified on the basis of 16S rRNA base sequences and serological, morphological, and biological characteristics. *International Journal of Systematic and Evolutionary Microbiology*, 45(2): 250-254.

Westmoreland, L.S., Stoskopf, M.K., Sheppard, E., DePerno, C.S., Gould, N.P., Olfenbuttel, C. and Maggi, R.G. (2019). Detection and prevalence of *Babesia* spp. in American black bears (Ursus americanus) from eastern and Western North Carolina, USA. *Journal of Wildlife Diseases*.

Wicki, R., Sauter, P., Mettler, C., Natsch, A., Enzler, T., Pusterla, N. and Lutz, H. (2000). Swiss Army Survey in Switzerland to determine the prevalence of *Francisella tularensis,* members of the *Ehrlichia phagocytophila* genogroup, *Borrelia burgdorferi* sensu lato, and tick-borne encephalitis virus in ticks. *European Journal of Clinical Microbiology and Infectious Diseases*, 19(6): 427-432.

World Organization for Animal Health – Oie (2008). Chapter 2.4.1 bovine. In: MCELWAIN, T.F. *Manual of diagnostic tests and vaccines for terrestrial animals*. 6th ed. Paris: World Organization for Animal Health. p. 599-610.

Wright, C.L., Gaff, H.D. and Hynes, W.L. (2014). Prevalence of *Ehrlichia chaffeensis* and *Ehrlichia ewingii* in *Amblyomma americanum* and *Dermacentor variabilis* collected from southeastern Virginia, 2010–2011. *Ticks and Tick-Borne Diseases*, 5(6): 978-982.

Yabsley, M.J., McKibben, J., Macpherson, C.N., Cattan, P.F., Cherry, N.A., Hegarty, B.C. and Perea, M.L. (2008). Prevalence of *Ehrlichia canis*, *Anaplasma platys*, *Babesia canis vogeli*, *Hepatozoon canis*, *Bartonella vinsonii berkhoffii*, and *Rickettsia* spp. in dogs from Grenada. *Veterinary Parasitology*, 151(2-4): 279-285.

Yamane, I., Gardner, I.A., Telford, S.R., Elward, T., Hair, J.A. and Conrad, P.A. (1993). Vector competence of *Rhipicephalus sanguineus* and *Dermacentor variabilis* for American isolates of Babesia gibsoni. *Experimental & Applied Acarology*, 17(12): 913-919.

Yang, Y., Mao, Y., Kelly, P., Yang, Z., Luan, L., Zhang, J., Li, J. El-Mahallawy, H.S. and Wang, C. (2014). A pan-Theileria FRET-qPCR survey for *Theileria* spp. in ruminants from nine provinces of China. *Parasites & Vectors*, 7(1): 413.

Yasini, S.P., Khaki, Z., Rahbari, S., Kazemi, B., Amoli, J.S., Gharabaghi, A. and Jalali, S.M. (2012). Hematologic and clinical aspects of experimental ovine anaplasmosis caused by *Anaplasma ovis* in Iran. *Iranian Journal of Parasitology*, 7(4): 91.

Ybanez, A.P., Sivakumar, T., Battsetseg, B., Battur, B., Altangerel, K., Matsumoto, K. and Inokuma, H. (2012). Specific molecular detection and characterization of *Anaplasma marginale* in Mongolian cattle. *Journal of Veterinary Medical Science*, 12: 0361.

Yin, H., Luo, J., Schnittger, L., Lu, B., Beyer, D., Ma, M., Guan, G., Bai, Q., Lu, C. and Ahmed, J. (2004). Phylogenetic analysis of *Theileria* species transmitted by *Haemaphysalis qinghaiensis*. *Parasitology research*, 92(1), pp.36-42.

Yin, H., Schnittger, L., Luo, J., Seitzer, U. and Ahmed, J.S. (2007). Ovine theileriosis in China: a new look at an old story. *Parasitology Research*, 101(2): 191-195.

Yitayew, D. and Samuel, D., (2015). Tick borne hemoparasitic diseases of ruminants: A review. *Adv. Biol. Res*, 9(4), 210-224.

Yousefi, A., Rahbari, S., Shayan, P., Sadeghi-dehkordi, Z. and Bahonar, A. (2017). Molecular detection of Anaplasma marginale and Anaplasma ovis in sheep and goat in west highland pasture of Iran. *Asian Pacific Journal of Tropical Biomedicine*, 7(5): 455-459.

Yunker, C.E. (1996). Heartwater in sheep and goats: a review. *Ondersteport Journal of Veterinary Research*, 63(2):159-70.

Zahler, M., Schein, E., Rinder, H. and Gothe, R. (1998). Characteristic genotypes discriminate between Babesia canis isolates of differing vector specificity and pathogenicity to dogs. *Parasitology Research*, 84(7): 544-548.

Zhan, L., Cao, W.C., Jiang, J.F., Zhang, X.A., Wu, X.M., Zhang, W.Y., Liu, W., Zuo, S.Q., Cao, Z-H, Yang, H, Richardus, J.H. and Habemma, J.D. (2010). *Anaplasma phagocytophilum* in livestock and small rodents. *Veterinary Microbiology*, 144(3-4): 405-408.

Zhang, X., Liu, Z., Yang, J., Chen, Z., Guan, G., Ren, Q., Liu, A., Luo, J., Yin, H. and Li, Y (2014). Multiplex PCR for diagnosis of *Theileria uilenbergi*, *Theileria luwenshuni*, and *Theileria ovis* in small ruminants. *Parasitology Research*, 113(2): 527-531.

Zhou, Z., Li, K., Sun, Y., Shi, J., Li, H., Chen, Y. and Wang, Z. (2019). Molecular epidemiology and risk factors of *Anaplasma* spp., *Babesia* spp. and *Theileria* spp. infection in cattle in Chongqing, China. *PloS One*, 14(7): e0215585.

Zintl, A., Mulcahy, G., Skerrett, H.E., Taylor, S.M. and Gray, J.S. (2003). *Babesia divergens*, a bovine blood parasite of veterinary and zoonotic importance. *Clinical Microbiology Reviews*, 16(4): 622-636.

In: The Entomological Guide to *Rhipicephalus*
Editors: Sachin Kumar, Raquel Cossio Bayugar et al.
ISBN: 978-1-53619-619-1

Chapter 3

TICKS AND TICK BORNE DISEASES IN PAKISTAN

***Asia Parveen*[1], *Sehrish Ashraf*[1], *Adil Khan*[2], *Muhammad Asif*[1] *and Furhan Iqbal*[1,*]**
[1]Institute of Pure and Applied Biology, Zoology Division.
Bahauddin Zakariya University Multan, 60800, Pakistan
[2]Department of Zoology, Abdul Wali Khan University Mardan, Pakistan

ABSTRACT

Ticks and tick-borne diseases (TBDs) are responsible for financial constrains for livestock sector in tropical and subtropical regions of the world and ultimately affects the livelihoods of poor farming communities as they are exclusively dependent on their animals. Livestock sector has become an integral part of Pakistan's economy and large number and variety of ruminants are reared, by the owners, in order to meet an increasing demand of milk, meat and hide products. It is timely to review current status of TBDs in Pakistan with a note on their vector ticks in order to identify the gaps in the knowledge of TBDs and their control. Anaplasmosis, babesiosis and theileriosis are the three major TBDs that are reported from various parts in Pakistan. This article will provide a brief background on key TBDs and ticks and reviews the current status of ticks and TBDs in Pakistan. This extensive review of literature will add to the understanding of these diseases and will propose areas for future research and will draw attention to the need for improved tools for the diagnosis and control of TBDs in Pakistan.

Keywords: tick borne diseases, ticks, theileriosis, anaplasmosis, babesiosis, Pakistan

LIVESTOCK SECTOR OF PAKISTAN

Livestock is our natural resource for the conversion of roughage into good quality-foods like milk and meat. These foods from animal resources are a rich source of essential amino acids and it has been reported that their quality is far better than those present in food from

[*] Corresponding Author's Email: furhan.iqbal@bzu.edu.pk.

plant sources. The other livestock products like hides are raw material for industries (Costa et al., 2011). Livestock sector is increasingly organized in long market chains that employ at least 1.3 billion people globally and directly support the livelihoods of 600 million poor small holder farmers in the developing world (Thornton et al. 2006).

In a country like Pakistan with rain-fed agricultural production system, livestock provides security against crop-failure. For low income countrymen in the villages, livestock is a form of social security that can be cashed at the time of need. Livestock are also sports and entertainment resource and are also considered as symbol of pride and prestige in rural areas of Pakistan (Zulfiqar et al., 2012). The livestock sector in Pakistan is represented mainly by small farm holders to meet the needs of nutrients and proteins, food security and income. In Pakistan 30–35 million people in the rural population rely on livestock sector for their livelihoods as 80% of the milk is produced by rural stock holders and commercial producers (de Castro, 1997). Livestock accounts for 37.5% of agricultural value-addition and about 9.4% of the Pakistan gross domestic product (GDP). In financial year 2000-2001, livestock earned 53 billion Rupees as net foreign exchange that was about 12.3% of all the export earnings of the country (Ensink et al., 2004). Livestock contributes 53.2% in agriculture sector of Pakistan and this contribution is higher than the crop sector (47.4% in agriculture and 10.3% in total gross domestic product GDP) that is major earning sector in Pakistan (Aslam et al., 2010). According to the 2013/14 Pakistan Livestock Census (Government of Pakistan, 2017), the livestock sector within the agricultural economy doubled from 25.3% in 1996 to 55%. The gross value of the livestock increased from $7.22 billion (Rs. 756.3 billion) in 2012/ 13 to $7.41 billion (Rs. 776.5 billion) in 2013/14, an increase of 2.7% as compared to the previous year. Gross production of milk in Pakistan increased from 47,895 million tons in 2011/12 to 50,990 million tons in 2013–14. Among the 8.4 million dairy-producing households, 51% own a herd of one to four animals, and 28% maintain five to ten animals (Karim et al., 2017)

Pakistan has 15 indigenous breeds of cattle belonging to zebu (one-humped) breed (*Bos indicus*), comprising 43% of the cattle population (Saeed et al., 2016). In addition to these local breeds Pakistan has exotic breeds (Friesian and Jersey) as well as cross breeds (Zulfiqar et al., 2012). Local breeds like Sahiwal and Cholistani are famous for heat and tick resistance while exotic dairy cattle like Holstein-Friesian has genetic potential of high milk production (Saeed et al., 2016). Pakistan is fortunate enough in having two best sub-tropical breeds of buffaloes such as N ili-Ravi and K undi both in the sense of meat and milk production (Bilal et al., 2006). In the livestock sector, water buffaloes and cattle, with an estimated population size of 35–40 million, are the main milk-producing animals and yielding approximately 18,000–31,000 million tons of milk annually (Pakistan Economic Survey, 2013). Out of total milk produced in the country, buffalo contributes about 68%, followed by cattle (27%) and sheep/ goat/camel (5%). Due to high fat contents of buffalo milk, it is the most preferred species in Pakistan (Bilal et al., 2006). Small ruminants like goat and sheep are the livestock of the choice of small farm holders due to their fast growth, economic and less hectic farming. Preference for goat meat is the major reason for its increased demand and hence increased forming (Khadim et al., 2003). Pakistan, at present, is having more than 60 million herds of goats, which consist of about 37 well recognized breeds found in different regions of the country (Abubakar et al., 2015). Equines (horses, donkeys and mules), family Equidae, are raised by man since long time for draught and game purposes and are playing a major role in the daily income of poor people. These animals are usually raised for transportation of people, carriage of goods and agricultural purposes. They are the most important draught animals, playing a key role in agriculture sector of the developing countries (Hussain et al., 2014).

In domesticated animals, Punjab Province of the Pakistan is richer in animal genetic resources as it is the home tract of world famous Nili-Ravi buffalo, Sahiwal cow, Kajli sheep and Beetal goat and most of the cows and buffaloes are also increasingly raised here to meet the growing demand for milk and meat (Khan 2003). Punjab province contributes 67% buffalo, 46% cattle, 37% goat and 26% sheep in livestock population of Pakistan. Livestock industry is economically more important in Khyber Pukhtoon Khwa (KPK) province as compared to Punjab and Sindh as these provinces have suitable land for crops and well established industrial sectors. According to an estimate in KPK, the livestock contributed 62.8 billion rupees to the national exchequer (Livestock census, 1996). Compound Annual Growth Rate (CAGR) of livestock population in Sindh for years 2006-2016 documents 2.397% cattle, 2.715% buffaloes, 0.65% sheep, 2.591% goats and 2.182% camels (Chandio et al., 2018). Raising livestock is one of the major activities and economic sources of livelihood for people living in the rural areas of Balochistan. A significant portion of national livestock population, particularly small ruminants, is reared in this part of the country (sheep 48% and goats 22% of the national flock). About 75% people of the province are dependent on agriculture and livestock, mixed farming being only sustainable (Raziq et al., 2010). Despite of the vital importance of livestock and dependency of farmers the productivity is far below than the actual potential. According to Arif et al. (2013), inadequate feed resources, unawareness of artificial insemination, finance shortage and limited health facilities are the major constraints among farmers. In this context, mostly small farmer or landless farmer suffer more because they fail to get the targeted outcome and in this way their livelihood becomes more diverse. Majority of farmers remain unaware of new practices and low adoption of recommended practices are also cause of poor dissemination of information through information sources (Ahmed et al. 2004).

TICKS OF PAKISTAN

Disease is a major threat to the livestock industry of developing countries like Pakistan because of the associated animal mortality, reduction in animal productivity and constraints to livestock improvement (Oura et al., 2004). Among the diseases, arthropod transmitted haemoparasitic diseases are of great economic impact on livestock affecting huge livestock population globally (Kasozi et al., 2014;. Jabbar et al., 2015). It has been estimated that economic losses due to vector transmitted diseases and cost of tick control worldwide is 7.0 billion dollars annually (Garcia, 2003). The direct production losses caused by ticks include loss of milk production, reduced weight (Sajid et al., 2018), low quality skins and hides. Besides these production losses, ticks also cause clinical and sub-clinical infections such as anaemia, dermatitis, paralysis, otoacariasis (Zeb et al., 2019). These voracious ectoparasites can infest wild, terrestrial, semi-aquatic vertebrates and human (Yu et al., 2015). Ticks transmit a wide variety of arbo viruses (Crimean-Congo hemorrhagic fever virus, tickborne encephalitis virus), bacteria (*Rickettsia* spp.), spirochetes (*Anaplasma* spp., *Borrelia* spp. and *Ehrlichia* spp.) and Protozoans (*Babesia* spp. and *Theileria* spp.)(Gondard et al., 2017; Schorderet-Weber et al., 2017).

The distribution of ticks in any region is closely associated with certain biotic and abiotic features. Tick dispersal in any area depends upon the environmental factors (Estrada-Peña, 2003) e.g., temperature of 27-39°C and relative humidity range of 60-80% render ticks more active however, ticks can also survive at temperatures down to - 18.5°C (Schulze et al., 2001; Mtshali et al., 2004). Being located in a subtropical zone (30° N, 70° E) within South Asia, most parts of Pakistan has favorable environmental conditions for the growth of various ticks (Shahzad et al., 2013; Karim et al., 2017; Rehman et al., 2017). Although ticks prevalent in various agrosystems of Pakistan are documented and are reported to be associated with economic losses, information is still missing about the tick's diversity, hosts they infest, seasonal dynamics and molecular phylogeny. Further large scale studies across the country are necessary to explore the molecular and cross breeding aspects at the geographical overlapping of various tick species and their associated pathogens to facilitate designing control strategies as well as awareness against tick infestation in the region (Ali et al., 2019).

Ghafar et al. (2020) had collected ticks from clinically healthy cattle (n = 116) and water buffaloes (n = 88) from five districts (Bahawalpur, Okara, Layyah, Jhelum in Punjab and Thatta district in Sindh) of Pakistan. The micro fluidic real time PCR was used to detect the genomic DNA of individual ticks for the presence of 27 bacterial and eight parasitic microorganisms.. The highest diversity of microorganisms was detected in *Hyalomma anatolicum* ticks, followed by *Rhipicephalus microplus*, *Hyalomma hussaini* and *Rhipicephalus annulatus*. Ticks collected from cattle were significantly more infected with piroplasms as compared to ticks collected from buffaloes (Table 1).

Ramazan et al. (2019) had collected and identified ticks from sheep (n = 25) and goats (n = 20) in Multan district of Punjab. They identified *Rhipicephalus sanguineus*, *Hyalomma anatolicum and Hyalomma marginatuma* from the enrolled animalsThey observed highest tick infestation in sheeps than in goats (Table 1).

Ali et al. (2019) collected large number of ticks infesting diverse hosts including cattle, buffalo, goat, sheep, domestic fowl, dog and horse from various regions of Khyber Pakhtunkhwa (KPK) Pakistan and morpho-taxonomically categorized ticks into six genera comprising of 17 species *(Rhipicephalus microplus*, *Hyalomma anatolicum*, *Argas persicus*, *Hyalomma impeltatum*, *Rhipicephalus turanicus*, *Rhipicephalus haemaphysaloides*, *Rhipicephalus annulatus*, *Hyalomma montgomeryi*, *Hyalomma marginatum*, *Rhipicephalus sanguineus* and *Hyalomma longicornis*, *Amblyomma gervaisi*, *Amblyomma exornatum*, *Amblyomma latum*, *Dermacentor marginatus* and *Haemaphysalis indica*, *Rhipicephalus microplus*, *Rhipicephalus annulatus*, *Hyalomma anatolicum*, *Hyalomma marginatum*, *and Haemaphysalis indica Punctata*). They observed highest tick infestation in cattle followed by buffalos, goats, sheep, domestic fowl, dogs, and horses. The phylogenetic analysis of cattle tick *Rhipicephalus microplus* based on partial mitochondrial cytochrome oxidase subunit I (*COX1*), 16S ribosomal RNA (*16S rRNA*) and internal transcribed spacer 2 (*ITS2*) sequences, revealed that *R. microplus* prevalent in this region belongs to clade which include ticks originating from Bangladesh, Malaysia, and India (Table 1).

Zeb et al. (2019) collected 123 ticks 123 from cattle in Bajaur Agency in KPK and 36 of them were used to compare two tick species (*Rhipicephalus microplus* and *Hyalomma anatolicum*) collected from these animals by molecular-phylogenetic analysis of two mitochondrial markers. All 11 *Rhipicephalus microplus* specimens were found identical for *cox1* haplotypes, whereas the 25 *Hyalomma anatolicum* specimens had nine *cox1* haplotypes. Among the 113 ticks molecularly analyzed for tick-borne protozoa, the sequence of *Babesia occultans* was successfully amplified from two specimens of *Hyalomma anatolicum*. *Theileria annulata* was present in both *Rhipicephalus microplus* (10.4%) and *Hyalomma anatolicum* (27.3%), with significantly higher prevalence rate in the latter species. They concluded, the above differences between *Rhipicephalus microplus* and *Hyalomma anatolicum* may be partly related to their life cycles involving one host or three hosts, respectively (Table 1).

Sajid et al. (2018) designed a study to report the point prevalence of ticks in the free range livestock population of the plain and hilly areas of Khyber Pakhtunkhwa (Mansehra, Haripur, Shangala, Kohistan), and Gilgit Baltistan (Diamer, Gilgit, Astor) along the Karakoram highway. Through convenient and snowball sampling techniques, a total of 813 animals, consisting of 232 sheep, 163 cattle, 365 goats and 53 buffaloes were screened for their tick burden. During the study, the two tick species identified were *Hyalomma anatolicum* and *Rhipicephalus microplus*. The overall prevalence of tick infestation among the screened livestock was 75.03%, with the highest distribution in sheep (81.47%) followed, in order, by cattle (77.91%), goats (72.05%) and buffalo (58.49%). The prevalence of ticks was found to be higher (85.67%) in younger livestock than adults (66.44%), and in females (80.33%) than males (66.44%) (Table 1).

Rehman et al. (2018) collected 3807 ticks from 369 ruminants from 108 livestock farms (semi-arid zone = 36, arid zone = 72) in Punjab. After morphological identification ticks were divided into 405 pools (*Hyalomma anatolicum* = 300, *Rhipicephalus microplus* = 89, *Hyalomma dromedarii* = 9, *Rhipicephalus turanicus* = 7) based on their species, locality of collection and the host. DNA from each pool was screened by a reverse line blot (RLB) hybridization assay for the presence of *Anaplasma*, *Ehrlichia*, *Rickettsia*, *Babesia* and *Theileria* species. The detected pathogens not only included species that were known to be endemic in Pakistan, such as *Theileria annulata* (6.7%), *Theileria orientalis* (3.5%), *Anaplasma marginale* (5.7%), *Anaplasma centrale* (2.7%), *Anaplasma ovis* (1.5%), *Babesia bigemina* (0.7%) and *Babesia bovis* (0.2%) but also several tick borne parasites that had not been reported to occur in Pakistan before were reported including *Ehrlichia minasensis* (3.2%), an *Anaplasma platys*-like organism (1.2%), *Babesia occultans* (0.2%) and *Rickettsia massiliae* (0.2%) as well as two previously uncharacterized species: *Ehrlichia* sp. Multan (18.0%) and *Anaplasma* sp. (BL 099-6) (2.22%). (Table 1).

Khalil et al. (2018) examined a total of 200 buffaloes at private farms from Jampur, district Rajanpur in Punjab and observed 44.5% prevalence of *Hyalomma* spp. in examined animals. Ticks were more prevalent in males (55.2%) as compared to females (42.7%). Prevalence of *Hyalomma* spp. was higher (48.78%) in younger animals having age between 1-5 years and lower (23.52%) in older animals with age range 16-20) years (Table 1).

Farooqi et al. (2017) collected 1200 ticks from 400 cattle and buffalo from districts Buner, Mardan and Bannu in KPK and reported *Rhepicephalus* as the most prevalent tick genus collected from these animals followed by *Heamaphysalis*, *Hyalomma*, *Dermacenter* and *Amblyomma*. A detailed account of tick species identified during this study is provided in Table 1.

Table 1. Studies from various districts of Pakistan reporting tick prevalence in livestock

Sampling locations	Host Animals	*Hyalomma* spp.	Haemaphysalis spp.	*Rhipicephalus* spp.	*Dermacentor* spp.	Other tick species	Parasites detected in ticks	References
Bahawalpur, Okara, Layyah, Jhelum Districts in Punjab and Thatta District in SIndh	Cattle and buffalo	*Hyalomma anatolicum*, *Hyalomma hussaini*, *Hyalomma scupense*	-	*Rhipicephalus microplus*, *Rhipicephalus annulatus*	-	-	Not applicable	Ghaffar et al., 2020
Multan District in Punjab	Sheep and goat	*Hyalomma anatolicum*, *Hyalomma marginatum*	-	*Rhipicephalus sanguineus*	-	-	-	Ramzan et al., 2019
Khyber Pakhtunkhwa (various parts)	Cattle, buffalo, sheep, goat, domestic fowl, dogs, horse	*Hyalomma impeltatum*, Hyalomma anatolicum, *Hyalomma marginatum*	*Haemaphysalis montgomeryi*, *Haemaphysalis indica*, *Haemaphysalis punctata*	*Rhipicephalus microplus*, *Rhipicephalus turanicus*, *Rhipicephalus haemaphysaloides*, *Rhipicephalus annulatus*, *Rhipicephalus sanguineus*, *Rhipicephalus haemaphysaloides*	*Dermacentor marginatus*	*Argas persicus* *Amblyomma gervaisi*, *Amblyomma exornatum*, *Amblyomma latum*	Not applicable	Ali et al., 2019
Bajaur Agency in KPK	Cattle	*Hyalomma anatolicum*	-	*Rhipicephalus microplus*	-	-	*Babesia occultans*, *Theileria annulata*	Zeb et al., 2019
Mansehra, Haripur, Shangala and Kohistan Districts in KPK and Gilgit Baltistan	Sheep, goats, cattle and buffaloes	*Hyalomma anatolicum*	-	*Rhipicephalus microplus*	8	-	Not applicable	Sajid et al., 2018
Punjab		*Hyalomma anatolicum*, *Hyalomma dromedarii*	-	*Rhipicephalus microplus*, *Rhipicephalus turanicus*	-	*Argas persicus*, *Ornithodoros tolozani*	*Theileria annulata*, *Theileria orientalis. Anaplasma marginale*, *Anaplasma centrale*, *Anaplasma ovis*, *Babesia bigemina Babesia bovis Ehrlichia minasensis Anaplasma platys*-like organism *Babesia occultans*, *Rickettsia massiliae*	Rehman et al., 2018
Rajanpur District in Punjab	Buffaloes	*Hyalomma* spp	-	-	-	-	Not applicable	Khalil et al., 2018

Sampling locations	Host Animals	***Hyalomma* spp.**	**Haemaphysalis spp.**	***Rhipicephalus* spp.**	***Dermacentor* spp.**	**Other tick species**	**Parasites detected in ticks**	**References**
Buner, Mardan and Bannu Districts in KPK	Cattle, buffaloes	*Hyalomma impeltatum, Hyalomma rufipes, Hyalomma egyptium, Hyalomma rufipes*	*Heamaphysalis aciculifer, Haemaphysalis excavatum, Haemaphysalis houyi, Haemaphysalis parmata,*	*Rhipicephalus microplus, Rhipicephalus appendiculatu, Rhipicephalus* decoloratus, *Rhipicephalus distinctus, Rhipicephalus longus, Rhipicephalus parvas*	*Dermacentor circumguttatus, Dermacentor rhinocerinus*	*Amblyomma pomposum*	Not applicable	Farooqi et al., 2017
Punjab, Sindh, KPK, Baluchistan Provinces and Gilgit-Baltistan	cattle, buffalo, sheep, goat, camel, poultry, cat, dog	*Hyalomma anatolicum*	*Haemaphysalis dromedarii*	*Rhipicephalus microplus*		*Ornithodorus Argas.*	Candidatus Rickettsia amblyommii, *Theilleria annulata*	Karim et al., 2017
Muzaffar Garh and Layyah Districts in Punjab	Goat	-	-	*Rhipicephalus sanguineus*	-	-	Not applicable	Sajid et al., 2011
Lahore District in Punjab	Cattle, buffalo	*Hyalomma* spp.	-	*Rhipicephalus* spp.	-	-	Not applicable	Ali et al., 2016
Toba Tek Singh District in Punjab	Buffaloes	*Hyalomma marginatum*	-	*Rhipicephalus microplus*	-	-	Not applicable	Iqbal et al., 2013
Faisalabad, Jhang, Khanewal Districts in Punjab	Cattle, buffaloes	*Hyalomma anatolicum, Hyalomma dromedary, Hyalomma marginatum*	-	*Boophilus annulatus,*	-	*Amblyomma varigatum*	*Theilleria annulata*	Ali et al., 2013
Lahore District in Punjab	Sheep	*Hyalomma* spp.	-	*Rhipicephalus* spp.	-	-	*Theileria ovis, Theileria lestoquardi*	Durrani et al., 2011
Kasur District in Punjab	Cattle	*Hyalomma* spp.	*Haemaphysalis* spp.	*Rhipicephalus* spp.			Not Applicable	Durrani and Kamal, 2008

Karim et al. (2017) had morphologically identified 3866 tick specimens from all four provinces and Gilgit-Baltistan of Pakistan and identified as 19 different tick species representing three hard tick genera, *Rhipicephalus*, *Haemaphysalis* and *Hyalomma*, and two soft tick genera, *Ornithodorus* and *Argas*. The bacterial diversity across these tick species was also assessed by amplifying bacterial *16S rRNA* gene sequences using a 454-sequencing platform in 10 of the different tick species infesting livestock. The notable genera detected include *Ralstonia*, *Clostridium*, *Staphylococcus*, *Rickettsia*, *Lactococcus*, *Lactobacillus*, *Corynebacterium*, *Enterobacter* and *Enterococcus*. Only three tick species *Rhipicephalus microplus*, *Hyalomma anatolicum* and *Hyalomma dromedarii* had evidence of infection with *Candidatus Rickettsia amblyommii* The *Hyalomma* ticks were also tested positive for *Theileria annulata* by using a qPCR assay (Table 1).

Sajid et al. (2017) had collected 800 tick samples from goats in Muzaffar Garh and Layyah districts in Punjab. Overall prevalence of goat tick infestation in both study districts was 60.1%. The prevalence was higher in district Muzaffar Garh than in district Layyah. *Rhipicephalus sanguineus* (24.1%) was the predominant species in both the districts. Highest tick infestation in goat was observed during July and the minimum was documented in November and December. Regarding host determinants, female animals were more heavily infested than males and younger animals were more burdened than older ones. Teddy goats are the most susceptible breeds followed in order by Beetal, cross-bred and Nachi (Table 1).

Ali et al. (2016) had enrolled 1258 tick infested bovines (cattle =726; buffaloes = 532) from Lahore district in Punjab to check determinants of tick infestation. Ticks belonging to *Hyalomma* genera were most prevalent followed by *Boophilus* spp. on animals under investigation. In cattle, crossbreds were the most affected breed followed by Sahiwal, Cholistani, Holstein Friesian and non-descript breed respectively, while in buffaloes, non-descriptive breed was more affected than Nili Ravi. Females were significantly the most affected gender than males in bovines. Calves were significantly more affected age group in both, followed by adult, young and old, respectively in cattle, while young buffaloes had highest tick infestation followed by adult and old respectively. Udder was the highly tick infested site in both type of animals followed by inner thighs, perineum, legs and tail, and neck. Summer was found to be the most tick favorable season, followed by winter, spring and autumn respectively in bovines (Table 1).

Iqbal et al. (2013) conducted a survey to define the epidemiological parameters related to the prevalence and associated risk factors of tick infestation in buffaloes in the Toba Tek Singh District of central Punjab. The prevalence of ticks on buffaloes was 31.21% (352/1,128). Among the species of ticks, the prevalence of *Hyalomma marginatum* (75.56% was higher than *Rhipicephalus microplus*. It was documented that female buffaloes and younger animals were more heavily infested with ticks than males and adult animals whereas breed was not a determinant. With regard to management and husbandry practices, the prevalence of ticks was higher in animals kept on un cemented flooring followed in order by partially cemented and fully cemented flooring. With regard to feeding systems, grazing animals were more burdened compared to stall-fed animals. The primary body area of infestation by ticks were head, neck, ear, dewlap, back, abdomen, foreleg, shoulder, hind leg, congenital areas, and tail (Table 1).

Ali et al. (2013) investigated the tick prevalence in dairy cattle (n = 710) and buffalo (n = 320) from Faisalabad, Jhang and Khanewal districts of Punjab. Overall prevalence of *Hyalomma* species was significantly higher (61%) than that of other genera of hard ticks. In sex-wise distribution, it was found that female *Hyalomma* species were the highest (85%) followed by *Amblyomma* species (81%), while prevalence of *Boophilus* and *Haemaphysalis* species were 77% each. The tick infestation rate in cattle (70%) was significantly higher than that of buffaloes (34%). *Theilleria annulata* was detected in 50% *Hyalomma anatolicum* and 40% *Hyalomma dromedary* ticks and they were not detected in *Hyalomma marginatum*, *Boophilus annulatus* and *Amblyomma varigatum* ticks when analysed by PCR (Table 1).

Durrani et al. (2011) had collected 100 ticks from sheep in District Lahore during spring and summer seasons The most commonly identified tick were belonging to *Hyalomma* (45%) followed by *Rhipicephalus* (41%) and *Boophilus* (14%). They found that 65.8% ticks were infected with *Theileria ovis* while *Theileria lestoquardi* was detected in 65.8% of collected ticks (Table 1).

Durrani and Kamal (2008) had conducted a tick infestation study in Friesian cattle from Kasur district of Pakistan and they had reported that *Hyalomma* ticks (15%) were most frequently collected from cattle under study followed by *Boophillus (*12%), *Haemaphysalis* (5%) and *Rhipicephalus* (3%) respectively (Table 1).

Tick Borne Diseases in Pakistan

The arid regions of Pakistan are notorious for the presence of the vector: tick (Acari: *Ixodidae*) and tick-borne diseases of domesticated animals as the hot and humid climate is highly conducive and favors the multiplication and survivability of ticks (Khan et al., 2004). Four main tick born diseases (anaplasmosis, babesiosis, theileriosis, and cowdriosis) affect animals throughout the world and the former three are of major economic importance in Pakistan (Jabbar et al., 2015; Hassan et al., 2018). In the next part of this chapter, we are going to briefly introduce theileriosis, babesiosis and anaplasmosis and their documented status in Pakistan.

Theileriosis

Theileriosis is caused by the protozoan parasite belonging to *Theileria* genera the members of whom are round, ovoid, rod like or irregular shaped and they are found in lymphocytes, histiocytes and erythrocytes (Bhatnagar et al., 2015). They infect both wild and domestic ruminants and have complex life cycles that involve both vertebrate and invertebrate hosts (OIE, 2014). *Theileria* spp. are transmitted by a variety of Ixodid ticks of the genera *Rhipicephalus*, *Hyalomma*, *Amblyomma* and *Haemaphysalis* (Bishop et al., 2009; Mans et al., 2015).

Theileria spp.

Globally, *Theileria annulata* (cause of tropical theileriosis) and *Theileria parva* (causes of East Coast fever) are the most economically important tick-transmitted pathogenic species (Bhatnagar et al. 2015; Gebrekidan et al., 2015). *Theileria annulata* occurs in cattle, yaks, water buffalo, camels (Pieszko, 2015), horses and donkeys (Asif et al., 2020). *Theileria mutans*, *Theileria Orientalis/buffeli*, *Theileria velifera* and *Theileria sergenti* can also infect domesticated ruminants and they are believed to cause milder and/or nonpathogenic theileriosis (Kalume et al., 2011). *Theileria* spp. has also been found in wild bovid in some continents. *Theileria lestoquardi*, *Theileria separata* and *Theileria ovis* are in sheep and goats (Iqbal et al., 2013).

Clinical Signs of Theileriosis

Infection due to *Theileria* spp. is characterized by a marked rise in body temperature (reaching 40-41.5°C), depression, lacrimation, nasal discharge, swelling of the superficial lymph nodes and anemia (Saeed et al., 2016).

Control of Theileriosis

Strategies to control tropical theileriosis are controlling tick infestation by acaricides, immunization using live vaccines and treatment of infected cattle (Akat et al., 2014). Specific anti-*Theileria* drugs (Buparvaquone and Parvaquone) can be used for treatment (Bishop et al., 2009; Mohamed et al., 2018).

Bovine Theileriosis in Pakistan

A number of studies regarding bovine, ovine and equine theileriosis has been reported from various parts of Pakistan and have used both conventional, modern as well as both types of methods for the detection of *Theileria* spp. in host animals. Conventional diagnostic methods used for parasite detection were microscopic examination of Giemsa stained blood films and serological tests, while modern techniques included PCR, real time PCR and reverse line blot (RLB) and recombinase polymerase amplification (RPA). Bovine theileriosis is the topic most extensively investigated topic among theileriosis in Pakistan and it has been reported from several districts in Punjab and KPK Provinces. In Pakistan, majority of the studies regarding theileriosis have been undertaken in apparently healthy animals either kept on private dairy/livestock farms or reared at public livestock research stations.

Rafiullah et al. (2019) collected 800 blood samples of cattle and buffalo from different areas in Districts Lakki Marwat and Peshawar in KPK and processed them for diagnosis of *Theileria parva* through different diagnostic techniques e.g., microscopy, ELISA and Real-time PCR. Overall prevalence for theileriosis was 27.8% through real-time PCR, followed by 24.6% through indirect ELISA and 11.4% through microscopic screening of Giemsa stained blood smear. Bovines of Lakki Marwat District were found more susceptible to *Theileria parva* than bovines from Peshawar. Smilarly, cattle were more prone to *Theileria* parva infection than buffalo (Table 2).

Hassan et al. (2018) had used recombinase polymerase amplification (RPA) and polymerase chain reaction (PCR) for thr detection of *Theileria annulata* in blood samples of 274 tick infested asymptomatic cattle from Chakwal, Faisalabad and Jhang District in Punjab.

17.5% samples were found positive by RPA while 7.7% samples amplified 281-bp fragment of the Enolase gene specific for *Theileria annulata* through PCR. Highest parasite prevalence was observed in Chakwal followed by Faisalabad and Jhang districts (Table 2).

Farooqi et al. (2017) determine the molecular prevalence of *Theileria annulata* in bovines of three distinct zones of KPK province through PCR. From 900 blood samples [cows (n = 479), buffaloes (n = 421)], 170 (18.9%) were found positive for *Theileria annulata* and highest prevalence of parasite (21.67%) was observed in central zone followed by southern (18.67%) and northern zone (16.33%)(Table 2).

Gebrekidan et al. (2017) had reported the first ever molecular characterization of *Theileria orientalis* in native as well as imported cattle and buffaloes from four districts of Punjab (Jhang, Qasur, Sheikhupura, Vehari) through conventional PCR amplifying the major piroplasm surface protein (MPSP) gene and multiplexed tandem PCR (MT-PCR). *Theileria orientalis* was detected in 15% of imported cattle by conventional PCR, whereas 24.5% of the imported cattle, 6% native Pakistani cattle and 6.1% buffaloes were test-positive for *Theileria orientalis* using MT-PCR. The prevalence of *Theileria orientalis* and copies of *chitose* and *ikeda* genotypes was higher in imported cattle than in native Pakistani cattle and buffaloes. Phylogenetic analysis revealed that the MPSP sequences of *buffeli*, *chitose* and *ikeda* from imported cattle were closely related to the MPSP sequences reported previously from Australia.

Saeed et al. (2016) screened 264 Cholistani cattle blood samples collected from Cholistan Desert in Bahawalpur by Giemsa stained blood smear screening and through PCR based amplification of 30 kDa merozoite surface antigen of *Theileria annulata.* They found through PCR that 19.3% apparently healthy cattle were positive for *Theileria annulata* while only 1.9% blood smears were positive for *Theileria* spp. (Table 2).

Khattak et al. (2012) had reported the prevalence of *Theileria annulata* in large ruminants from two districts, Peshawar and Kohat, in KPK. Blood samples were collected from 95 cattle and they were screened through two different parasite detection techniques [Giemsa stained blood smear screening (conventional) and PCR amplification (modern)]. They reported that PCR based detection was more sensitive tool as it detected *Theileria annulata* in 33.7% blood samples through amplification of 30 kDa merozoite surface antigen of *Theileria annulata* as compared to blood smear scanning that resulted in detection of *Theileria* spp. in 5.2% of samples (Table 2).

Ovine Theileriosis in Pakistan

Khan et al. (2020) presented the first report of distribution and prevalence of theileriosis caused by *Theileria luwenshuni* in small ruminants from the Lower Dir district of the KPK Province. A total of 200 blood samples from healthy sheep and goats were screened through nested PCR using universal primers to amplify 492–498 bp amplicon from hyper variable of the 18S rRNA gene and subjected the products to reverse line blotting (RLB) in order to report *Theileria* spp. in the collected blood samples. Nested PCR results indicated 53.5% prevalence of one or more *Theileria* genotypes in the blood of the host animal while RLB assay revealed that 13.5% animals were infected with a single species of *Theileria* while 40% animals were co infection with multiple *Theileria* species. The most common species was

Theileria luwenshuni and based on the 18S rRNA phylogeny, they were closely related to Chinese isolates of *Theileria luwenshuni* (Table 2).

Riaz et al. (2017) conducted a molecular survey to determine the prevalence of *Theileria* spp. in 250 sheep (n = 156) and goat (n = 94) blood samples collected during 2013 from Multan district in southern part of Punjab Province Punjab, Pakistan. Overall incidence of theileriosis in small ruminants was found to be 12.5% and 36% through blood smears and PCR assay, respectively. They reported high prevalence rate of *Theileria lestoquardi* in investigated small ruminants as compared to *Theileria ovis* (Table 2).

Riaz et al. (2016) collected a total of 200 blood samples from apparently healthy small ruminants comprising sheep (n = 161) and goats (n = 39) from different sampling sites in Multan District of Punjab and reported that 7% blood samples were *Theileria* spp. positive through microscopic examination of Giemsa stained blood smears while 22.5% samples were positive for *Theileria* spp. by PCR amplification of 18S ssu rRNA gene. Further, 12.5% blood samples were found positive for *Theileria lestoquardi* while 6% samples were infected with *Theileria ovis.* They also documented that sheep were more prone to theileriosis than goats (Table 2).

Fatima et al. (2015) reported the prevalence of *Theileria lestoquardi* in 115 blood samples small ruminants [goat (n = 66) and sheep (n = 49)] from randomly selected herds collected from five sampling sites (Bahawalnagar, Dera Ghazi Khan, Layyah, Multan and Muzaffargarh Districts) in Southern Punjab. They have documented that 3.74% animals amplified 730 base pairs of 18S rRNA gene specific for *Theileria* lestoquardi through PCR and sheep from Multan were found to be more prone to theileriosis than other sampling sites (Table 2).

Iqbal et al. (2013) collected 196 blood samples from apparently healthy sheep and goats from two provinces, Punjab (from Multan District) and KPK (from Kohat District), detected the presence of tick-borne parasites (*Theileria* and *Babesia* spp.) by the amplification of hyper variable V4 region of the 18S ribosomal RNA (rRNA) gene through Reverse line blot (RLB) assay. Parasites were detected in 16% of the ruminant blood samples under study. Two *Theileria* species, *Theileria lestoquardi* and *Theileria ovis*, were identified in samples. Blood samples collected from KPK were found significantly more infected with *Theileria* spp. than samples from Punjab.

Shahzad et al. (2013) investigated the prevalence of *Theileria* spp. by microscopy and that of *Theileria ovis* through PCR amplification of 18S rRNA gene in 200 blood samples from Lohi sheepcollected from Okara District in Punjab. 24% of sheep blood samples were found positive for *Theileria* spp. by microscopy while 37% were found positive for *Theileria ovis* through PCR (Table 2).

Durrani et al. (2011) had collected 200 blood samples from twenty sheep flocks along with 100 ticks from these sheep from District Lahore during spring and summer seasons. 22% blood samples were found positive through blood smear screening while 35% were found for *Theileria* spp. through PCR. Among the PCR positive samples, 79% were positive for *Theileria ovis* and 21% were found to be *Theileria lestoquardi* (Table 2).

Equine theileriosis in Pakistan

Asif et al. (2020) has recently reported the simultaneous detection of *Anaplasma marginale* and *Theileria annulata* in 195 blood samples of apparently healthy horses (n = 141) and donkeys (n = 54) collected from Dera Ghazi Khan District in Punjab through multiplex PCR. They have reported 4.9% horses and 9.2% donkey were infected with *Anaplasma marginale* while 5.6% horses and 11.1% donkey were infected with *Theileria annulata*. 1.4% horses and 3.7% donkeys were found co-infected with both parasites (Table 2 and 4).

Table 2. Studies from various Districts of Pakistan reporting bovine, ovine equine theileriosis

Causative Parasite(s) Screened	Host Animals	Sampling Site(s)	Diagnostic Methods (% Prevalence)	Reference
Bovine theileriosis				
Theileria parva	Cattle and buffalo	Lakki Marwat and Peshawar Districts in KPK	Real-time PCR (27.8%) Indirect ELISA (24.6%) Microscopy (11.4%)	Rafiullah et al., 2019
Theileria annulata	Cattle	Chakwal, Faisalabad and Jhang Districts in Punjab	Recombinase polymerase amplification (17.5%) PCR (7.7%)	Hassan et al., 2018
Theileria annulata	Cattle and buffalo	Central, southern and northern zones in KPK	PCR (18.9%)	Farooqi et al., 2017
Theileria orientalis	Cattle and buffalo	Jhang, Qasur, Sheikhupura, Vehari Districts in Punjab	multiplexed tandem PCR () PCR (15%)	Gebrekidan et al., 2017
Theileria annulata	Cattle	Bahawalpur District in Punjab	PCR (19.3%) Microscopy (1.9%)	Saeed et al., 2016
Theileria annulata	Cattle	Peshawar and Kohat Districts in KPK	PCR (33.7%) Microscopy (5.2%)	Khattak et al., 2012
Ovine theileriosis				
Theileria luwenshuni	Sheep and goat	Lower Dir District in KPK	Nested PCR (53%)	Khan et al. (2020)
Theileria ovis and *Theileria lestoquardi*	Sheep and goat	Multan District in Punjab	Microscopy (12.5%) PCR (36%)	Riaz et al., 2017
Theileria ovis and *Theileria lestoquardi*	Sheep and goat	Multan District in Punjab	Microscopy (7%) PCR (22.5%)	Riaz et al., 2016
Theileria lestoquardi	Sheep and goat	Bahawalnagar, Dera Ghazi Khan, Layyah, Multan and Muzaffargarh Districts in Punjab	PCR (3.7%)	Fatima et al., 2015
Theileria ovis and *Theileria lestoquardi*	Sheep and goat	Multan District in Punjab, Kohat District in KPK	PCR (16%)	Iqbal et al., 2013
Theileria ovis	Sheep	Okara District in Punjab	PCR (37%) Microscopy (24%)	Shahzad et al., 2013
Theileria ovis and *Theileria lestoquardi*	Sheep	Lahore District in Punjab	PCR (35%) Microscopy (22%)	Durrani et al., 2011
Equine theileriosis				
Theileria annulata	Horse and donkey	Dera Ghazi Khan District in Punjab	PCR (7.2%)	Asif et al., 2020
Theileria equi	Donkey and mule	Mianwali District in Punjab	PCR (17.3%)	Shah et al., 2020
Theileria equi	Horse	Various localities in Punjab	PCR (9%)	Ali et al., 2019
Theileria equi	Horse, donkey and mule	Charsada and Peshawar Districts in KPK	cELISA (38.2%)	Afridi et al., 2017
Theileria equi	Horse, donkey and mule	Bahawalpur, Faisalabad, Gujranwala, Lahore and Multan Districts in Punjab	cELISA (41.2%)	Hussain et al., 2014

Shah et al. (2020) has investigated the prevalence of theileriosis in 150 equine blood sample [donkeys (n = 75); mules (n = 75)] collected from District Mianwali in Punjab along with the effect of parasite on hematological parameters of equine. They used PCR based amplification of the *18S rRNA* gene for the detection of *Theileria equi* in enrolled blood samples. They reported an overall 17.3% prevalence of *Theileria equi* and were observed that donkey is more susceptible to parasite than mule. They sequenced selected amplified PCR products that showed 99% homology with isolates from the United States of America, Spain, Brazil, Israel, Cuba, France, South Africa, Korea, Turkey, Tunisia, India and Nigeria (Table 2).

Ali et al. (2019) collected 500 blood samples of horses from various localities in Punjab Province and amplified them for 18S rRNA gene of *Theileria equi* through PCR followed by DNA sequencing and found that 9% of enrolled animals were positive for *Theileria equi*. The isolates of this study showed high homology with Cuban, Russian and Brazilian isolates of *Theileria equi*. They also analyzed the complete blood count and selected serum parameters in all enrolled animals and found decreased granulocytes and erythrocytes while increased lymphocytes, monocytes, mean corpuscular volume, mean corpuscular hemoglobin, mean platelet volume, glucose, phosphorus and aspartate aminotransferase in *Theileria equi* positive horses when compared with parasite negative horses (Table 2).

Afridi et al. (2017) collected a total of 393 equine [horses (n = 195), donkey (n = 194) and mule (n = 4) serum samples from randomly selected localities in Charsadda (n = 193) and Peshawar (n = 200) Districts in KPK and determined the presence of antibodies for *Theileria equi* by using competitive enzyme-linked immunosorbent assay. An overall seroprevalence of 38.2% was observed among all the tested animals. Higher sero positivity for *Theileria equi* was observed in equine belonging to Charsada (50.3%) as compared to equine from Peshawar District (27.5%) (Table 2).

Hussain et al. (2014) had collected a total of 430 equine blood samples [mules (n = 33), horse (n = 65) and donkeys (n = 332) from five metropolises (Bahawalpur, Faisalabad, Gujranwala, Lahore and Multan Districts) of Punjab and they were serologically tested for the presence of antibodies directed against *Theileria equi* by using a competitive enzyme-linked immune sorbent assay (cELISA). Out of 430 equine serum samples tested, 52.6% were found cELISA positive for equine piroplasm (*Theileria equi* and/or *Babesia caballi*) infections. Seroprevalence of for *Theileria equi* was 41.2% and 10.2% equine were sero positive for both *Theileria equi* and *Babesia caballi*. Highest seroprevalence of *Theileria equi* was observed in horses out of enrolled equines and equines from the metropolis of Lahore had higher sero prevalence than samples collected from other four Districts of Punjab (Table 2).

BABESIOSIS

Babesiosis is a haemolytic disease of small and large ruminants is caused by multiple species of intraerythrocytic Protozoan parasites of the genus *Babesia* (El Moghazy et al. 2014). *Babesia* spp. can be pear, round, oval, leaf, amoeboid and irregularly shaped (Laha et al., 2015). Although babesiosis occurs throughout the world however, the distribution of the causative protozoa is governed by the geographical and seasonal distribution of their vectors

(Radostits et al., 2007; Fakhar et al., 2012). A number of tick genera including *Hyalomma*, *Boophilus* and *Rhipicephalus* are reported as vector for *Babesia* species (Uilenberg, 2006).

Babesia spp.

The causative agents of babesiosis are specific for particular species of animals. Babesiosis in cattle is reported to be caused by *Babesia bovis*, *Babesia bigemina*, *Babesia divergens and Babesia major* (Radostits et al. 2007). *Babesia ovis* and *Babesia motasi* are known to be pathogenic agents in sheep and goats (Fakhar et al. 2012). *Babesia bovis* is considered to be the most pathogenic species, followed by *Babesia bigemina* and *Babesia divergens* and cause enormous economic losses in the livestock industry (Bock et al., 2004).

Clinical Signs of Babesiosis

Babesia spp. causes rapid destruction of the erythrocytes accompanying haemoglobinaemia, haemoglobinuria and fever (40-42°C) which may be sudden in onset and is associated with anemia, icterus, anorexia, jaundice and may lead to death (El Moghazy et al., 2014).

Control of Babesiosis

Two drugs, Diminazine, and Imidocarb are available for treatment and prophylaxis of babesiosis (Figueroa et al., 2010). Mixture of Atovaquone with Azithromycin is proved to be more beneficial for the treatment of babesiosis than mixture of Quinine and Clindamycin (Thalmann et al., 2013). Clofazimine is also discovered to be significantly effective for treatment of babesiosisis that hinders the growth and development parasites (Tuvshintulga et al., 2016).

Bovine Babesiosis in Pakistan

A fair number of studies have been reported regarding babesiosis in Pakistan. Majority of investigations are focused on bovine (cattle, buffaloes) and ovine (sheep and goat) whereas only few studies are reporting babesiosis in equines (mule, horses and donkey) and canines (dogs, cat, fox). Maximum numbers of studies were conducted in Punjab followed by KPK, Sindh and Balochistan provinces respectively.

Siddique et al. (2020) collected 2176 cattle and buffalos blood samples from three Districts in Punjab: Khanewal (southern sandy zone), Faisalabad (central plain zone) and Chakwal (northern arid zone) and screened them for *Babesia bovis* through both microscopy of Giemsa stained blood films and PCR. They had reported 17.2% prevalence of *Babesia* spp. through optical microscopy and PCR based prevalence of *Babesia bovis* was 26.9. Highest parasite prevalence was observed in bovine samples from District Khanewal followed by Faisalabad and Chakwal (Table 3).

Farooq et al. (2019) had collected 200 cattle blood samples from District Narowal in Punjab and screened them for *Babesia bigemina*. They reported 18% prevalence of *Babesia bigemina* through microscopy while 8.5% by real time PCR (Table 3).

Khetran et al. (2019) collected blood samples from water buffaloes (n = 100), including tick-free and tick-infested animals (50 animals each group), from Southeastern region of Pakistan and screened *Babesia bigemina* in these samples by using both traditional blood smear screening and PCR. The overall prevalence was 20% and 17% through microscopy and PCR respectively. *Babesia bigemina* infection was found to be higher in tick infested animals as compared to tick free animals (Table 3).

Zia and Nazir (2019) examined Giemsa stained blood samples of 200 cattle collected from various villages of Taunsa Sharif in District Dera Ghazi Khan for the presence of *Babesia* spp. They had reported 35% overall prevalence of *Babesia* spp. in the analyzed samples. Parasite was more common in female and young animals than in male and adult cattle (Table 3).

Akhtar et al. (2018) collected 100 cattle blood samples (60 samples from Government dairy farm and 40 samples from private dairy farms) from different areas of Quetta City and screened them for the presence of *Babesia* spp. by conventional microscopic examination and through PCR. Microscopic screening revealed 19% while molecular detection through PCR confirmed 33% samples positive for *Babesia* spp. A duplex species level PCR analysis revealed that 60.6% of *Babesia* spp. samples were positive for *Babesia bigemina* while remaining 39.4% were positive for *Babesia bovis* (Table 3).

Farooqi et al. (2017) conducted an epidemiological study on 900 blood samples (cattle = 479, buffaloes = 421) collected from three distinct temporal zones of KPK Province and screened the spherical body protein-4 (BbSBP-4) gene of *Babesia bovis* in them. Higher prevalence of *Babesia bovis* was recorded in cattle (11.9%) as compared to buffaloes (8.1%). Phylogenetic analysis of sequenced isolates presented their homology with the reported Syrian strains (Table 3).

Hussain et al. (2017) reported the incidence of *Babeisa bovis* infection in clinically ill crossbred and indigenous cattle from Lahore District in Punjab by using microscopic examination of blood smears and PCR. Overall *Babesia bovis* prevalence of 8.3% by PCR and 6. 7% by microscopy were documented for enrolled cattle. The crossbred cattle were significantly more infected with the parasite compared to indigenous cattle (Table 3).

Khan et al. (2016) collected blood samples of house hold dairies from southern part of KPK (Bannu and Lakki Marwat Districts) and screened their blood smear for the presence of *Babesia bigemina* and *Babesia bovis*. They had reported 20.7% and 34% prevalence for *Babesia bigemina* and for *Babesia bovis* respectively (Table 3).

Saad et al. (2015) analyzed 100 blood samples of tick infested cattle (cow = 73, bulls = 27) collected from Peshawar District in KPK and screened them for *Babesia bovis* and *Babesia bigemina* through microscopy and PCR. 67% samples through microscopy and 57% through PCR were found positive for *Babesia bigemina*, while *Babesia bovis* was detected in 33% samples by microscopy and in 3o% samples through PCR (Table 3).

Ahmad et al. (2014) determined the prevalence of babesiosis in cows and calves from districts Charsadda and Sawabi in KPK. A total of 100 ticks and 100 blood samples were randomly collected from enrolled animals and blood smear were microscopically for the presence of *Babesia* species. Overall 30% samples were found infected with *Babesia* spp. out of which 19% samples were positive for *Babesia bigemina* and 11% were positive for *Babesia bovis*. In district Charsadda, the prevalence rate of *Babesia bigemina* was 18% while in Swabi district, it was 20%. Similarly the prevalence rate of *Babesia bovis* in was 14% and 8% in animals from district Charsadda and Swabi respectively (Table 3).

Ayaz et al. (2013) examined 2400 cattle blood samples (1200 calves and 1200 cows) collected from Karak and Kohat Districts in KPK through PCR and microscopy. The overall prevalence of babesiosis in cattle by microscopy was 9.9% (Karak District 9.25% and Kohat District 10.5%). PCR based prevalence of babesiosis was 27.5%. Prevalence of *Babesia bovis* was higher than that of *Babesia bigemina* in enrolled animals (Table 3).

Shams et al. (2013) had collected 600 blood samples of clinically suspected cattle from veterinary hospitals of District Karak and Kohat in KPK. Thick and thin blood smear slides were examined under microscope: PCR was also applied for detection of *Babesia* spp. in collected blood samples. Overall prevalence of *Babesia* spp. by PCR was 27.5% while through microscopy it was 9.8%. 6.3% of calves and 10.6% of cows were found positive for *Babesia bovis*. While 4% calves and 6.6% cows were found infected with *Babesia bigemina* (Table 3).

Chaudhry et al. (2010) screened *Babesia bovis* and *Babesia bigemina* in cross bred carrier cattle. Screening by light microscopy showed that 18% of samples were *Babesia* spp. positive. While PCR based analysis diagnosed 29% of samples positive for *Babesia* spp. out of which 11% were positive for *Babesia bovis* and 18% were found infected with *Babesia bigemina* (Table 3).

Durani and Kamal (2008) screened cattle blood samples from district Kasur in Punjab for *Babesia* spp. through microscopy and PCR. Prevalence of *Babesia* spp. was recorded to be 42.8% by microscopy and 33.3% by PCR (Table 3).

Ovine Babesiosis in Pakistan

Shahzad et al. 2013 investigated the prevalence of *Babesia ovis* in 200 Lohi sheep reared at livestock experiment station Bahadurnagar in Okara District of Punjab by using PCR and reported that 16% animals were positive for *Babesia ovis* (Table 3). Iqbal et al. (2011) established a PCR based protocol for the detection of *Babesia* spp. in small ruminants from seven Districts (Multan, Muzaffar Garh, Layyah, Dera Ghazi Khan, Khanewal, Vehari and Bahawalnagar) in Southern Punjab. A total of 107 blood samples (40 sheep and 67 goats) were screened through PCR and 34% blood samples produced the DNA fragment specific for 18S rRNA gene of *Babesia* spp. Samples from all seven districts of Southern Punjab were foundpositive for *Babesia* spp. (Table 3).

Equine Babesiosis in Pakistan

Hussain et al. (2014) serologically screened prevalence 430 equine (33 mules, 65 horses and 332 donkeys) of different age and sex from Bahawalpur, Faisalabad, Gujranwala, Lahore and Multan Districts of Punjab (Pakistan) for *Babesia caballi* Antibodies directed against

Babesia caballi, using a competitive enzyme-linked immunosorbent assay (cELISA), were detected in 21.6% of enrolled equines (Table 3).

Table 3. Studies from various Districts of Pakistan reporting bovine, ovine, equine and canine babesiosis

Causative Parasite(s) Screened	Host Animals	Sampling Site(s)	Diagnostic Methods (% Prevalence)	Reference
Bovine babesiosis				
Babesia bovis	Cattle, buffalo	Khanewal, Faisalabad and Chakwal Districts in Punjab	Microscopy (17.23%) PCR (26.9%)	Siddique et al., 2020
Babesia bigemina	Cattle	Narowal District in Punjab	Microscopy (18%) Real time PCR (8.5%)	Farooq et al., 2019
Babesia bigemina	Water buffalo	South eastern region of Pakistan	Microscopy (20%) PCR (17%)	Khetran et al., 2019
Babesia spp.	Cattle	Dera Ghazi Khan District in Punjab	Microscopy (35%)	Zia and Nazir, 2019
Babesia spp. *Babesia bigemina* *Babesia bovis*	Cattle	Quetta District in Balochistan	Microscopy (19%) PCR (33%)	Akhtar et al., 2018
Babesia bovis	Cattle, buffalo	Distinct temporal zones of KPK	PCR (cattle 11.9%; buffalo 8.1%).	Farooqi et al., 2017
Babesia bovis	Cattle	Lahore District in Punjab	Microscopy (6. 7%) PCR (8.3%)	Hussain et al., 2017
Babesia bigemina *Babesia bovis*	Cattle	Bannu and Lakki Marwat Districts in KPK	Microscopy	Khan et al., 2016
Babesia bigemina and Babesia bovis	Cattle	Peshawar	Microscopy and PCR	Saad et al., 2015
Babesia bigemina and *Babesia bigemina*	Cattle	Charsadda and Sawabi Districts in KPK	Micoscopy (30%)	Ahmad et al., 2014
Babesia bigemina *Babesia bovis*	Cattle	Kohat and Karak Districts in KPK	Microscopy and PCR	Ayaz et al., 2013
Babesia bigemina *Babesia bovis*	Cattle	Kohat and Karak Districts in KPK	PCR (27.5%) Microscopy (9.8%)	Shams et al., 2013
Babesia bigemina *Babesia bovis*	Cattle	Qadirabad in Punjab	PCR (29%) Microscopy (18%)	Chaudhry et al., 2010
Babesia spp.	Cattle	Kasur District in Punjab	Microscopy (42.8%) PCR (33.3%)	Durani and Kamal, 2008
Ovine babesiosis				
Babesia bovis	Lohi sheep	Okara District in Punjab	PCR (16%)	Shahzad et al., 2013
Babesia spp	Sheep, goat	Multan, Muzaffar Garh, Layyah, Dera Ghazi Khan, Khanewal, Vehari and Bahawalnagar Districts in Punjab	PCR (34%)	Iqbal et al., 2011
Equine babesiosis				
Babesia caballi	Mules, horses, donkey	Bahawalpur, Faisalabad, Gujranwala, Lahore, Multan Districts in Punjab	cELISA (21.6%)	Hussain et al., 2014
Canine babesiosis				
Babesia microti	Dogs, cats	Multan, Islamabad/Rawalpindi and Lahore	PCR (dog 1.3%; cat 13.2%)	Akram et al., 2019
Babesia canis Babesia gibsoni	Dogs	Lahore and Narowal Districts in Punjab	Microscopy (42%) PCR (39%)	Tayyub et al., 2019

Canine Babesiosis in Pakistan

Akram et al. (2019) had collected 150 blood samples of dogs from three districts in Punjab including Multan (n = 49), Islamabad/Rawalpindi (n = 49) and Lahore (n = 52) and 159 blood samples of cats from Lahore and analyzed them by PCR to report the presence of *Babesia microti* in them. They reported that two dogs (1.3%) and 21 cats (13.2%) amplified 238 base pair amplicon from 18S rRNA gene of *Babesia microti* (Table 3).

Tayyub et al. (2019) collected 200 blood samples of dogs three from three clinics in Lahore and one clinic in Narowal District and screened them through microscopy for the presence of *Babesia* spp. and reported that 84 (42%) were positive for *Babesia* spp. 39% of these samples were confirmed for *Babesia* spp. through PCR and positive samples were confirmed by DNA sequencing and both *Babesia canis* and *Babesia gibsoni* were confirmed in enrolled dogs (Table 3).

ANAPLASMOSIS

Anaplasmosis is disease reported in human as well as in domestic and wild ruminants and it is caused by obligate intraerythrocytic Rickettsiae of the genus *Anaplasma* (Rickettsiales: Anaplasmatacea) and is distributed in tropical and subtropical regions of the world (Jabbar et al., 2015). Anaplasmosis is biologically transmitted by about twenty types of ticks including *Rhipicephalus* spp., *Hyalomma* spp., *Boophilus*, *Demacentor* spp. and *Ixodes* spp. (Kocan et al., 2003). Mechanical transmission has been reported through a number of biting flies of the family Tabanidae (horseflies) and by mosquitoes of the genus *Psorophora* (Rajput et al., 2005) and through fomites by infected hypodermic needle, castrating and dehorning instruments, blood transfusion and during embryo transplant (Demessie and Derso 2015).

Anaplasma spp. are often coccoid to ellipsoidal and pleomorphic residing in vacuoles of erythrocytes of mammalian hosts (Dumler et al., 2001). The organisms also develop tightly packed formations of cells termed morulae. These inclusions are seen in the monocytes, granulocytes and in neutrophils (Rikihisa, 2011). *Anaplasma marginale* was the first among *Anaplasma* species amd it was first described by Theiler in South Africa in 1910 and he described them as "marginal points" in bovine erythrocytes and hence the name "marginale" derived from its location within the host cell (Theiler, 1910). Additionally, there is a high similarity between the *Anaplasma* inclusions and other intra-erythrocyte structures like Heinz bodies, Howell-Jolly bodies, basophilic stippling in reticulocytes or staining artifacts (Noaman, 2013).

Anaplasma spp.

Anaplasmosis is caused by a number of *Anaplasma* species including *Anaplasma marginale*, *Anaplasma centrale*, *Anaplasma bovis*, *Anaplasma ovis* that cause disease in ruminants, *Anaplasma platys* infects canines and *Anaplasma pagacytophilum* is known to cause anaplasmosis in human and horse (Kocan et al., 2010).

Clinical Signs of Anaplasmosis

Common clinical signs of anaplasmosis are fever, dyspnoea, lethargy, icterus, hyper excitability, abortion due to hypoxia (Aubry and Geale 2011). The most marked clinical signs of anaplasmosis are anemia and jaundice, the latter occurring late during the disease. Affected animals may die suddenly from hypoxia if they are handled roughly (Quinn et al., 2002). Recovered animals from the acute stage of anaplasmosis will develop lifelong low level of bacteraemia. These carrier animals act as a reservoir of infection for the susceptible host (Ashraf et al., 2013).

Control of Anaplasmosis

Methods that are used for the control of anaplasmosis include use of antibiotics, vaccination and arthropod control. Treatment of *Anaplasma* carrier animals with Oxytetracycline is normally successful for the elimination of infection (Atif, 2015).

Bovine Anaplasmosis in Pakistan

Bovine anaplasmosis is endemic in Pakistan and it reduces livestock productivity and leads to high mortality, especially in young animals. Anaplasmosis is highly prevalent hemo-rickettsial bovine disease in Pakistan with its prevalence ranging from 4 to 75.5% (Khan et al., 2004). The major risk factors assumed to affect the dynamics of this disease include species of animal, sex of the animal, breed, age, tick burden, flies, management system and access to various infected animals and non-living fomites (Ashraf et al., 2013). *Anaplasma* spp. that are known to cause bovine anaplasmosis in Pakistan are *Anaplasma marginale* and *Anaplasma centrale.* Some studies have also reported *Anaplasma phagocytophilum* and *Anaplasma bovis* as causative agent of bovine anaplasmosis.

Iqbal et al., 2019 Collected 450 blood samples on FTA cards from asymptomatic, tick-infested cattle from Jhang District in Punjab and tested them for the presence of *Anaplasma* spp. by using nested PCR.. The nested-PCR amplified 16S ribosomal RNA gene r of two *Anaplasma* species with an overall prevalence rate of 10.4%, where the prevalence *Anaplasma phagocytophilum* was 2.66% and that of *Anaplasma bovis* was 7.8% (Table 4).

Farooqi et al. (2018) had collected a total of 900 blood samples (cattle = 479, buffalo = 421) from three distinct zones of KPK province and applied PCR for the molecular detection of *Anaplasma* spp. The 611 bp DNA fragment specific for 16S rRNA gene of *Anaplasma marginale* was amplified in 18.3% of analyzed samples. Parasite positive samples were confirmed by DNA sequencing and BLAST analysis. The *Anaplasma marginale* isolates o showed resemblance with previously reported Chinese isolates (Table 4).

Khan et al. (2017) had collected a total of 234 blood samples and 100 fecal samples of cattle from cattle breeding and dairy farm Harichand in Charsada District of KPK and screened them for endo-parasites. Giemsa stained thin blood smear screening was used for the detection of haemo-protozoan while for gastrointestinal parasites direct examination and floatation technique was used. Out of total 234 blood samples, 48% were found positive for haemo-protozoans ou of which 28% were *Anaplasma* species. They had documented that parasitic diseases have strong seasonal correlation as hot and humid weather provide conducive environment for the propagation of ticks that act as vector for haemo-parasitic diseases (Table 4).

Hussain et al., (2017) had reported the prevalence of *Anaplasma* spp. in 148 blood samples of apparently healthy Cholistan breed of cattle collected from Bahawalpur District in Punjab. 47 blood samples (31.8% of total) amplified 577 base pairs DNA fragment specific for 16S rRNA gene of *Anaplasma* spp. by PCR. Out of 47 *Anaplasma* spp. positive cattle, 9 were confirmed to be *Anaplasma marginale* and 9 were confirmed to be *Anaplasma phagocytophilum* upon restriction with *BssNa1*. They had also documented that *Anaplasma* spp. positive cattle had significantly reduced levels of mean corpuscular volume and eosinophils than in parasite negative animals. While studied serum biochemical profile remain unaffected when compared between the two groups (Table 4).

Atif et al. (2013) designed a study to report seroprevalence of *Anaplasma marginale* in cattle from Sargodha, Khushab and Rawalpindi districts of the Punjab. A total of 1050 serum samples were collected from livestock farms by using multistage cluster random sampling technique. Antibodies against *Anaplasma marginale* were detected by MSP-5 competitive enzyme linked immunosorbent assay. They reported that cattle from Sargodha district were more susceptible to *Anaplasma marginale* infection as compared to Khushab and Rawalpindi districts (Table 4).

A survey on tick-borne diseases of cattle was conducted by Atif et al. (2012) at Sargodha district in Punjab. A total of 350 blood samples were collected from selected small holders having 1 to 10 cattle ($n = 30$) and private livestock farms ($n = 4$) having ≥ 50 cattle using multistage cluster random sampling technique. Out of 350 samples, 94 were positive for tick-borne pathogens with an overall prevalence of 26.9% through blood smear screening. *Anaplasma marginale* was the most prevalent (9.7%) (Table 4).

Talat et al. (2005) collected a total of 118 blood samples from cows and buffalos that were slaughtered in slaughter houses in Peshawar city and screened their Giemsa stained blood smears. They reported that 15% cows and 26% buffalos were infected with *Anaplasma marginale* (Table 4).

Rajput et al. (2005) had collected five hundred blood samples of buffaloes and cattle (250 of each) from Hyderabad district in Sindh and screened the Giemsa stained blood smears. Microscopic examination revealed that 205 (41%) animals were positive for *Anaplasma* species, out of which 89 animals had *Anaplasma marginale and* 44 animals had *Anaplasma centrale* infection. Comparative study revealed that the cattle were more susceptible to *Anaplasma* infection than buffaloes (Table 4).

Afridi and Ahmed (2005) had conducted a survey in Watpaga, Nasirpur, lahgra, Chamkani/Surezai villages in Peshawar district during 2001. They analyzed 285 blood smears of cattle and found that 68 (23.9%) were positive for blood parasites. Out of these 68 parasite positive samples, 12 (4.2%) were harboring *Anaplasma marginale* infection and 11 samples (3.9%) were positive for *Anaplasma centrale* (Table 4)

Khan et al. (2004) had collected blood samples from 307 cattle and 155 buffaloes during September 1999 till May 2000. Of these, 65 cattle were maintained at Animal Sciences Institute (ASI), National Agricultural Research Centre (NARC), Islamabad and 242 (of three breeds) were kept at Barani livestock production research institute (BLPRI, Kherimurat in District Attock. Prevalence of blood parasites in cattle and buffaloes maintained at NARC and BLPRI was 27.7 and 16.1%, respectively through blood smear examination. Among the infested animals, 75.1% cattle and 80% buffaloes were found infected with *Anaplasma marginale* (Table 4).

Table 4. Studies from various districts of Pakistan reporting bovine, ovine and equine theileriosis

Causative Parasite(s) Screened	Host Animals	Sampling Site(s)	Diagnostic Methods (% Prevalence)	Reference
Bovine anaplasmosis				
Anaplasma bovis *Anaplasma phagocytophilum*	Cattle	Jhang District in Punjab	nested-PCR (10.4%)	Iqbal et al., 2019
Anaplasma marginale	Cattle and buffalo	Khyber Pakhtunkhhwa Province	PCR (18.3%)	Farooqi et al., 2018
Anaplasma spp.	Cattle	Charsada District in KPK	PCR (28%)	Khan et al., (2017)
Anaplasma spp. *Anaplasma marginale* *Anaplasma phagocytophilum*	Cattle	Bahawalpur District in Punjab	PCR (31.8%)	Hussain et al., (2017)
Anaplasma marginale	Cattle	Sargodha, Khushab and Rawalpindi Districts of Punjab	cELISA	Atif et al., 2013,
Anaplasma marginale	Cattle	Sargodha district from Pakistan	Microscopy (9.7%)	Atif et al., 2012
Anaplasma marginale	Cattle and buffalo	Peshawar District in PK NWFP	Microscopy (21.2%)	Talat et al., 2005
Anaplasma spp. *Anaplasma marginale* *Anaplasma centrale*	Cattle and buffalo	Hyderabad District in Sindh	Microscopy (41%)	Rajput et al., 2005
Anaplasma marginale *Anaplasma centrale*	Cattle	Peshawar District in KPK	Microscopy (4.21%) and (3.85%)	Afridi and Ahmed 2005
Anaplasma marginale	Cattle and buffalo	Islamabad and Attock Districts in Punjab	Microscopy	Khan et al., 2004
Ovine anaplasmosis				
Anaplasma ovis	Sheep and goat	Charsadda District in KPK	cELISA	Khan et al. (2019)
Anaplasma ovis	Bulkhi sheep	Peshawar District in KPK	Microscopy (28%) PCR (40%)	Khan et al., 2015
Anaplasma marginale	Sheep	Peshawar District in KPK	indirect ELISA (24.5%)	Kashif et al. (2014)
Anaplasma ovis	Sheep and goat	different areas of NWFP	Microscopy (5.8%)	Talat et al., 2005
Equine anaplasmosis				
Anaplasma marginale	Horse and donkeys	Dera Ghazi Khan District in Punjab	PCR (4.9% in horses, 9.2% in donkey)	Asif et al., (2020)
Anaplasma phagocytophilum	Horse, mule and donkey	Mianwali District in Punjab	PCR (10.7%)	Saleem et al., (2018)
Anaplasma spp.	Camel	Mianwali District -in Punjab	PCR (13.3%)	Azmat et al., 2018
Anaplasma phagocytophilum	Donkey and horse	Dera Ghazi Khan and Khanewal districts District in Southern Punjab	PCR (4.3%)	Razzaq et al., 2014

Bovine Anaplasmosis in Pakistan

Khan et al. (2019) used commercially available competitive enzyme-linked immunosorbent assay (cELISA) to investigate the seroprevalence of *Anaplasma* antibodies in small ruminants from the three tehsils of the Charsadda District in KPK. A total of 600 blood samples (300 sheep and 300 goats) were collected from small farms and it was observed that seroprevalence of *Anaplasma* antibodies was higher in goats than in sheep. A significantly higher infection rate was observed in the animals from Charsadda tehsil as compared to the animals from Shabqadar and Tangi tehsils (Table 4).

Khan et al. (2015) investigated the prevalence of anaplasmosis in Bulkhi sheep from district Peshawar in KPK during 2013-2014. They collected 100 blood samples and prepared thin blood smear, stained them with Giemsa and observed under microscope. Out of 100 animals, 40 showed clinical signs of anaplasmosis 28/40 were found positive for *Anaplasma* spp. through microscopy. All these microscopy positive samples were also confirmed through PCR. While the 12 samples having clinical signs of anaplasmosis but were found negative for Anaplasma spp. under microscope were also found positive for *Anaplasma ovis* through PCR (Table 4).

Kashif et al. (2014) reported geographical seroprevalence of *Anaplasma marginale* in sheep from district Peshawar in KPK. For this purpose, 376 serum samples were collected from 4 different sheep breeds from different geographical areas of Peshawar. An indirect ELISA using recombinant MSP-5 as antigen of *Anaplasma marginale* was performed. 24.5% of sheep sera were found positive for parasite. In 6 areas of Peshawar, Peshthakhara and Mashokhel area were found highly infected while Ghazi baba area was found least infected with *Anaplasma marginale* (Table 4).

Talat et al. (2005) had conducted a study in KPK during 2003 in which they had collected 209 blood samples of goats and sheep from the slaughter houses. 25/209 samples were found infected with *Anaplasma ovis* and over all incidence of infection was 5.6% (Table 4).

Eqine Anaplasmosis in Pakistan

Saleem et al., (2018) reported a study to assess the molecular prevalence of equine granulocytic anaplasmosis in donkey, horse and mule samples collected from Mianwali district in Punjab. They have reported an overall 10.7% prevalence of *Anaplasma phagocytophilum* in equine. Horses (11.9%) showed highest prevalence of this parasite followed by mule (10.5%) and donkeys (9.43%) respectively. Parasite positive samples were confirmed through DNA sequencing. The BLAST queries confirmed very high homology of these isolates with Chinese and Japanese isolates of *Anaplasma phagocytophilum*. The hematological parameters like hemoglobin, total leukocyte count, total erythrocytes count and granulocyte count were significantly decreased in infected equine as compared to parasite negative animals (Table 4).

Azmat et al. (2018) has recently investigated prevalence of anaplasmosis in camel from district Mianwali in Punjab. The study found an overall 13.3% prevalence of *Anaplasma* spp. in camels. They have reported the presence of *Anaplasma* spp., i.e., *Anaplasma phagocytophilum*, *Anaplasma ovis* and *Anaplasma platys* for the first time in camels from

Pakistan. The PCR products (16S rRNA gene fragment) of parasite positive samples were sequenced (Table 4).

Razzaq et al. (2014) used a PCR-RFLP protocol for the molecular detection of *Anaplasma* spp. in blood samples of equine from Southern Punjab. They collected a total of 210 equine blood samples (donkey and horse) from Dera Ghazi Khan and Khanewal districts in Punjab. PCR amplified the 577bp product specific for 16S rRNA gene of *Anaplasma* spp. in 9 blood samples (4.3% of total). These *Anaplasma* spp. positive blood samples were used for the detection of *Anaplasma phagocytophilum* in a second round of PCR by using species specific primers and all 9 samples were found *Anaplasma phagocytophilum* positive. Among the studied hematological parameters, lymphocyte count was significantly lowered in *Anaplasma phagocytophilum* positive samples as compared to uninfected equine (Table 4).

Abbreviation

16S rRNA	16S ribosomal RNA
ASI	Animal Sciences Institute
BLPRI	Barani livestock production research institute
cELISA	Competitive enzyme-linked immune sorbent assay
COX1	Cytochrome oxidase subunit I
GDP	Gross domestic product
ITS2	Internal transcribed spacer 2
KPK	Khyber Pukhtoon Khwa
MPSM	Major piroplasm surface protein
MT- PCR	Multiplexed tandem PCR
NARC	National Agricultural Research Centre
PRC	Polymerase chain reaction
RLB	Reverse line blot
RPA	Recombinase polymerase amplification
TBDs	Ticks and tick-borne diseases

References

Abubakar, M., Manzoor, S. & Ali, Q. (2015). Evaluating the role of vaccine to combat peste des petits ruminants outbreak in endemic disease situation. *Journal of Animal Science and Technology*, 57, 2-7.

Afridi, M. J. K., Mian, A. H., Saqib, M., Abbas, G., Ali, J., Mansoor, M. K., Rahman, Sial., A. U. R., Rasheed, I. & Hussain, M. H. (2017). Seroprevalence and Risk Factors for *Theileria equi* Infection in Equines from Khyber Pakhtunkhwa Province, Pakistan. *Iranian Journal of Parasitolgy*, 12(4), 597-605.

Afridi, Z. K. & Ahmad, I. (2005). Incidence of anaplasmosis, babesiosis and theileriosis in dairy cattle in Peshawar (Pakistan). *Sarhad Journal of Agriculture (Pakistan)*, 1016-4383.

Ahmed, A. M., Simeon, E. & Yemesrach, A. (2004). *Dairy development in Ethiopia*, International Food Policy Research Institute, 2033 K Street, Washington DC 20006 U.S.A.

Ahmad, I., Khwaja, A., Shams, S., Ayaz, S., Khan, S., Akbar, N. U., Waqar M., Alam, S., Khan, M. A., Rehman, A. & Zakir, M. (2014). Detection of babesiosis and identification of associated ticks in cattle. *International Journal of Bioassays*, 3(07), 3195-3199.

Akhtar, W., Jan, S., Ahmed, Z., Samad, A., Bugti, F. S., Naeem, M., Rizwan, M., Achakzai, S. K. Hamdullah, Yousaf, M., Tayyeb, T., Asim, M. & Lodhi, M. (2018). PCR based detection of Babesia spp. from cattle of Quetta Pakistan. *Indo American Journal of Pharmaceutical Sciences*, 05(11), 11283-11287.

Akat, A., Aktas, M., Dumanli, N. & Turgut-Balik, D. (2014). Isolation, cloning and sequence analysis of enolase enzyme encoding gene from *Theileria* annulata for assessment of important residues of this enzyme. *Kafkas Universitesi Veteriner Fakultesi Dergisi*, 20, 243–248.

Akram, I. N., Parveen, T., Abrar, A., Mehmood, A. K. & Iqbal, F. (2019). Molecular detection of *Babesia microti* in dogs and cat blood samples collected from Punjab (Pakistan). *Tropical Biomedicine*, 36(1), 304–309.

Ali, A., Khan, M. A., Zahid, H., Yaseen, P. M., Khan, M. Q., Nawab, J., Rehman, Z. U., Ateeq, M., Khan, S. & Ibrahim, M. (2019). Seasonal dynamics, record of ticks infesting humans, wild and domestic animals and molecular phylogeny of *Rhipicephalus microplus* in Khyber Pakhtunkhwa Pakistan. *Frontiers in Physiology*, 10, 793-798.

Ali, S., Ijaz, M., Farooqi, S. H., Durrani, A. Z., Rashid, M. I., Ghaffar, A., Ali, A., Aslam, A., Rehman, A., Khan, I., Masud, A. & Mehmood, K. (2019). Molecular characterization of *Theileria equi* and risk factors associated with the occurrence of theileriosis in horses of Punjab (Pakistan). *Equine Veterinary Education*, https://doi.org/10.1111/eve.13161.

Ali, S., Ijaz, M., Zameer, A., Durrani, A., Maqbool, A., Ali, M. & Mehmood, K. (2016). Epidemiological aspects of bovine tick infestation in the river Ravi Region, Lahore. Pakistan. *Journal of Zoology*, 48(2), 563-567.

Ali, Z., Maqbool, A., Muhammad, K., Khan M. S. & Younis. M. (2013). Prevalence of *Theileria annulata* infected hard ticks of cattle and buffalo in Punjab, Pakistan. *Journal of Animal and Plant Science*, 23(1), 17-23.

Ashraf, Q. U. A., Khan, A. U., Khattak, R. M., Ali, M., Shaikh, R. S., Ali, M. & Iqbal, F. (2013). A report on high prevalence of Anaplasma sp. In buffaloes from two provices in Pakistan. *Tick and Tick - Borne Diseases*, 4, 395-398.

Arif, M., Iqbal, A., Younas, M., Khan, B. B., Sarwarand, M. & Ahmad, S. (2013). Prospects and limitations of dairying in Gujranwala district (Punjab-Pakistan). Proceedings of International Workshop on Dairy Sciences. Park. Nov 21-23, 2011, Agriculture University Peshawar, Pakistan. *Journal of Animal and Plant Sciences*, 23(Sup. 1), 34-37.

Asif, M., Parveen, A., Sehrish Ashraf, S., Hussain, M., Aktas, M., Ozubek, S., Shaikh, R. S. & Iqbal, F. (2020). First report regarding the simultaneous molecular detection of *Anaplasma marginale* and *Theileria annulata* in equine blood samples collected from Southern Punjab in Pakistan. *Acta Parasitologica*, 65(1), 259-263.

Aslam, A., Chaudhary, Z. I., Rehman, H., Ashraf, K., Ahmad, N., Yaqub, T., Maqbool, A. & Shakori, A. R. (2010). Comparative evaluation of parasitological, serological and DNA amplification methods for diagnosis of natural trypanosomal Infection in equines. *Pakistan Journal of Zoology*, 27, 115-120.

Atif, F., Khan, M. S., Muhammad, F. & Ahmad, B. (2013). Sero-epidemiological study of *Anaplasma marginale* among cattle. *Journal of Animal and Plant Sciences*, 23(3), 740-744.

Atif, F. H., Khan, M. S., Iqbal, H. J., Arshad, G. M., Ashraf, E. & Sami Ullah (2012). Prevalence of *Anaplasma marginale*, Babesia *bigemina* and *Theileria annulata* infections among cattle in Sargodha District, Pakistan. *African Journal of Agricultural Research*, 7(22), 3302-3307.

Atif, F. H. (2015). *Anaplasma marginale* and *Anaplasma phagocytophilum*: Rickettsiales pathogens of veterinary and public health significance. *Parasitology Research*, doi: 10. 1007/s00436-015-4698-2.

Aubry, P. & Geale, D. W. (2011). A review of bovine anaplasmosis. *Transboundary and Emerging Diseases*. 58(1), 1-30.

Ayaz, S., Shams, S. A. T., Abdel-Reheem, M. A. T., Khan, S. & Ullah, R. (2013). Epidemiology and Molecular Detection of Babesiosis in Household Dairies in Districts Kohat and Karak, Khyber Pakhtunkhwa Pakistan. *Life Science Journal*, 10(10s), 65-67.

Azmat, M., Ijaz, M., Farooqi, S. H., Ghaffar, A., Ali, A., Masud, A. & Zhang, H. (2018). Molecular epidemiology, associated risk factors, and phylogenetic analysis of anaplasmosis in camel. *Microbial Pathogenesis*, 123, 377–384. doi: 10.1016/j.micpath. 2018.07.034.

Bhatnagar, C. S., Bhardawaj, B., Sharma, D. K. & Meena, S. K. (2015). Incidence of haemoprotozoan diseases in cattle in Southern Rajasthan, India. *International Journal of Current Microbiology and Applied Sciences*, 4(3), 509-514.

Bilal, M. Q., Suleman, M. & Raziq, A. (2006). Buffalo: Black gold of Pakistan. *Live Stock Research of Rural Development*, 18(9), 1-15.

Bishop, R. P., Odongo, D. O., Mann, D. J., Pearson, T. W., Sugimoto, C. & Haines, L. R. (2009). *Theileria*. In: Nene V, Kole C, editors. *Genome mapping and genomics in animal-associated microbes*. Springer-Verlag Berlin Heidelberg, 191-231.

Bock, R., Jackson, L., de Vos, A. & Jorgensen, W. (2004). Babesiosis of cattle. *Parasitology*, 129, 247–269.

Chandio, S. P., Talpur, G. H. & Rajput, M. I. (2018). Empirical Study of Livestock Population Growth and Dairy Farming in Sindh, Pakistan. *Journal of Applied Environmental and Biological Science*, 8(5), 41-48.

Costa, V. M., Rodrigues, A. L., Medeiros, J. M. D. A., Labruna, M. B., Simões, S. V. & Riet-Correa, F. (2011). Tristeza parasitária bovina no Sertão da Paraíba [Bovine parasitic sadness in the Paraíba Hinterland]. *Pesq Vet Bras*, 31(3), 239-243.

Chaudhry, Z. I., Suleman, M., Younus, M. & Aslim, A. (2010). Molecular detection of Babesia *bigemina* and *Babesia bovis* in crossbred carrier cattle through PCR. *Pakistan Journal of Zoology*, 42(2), 201-204.

de Castro, J. J. (1997). Sustainable tick and tickborne disease control in livestock improvement in developing countries. *Veterinary Parasitology*, 71(2-3), 77-97.

Demessie, Y. & Derso, S. (2015). Tick borne hemoparasitic diseases of ruminants: A review. *Advances in Biological Research*, 9(4), 210-224.

Dumler, J. S., Barbet, A. F., Bekker, C. P., Dasch, G. A., Palmer, G. H., Ray, S. C., Rikihisa, Y. & Rurangirwa, F. R. (2001). Reorganization of genera in the families Rickettsiaceae and Anaplasmataceae in the order Rickettsiales: unification of some species of *Ehrlichia* with *Anaplasma*, *Cowdria* with *Ehrlichia* and *Ehrlichia* with *Neorickettsia*, descriptions of six new species combinations and designation of *Ehrlichia equi* and 'HGE agent'as subjective synonyms of *Ehrlichia phagocytophila. International Journal of Systematic and Evolutionary Microbiology*, 51(6), 2145-2165.

Durrani, A. & Kamal, N. (2008). Identification of ticks and detection of blood protozoa in Friesian cattle by polymerase chain reaction test and estimation of blood parameters in district Kasur Pakistan. *Tropical Animal Health & Production*, 40(6), 441–447.

Durrani, A. Z., Younus, M., Kamal, N., Mehmood, N. & Shakoori, A. R. (2011). Prevalence of Ovine *Theileria* Species in District Lahore, Pakistan. *Pakistan Journal of Zoology*, 43(1), 57-60.

El Moghazy, H. M., Ebied, M. M., Abdelwahab, M. G. & El Sayed, A. A. (2014). Epidemiological studies on bovine Babesiosis and Theileriosis in Qalubia governorate. *Benha Veterinary Medical Journal*, 27, 36-48.

Ensink, J. H., Mahmood, T., Van der Hoek, W., Raschid-Sally, L. & Amerasinghe, F. P. (2004). A nationwide assessment of wastewater use in Pakistan: An obscure activity or a vitally important one. *Water policy*, 6(3), 197-206.

Estrada-Peña, A. (2003). The relationships between habitat topology, critical scales of connectivity and tick abundance *Ixodes ricinus* in a heterogeneous landscape in northern Spain. *Ecography*, 26, 661-671.

Fakhar, M., Hajihasani, A., Maroufi, S., Alizadeh, H., Shirzad., P. F. & Pagheh, A. S. (2012). An epidemiological survey on bovine and ovine Babesiosis in Kurdistan Province, western Iran. *Tropical Animal Health and Production*, 44, 319-322.

Farooq, R., Hafeez, M. A. & Oneeb, M. (2020). Molecular Characterization and Phylogenetic Analysis of Babesia Species Isolated from Domestic Cattle. *Pakistan Veterinary Journal*, 40(2), 224-228.

Farooqi, S. H., Ijaz, M., Rashid, M. I., Nabi, H., Islam, S., Aqib, A. I. & Zhang, H. (2018). Molecular epidemiology of bovine anaplasmosis in Khyber Pakhtunkhwa, Pakistan. *Tropical Animal Health and Production*, doi: 10.1007/s11250-018-1599-2.

Farooqi, S. H., Ijaz, M., Saleem, M. H., Imran, M., Oneeb, M., Khan, A., Aqib, A. I. & Mahmood, S. (2017). Distribution of ixodid tick species and associated risk factors in temporal zones of Khyber Pakhtunkhwa Province, Pakistan. *Pakistan Journal of Zoology*, 2017, 49(6), 2011-2017.

Farooqi, S. H., Ijaz, M., Saleem, M. H., Rashid, M. I., Ahmad, S. S., Islam, S., Aqib, A. I., Khan, A., Hussain, K. & Khan, N. U. (2017). Prevalence And Molecular Diagnosis of *Theileria annulata* in Bovine from three Distincts zones of Khyber Pakhtunkhwa Province, Pakistan. *Pakistan Journal of Zoology*, 49(6), 2011-2017.

Fatima, M., Saeed, S., Shaikh, R. S., Ali, M. & Iqbal, F. (2015). A Study on Molecular Detection of *Theileria lestoquardi* by PCR Amplification in Apparently Healthy Small Ruminants from Five Districts of Southern Punjab. *Pakistan Journal of Zoology*, 47(2), 441-446.

Figueroa, J. V., Chieves, L. P., Johnson, G. S. & Buening, G. M. (1993). Multiplex polymerase chain reaction based assay for the detection of Babesia bigemina, Babesia bovis and Anaplasma marginale DNA in bovine blood. *Veterinary Parasitology*, 50, 69–81.

Gebrekidan, H., Abbasb, T., Wajidc, M., Alic, A., Gassera, R. B. & Jabbara, A. (2017). Molecular characterisation of *Theileria* orientalis in imported 9 and native bovines from Pakistan. *Infection, Genetics and Evolution*, 47, 19-25.

Gebrekidan, H., Gasser, R. B., Perera, P. K., McGrath, S., Stevenson, M. A. & Jabbar, A. (2015). Investigating the first outbreak of oriental theileriosis in cattle in South Australia using multiplexed tandem PCR (MT-PCR). *Tick and Tick Borne Diseases*, 6(5), 574-578.

Ghafar, A., Robin, A., Gassera, B., Rashidb, I., Ghafoorc, A. & Jabbara, A. (2020). Exploring the prevalence and diversity of bovineticksin fiveagro-ecological zones of Pakistan using phenetic and genetic tools. *Ticks and Tick-borne Diseases Ticks*, 11, 101472.

Gondard, M., Cabezas-Cruz, A., Charles, R. A., Vayssier-Taussat, M., Albina, E. & Moutailler, S. (2017). Ticks and tick-borne pathogens of the Caribbean: current understanding and future directions for more comprehensive surveillance. *Frontier Cell Infection and Microbiology*, 7, 490.

Government of Pakistan. *Ministry of Finance*. (2017). http://finance.gov.pk/survey_1314. html.

Garcia, Z. (2003). Integrated control of *Boophilus microplus* in cattle. In: *Proc. 11th International Congress*. International Society for Animal Hygiene, Mexico City, Mexico.

Hassan, M. A., Liu, J., Rashid, M., Iqbal, N., Guan, G., Yin, H. & Luo, J. (2018). Molecular survey of piroplasm species from selected areas of China and Pakistan. *Parasites & Vectors*, 11, 457.

Hussain, H. M., Saqib, M., Raza, F., Muhammad, G., Asi., M. M., Mansoon, M. K., Saleem, M. & Jabbar, A. (2014). Seroprevalence of *Babesia caballi* and *Theileria equi* in five draught equine populated metropolises of Punjab, Pakistan. *Veterinary Parasitology*, 202(2), 3–4.

Hussain, M. F., Qamar, M., Malik, M. I., Hussain, M., Saeed, Z., Shaikh, R. S. & Iqbal, F. (2017). Molecular detection of *Anaplasma* in apparently healthy Cholistan breed of cattle from the Bahawalpur district, Pakistan. *Tropical Biomedicine*, 34(1), 37–44, (2017).

Iqbal, A., Sajid, M. S., Khan, M. N. & Khan, M. K. (2013). Frequency distribution of hard ticks (Acari: Ixodidae) infesting bubaline population of district Toba Tek Singh, Punjab, Pakistan. *Parasitology Research*, 112, 535–541.

Iqbal, F., Khattak, R. M., Ozubek, S., Khattak, M. N. K., Rasul, A. & Aktas, M. (2013). Application of the reverse line blot assay for the molecular detection of *Theileria* and Babesia sp. in sheep and goat blood samples from Pakistan. *Iranian Journal of Parasitology*, 8(2), 289-295.

Iqbal, F., Fatima, M., Shahnawaz, S., Naeem, M., Shaikh, R. S., Ali, M., Shaikh, A. S., Aktas, M. & Ali, M. (2011). A study on the determination of risk factors associated with Babesiosis and prevalence of *Babesia* sp., by PCR amplification, in small ruminants from Southern Punjab (Pakistan). *Parasite*, 18, 229-234.

Iqbal, N., Mukhtar, M. U., Yang, J., Sajid, M. S., Niu, Q., Guan, G., Liu, Z. & Hong Yin, H. (2019). First Molecular Evidence of *Anaplasma bovis* and *Anaplasma phagocytophilum* in bovine from Central Punjab, Pakistan. *Pathogens*, 8, 155 doi: 10.3390/ pathogens8030155.

Jabbar, A., Abbas, T., Saddiqi, H. A., Qamar, M. F. & Gasser, R. B. (2015). Tick borne diseases of bovines in Pakistan: major scope for future research and improved control. *Parasite & Vectors*, 8, 283. doi: 10.1186/s13071-015-0894-2.

Kalume, M. K., Losson, B. & Saegerman, C. (2011). Epidemiologie et controle de la theileriose bovinea *Theileria* parva en Afrique: une revue de la litterature [Epidemiology and control of bovine theileriosis *Theileria* parva in Africa: a review of the literature]. *Annales de médecine vétérinaire*, 155, 88–104.

Karim, S., Budachetri, K., Mukherjee, N., Williams, J., Kausar A., Hassan, M. J., Adamson, S., Dowd, S. E., Apanskevich, D., Arijo, A., Sindhu, Z. U., Kakar, M. A., Khan, R. M. D., Ullah, S., Sajid, M. S., Ali, A. & Iqbal, Z. (2017). A study of ticks and tickborne livestock pathogens in Pakistan. *PLoS Neglected Tropical Diseases*, 11(6), e0005681.

Kashif, M., Ahmad, M. S., Ullah, G., Fareed, A. I. & Shah, S. T. (2014). Seroprevalence of *Anaplasma sp.* in sheep (ovis aries) Peshawar, Pakistan. *International Journal of Current Research*, 6(01), 4502-4504

Kasozi, K. I., Matovu, E., Tayebwa, D. S., Natuhwera, J., Mugezi, I. & Mahero, M. (2014). Epidemiology of increasing hemo-parasite burden in Ugandan Cattle. *Open Journal of Veterinary Medicine*, 4(10), 220.

Khetran, I. B., Arijo, A. G., Buhtto, M. B., Gadahi, J. A. & Laghari, Z. A. (2019). Prevalence and molecular detection of *Babesia bigemina* in water buffalo from southeastern region of Pakistan. *Veterinaria*, 68(1), 121-126.

Khadim, I., Mahgoub, O., Al-Ajmi, D., Al-Maqbaly, R., Al-saqri, N. & Ritchie, A. (2003). An evaluationof the growth, carcass and meat quality characteristics of Omani goat breeds. *Meat Science*, 66, 203-210.

Khalil, M. I., Lashari, M. H., Akhtar, M. S. & Tasawar, T. (2018). Prevalence of ticks infesting Buffaloes in and around Jampur District Ranjanpur, Pakistan. *FUUAST Journal of Biology*, 8(2), 327-330.

Khan, A., Nasreen, Mitchell, R. D., Niaz, S., Ayaz, S., Khattak, I., Naeem, H., PérezdeLeón, A. A. & Zaman, M. A. (2019). Seroprevalence of Anaplasma spp. among sheep and goats in Charsadda District, Pakistan. *Small Ruminant Research*, 165, 5-10.

Khan, A., Khan, M. R., Tufail, M. G. Saleem, M. G. & Masood, S. (2015). Pathogenesis of anaplasmosis in bulkhi sheep and their confirmation with polymerase chain reaction. *Scholar's Advances in Animal Veterinary Research*, 2(2), 107-114.

Khan, A., Rehman, A. U., Hisham, M., Khan, A. Z., Rehman, H. U., Khan, M. I., Quershi, M. F. & Ameen, M. A. (2016). Burden of babesiosis among domestic cattle of southern Khyber Pakhtunkhwa. *Pakistan Journal of Entomology and Zoology Studies*, 4(5), 305-307.

Khan, M. I., Shah, S. S. A., Khan, H., Irshad, M. & Ullah, A. (2017). Determination of parasitic load in government cattle breeding and dairy farm, Charsadda, Khyber Pakhtunkhwa-Pakistan. *Advances in Animal and Veterinary Sciences*, 5(4), 174-178.

Khan, M. Q., Zahoor, A., Jahangir, M. & Mirza, M. A. (2004). Prevalence of Blood Parasites in cattle and Buffaloes. *Pakistan Veterinary Journal*, 24(4), 16-20.

Khan, M. S. (2003). "*Livestock Action Plan: Breeding Policy for Pakistan*", Draft Report Prepared for FAO, Pakistan under the Project TCP/PAK/0168.

Khan, N. A., Niaz, S., Shah, M. H., Khan, A., Ahmad, H., Khattak, I., Zeb, J., Naeem, H., Hassam, A. D., Ulucesme, M. C., Ozubek, S. & Aktas, M. (2020). Molecular detection of small ruminant piroplasmosis and first report of *Theileria luwenshuni* (Apicomplexa: Theileridae) in small ruminants of Pakistan. *Experimental Parasitology*, 212, 107872.

Khattak, R. M., Rabib, M., Khan, Z., Ishaq, M., Hameed, H., Taqddus, A., Faryal, M., Durrani, S., Gillani, Q. U. A., Allahyar, R., Shaikh, R. S., Khan, M. A., Ali, M. & Iqbal, F. (2012). A comparison of two different techniques for the detection of blood parasite *Theileria annulata* in cattle from two districts in Khyber Pukhtoon Khwa province (Pakistan). *Parasites*, 19, 91-95.

Kocan, K. M., De la Fuente, J., Guglielmone, A. A. & Meléndez, R. D. (2003). Antigens and alternatives for control of *Anaplasma* marginale infection in cattle. *Clinical Microbiology Reviews*, 16(4), 698-712.

Kocan, K. M., de la Fuente, J., Bouin, E. F., Coetzee, J. F. & Ewing, S. A. (2010). The natural history of *Anaplasma marginale*. *Veterinary Parasitology*, 167, 95107.

Laha, R., Das, M. & Sen, A. (2015). Morphology, epidemiology, and phylogeny of *Babesia*: An overview. *Tropical Parasitology*, 5(2), 94-98.

Livestock census: data on Punjab livestock. (1996). *Livestock Census Organization*, Government of the Punjab.

Mans, B. J. R., Pienaar, A. & Latif, A. (2015). A review of *Theileria* diagnostics and epidemiology. *International Journal of Parasitology: Parasites Wild life*, 4(1), 104-118.

Mohamed, S. B., Alagib, A., AbdElkareim, T. B., Hassan, M. M., Johnson, W. C., Hussein, W. H., Taus, N. S. & Ueti, M. W. (2018). Molecular detection and characterization of *Theileria* spp. infecting cattle in Sennar State, Sudan. *Parasitology Research*, 117, 1271–1276.

Mtshali, M., Mbati, P. & De Waal, D. (2004). A sero-epidemiological survey of blood parasites in cattle in the north-eastern Free State, South Africa. *Onderstepoort Journal of Veterinary Research*, 71(1), 67-75.

Noaman, V. (2013). Discrimination between *Anaplasma marginale* and *Anaplasma ovis* by PCR-RFLP. *World Applied Science Journal*, 21(2), 190-195.

OIE. (2014). 7th ed. volume 1–2. *Office International Des Epizooties*; Paris: Manual of Diagnostic Tests and Vaccines for Terrestrial Animals.

Oura, C., Bishop, R., Wampande, E., Lubega, G. & Tait, A. (2004). The persistence of component *Theileria parva* stocks in cattle immunized with the 'Muguga cocktail' live vaccine against East Coast fever in Uganda. *Parasitology*, 129(1), 27-42.

Pakistan Economic Survey. (2013–14). Ministry of Finance, Islamabad: Government of Pakistan, 2014, 23–41.

Pieszko, M. (2015). *Molecular regulation of the macroschizont to merozoite differentiation in Theileria annulata*. PhD Thesis, University of Glasgow, UK.

Quinn, P. J., Markey, B. K., Carter, M. E., Donnelly, W. J. C. & Leonard, F. C. (2002). *Veterinary microbiology and microbial diseases*. 1st Iowa State University Press Blackwell Science.

Radostits, O. M., Gay, G. C., Hinchiff, K. W. & Constable, P. O. (2007). *Veterinary Medicine: A text book of the disease of cattle, sheep, goat, pigs and horses*. 10th edn. Saunders Elsevier, London, pp. 1110-1489.

Rafiullah, A., Rahman, K., Khan, A., Ali, A., Khan, A. & Sajid, N. K. (2019). Prevalence of *Theileria parva* in large ruminants through conventional and molecular techniques in district Lakki Marwat and Peshawar (Pakistan). *Sarhad Journal of Agriculture*, 35(2), 320-329.

Rajput, Z. I., Hu, S. H., Arijo, A. G., Habib, M. & Khalid, M. (2005). Comparative study of *Anaplasma* parasites in tick carrying buffaloes and cattle. *Journal of Zhejiang University Science B*, 6(11), 1057-1062. doi: 10.1631/jzus.2005.B1057.

Ramzan, M., Unsar, N. U., Abbas, H., Adnan, M., Rasheed, Z. & Khan, S. (2019). Diversity of hard ticks in goats and sheep in Multan, Punjab, Pakistan. *Agricultural and Biological Research*, 35(2), 7-9.

Razzaq, F., Khosa, T., Ahmad, S., Hussain, M., Saeed, Z., Khan, M., Shaikh, R., Ali, M. & Iqbal, F. (2015). Prevalence of *Anaplasma phagocytophilum* in horses from Southern Punjab (Pakistan). *Tropical Biomedicine*, 32, 233–239.

Raziq, A., Younas, M. & Rehman, Z. (2010). Prospects of livestock production in Balochistan. *Pakistan Veterinary Journal*, 30(3), 181-186.

Rehman, A., Conraths, F. G., Sauter-Louis, C., Krücken, C. & Nijhof, A. M. (2018). Epidemiology of tick borne pathogens in the semi arid and the arid agro-ecological zones of Punjab province, Pakistan. *Transbound and Emerging Diseases*, 66(1), 526-536.

Rehman, A., Nijhof, A. M., Sauter-Louis, C., Schauer, B., Staubach, C. & Conraths, F. J. (2017). Distribution of ticks infesting ruminants and risk factors associated with high tick prevalence in livestock farms in the semi-arid and arid agro-ecological zones of Pakistan. *Parasites & Vectors*, 10, 190.

Riaz, M., Tasawar, Z. & Ullah, M. Z. (2017). A Study on Molecular Prevalence, Intensity and Associated Risk Factors for Ovine and Caprine Theileriosis from Southern Punjab, Pakistan. *Pakistan Journal of Life and Social Sciences*, 15(3), 150-157.

Riaz, M. & Tasawar, Z. (2016). Detection and discrimination of *Theileria* species infection by using PCR amplification in small ruminants in and around Multan, Pakistan. *International Journal of Bioscience*, 9(4), 61-71.

Rikihisa, Y. (2011). Mechanisms of obligatory intracellular infection with *Anaplasma phagocytophilum. Clinical Microbiology Reviews*, 24(3), 469-489.

Saad, F., Khaisroon, M., Khan, K. & Akbar, N. L. (2015). Prevalence and molecular detection of babesiosis in the slaughter animals of Peshawar Khyber Pakhunkhawa Pakistan. *Pakistan International Journal of Current Microbiology and Applied*, 4(8), 1030-1036.

Saeed, Z., Iqbal, F., Hussain, M., Shaikh, R. S., Farooq, U., Akbar, A., Gulsher, M., Ayaz, M. M., Sajid, M., Ali, M. & Aktas, M. (2016). Molecular prevalence and haematology of tropical Theileriosis in Cholistani cattle from nomadic herds of the Cholistan Desert, Pakistan. *Kafkas Universitesi Veteriner Fakultesi Dergisi*, 22(2), 281-286.

Sajid, M. S., Iqbal, Z., Shamim, A., Saddique, R. M., Rizwan, H. M. & Hassan, M. J. U. (2018). Distribution and abundance of ticks infesting livestock population along Karakorum highway from Mansehra to Gilgit, Pakistan. *Journal of the Hellenic Veterinary Medical Society*, 68(1), 51-55.

Sajid, M. S., Iqbal, Z., Khan, M. N., Muhammad, G., Needham, G. & Khan, M. K. (2011). Prevalence, associated determinants, and *in vivo* chemotherapeutic control of hard ticks (Acari: Ixodidae) infesting domestic goats (*Capra hircus*) of lower Punjab, Pakistan. *Parasitology Research*, 108, 601–609.

Saleem, S., Ijaz, M,, Farooqi, S. H., Rashid, M. I., Khan, A., Masud, A., Aqib, A. I., Hussain, K., Mehmood, K. & Zhang, H. (2018). First molecular evidence of equine granulocytic anaplasmosis in Pakistan. *Acta Tropica,* 180, 18-25.

Schorderet-Weber, S., Noack, S., Selzer, P. M. & Kaminsky, R. (2017). Blocking transmission of vector-borne diseases. *International Journal of Parasitology: Drugs Drug Resistance*, 7, 90–109.

Schulze, T. L., Jordan, R. A. & Hung, R. W. (2001) Effects of selected meteorological factors on diurnal questing of *Ixodes scapularis* and *Amblyomma americanum* (Acari :Ixodidae). *Journal of Medical Entomology*, 38, 318–324.

Shah, M. H., Ijaz, M., Ahmed, A., Aziz, M. U., Hammad, A. G., Ghauri, N. & Naveed, M. (2020). Molecular analysis and risk factors associated with *Theileria equi* Infection in

Domestic Donkeys and Mules of Punjab, Pakistan. *Journal of Equine Veterinary Science*, 92, 103164.

Shahzad, M., Noor, H., Ahmad, M. U. D., Rashid Munir, R., Saghar, M. S., Mushtaq, M. H., Ahmad, N., Akbar, G. & Mehmood, F. (2013). Prevalence and Molecular Diagnosis of *Babesia ovis* and *Theileria ovis* in Lohi Sheep at Livestock Experiment Station (LES), Bahadurnagar, Okara, Pakistan. *Iranian Journal of Parasitolgy*, 8(4), 570-578.

Shams, S., Ayaz, S., Ali, I., Khan, S., Gul, I., Gul1, N. & Khan, S. N. (2013). Sensitivity and specificity of PCR & microscopy in detection of Babesiosis in domesticated cattle of Khyber Pakhtunkhwa, Pakistan. *International Journal of Advancements in Research and Technology*, 2(5), 345-347.

Siddique, R. M., Sajid1, M. S., Iqbal, Z. & Saqib, M. (2020). Association of different risk factors with the prevalence of Babesiosis in cattle and buffalos. *Pakistan Journal of Agriculture Sciences*, 57(2), 517-524.

Talat, R., Khanum, T. & Hayat, A. (2005). Studies on mammalian haematozoan parasites of NWFP Pakistan. *Pakistan Journal of Biological Sciences*, 8, 726–9.

Tayyub, M., Ashraf, K., Lateef, M., Anjum, A. A., Ali, M. A., Ahmad, N., Nawaz, M. & Nazir, M. M. (2019). Genetic diversity of canine *Babesia* species prevalent in pet dogs of Punjab, Pakistan. *Animals*, 9, 439.

Thalmann, O., Shapiro, B., Cui, P., Schuenemann, V. J., Sawyer, S. K., Greenfield, D., Germonpre, M. B., Sablin, M. V., Lopez-Giraldez, F. & Domingo-Roura, X. (2013). Complete mitochondrial genomes of ancient canids suggest a European origin of domestic dogs. *Science*, 342, 871-874.

Theiler, A. (1911). Further investigations into anaplasmosis of South African cattle, p7-46. In *First Report of the Director of Veterinary Research*, Union of South Africa. Johannesburg, South Africa.

Thornton, P. K. & Gerber, P. (2010). Climate change and the growth of the livestock sector in developing countries. *Mitigation Adapt Strategic Globe Change*, 15, 169–184.

Tuvshintulga, B., AbouLaila, M., Davaasuren, B., Ishiyama, A., Sivakumar, T., Yokoyama, N., Iwatsuki, M., Otoguro, K., Ōmura, S. & Igarashi, I. (2016). Clofazimine inhibits the growth of *Babesia* and *Theileria* parasites *in vitro* and *in vivo*. *Antimicrobial Agents and Chemotherapy*, 60, 2739-2746.

Uilenberg, G. (2006). Babesia -a historical overview. *Veterinary Parasitology*, 138(1-2), 3-10.

Yu, Z., Wang, H., Wang, T., Sun, W., Yang, X. & Liu, J. (2015). Tick-borne pathogens and the vector potential of ticks in China. *Parasites & Vectors*, 8, 24-26.

Zeb, J., Takács, N., Kontschán, J., Shams, S., Ayaz, S. & Hornok, S. (2019). Genetic diversity, piroplasma and trypaosoms in *Rhipicephalus microplus* and *Hyalomma anatolicum* collected from cattle in Northern Pakistan, *Experimental and Applied Acarology* 79, 233–243

Zia, A. & Nazir, Q. (2019). Investigation on prevalence of Babesiosis in cattle in various areas of Taunsa Sharif District Dera Ghazi Khan, Punjab, Pakistan. *Buffalo Bulletin*, 38(3), 78-82.

Zulfiqar, S., Shahnawaz, S., Ali, M., Bhutta, A. M., Iqbal, S., Hayat, S. & Saeed, A. (2012). Detection of Babesia bovis in blood samples and its effect on the hematological and serum biochemical profile in large ruminants from Southern Punjab. *Asian Pacific Journal of Tropical Biomedicine*, 2(2), 104-108.

In: The Entomological Guide to *Rhipicephalus*
Editors: Sachin Kumar, Raquel Cossio Bayugar et al. ISBN: 978-1-53619-619-1

Chapter 4

Current Status of Tick-Borne Diseases in India

***Sachin Kumar*[1,*], *Anil Kumar Sharma*[1], *Gajanan M. Chigure*[2], *Ashok K. Chaubey*[3] *and Srikanta Ghosh*[1]**
[1]Division of Parasitology, ICAR-Indian Veterinary Research Institute, Izatnagar, Bareilly (UP), India
[2]Department of Veterinary Parasitology, College of Veterinary and Animal Sciences, MAFSU, Parbhani, (Maharashtra), India
Department of Zoology, Ch. Charan Singh University, Meerut, (Uttar Pradesh), India

Abstract

Ticks and tick borne diseases are a major threat to the livestock and economy of developing countries including India. Ticks, as vectors of several zoonotic diseases, are second only to mosquitoes as vectors. The diseases spread by ticks are a major constraint to animal productivity while causing morbidity and mortality in both animals and humans. *Rhipicephalus (Boophilus) microplus* and *Hyalomma anatolicum* have been recognized since long as vectors of importance for transmitting diseases like babesiosis, theilieriosis and CCHF. There is need to recognize the importance of ticks as vectors in India and devise better methods for their control. This review is focused on the major tick borne animal and human diseases in India and the current situation in India regarding control of ticks including the development of resistance, progress in tick vaccine and the development of potential phytoacaricides as alternatives to chemical acaricides.

Keywords: ticks, zoonotic diseases, *Rhipicephalus (Boophilus) microplus*, babesiosis, theilieriosis

---

* Corresponding Author's Email: sachin.amroha@gmail.com.

INTRODUCTION

Ticks are highly specialized obligate haematophagous ectoparasite of mammals, birds and reptiles, distributed worldwide and are of enormous medical and veterinary relevance owing to the direct damage they cause to their hosts and as vectors of a large variety of human and animal pathogens. Today, most emerging infectious diseases arise from zoonotic pathogens, and many of them are transmitted by tick vectors. Ticks are among the most competent and versatile vectors of pathogens and are second to mosquitoes as vectors of number of human pathogens and the most important vector of pathogens affecting cattle worldwide (Peter et al., 2005). Several characteristics of ticks make them outstanding vectors of pathogenic agents. Their wide host range and tendency to feed on several hosts during their life cycle ensures ample opportunity to acquire and transmit pathogens. Their hardiness and longevity enable them to survive long periods in unfavorable environmental conditions. They have a high reproductive potential, ensuring maintenance of large populations and a high frequency of host-vector contact. They also feed slowly and, in the case of ixodids, attach to hosts for relatively long periods, which allows sufficient time for pathogen acquisition and transmission, as well as tick dispersal by migrating or wandering hosts. In humans, tick infestations typically involve few specimens and the greatest risk for people bitten by a tick lies in infection due to a tickborne pathogen (Parola & Raoul, 2001). Such pathogens are diverse and include viruses, bacteria, and protozoa (Jongejan & Uilenberg, 2004; de la Fuente et al., 2008). In animals, tick infestations are much more severe than in humans. Animals can be parasitized by hundreds or even thousands of ticks, which obviously multiplies the effect on the host, either by direct injuries or disease transmission. Direct injuries to animals can be very serious, especially in tropical climates, and are mainly observed in infestations with ixodid ticks. Tick-borne infectious diseases are growing steadily partly due to the establishment of the tick vector into urban areas/new areas and posing serious threat to world health problem and also a major obstacle for animal health and production (Rajput et al., 2006). The range of pathogens transmitted by ticks (Table 1) and its consequences to human and animal health signifies the involvement of interdisciplinary research team in the area of research. The complex triangular interactions between pathogen-host- vectors complicated the subject and multiple pathways were targeted to control the ticks and tick borne diseases (TTBDs).

Distribution of Important Tick Vectors in India

India is predominantly an agricultural country with about 70% of its population dependent on income from agriculture. Farmers are keeping animals for milk, meat, wool, and hide production and also for various farm operations. India accounts for a significant share of the world's livestock resources with nearly 57% of world's buffaloes, 16.5% cattle, 16.3% goats, and 5.7% sheep (FAO 2004). Livestock is an important source of animal protein for farm families and also used for draught power in agriculture and transport, and their dung is used to increase soil fertility. In India, the damage caused by ticks and tick borne diseases (TTBDs) to livestock is considered very high (Ghosh et al., 2006). An estimate of US$ 498.7 million per annum has been calculated as the cost of TTBDs in India (Minjauw and McLeod

2003). India has a wide variety of tick species spread throughout country as documented by Ghosh et al., (2007). However, a few of them have been experimentally established as the principal vector of pathogens (Table 2).

Table 1. Important tick borne diseases of man and livestock

Vector	Vector Borne Diseases	Parasite/pathogen
Viral diseases		
Haemophysalis spinigera	Kyasanur Forest Disease	Flavidiridae (KFD virus)
Hyalomma spp.	Crimean-Congo Haemorragic Fever	Bunyaviridae(CCHF virus)
Hyalomma dromedarii	African Horse Sickness	Reoviridae(African Horse Sickness virus)
Ornithodorus mobuta	African Swine Fever	African Swine Fever virus
Rhipicephalus appendicul	Nairobi sheep disease	Bunyaviridae
Rickettsial diseases		
Rhipicephalus sanguineus	Ehrlichiosis Human monocytic ehrlichiosis	*Ehrlichia canis, E. equi* *E, senetsu, E. chaffeensis, E. phgocytophilia*
Amblyoma variegatum	Cowdriosis Anaplasmosis	*Cowdria ruminantium Anaplasma marginale*
R. sanguineus, Dermacenter andersoni, Boophilus decoloratus	Epidemic typhus	*Rickettsia prowazekii* *Rickettsia conorii*
Spirochete diseases		
Ixodes ricinus	Lyme disease	*Borrelia burgdoferi*
Bacterial diseases		
Dermacentor spp.	Tularemia	*Francisella tularensis*
Protozoan diseases		
H.a. anatolicum, R. appenticulatus	Theileriosis	*Theileria annulata, T. parva,T. hirci*
Boophilus spp	Babesiosis	*B. bigemina, B. bovis, B, ovis, B. motasi, B. equi*
Ixodes spp.	Human babesiosis	*B. microti*

Table 2. Important tick borne diseases in India, distribution of vectors and status of vaccine development

Tick vector	Pathogen/Parasite	Distribution of vectors [a]	Host	Vaccine status
Haemophysalis spinigera	KFD virus	1, 2, 6, 8, 14 -17, 19, 22, 26	Man	Chick embryo tissue culture vaccine *
Rhipicephalus sanguineus?	*Rickettsia conorii*	2- 6, 9 - 17, 20, 22 -, 24, 26, 29, 30	Man	Nil
Hyalomma anatolicum	*Theileria annulata* *CCHF*	2, 4 - 12, 14, 16, 17, 22 - 24, 26, 28 - 30	Cattle Man	Attenuated macroschizont vaccine**
Rhipicephalus (Boophilus) microplus *Haemaphysalis* sp *R. sanguineus* *Rhipicephalus* sp. *Hyalomma* sp	*B. bigemina* *B. motasi,* *B. canis* *B.ovis* *B. equi*	2 – 14, 16, 19, 20, 22 – 26, 28-30 2- 6, 9 - 17,20,22,24,26,29,30 ***	Cattle, Buffalo Goat Dog Sheep Horse	Nil
B. microplus	*Anaplasma marginale*	2 – 14, 16, 19,20, 22 – 26, 28-30	Cattle, Bufallo, sheep	Nil
Hyalomma sp. *R. sanguineus*	*Ehrlichia bovis* *E. canis*	2- 6, 9 - 17,20,22 - ,24,26,29,30	Cattle Dog	Nil

*Not fully effective in field situation6; **Effective but not in large-scale use; *** Not known;? Tick species not yet confirmed. [a]1, Andaman and Nicobar Islands; 2, Andhra Pradesh, 3, Arunachal Pradesh; 4, Assam; 5, Bihar; 6, Chattisgarh; 7, Delhi; 8, Goa;9, Gujarat; 10, Haryana; 11, Himachal Pradesh; 12, Jammu and Kashmir; 13, Jharkhand; 14, Karnataka; 15, Kerala; 16, Madhya Pradesh; 17, Maharashtra;18, Manipur; 19, Meghalaya; 20, Mizoram; 21, Nagaland; 22, Orissa; 23, Punjab; 24, Rajasthan; 25, Sikkim; 26, Tamil Nadu; 27, Tripura; 28, Uttaranchal; 29, Uttar Pradesh; West Bengal.

MAJOR TBDS PREVALENT IN INDIA

Indian Tick Typhus

Indian tick typhus (ITT) is a re-emerging zoonotic disease in India. It is a type of rickettsial spotted fever similar to Rocky Mountain spotted fever (SF) prevalent in many parts of North and South America. Indian tick typhus is caused by *Rickettsia conorii* and similar strain was isolated from the Mediterranean region, Kenya, South Africa. The disease is reported from Maharashtra, Tamil nadu, Karnataka, Kerala, Jammu and Kashmir, Uttaranchal, Himachal Pradesh, Rajasthan, Assam and West Bengal (Mahajan et al., 2007; Mathai et al., 2001; Sundhindra et al., 2004). The dog tick *Rhipicephalus sanguineus* is the most important vector of ITT although some species of *Haemaphysalis* and Hyalomma ticks may also transmit the infection (CD Alert, May-July, 2009). ITT in India has been recognized clinically but cases have not been documented frequently maybe due to the lack of definitive diagnostic tools. Between 1996 and 1998, serological testing amongst residents of south India confirmed that spotted fever continues to occur in south India (Mathai et al., 2001). An extensive study on tick borne rickettsiosis in Pune district of Maharashtra revealed that Indian tick typhus exists as zoonosis (Padbidri et al., 1984). Subsequently, ITT has been reported in Mumbai (Shah et al., 2009), Himachal Pradesh (Mahajan et al., 2007), in a French traveller returning from India (Parola et al., 2001) and in Haryana (Chaudhry et al., 2009). Recently in June, 2013 Manish Kumar et al., reported a case of ITT with gangrene in a 10 year old boy from Delhi.

No rapid laboratory tests are available to diagnose rickettsial infection early in the course of disease. Indirect immuno fluorescence Assay (IFA) is the preferred diagnostic assay for detection of rickettsial infection but its availability and cost are major constrains in India and other developing countries. ELISA techniques, particularly immunoglobulin M (IgM) capture assays, are probably the most sensitive tests available for rickettsial diagnosis, and the presence of IgM antibodies, indicate recent infection with rickettsial diseases. The weil felix test is a cheap and easily available diagnostic assay for rickettsial diseases when other means of diagnosis are not available but the downside is its poor reliability and specificity. Rickettsial diseases can be easily confused with a variety of viral (measles, enteroviral exanthems, dengue, infectious mononucleosis), protozoal (malaria), bacterial (meningococcemia, typhoid, leptospirosis, toxic shock syndrome, scarlet fever) and collagen vascular (Kawasaki disease, other vasculitis) diseases, and adverse drug reactions. Invasive meningococcal disease may not be reliably distiguished from rickettsial disease clinically, hence one may need to treat for both conditions, after sending cerebrospinal fluid and blood for appropriate studies (Chapman et al., 2006). The possibility of rickettsial disease should be considered in those leptospirosis patients who present with atypical features or respond poorly to therapy (Watt et al., 2003). The treatment of rickettsial infection is relatively easy after diagnosis and the commonly used antibiotics for treatment are Tetracyclines, Chloramphenicol, Macrolides and Rifampicin. Among tetracyclines, Doxycycline is considered as a drug of choice for SFG rickettsioses.

Crimean Congo Hemorrhagic Fever (CCHF)

Crimean-Congo hemorrhagic fever (CCHF) was described in the Crimea in 1944 during an outbreak and was called Crimean hemorrhagic fever. A later virus isolate from Congo was noted to be the same pathogen and the nomenclature was changed to Crimean-Congo hemorrhagic fever virus (CCHFV) (Wallace et al., 2002). The CCHF virus of *Nairovirus* group circulates in an enzootic tick-vertebrate-tick cycle. However, there is no evidence that the virus causes disease in animals. Wide range of domestic and wild animals may get CCHF virus infection. The virus infection has been commonly demonstrated among smaller wildlife species-eg, hares and hedgehogs that harbor the tick vectors that are at immature stages (Hoogstraal, 1979; Watts et al., 1988).

CCHF is one of the severe forms of hemorrhagic fever endemic in Africa, Asia, Eastern Europe and the Middle East with a near fatal mortality rate. Since its discovery in 1960s, nearly 140 outbreaks involving more than 5,000 cases have been reported all over the world. A total of 52 countries have been recognized as endemic or potentially endemic regions, reporting substantial number of cases every year. The geographic range of CCHF virus is the most extensive one among the medically important tick borne viruses. The distribution of *Hyalomma* spp., the principal tick vector, also has wide distribution. CCHFV is mainly seen in the Middle East and Asia and parts of Europe including southern portions of the former Union of Soviet Socialist Republics (USSR) (Appannanavar and Mishra, 2011). In the recent past, most cases have been reported from Pakistan (Sheikh et al, 2005), Iran (Sharifi et al., 2009; Nadiri et al., 2010), Sudan (Aradaib et al., 2010), Bulgaria (Papa et al., 2004) and from Turkey (Yilmaz et al., 2009). In India the CCHF had not been reported before outbreak of CCHF viral infection in Gujarat in January 2011. This was the first report of CCHF virus in India (Appannanavar and Mishra, 2011; Patel et al., 2011).

Diagnosis of CCHF virus infection is possible with serological assays and molecular assays. Drosten et al., (2003) developed a one-step real-time RT-PCR assay for detecting CCHFV using primers to the nucleoprotein gene. Later, a real-time RT-PCR assay was developed using TaqMan-Minor Groove Binding Protein (MGB) probe, which had higher specificity and a shorter probe length (Whitehouse, 2004). A CCHFV recombinant nucleoprotein (rNP) based IgG ELISA and IgM-capture ELISA have also been shown to be useful methods for diagnosis of CCHFV infections (Saijo et al., 2002, 2005). Treatment options for CCHF are limited as there is currently no specific antiviral therapy approved for use in humans by the FDA. Although nucleoside analog ribavirin is shown to be effective for inhibiting CCHFV *in vitro* but its efficacy is not very well documented in randomized control study. However, Tasdelen et al., (2009) have shown the beneficial effect of ribavirin if given at an early phase of the CCHF. According to World Health Organization (WHO), ribavirin is the anti-viral medication of choice for CCHF and the recommended dose is, an initial dose of 30 mg/kg followed by 15 mg/kg for four days and then 7.5 mg/kg for six days for a total of 10 days. A vaccine for CCHF has been in use in Bulgaria since many years but its efficacy and safety is not well quantified.

Kyasanur Forest Disease

Kyasanur forest disease (KFD) is caused by Kyasanur forest disease virus (KFDV), a member of the virus family Flaviviridae. KFDV was identified in 1957 when it was isolated from a sick monkey from the Kyasanur forest in the Karnataka State, India. The disease is localised in five districts (Shimoga, Chikamagalur, Uttar Kannada, Dakshina Kannada and Udupi) of the state and occurs as seasonal outbreaks during December to May when the nymphal activity of the vector, *Haemaphysalis spinigera*, in the forest is maximum (Kasabi at al., 2013)

The main hosts of KFDV are small rodents, but shrews, bats, and monkeys may also carry the virus and is transmitted through the bite of an infected tick . Humans can get this disease from a tick bite or by contact with an infected animal, such as sick or recently dead monkey. Larger animals such as goats, cows, and sheep may become infected with KFD, but they do not have a role in the transmission of the disease.

Diagnosis can be made by serological testing during the convalescent phase using hemagglutination inhibition, complement fixation, and through mass tag polymerase chain reaction. There is no specific treatment for KFD, but supportive therapy is important. Supportive therapy includes the maintenance of hydration and the usual precautions for patients with bleeding disorders. Due to lack of specific treatments, prophylaxis by vaccination is advised. National Institute of Virology, Pune (India) developed a formalin inactivated chick embryo tissue culture vaccine against KFD. This vaccine evokes neutralizing antibodies response in about 70% of the vaccinated persons. The technology has been transferred to Karnataka Public Health Department for production and vaccination. Regular vaccination and booster campaigns are run by the State authorities in the KFD affected areas. There are reports of the reduced efficacy of vaccine in recent years (Pattnaik, 2006; Kasabi et al., 2013) that necessitates a review of the current vaccination protocol including the storage and administration. The antigenic variations in the current strains of KFD virus and the strain used for vaccine preparation (isolated in 1950s) have to be determined and a new vaccine with current strain of virus should be developed.

Theileriosis

Bovine theileriosis is a tick borne protozoan infection that affects many animal species around the globe. It has a significant adverse effect on the productivity of animals and also proves to be fatal if left untreated. Latest reports cited an annual loss of US$ 239.5 million alone due to bovine tropical theileriosis in India (Minjauw and McLeod, 2003). Bovine theileriosis in India is caused by an intracellular protozoan, *Theileria annulata,* and is transmitted by three hosts tick, *Hyalomma anatolicum*. Approximately ten million crossbred cattle in India are at risk to this disease and several reports of subclinical infections and severe outbreaks have been documented (Chandra et al., 2000, Bansal et al., 1987). In a six year survey in project area of cooperative milk producers union, Mysore, a district of Southern India, a total of 17.7% cattle showed infection of *T. annulata* in blood smears. The incidence of theileriosis was highest in southwest monsoon period and lowest in cold weather. (Muraleedharan et al., 1994). High parasitemia (40-50%) was detected in the blood smears of the affected animals and seventeen cross-bred calves died due to the infection. The authors also detected 18.9% carrier cattle showing low level of *T. annulata* infection in100 smears from the state of Punjab, Hariana, Uttar Pradesh (Northern state), Rajasthan (Western state) and in Orissa (Eastern state) Chandra et al., 2000. Serological screening indicated that 30-60% of cattle maintained in unorganized cattle farms had antibodies to *T. annulata* (Singh, 1991). Haque *et al.,* (2010) carried out a prevalence study of *T. annulata* in *Hyalomma anatolicum* ticks in Punjab and detected 15.45% prevalence of *T. annulata* in female ticks. Diagnosis of theileriosis is mainly dependent on clinical signs as well as on microscopic examination of Giemsa-stained lymph nodes and blood smears.

Polymerase chain reaction (PCR) amplification of small subunit ribosomal RNA (SSU rRNA) gene of *T. annulata* was used to detect infection in carrier cattle (Roy *et al.,* 2000). Antibodies to *T. annulata* can be detected with an enzyme-linked immunosorbent assay (ELISA) using TASP and Tams1 antigens (Bakheit et al., 2004; Seitzer et al., 2008; Rajendran and Ray, 2012). Developments of sensitive and specific diagnostic assays have become possible with the advent of the polymerase chain reaction (PCR) for the detection numerous *theileria* species, in animals. Antiparasitic drugs are effective in animals with clinical signs, but in most of the cases the animals may remain carriers. For treatment of animals against theileriosis "Menoctone," a hydroxynethoquinone is found to be effective against *T parva* infection in a single injection at a dose rate of 10mg per kg. body weight. A cheaper analog with generic name Parvaquone (Clexon, Coopers Animal Health Ltd., UK) is effective against theileriosis at a dose rate of 20 mg. per kg. body weight of cattle. Another analog, "Buparvaquone" (Butalex- Coopers Animal Health Ltd., UK) is more effective in the treatment of both experimental and natural infections of *T. annulata* and *T parva* in cattle. The drug is found to protect cattle for up to a week from *Theileria annulata* or *T. parva* infection. Milk and meat of the treated cattle are recommended as unfit for human consumption for 48 and 42 days, respectively, post-treatment.

The live attenuated vaccine of *T. annulata* was produced by prolonged *in vitro* cultivation of lymphoblasts of cattle infected with macroschizonts of *T. annulata* (Pipano and Tsur, 1966). Complete attenuation of schizonts has been reported after *in vitro* cultivation for 600-900 days depending on the nature of the isolates (Subramanian et al., 1988; Singh, 1990). The protection engendered by the attenuated schizont vaccine has been evaluated by laboratory challenge with live infected ticks or with ground up tick sporozoites (GUTS) inoculated through syringe passage. The results of such challenge ranged from no clinical response at all (Gill et al., 1976; Subramanian et al., 1988; Khatri et al., 2001) through mild transient clinical reactions with low parasitaemia (Khdier and Latif, 1991), to death from acute theileriosis (Shukla and Sharma, 1991). In endemic areas, single vaccination appeared to be adequate for protection when the animals receive continuous challenge from natural tick infestation (Ray and Bansal, 1997). The attenuated schizont vaccine of *T. annulata* was commercialized under the trade name of "Raksha Vac-T" and is produced and marketed by Indian Immunologicals, Hyderabad, India.

Babesiosis

Babesiosis, or tick fever, is a febrile disease of domestic and wild animals characterized by extensive erythrocytic lysis leading to anaemia, icterus and haemoglobinuria. The disease is caused by protozoan parasites of the genus *Babesia,* an intraerythrocytic protozoan parasites transmitted mainly by R. (B.) *microplus*. There are at least 14 distinct species of *Babesia* from various vertebrate hosts (Riek, 1968). In cattle, *Babesia bigemina* is the primary pathogen and its incidence in indigenous, cross-bred cattle and in buffaloes has been reported frequently since long (Dwivedi et al., 1976) Depending on the sensitivity of serological tests, different authors reported upto 86% seroprevalence of the pathogen in Indian dairy animals (Bhikane at al. 2001; Singh et al., 2007).

Diagnosis of babesiosis usually includes examination of stained blood smears as well as serologic evaluation with indirect (immuno)fluorescent antibody tests (IFATs) (Chisholm et al., 1978, Singh et al., 2007) and also by PCR. Because of improved sensitivity, PCR has become the test of choice for confirmation of actual infection in antibody-reactive individuals and for monitoring therapeutic responses (Persing et al., 1992; Ravindran et al., 2007). Live, attenuated strains of *B. bovis, B. bigemina* or *B. divergens* are used to vaccinate cattle in some countries but some safety issues issues including the potential for virulence in adult animals, possible contamination with other pathogens, and hypersensitivity reactions to blood proteins has been noted. Clinically affected animals treated with an antiparasitic drug (diminazene diaceturate, imidocarb, amicarbalide); efficacy depends on timely detection early in disease. Imidocarb has been reported to protect animals from disease but allows development of immunity and there is concern regarding its residues in milk and meat.

Human babesiosis mostly occurs in the USA, but cases have also been reported in several European countries. Human babesiosis is caused by one of several babesial species that have distinct geographic distributions based on the presence of competent hosts. In North America, babesiosis is caused predominantly by *Babesia microti (*Dammin et al., 1981*)*. In Europe, babesiosis is considerably rarer but more lethal; it is caused by the bovine pathogen *Babesia divergens* (Homer et al., 2000). In Indian situation there is only one documented report of a *Babesia microti* infection in a 51 year old male patient (Marathe et al., 2005). However, as babesiosis in humans can be confused with plasmodium infection on examination of blood smears, the actual incidence of babesiosis in humans maybe higher than reported. Careful examination of peripheral smears and surveillance studies may be necessary to know the actual prevalence of human babesiosis in India.

Bovine Anaplasmosis

Bovine anaplasmosis is an infectious, noncontagious haemotropic disease of cattle characterized in the acute form by fever, anaemia, weakness, constipation, yellowing of the mucous membranes, lack of appetite, depression, dehydration and labored breathing. Animals surviving an acute attack often make a slow recovery, resulting in losses in milk or meat production. Generally, mortality is between 5 and 40 percent, but may reach 70 percent during a severe outbreak. Anaplasmosis is considered as one of the top ten economically important rickettsial diseases affecting ruminants in India (PDADMAS Annual Report, 2005-06) and is principally transmitted by *R. (B.) microplus*. The epizootiology of anaplasmosis is complicated by the life-long carrier state which occurs in animals that have recovered from the clinical disease.

Clinical anaplasmosis was first recorded in Indian cattle from the state of Orrisa (Patnaik, 1963). Subsequently *A. marginale* infection was detected in livestock of Uttar Pradesh, Punjab, Haryana, Tamilnadu and parts of north and central India. (Mohan and Pathak 1966,

Sinha and Pathak, 1966, Misraulia et al., 1988, Srinivasan et al., 1995). However, there is a paucity of recent information about the presence of *A. marginale* infections in cattle from India. Recently anaplasmosis has been recorded in Jammu (Raina et al., 2005) Karnataka (Muraleedharan, 2005), Haryana (Kumar and Sangwan, 2010) and Tamil Nadu (Arunkumar and Nagarajan, 2013) With the advent of molecular tools, the sensitivity of detection of Anaplasma infection in diseased and carrier animals has improved. Nair at al. 2013 detected *A. marginale* in 25 samples and *A. bovis* in one sample out of total 150 samples using PCR while the microscopic examination could detect *A. marginale* infection in only 3 samples. Singh et al., (2012) reported 45.2 to 73.1 infectivity in cattle of Punjab using semi nested PCR assay. Oxytetracycline at 20mg/kg body weight is the drug of choice. It was believed earlier that repeated treatments with oxytetracycline eliminated the carrier status of treated animals but with sensitivity of PCR it has been proved to be wrong.

Vector Control

Control of TTBDs has primarily focused on the treatment of the disease and development of a suitable prophylaxis system for protection against the disease in future. However, the treatment of disease does not necessarily mean the complete removal of the infecting parasite which persists in treated animals in latent form. In the absence of a suitable vaccine there is continued risk of spread of the infection at a later stage. Since effective vaccines or drugs are not always available for the prevention or treatment of these diseases, control of transmission has to rely on control of vectors. Large scale vector control has been shown to be possible in U.S.A with eradication of *R.(B.)microplus* and *R.(B.) annulatus*, the vectors of babesiosis. But in India the large scale campaign for tick control is debatable for both financial and practical reasons as the country shares a porous border for livestock movement with 4-5 countries and has a highly unorganized dairy sector. Therefore, there is need for tick control methods that can be applied by individuals and communities.

References

Aguirre, J., Sobrino, A., Santamaria, M., Aburto, A., Roman, S., Hernandez, M., Ortiz, Y. A. *Resistancia de garrapatas en Mexico*, 1986 pp. 282D306. In A. H. Cavazzani and Z. Garcia [eds.], *Seminario Internacional de Parasitologia Animal*, Cuernavaca, Morelos, Mexico.

Almazán, C., Blas-Machado, U., Kocan, K. M., Yoshioka, J. H., Blouin, E. F., Mangold, A. J. and de la Fuente, J. (2005). Characterization of three *Ixodes scapularis* cDNAs protective against tick infestations. *Vaccine,* 23 (35): 4403-16.

Almazan, C., Lagunes, R., Villar, M., Canales, M., Rosario-Cruz, R., Jongejan, F. and de la Fuente, J.(2010). Identification and characterization of *Rhipicephalus (Boophilus) microplus* candidate protective antigens for the control of cattle tick infestations. *Parasitology Research,* 106: 471–479.

Almazan, C., Moreno-Cantu O, Moreno-Cid JA, Galindo RC, Canales M, Villar, M. and de la Fuente J. (2012). Control of tick infestations in cattle vaccinated with bacterial membranes containing surface-exposed tick protective antigens. *Vaccine,* 30: 265–272.

Andreotti, R., Cunha, R. C., Soares, M. A., Guerrero, F. D., Leite, F. P. and de León, A. A. (2012). Protective immunity against tick infestation in cattle vaccinated with recombinant trypsin inhibitor of *Rhipicephalus microplus*. *Vaccine,* 30 (47): 6678-85.

Appannanavar, S. B. and Mishra, B. (2011). An update on Crimean Congo Hemorrhagic Fever. *Journal of Global Infection Diseases,* 3: 285-92.

Aradaib, I. E., Erickson, B. R., Mustafa, M. E., Khristova, M. L., Saeed, N. S., Elageb, R. M. and Nichol, S. T. (2010). Nosocomial outbreak of Crimean-Congo hemorrhagic fever, Sudan. *Emerging Infectious Diseases,* 16: 837-9.

Arunkumar, S. and Nagarajan, K. (2013). A study on prevalence status of *Anaplasma marginale* infection among cattle population of Kancheepuram and in and around Chennai districts of Tamil Nadu. *International Journal of Food, Agriculture & Veterinary Sciences,* 3(1): 155-157.

Azhahianambi, P., de La Fuente, J., Suryanarayana, V. V. S. and Ghosh, S. (2009a). Cloning, expression and immunoprotective efficacy of rHaa86, the homologue of the Bm86 tick vaccine antigen, from *Hyalomma anatolicum anatolicum*. *Parasite Immunology,* 31: 111–122.

Azhahianambi, P., Ray, D. D., Pallab, C., Gupta, R. and Ghosh, S. (2009b). Vaccine efficacy of Bm86 ortholog of *H. a. anatolicum*, rHaa86 expressed in prokaryotic expression system. *Journal of Parasitology Research,* 7.

Bakheit, M. A., Schnittger, L., Salih, D. A., Boguslawski, K., Beyer, D., Fadl, M. and Ahmed, J. S. (2004). Application of the recombinant *Theileria annulata* surface protein in an indirect ELISA for the diagnosis of tropical theileriosis. *Parasitology Research,*; 92(4): 299-302.

Bansal, G. C., Ray, D., Srivastava, R. V. N. and Subramanian, G. (1987). Seroprevalence of bovine theileriasis in some farms in India. *Indian Journal of Animal Science,* 57: 366-368.

Bhikane, A. U., Narladkar. B. W., Anantwar, L G. and Bhokre, A. P. (2001). Epidemiology, Clinico-pathology and treatment of Babesiosis in Cattle. *Indian Veterinary Journal,* 78: 726-729.

Carreón, D., de la Lastra, J. M. P., Almazán, C., Canales, M., Ruiz-Fons, F., Boadella, M., Moreno-Cid, J. A., Villar, M., Gortázar, C., Reglero, M., Villarreal, R. and de la Fuente, J. (2012). Vaccination with BM86, subolesin and akirin protective antigens for the control of tick infestations in white tailed deer and red deer. *Vaccine*, 30: 273–279.

Cetin, H., Cilek, J. E., Aydin, L. and Yanikoglu, A. (2009). Acaricidal effects of the essential oil of *Origanum minutiflorum* (Lamiaceae) against *Rhipicephalus turanicus* (Acari: Ixodidae). *Veterinary Parasitology,* 160 (3-4): 359-61.

Chandra, D., Srivastava, V. K., Singh, R. and Bansal, G. C. (2000). *Journal of Veterinary Parasitology,* 14: 181-182.

Chapman, A. S., Bakken, J. S., Folk, S. M., Paddock, C. D., Bloch, K. C. and Krusell, A. (2006). Diagnosis and management of tickborne rickettsial diseases. *MMWR Recommendations and Reports*, 55: 1-27.

Chaudhry, D., Garg, A., Singh, I., Tandon, C., Saini, R. (2009). Rickettsial diseases in Haryana: not an uncommon entity. *Journal of the Association of Physicians of India, 57:* 334–7.

Chisholm, E. S., Ruebush, T. K. 2nd, Sulzer, A. J. and Healy, G. R. (1978). *Babesia microti* infection in man: evaluation of an indirect immunofluorescent antibody test. *American Journal of Tropical Medicine and Hygiene,* 27: 14–19.

Choudhury, R. T., Vasanthi, C., Latha, B. R. and John, L. (2004). *In vitro* effect of *Nicotina tabacum* aqueous extract on *Rhipicephalus haemaphysaloides* ticks. *Indian Journal of Animal Science,* 74: 730–731.

Dammin, G. J., Spielman, A., Benach, J. L. and Piesman, J. (1981). The rising incidence of clinical *Babesia microti* infection. *Human Pathology,* 12: 398-400.

Davey, R. B., George, J. E. and Miller, R. J. (2006). Comparison of reproductive biology between acaricide-resistant and acaricide-susceptible *Rhipicephalus* (*Boophilus*) *microplus* (Acari: Ixodidae). *Veterinary Parasitology,* 139 (1–3): 211–220.

de la Fuente, J., Almazán, C., Canales, M., Pérez de la Lastra, J. M., Kocan, K. M. and Willadsen, P.(2007). A ten-year review of commercial vaccine performance for control of tick infestations on cattle. *Animal Health Research Review,* 8 (1):23-8.

de la Fuente, J., Estrada-Peña, A., Venzal, J. M., Kocan, K. M. and Sonenshine, D. E. (2008). Overview: ticks as vectors of pathogens that cause disease in humans and animals. *Frontiers in Bioscience,* 1 (13): 6938-6946.

de La Fuente, J., Rodríguez, M. and García-García, J. C.(2000). Immunological controlof ticks through vaccination with *Boophilus microplus* gut antigens. *Annals of the New York Academy of Sciences,* 916:617-21.

de la Fuente, J., Rodríguez, M., Montero, C., Redondo, M., García-García, J. C., Méndez, L. and Serrano, E. (1999). Vaccination against ticks (Boophilus spp.): the experience with the Bm86-based vaccine Gavac. *Genetic Analalysis,* 15 (3-5): 143-8.

de la Fuente, J. and Kocan, K. M. (2006). Strategies for development of vaccines for control of ixodid tick species. Parasite Immunology, 28 (7): 275-83.

de Vos, S., Zeinstra, L., Taoufik, O., Willadsen, P., Jongejan, F. (2001). Evidence for the utility from *Boophilus microplus* in vaccination against other tick species. *Experimantal Applied Acarology,* 25: 245–261.

Drosten, C., Kummerer, B. M., Schmitz, H. and Gunther, S. (2003). Molecular diagnostics of viral hemorrhagic fevers. *Antiviral Research,* 57: 61-87.

Dwivedi, S. K., Sharma, S. P. and Gautarn, O. P. (1976). Babesiosis clinical cases in exotic and cross bred cattle. *Indian Veterinary Journal*, 53: 469-472.

FAO, (2004). *Guidelines Resistance Management and Integrated Parasite Control in Ruminants*. Module 1: Ticks: Acaricide Resistance: Diagnosis, Management and Prevention. Food and Agriculture Organization, Animal Production and Health Division, Rome, pp. 1–53.

Fragoso, H. and Hoshmand, R. P. (1998). Protection against *Rhipicephalus (Boophilus) annulatus* infestation in cattle vaccinated with *Boophilus microplus* Bm86 containing vaccine Gavac. *Vaccine,* 16: 1990–1992.

Garcia-Garcia, J. C., Gonzalez, I. L. and Gonzalez, D. M. (1999). Sequence variation in the *Rhipicephalus (Boophilus) microplus* Bm86 locus and implication for immunoprotection in cattle vaccinated with this antigen. *Experimantal Applied Acarology,* 23: 883–895.

Garcia-Garcia, J. C., Montero, C. and Redondo, M. (2000). Control of ticks resistance to immunization with Bm86 in cattle vaccinated with the recombinant antigen Bm95 isolated from the cattle ticks, *Rhipicephalus (Boophilus) microplus*. *Vaccine,* 18: 2275–2287.

Ghosh, S., Azhahianambi, P. and de la Fuente, J. (2006). Control of ticks of ruminants, with special emphasis on livestock farming system in India—present and future possibilities for integrated control: A review. *Experimantal Applied Acarology*, 40:49–66.

Ghosh, S., Bansal, G. C., Gupta, S. C., Ray, D. D., Khan, M. Q., Irshad, H., Shahiduzzaman, M., Seitzer, U. and Ahmed, J. S. (2007). Status of tick distribution in Bangladesh, India and Pakistan. *Parasitology Research,* 101: 207–216.

Ghosh, S., Khan, M. H. and Ahmed, N. (1999). Immunization of cattle against *H. anatolicum anatolicum* using antigen purified by immunoaffinity chromatography. *Tropical Animal Health Production,* 31: 263–273.

Ghosh, S., Khan, M. H. (998). Immunization of rabbits against *Hyalomma anatolicum anatolicum* using homogenate from unfed immature ticks. *Indian Journal of Experimental Biology,* 36: 167–170.

Ghosh, S. and Khan, M. H. (1996). Studies on immunological control of ticks. II. Immunization of cattle against *Boophilus microplus* using tick extract supernatant antigen. *Journal of Veterinary Parasitology,* 10: 33–37.

Ghosh, S., Ray, D. D., Vanlahmuaka, Das, G., Singh, N. K., Sharma, J. K. and Azhahianambi, P. Progress in development of vaccine against *Hyalomma anatolicum anatolicum*-Indian scenario. *Vaccine,* 26 (6): G40-7.

Ghosh, S., Tiwari, S. S., Srivastava, S., Sharma, A. K., Kumar, S., Ray, D. D. and Rawat, A. K. (2013). Acaricidal properties of *Ricinus communis* leaf extracts against organophosphates and pyrethroids resistant *Rhipicephalus (Boophilus) microplus. Veterinary Parasitology,* 192 (1-3): 259-67.

Ghosh, S., Sharma, A. K., Kumar, S., Rastogi, S., Singh, M., Kumar, R., Paul, S., Ray, D. D. and Rawat, A. K. (2011). In vitro and in vivo efficacy of *Acorus calamus* extract against *Rhipicephalus (Boophilus) microplus*. *Parasitology Research,* 108 (2): 361-370.

Gill, B. S., Bhattachayulu, Y., Kaur, D. and Singh A. (1976). Vaccination against bovine tropical theileriosis (*Theileria annulata*). *Nature,* 264: 355–356.

Guerrero, F. D., Davey, R. B., Miller, R. J. (2011). Use of an allele-specific polymerase chain reaction assay to genotype pyrethroid resistant strains of *Boophilus microplus* (Acari: Ixodidae). *Journal of Medical Entomology,* 38 (1): 44-50.

Guerrero, F. D. and Leonore, L. (2012). Martins JR. Acaricide resistance mechanisms in *Rhipicephalus (Boophilus) microplus*. *Revista Brasileira de Parasitologia Veterinária,* 21 (1): 1–6.

Hajdusek, O., Almazan, C., Loosova, G., Villar, M., Canales, M., Grubhoffer, L., Kopacek, P. and de la Fuente, J. (2010). Characterization of ferritin 2 for the control of tick infestations. *Vaccine,* 28: 2993-2998.

Haque, M., Jyoti, Singh, N. K. and Rath, S. S. (2010). Prevalence of *Theileria annulata* infection in *Hyalomma anatolicum anatolicum* in Punjab state, India. *Journal of Parasitic Diseases,* 34 (1): 48–51.

Hill, C. A. and Wikel, S. K. (2005). The *Ixodes scapularis* Genome Project: an opportunity for advancing tick research. *Trends in Parasitology,* 21 (4): 151-3.

Homer, M. J., Aguilar-Delfin, I., Telford SR 3rd, Krause, P. J. and Persing, D. H. (2000). Babesiosis. *Clinical Microbiology Review,* 13: 451-69.

Hoogstraal, H. (1979). The epidemiology of tick borne Crimean-Congo hemorrhagic fever in Asia, Europe and Africa. *Journal of Medical Entomology,* 15: 307-417.

Jeyabal, L., Kumar, B., Ray, D., Azahahianambi, P., Ghosh, S. (2012). Vaccine potential of recombinant antigens of *Theileria annulata* and *Hyalomma anatolicum anatolicum* against vector and parasite. *Veterinary Parasitology,* 188 (3-4): 231-8.

Jongejan, F., Uilenberg, G. (2004). The global importance of ticks. *Parasitology*, 129 Suppl: S3-14.

Jonsson, N. N., Mayer, D. G., Green, P. E. (2000). Possible risk factors on Queensland dairy farms for acaricide resistance in cattle tick (*Boophilus microplus*). *Veterinary Parasitology,* 88: 79–92.

Jyoti, Singh, N. K., Singh, H. and Rath, S. S. (2013). Malathion resistance in *Rhipicephalus (Boophilus) microplus* from Ludhiana district, Punjab. *Journal of Parasitic Diseases,* doi 10.1007/s12639-013-0322-5.

Kasabi, G. S., Murhekar, M. V., Sandhya, V. K., Raghunandan, R., Kiran, S. K., Channabasappa, G. H. and Mehendale, S. M. (2005). Coverage and Effectiveness of Kyasanur Forest Disease (KFD) Vaccine in Karnataka, South India, 2005–10. *PLOS Neglected Tropical Diseases,* 7 (1): e2025.

Khatri, N., Nichani, A. K., Sharma, R. D., Khatri, M. and Malhotra, D. V. (2001). Effect of vaccination in the field with the *Theileria annulata* (Hisar) cell culture vaccine on young calves born during the winter season. *Veterinary Research Communication,* 25: 179-188.

Khdier, I. A. and Latif, B. M. A. (1991). Immunization of cattle against *Theileria annulata* infection using schizont's propagated in the tissue culture. *Indian Veterinary Journal,* 68 (pt 3): 209-214.

Khudrathulla, M. and Jagannath, M. S. (2000). Effect of methonolic extract of *Stylosanthes scabra* on ixodid ticks of animals. *Indian Journal of Animal Sciences,* 70: 1057–1058.

Klafke, G. M., Sabatini, G. A., de Albuquerque, T. A., Martins, J. R., Kemp, D. H., Miller, R. J. and Schumaker, T. T. (2006). Larval immersion tests with ivermectin in populations of the cattle tick *Rhipicephalus (Boophilus) microplus* (Acari: Ixodidae) from State of Sao Paulo, Brazil. *Veterinary Parasitology,* 142 (pt 3-4): 386-390.

Kumar, B., Kumar, R., Azhahianambi, P., Ray, D. D., Chaudhuri, P., de la Fuente, J. and Ghosh, S. (2012). Comparative protective efficacy of rHaa86 and rBm86 against *Hyalomma anatolicum anatolicum* and *Rhipicephalus (Boophilus) microplus* infestation in cattle. *Parasite Immunology,* 34 (pt 6): 297-301.

Kumar, B., Murugan, K., Ray, D. D. and Ghosh, S. (2012). Efficacy of rBm86 against *Rhipicephalus (Boophilus) microplus* (IVRI-I line) and *Hyalomma anatolicum anatolicum* (IVRI-II line) infestations on bovine calves. *Parasitology Research,* 111 (pt 2): 629-635.

Kumar, M., Singh and R., Yadav, M. (2013). Indian Tick Typhus Presenting with Gangrene: A Case Report from an Urban Slum of Delhi. *Indian Journal of Pediatrics.* doi 10. 1007/s12098-013-1095-7.

Kumar, P. P. and Sangwan, A. K. (2010). Comparative prevalence of subclinical bovine anaplasmosis under different cattle management systems in Haryana. *Haryana Veterinary,* 49: 1-5.

Kumar Rinesh, Nagar, G., Sharma, A. K., Kumar, S., Ray, D. D., Chaudhuri, P. and Ghosh, S. (2013). Survey of pyrethroids resistance in Indian isolates of *Rhipicephalus (Boophilus)* microplus: Identification of C190A mutation in the domain II of the para-sodium channel gene. *Acta Tropica,* 125*:* 237-245.

Kumar, S., Paul, S., Sharma, A. K., Kumar, R., Tewari, S. S., Chaudhuri, P., Ray, D. D., Rawat, A. K. S. and Ghosh, S. (2011). Diazinon resistant status in *Rhipicephalus (Boophilus) microplus* collected from different agroclimatic zones of India. *Veterinary Parasitology,* 181*:* 274-281.

Kunz, S. E. and Kemp, D. H. (1994). Insecticides and acaricides resistance and environmental impacts. *Scientific and Technical Review of the Office International des Epizooties (Paris)* 13*:* 1249–1286.

Labuda, M., Trimnell, A. R., Licková, M., Kazimírová, M., Davies, G. M., Lissina, O., Hails, R. S. and Nuttall, P. A. (2006). An antivector vaccine protects against a lethal vector-borne pathogen. *PLoS Pathogen,* 2 (pt 4): e27.

Li, A. Y., Davey, R. B., Miller, R. J. and George, J. E. (2004). Detection and Characterization of Amitraz Resistance in the Southern Cattle Tick, *Boophilus microplus* (Acari: Ixodidae). *Journal of Medical Entomology,* 41(2):193-200.

Lyndon, J., Lawrence, W. and Earle, V. R. (1997). An insecticidal and acaricidal polysulfide metabolite from the roots of *Petiveria alliacea. Pest Science,* 50: 228-232.

Mahajan, S. K., Kashyap, R., Sankhyan, N., Sharma, V., Rolain, J. M., Prasher, B. S. and Raoult, D. (2007). Spotted fever group rickettsioses in Himachal Pradesh. *Journal of Association of Physicians India,* 55: 868.

Marathe, A., Tripathi, J., Handa, V. and Date, V. (2005). Human babesiosis–a case report. *Indian Journal of Medical Microbiology,* 23: 267-269.

Martin, C. J., Torres, F., Quinones, W. and Cheverri, F. (2001). Phytochemistry and evaluation of biological action of *Polygonum punctatum. Revista Latinoamericana de Química,* 29 (pt 2): 100-107.

Martins, J. R. and Furlong, J. (2001). Avermectin resistance of the cattle tick *Boophilus microplus* in Brazil. *Veterinary Record,* 149 (pt 2): 64.

Mathai, E., Lloyd, G., Cherian, T., Abraham, O. C. and Cherian, A. M. (2001). Serological evidence for the continued presence of human rickettsioses in southern India. *Annals of Tropical Medicine and Parasitology,* 95*:* 395–8.

Miller, R. J., Davey, R. B. and George, J. E. (2005). First Report of Organophosphate-resistant *Boophilus microplus* (Acari: Ixodidae) within the United States. *Journal of Medical Entomology,* 42 (pt 5): 912-917.

Miller, R. J., Davey, R. B. and George, J. E. (2007). First report of permethrin-resistant *Boophilus microplus* (Acari: Ixodidae) collected within the United States. *Journal of Medical Entomology,* 44 (pt 2): 308-15.

Minjauw, B. and McLeod, A. (2003). *Tick borne disease and poverty. The impact of tick and tick borne disease on the livelihoods of small scale and marginal livestock in India and eastern and southern Africa.* Research report, DFID Animal Health Programme, Centre for Tropical Veterinary Medicine, University of Edinburgh, UK.

Misraulia, K. S., Sisodia, R. S. and Mehta, H. K. (1988). Incidence of anaplasmosis in exotic and crossbred cattle in and around Indore and Mhow, Madhya Pradesh. *Indian Veterinary Journal,* 65: 574-576.

Mohan, R. N. and Pathak, R. C. (1966). Prevalence of anaplasmosis in ruminants in Northern India. *Indian Veterinary Journal,* 43 (pt 8): 685–689.

Morais-Urano, R. P., Chagas, A. C. and Berlinck, R. G. (2012). Acaricidal action of destruxins produced by a marine-derived *Beauveria felina* on the bovine tick *Rhipicephalus (Boophilus) microplus. Experimental Parasitology,* 132(3):362-6.

Morgan, J. A. T., Corley, S. W., Jackson, L. A., Lew-Tabor, A. E., Moolhuijzen, P. M. and Jonsson, N. N. (2009). Identification of point mutation in the para-sodium channel gene of the cattle *Rhipicephalus (Boophilus) microplus* associated with resistance to synthetic pyrethroids acaricides. *International Journal of Parasitology,* 39: 775–779.

Mulenga, A., Sugimoto, C., Sako, Y., Oaci, K., Musoke, A., Mozaria, S. and Onuma, M. (1999). Molecular characterization of a *Haemaphysalis longicornis* tick salivary gland associated 29-kilodalton protein and its effect as a vaccine against tick infestation in rabbits. *Infection Immunity,* 67:1652–1658.

Muraleedharan, K., Syed Ziauddin, K., Mallikarjun, G. B. and Seshadri, S. J. (2005). Incidence of *Anaplasma spp*, *Babesia spp* and *Trypanasoma spp* in cattle of Karnataka. *Journal of Veterinary Parasitogy,* 19 135-137.

Muraleedharan, K., Ziauddin, K. S., Hussain, P. M., Seshadri, S. J., Mallikarjun, G. B. and Puttabyatappa, B. (1994). Observations on theilerial infection of cattle in project area of Mysore cooperative milk producers union, Karnataka State. *Cheiron,* 23 (pt 3): 130-139.

Naderi, H. R., Sarvghad, M. R., Bojdy, A., Hadizadeh, M. R., Sadeghi, R. and Sheybani, F. (2010). Nosocomial outbreak of Crimean-Congo haemorrhagic fever. *Epidemiology and Infections,* 27:1-5.

Nogueira Domingues, L., dos Santos Alves Figueiredo Brasil, B., Passos de Paiva Bello, A. C., Pinto da Cunha, A., Thadeu Medeiros de Barros, A. and Cerqueira Leite, R. (2012). Survey of pyrethroid and organophosphate resistance in Brazilian field populations of *Rhipicephalus (Boophilus) microplus*: Detection of C190A mutation in domain II of the para type sodium channel gene. *Veterinary Parasitology,* 189 (pt 2-4): 327-32.

Padbidri, V. S., Rodrigues, J. J., Shetty, P. S., Joshi, M. V., Rao, B. L. and Shukla, R. N. (1984). Tick-borne rickettsioses in Pune district, Maharashtra, India. *Indian Journal of Zoonoses,* 11*:* 45–52.

Papa, A., Christova, I., Papadimitriou, E. and Antoniadis, A. (2004). Crimean-Congo hemorrhagic fever in Bulgaria. *Emerging Infectious Diseases,* 10*:*1465-7.

Parmar, A. and Grewal, A. S. (1996). Characterization of immunogenic components of *Hyalomma anatolicum anatolicum*: the vector of bovine tropical theileriosis. *Current Science,* 70: 161–164.

Parola, P., Fenollar, F., Badiaga, S., Brouqui, P. and Raoult, D. (2001). First documentation of *Rickettsia conorii* infection (strain Indian tick typhus) in a traveler. *Journal of Emerging Infectious Diseases,* 7*:* 909–10.

Parola, P. and Raoult, D. (2001). Ticks and tickborne bacterial disease in humans: an emerging infection threat. *Clinicial Infectious Diseases,* 32: 897-928.

Patarroyo, J. H., Portela, R. W., De Castro, R. O., Couto Pimentel, J., Guzman, F., Patarroyo, M. E., Vargas, M. I., Prates, A. A. and Dias Mendes, M. A. (2002). Immunization of cattle with synthetic peptides derived from the *Boophilus microplus* gut protein (Bm86). *Veterinary Immunology and Immunopathology,* 88: 163-172.

Patel, A. K., Patel, K. K., Mehta, M., Parikh, T. M., Toshniwal, H. and Patel, K. (2011). First Crimean-Congo hemorrhagic fever outbreak in India. *Journal of Association of Physicians India,* 59:585-9.

Pathak, D., Mathur, V. C., Latha, B. R. and John, L. (2004). *In vitro* effect of indigenous plant extract on ixodid ticks of small ruminants. *Indian Journal of Animal Sciences,* 74: 616–617.

Patnaik, M. M. (1963). A Note on Bovine Anaplasmosis. *Indian Veterinary Journal,* 40: 655-657.

Pattnaik, P. (2006). Kyasanur forest disease: an epidemiological view in India. *Review of Medical Virology,* 16: 151–165.

Perez-Cogollo, L. C., Rodriguez-Vivas, R. I., Ramirez-Cruz, G. T. and Rosado-Aguilar, J. A. (2010). Survey of *Rhipicephalus microplus* resistance to ivermectin at cattle farms with history of macrocyclic lactones use in Yucatan, Mexico. *Veterinary Parasitology,* 172 (pt1-2):109-13.

Perez-Perez, D., Bechara, G. H., Machado, R. Z., Andrade, G. M., Del Vecchio, R. E., Pedroso, M. S., Hernández, M. V. and Farnós, O. (2010). Efficacy of Bm86 antigen against immature instars and adults of the dog tick *Rhipicephalus sanguineus* (Latreille, 1806). *Veterinary Parasitology,* 167:321–326.

Persing, D. H., Mathiesen, D., Marshall, W. F., Telford, S. R. 3rd, Spielman, A., Thomford, J. W. and Conrad, P. A. (1992). Detection of *Babesia microti* by polymerase chain reaction. *Journal of Clinical Microbilogy,* 30: 2097–2103.

Peter, R. J., Van den Bossche, P., Penzhorn, B. L. and Sharp, B. (2005). Tick, fly, and mosquito control-lessons from the past, solutions for the future. *Veterinary Parasitology,* 132(3): 205-15.

Pipano, E., Alekceev, E., Galker, F., Fish, L., Samish, M. and Shkap, V. (2003). Immunity against *Boophilus annulatus* induced by the Bm86 (Tick-GARD) vaccine. *Experimental and Applied Acarology,* 29: 141–149.

Pipano, E. and Tsur, I. (1966). Experimental immunization against *Theileria annulata* with a tissue culture vaccine. *Refuah Veterinarith,* 23:186–194.

Porter, R. B., Reese, P. B., Williams, L. A. and Williams, D. J. (1995). Acaricidal and insecticidal activities of cadina-4, 10 (15)-dien-3-one. *Phytochemistry, 4 0* (pt 3), p. 735-738.

Project directorate on animal disease monitoring and surveillance Annual report (2005). Indian Council of Agricultural Research, New Delhi.

Raina, A. K., Yadav, A., Gupta, S. K., Singh, R., Agrawal, R. and Hussain, K. (2005). Occurrence of haemoprotozoan diseases and anaplasmosis in Jammu district. *Indian Journal of Veterinary Medicine,* 25: 110-111.

Rajendran, C. and Ray, D. D. (2012). Diagnosis of tropical bovine theileriosis by ELISA with recombinant merozoite surface protein of *Theileria annulata* (Tams1). *Journal of Parasitic Diseases.* doi 10.1007/s12639-012-0183-3.

Rajput, Z. I., Hu, S. H., Chen, W. J., Arijo, A. G. and Xiao, C. W. (2006). Importance of ticks and their chemical and immunological control in livestock. *Journal of Zhejiang University of Science B 7* (pt 11): 912-921.

Ravindran, R., Juliet, S., Sunil, A. R., Kumar, K. G., Nair, S. N., Amithamol, K. K., Shynu, M., Rawat, A. K. and Ghosh, S. (2011). Eclosion blocking effect of ethanolic extract of

Leucas aspera (Lamiaceae) on *Rhipicephalus (Boophilus) annulatus*. *Veterinary Parasitology,* 179 (1-3): 287-90.

Ravindran, R., Juliet, S., Sunil, A. R., Ajith Kumar, K. G., Nair, S. N., Amithamol, K. K., Bandyopadhyay, A., Rawat, A. K. and Ghosh, S. (2012). Acaricidal activity of *Cassia alata against Rhipicephalus (Boophilus) annulatus*. *Experimental and Applied Acarology,* 56 (1): 69-74.

Ravindran, R., Saravanan, B. C., Rao, J. R., Mishra, A. K., Bansal, G. C. and Ray, D. D. (2007). A PCR-RFLP method for the simultaneous detection of *Babesia bigemina* and *Theileria annulata* infections in cattle. *Current Science,* 93(12): 1840-1843.

Ray, D. and Bansal, G. C. (1997). Duration of immunity in calves vaccinated with attenuated macroschizonts of *Theileria annulata*. *Journal of Veterinary Parasitology,* 11 (pt 2): 197-199.

Riek, R. F. (1968). Babesiosis, p. 219–268. In D. Weinman and M. Ristic (ed.), *Infectious blood diseases of man and animals*, vol. 2. Academic Press, New York, N. Y.

Rosado-Aguilar, J. A., Aguilar-Caballero, A., Rodriguez-Vivas, R. I., Borges-Argaez, R., Garcia-Vazquez, Z. and Mendez-Gonzalez, M. (2010). Acaricidal activity of extracts from *Petiveria alliacea* (Phytolaccaceae) against the cattle tick, *Rhipicephalus (Boophilus) microplus* (Acari: ixodidae). *Veterinary Parasitology,* 168 (3-4): 299-303.

Roy, K. C., Ray, D., Bansal, G. C. and Singh, R. K. (2000). Detection of *Theileria annulata* carrier cattle by PCR. *Indian Journal of Experimental Biology*; 38 (pt 3): 283-4.

Saijo, M., Qing, T., Niikura, M., Maeda, A., Ikegami, T. and Prehaud, C. (2002). Recombinant nucleoprotein-based enzyme-linked immunosorbent assay for detection of immunoglobulin G antibodies to Crimean-Congo hemorrhagic fever virus. *Journal of Clinical Microbiology,* 40: 1587-91.

Saijo, M., Tang, Q., Shimayi, B., Han, L., Zhang, Y., Asiguma, M. (2005). Recombinant nucleoprotein-based serological diagnosis of Crimean-Congo hemorrhagic fever virus infections. *Journal of Medical Virology,* 75: 295-9.

Scrub typhus & other rickettsioses. *CD Alert Monthly Newsletter of National Centre for Disease Control*, Directorate General of Health Services, Government of India. May-July, 2009 Vol. 13. No. 1.

Seitzer, U., Beyer, D., Kullmann, B., Bakheit, M. A. and Ahmed, J. S. (2008). Evaluation of *Theileria annulata* recombinant immunodominant proteins for the development of ELISA. *Transbound Emerging Diseases,* 55 (pt 5-6): 244-8.

Shah, V., Vaidya, V., Bang, V. and Shah, Ira. (2009). Spotted fever in a child in Mumbai, India. *Journal of Vector Borne Diseases,* 46: 310–312.

Sharifi Mood, B., Alavi-naini, R. and Metanat, M. (2009). Ten years after the beginning of Crimean-Congo hemorrhagic fever outbreak in Iran: A promising report. *Iranian Journal of Clinical Infectious Diseases,* 4:189-93.

Sharma, A. K., Kumar, R., Kumar, S., Nagar, G., Singh, N. K., Rawat, S. S., Dhakadd, M. L., Rawat, A. K. S., Ray, D. D. and Ghosh, S. (2012). Deltamethrin and cypermethrin resistance status of *Rhipicephalus (Boophilus) microplus* collected from six agro-climatic regions of India. *Veterinary Parasitology,* 188: 337– 345.

Sharma, J. K., Ghosh, S., Khan, M. H. and Das, G. (2001). Immunoprotective efficacy of a 39 kDa purified nymphal antigen of *Hyalomma anatolicum anatolicum*. *Tropical Animal Health Production,* 33: 103–116.

Sheikh, A. S., Sheikh, A. A., Sheikh, N. S., Rafi-U-Shan, Asif, M. and Afridi, F. (2005). Bi-annual surge of Crimean-Congo haemorrhagic fever (CCHF): A five-year experience. *International Journal Infectious Diseases*, 9:37-42.

Shukla, P. C. and Sharma, R. D. (1991). Immunization of bovine calves with cell culture vaccine against *Theileria annulata. Acta Veterinaria,* 60: 79-86.

Shyma, K. P., Kumar, S., Sangwan, A. K., Sharma, A. K., Nagar, G., Ray, D. D. and Ghosh, S. (2013). Acaricide resistance status of *Rhipicephalus (Boophilus) microplus* and *Hyalomma anatolicum* collected from Haryana. *Indian Journal of Animal Science,* 83 (pt 6).

Shyma, K. P., Kumar, S., Sharma, A. K., Ray, D. D. and Ghosh, S. (2012). Acaricide resistance status in Indian isolates of *Hyalomma anatolicum. Experimantal Applied Acarology,* 58(4): 471-81.

Singh, D. K. (1991). "Theileriosis in India." In *Orientation and Coordination of Research on Tropical Theileriosis.* p. 23. Eds. Singh, D. K. and Varshney, B. C., National Dairy Development Board, Anand, India.

Singh, D. K. (1990). Methods currently used for the control of *Theileria annulata*: their validity and proposals for future control strategies. *Parassitologia,* 32 (pt 1): 33-40.

Singh, H., Mishra, A. K., Rao, J. R. and Tewari, A. K. (2007). Seroprevalence of babesiosis in cattle and buffaloes by indirect fluorescent antibody test. *Journal Veterinary Parasitogy,* 21 (pt 1): 1-4.

Singh, H., Jyoti, Haque, M., Singh, N. K., Rath, S. S. (2012). Molecular detection of *Anaplasma marginale* infection in carrier cattle. *Ticks Tick Borne Diseases*, 3 (pt 1): 55-8.

Singh, N. K., Gelot, I. S., Jyoti, Singh, V. and Rath, S. S. (2013). Detection of amitraz resistance in *Rhipicephalus (Boophilus) microplus* from North Gujarat, India. *Journal of Parasitic Diseases*, 24: 207–208.

Singh, N. K., Haque, M., Jyoti and Rath, S. S. (2012). Deltamethrin resistance in *Rhipicephalus microplus* in Ludhiana. *Indian Veterinary Journal,* 89: 23–25.

Singh, N. K., Jyoti, Haque, M. and Rath, S. S. (2010). Studies on acaricide resistance in *Rhipicephalus (Boophilus) microplus* against synthetic pyrethroids by adult immersion test with a discriminating dose. *Journal of Veterinary Parasitology*, 24: 207–208.

Sinha, G. K. and Pathak, R. C. (1966). Anaplasmosis in goat and sheep. *Indian Veterinary Journal,* 43(6): 490–493.

Srinivasan, E. V. R., Rajendran, M. P. and Anbumani, S. P. (1995). A report on the incidence of anaplasmosis in sheep in Tamil Nadu. *Indian Veterinary Journal,* 72: 1307.

Subramanian, G., Ray, D., Bansal, G. C. and Srivastava, R. V. N., (1988). A field trial with a live schizont vaccine (*Theileria annulata*) in adolescent cross-bred cattle in India. *Indian Journal Animal Science*, 58: 529–533.

Sundhindra, B. K., Vijaykumar, S. and Kutti, A. K. (2004). Rickettsial spotted fevers in Kerala. *National Medical Journal of India,* 17: 51-52.

Sunil, A. R., Amithamol, K. K., Juliet, S., Nair, S. N., Ajithkumar, K. G., Soorya, V. C., Divya, T. M., Jyothymol, G., Ghosh, S. and Ravindran, R. (2013). Acaricidal effect of *Cassia fistula* Linn. leaf ethanolic extract against *Rhipicephlaus (Boophilus) annulatus. Tropical Biomedicine,* 30 (2): 231–237.

Tasdelen, Fisgin, N., Ergonul, O., Doganci, L. and Tulek, N. (2009). The role of ribavirin in the therapy of Crimean-Congo hemorrhagic fever: Early use is promising. *European Journal of Clinical Microbiology & Infectious Diseases,* 28: 929-33.

Tedonkeng, Pamo, E., Tendonkeng, F., Kana, J. R., Khan, Payne, V., Boukila, B., Lemoufouet, J., Miegoue, E. and Nanda, A. S. (2005). A study of the acaricidal properties of an essential oil extracted from the leaves of *Ageratum houstonianum. Veterinary Parasitology,* 128 (3-4): 319-23.

Tellam, R. L., Kemp, D., Riding, G., Briscoe, S., Smith, D., Sharp, P., Irving, D. and Willadsen, P. (2002). Reduced oviposition of *Boophilus microplus* feeding on sheep vaccinated with vitellin. *Veterinary Parasitology,* 103 (1-2): 141-56.

Trimnell, A. R., Hails, R. S. and Nuttall, P. A. (2002) Dual action ectoparasite vaccine targeting exposed and concealed antigens. *Vaccine,* 20: 3560–3568.

Ullmann, A. J., Lima, C. M., Guerrero, F. D., Piesman, J. and Black, W. C. (2005). Genome size and organization in the blacklegged tick, *Ixodes scapularis* and the Southern cattle tick, *Boophilus microplus. Insect Molecular Biology,* 14: 217–22.

Vatsya, S., Yadav, C. L. and Garg, R. (2012). Detecting mutation in sodium channel gene in *Rhipicephalus (Boophilus) microplus* collected from northern India for genotyping deltamethrin-susceptible and deltamethrin-resistant alleles. *International Journal Acarology,* 38 (8): 726–730.

Vatsya, S. and Yadav C. L. (2011). Evaluation of acaricide resistance mechanisms in field populations of *Rhipicephalus (Boophilus) microplus* collected from India. *International Journal of Acarology*, 37: 405–410.

Veiga, L. P., Souza, A. P., Bellato, V., Sartor, A. A., Nunes, A. P. and Cardoso, H. M. (2012). Resistance to cypermethrin and amitraz in *Rhipicephalus (Boophilus) microplus* on the Santa Catarina Plateau, Brazil. *Brazilian Journal of Veterinary Parasitology,* 21 (2): 133-136.

Wallace, M. R., Hale, B. R., Utz, G. C., Olson, P. E., Earhart, K. C. and Thornton, S. A. (2002). Endemic infectious diseases of Afghanistan. *Clinical Infectious Diseases*, 34(suppl 5): S171 – S207.

Watt, G., Jongsakul, K. and Suttinont, C. (2003). Possible scrub typhus coinfection in Thai agriculture worker hospitalised with leptospirosis. *American Journal of Tropical Medicine and Hygiene,* 68: 89-91.

Watts, D. M., Ksiazeck, T. G., Linthicum, K. J., Hoogstraal, H. (1988). Crimean-Congo hemorrhagic fever. In: Monath TP, editor. *The Arboviruses: Epidemiology and Ecology.* Florida: CRC Press, 177-222.

Whitehouse, C. A. (2004) Crimean-Congo hemorrhagic fever. *Antiviral Research,* 64:145-60.

Willadsen, P. Anti-tick vaccines. In: Bowman AS, Patricia A. (2008). *Ticks: Biology, Disease and Control*, eds. Nuttall. Cambridge: Cambridge University Press, 424-446.

Yilmaz, G. R., Buzgan, T., Irmak, H., Safran, A., Uzun, R., Cevik, M. A. and Torunoglu, M. A. (2009). The epidemiology of Crimean-Congo hemorrhagic fever in Turkey, 2002–2007. *International Journal Infectious diseases,* 13: 380–386.

In: The Entomological Guide to *Rhipicephalus* ISBN: 978-1-53619-619-1
Editors: Sachin Kumar, Raquel Cossio Bayugar et al.

Chapter 5

INTEGRATED TICK MANAGEMENT

Muhammad Arfan Zaman * ***and Uzma Mehreen***
Section of Parasitology, Department of Pathobiology, College of Veterinary and Animal Sciences, Jhang-Pakistan

ABSTRACT

Ticks are the global problems, most important vectors of Rickettsia and protozoa (*Anaplasma*, *Babesia* and *Theileria*), rickettsiae (*Ehrlichia, Yersinia pestis*, *Francisella tularensis, Coxiella burnetii)*, viruses (Tick-borne encephalitis, Nairobi sheep disease virus, louping ill virus, Bhanja viruses, west Nile virus, African swine fever virus) and spirochaetes (*Borrelia burgdogferi* and *B. hermsii*). Several efforts have been made to eradicate ticks and tick-borne diseases by using various chemical control strategies in many developing countries. However, problems associated with the continuous usage of these acaricides (Organochlorides, macrocyclic lactones, pyrethroides, carbamates, organophosphates, formamidines) Integrated tick management is recommended. Various plants e.g., neem tree (*Azadiratcha indica*), African spider (*Gynandropsis gynandra*), tamarind tree (*Tamarindus Indica*), Jamaica pepper (*Pimenta dioica*), and wormwood (*Artmisia brevifolia*) have been evaluated for their anti-tick properties. So due to having folded benefits there are more opportunities for testing and applications of these plants as anti-tick. Futuristic approach involves the development of combination of chemical and vaccine strategies, directed towards the both tick eradication and transmission. New emerging technologies should be used for gene discovery, cocktail vaccines formulation and delivery. Usage of endosymbiotic approach, role played by *Buphagus erythrorhynchus* for tick control and fungal biocontrol agents will be used in near future research for the control of tick vectors.

Keywords: Ticks, chemicals, plants, vaccine, endosymbiont

Ticks are the globally well-known vectors, belongs to the phylum arthropod having great medical and veterinary importance. They are the Jointed legged, obligate blood-feeding parasites of terrestrial animals (humans, domestic, and wildlife), responsible for the

* Corresponding Author's Email: arfan.zaman@uvas.edu.pk.

transmission of numerous pathogens like protozoa, helminths bacteria (including rickettsial) viruses, and fungal pathogens. (Nicholson et al., 2019). A thousand cases of these tick-borne diseases occur annually. They are major constraints to livestock production, public health, and livestock suspension.

1. The Genus of *Rhipicephalus*

The genus *Rhipicephalus* (*R*) is the 4th largest hard tick in the family Ixodidae comprises approximately 84 species. The most important species in veterinary sciences are *R. appendiculatus* and *R. sanguineus* and to some extent *R. evertsi* also included. *Rhipicephalus* also divided into further subgenus called *Boophilus* (*B*). Its maybe due to having different sizes and morphological features like ornamentation which is absent in *Boophilus* species. They are small in size as compared to *Rhipicephalus*. Among the most important species of *Boophilus* are (*B*) *R. microplus*, (*B*) *R. annulatus*, and (*B*) *R. decoloratus* (Nicholson et al., 2019).

1.1. Common Name

Important species of domestic animals include *R. sanguineus* (Brown dog tick/kennel tick), *R. appendiculatus* (brown ear tick), *R. evertsi* (red-legged tick), Ticks previously classified in the genus *Boophilus* (*B*) are now incorporated in *Rhipicephalus* including the economically important *R. annulatus* (cattle fever tick), *R. microplus* (tropical cattle tick) and *R. decoloratus* (blue tick). It's called blue tick due to the color of the engorged female (Paikade et al., 2019).

1.2. Taxonomy

Family	Sub family	Genera	Species	References
Ixodidae	Rhipicephalinae	*Rhipicephalus*	*R. sanguineus*	Nicholson et al., 2019)
			R. appendiculatus	
			R. evertsi	
		Sub genera:	Species	
		Rhipicephalus (*Boophilus*)	*R. microplus*	
			R. annulatus	
			R. decoloratus	

1.3. Geographical Distribution

R. sanguineus have a cosmopolitan in distribution. It's the most spread tick common in warmer climates. *R. appendiculatus* and *R. evertsi* found on the livestock in Africa. *Boophilus* subgenus (*R. microplus*) are found worldwide (tropical and sub-tropical areas). *R. annulatus* is found in the western hemisphere. *R. decoloratus* is an African tick (Zajac et al., 2012).

1.4. Morphology

Genus and subgenus of *Rhipicephalus* are recognized by the hexagonal shape. Basis capituli have a hexagonal shape, it's the portion of capitulum with broad lateral angles on which mouthparts are attached. The shape of the basis capituli may vary at different developmental stages of the tick. (Nava et al., 2018) Mouthparts are short to medium in size with flat eyes which are clearly visible for the genus *Rhipicephalus* but in the case of subgenus (*Boophilus*) mouthparts are very short with invisible eyes. Have different teeth arrangements e.g., *R. decoloratus* have teeth arranged on the hypostome into two columns (3 + 3) and *R. microplus* have teeth arrangement on the hypostome in two columns (4 + 4). Scutum is sclerotized which covers the entire dorsum and present in all life stages of male tick (larvae, nymphs, and adults), but in female and immature stages it's just like a small shield present just behind the capitulum. Festoons are present in the genus *Rhipicephalus* with 4 coxae of normal size in male. But in the case of subgenus festoons are absent, instead male has caudal appendages. Genus *Rhipicephalus* have an Anal groove surrounding the anus posteriorly with ventral plates but in *Boophilus* subgenus, the anal groove is absent or indistinct in females or faint in males. A pair of curved and broad Adanal shields with accessary plates are present in the *Rhipicephalus* genus but the subgenus has well developed adanal shield and accessary plates which tips can be seen from the dorsal sides (Paikade et al., 2019). Respiratory pores are oval in the male of *Rhipicephalus* spp. and females have short coma shaped. While the *Boophilus* spp. have oval or circular respiratory pores in male and coma shape in females (Figure 1 & Figure 2) (Ali et al., 2014). Palps located anterior to the basis capituli, present parallel to the hypostomes, are short in *Rhipicephalus* species, in *Boophilus* spp. these are ridged dorsally and ventrally. Ornamentation is an Enamel-like reticulated color pattern present usually on the scutum in the hard tick. This structure also used as a taxonomic character for Ixodidae species. It's known to perform some specific function. It superimposes the integument base color of some male *Rhipicephalus* species but it's absent in subgenus *Boophilus*. (Schachat et al., 2018). A pair of long and coiled Malpighian tubules present which is part of the excretory structure and empty into the rectal sac. The caudal process presents to the medial posterior end of many engorged males. It's a distinct posteromedial projection. Many *Rhipicephalus* species have this structure (Horak et al., 2013). Genital organs have an external opening called a Genital aperture present on the ventromedial line. But this structure has taxonomic value only in females belongs to the *Rhipicephalus* genus. (Figure 3 & 4). This structure also plays an important role during copulation because of the male extruding spermatophore from his genital aperture. (Oliver et al., 1974). The female genital aperture has a lateral margin called the genital apron. There are different segmented appendages called as legs. It may be 3 pairs in larvae and 4 pairs in adults. Legs identification can be done from anterior to posterior and given Roman numerals I, II, III, and IV, etc. These segments are coxa, trochanter, femur, genu, tibia, metatarsus, and tarsus designated from proximal to distal.

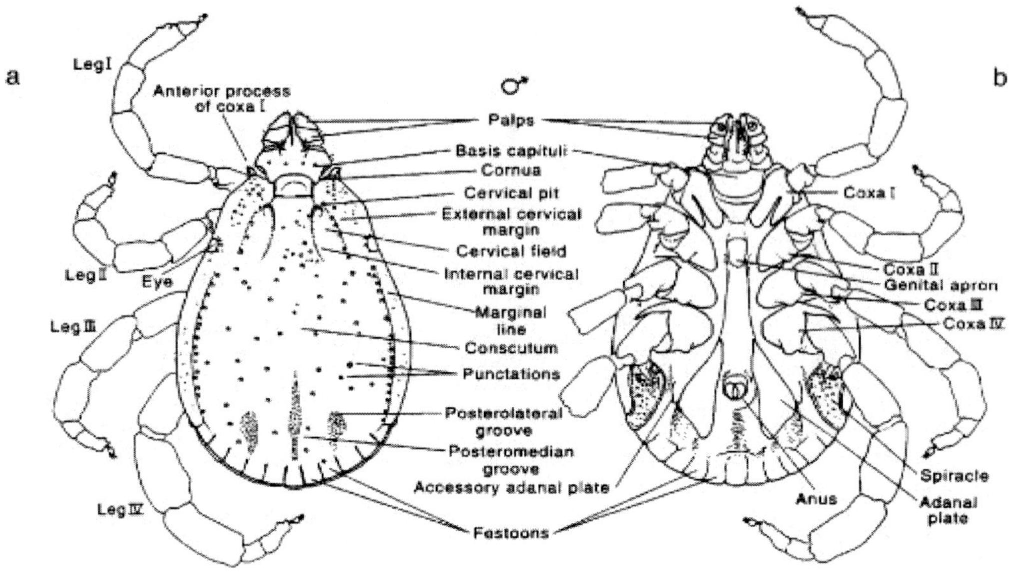

Figure 1. Morphological features of *Rhipicephalus* spp. adult's male a) Posterior b) Anterior.

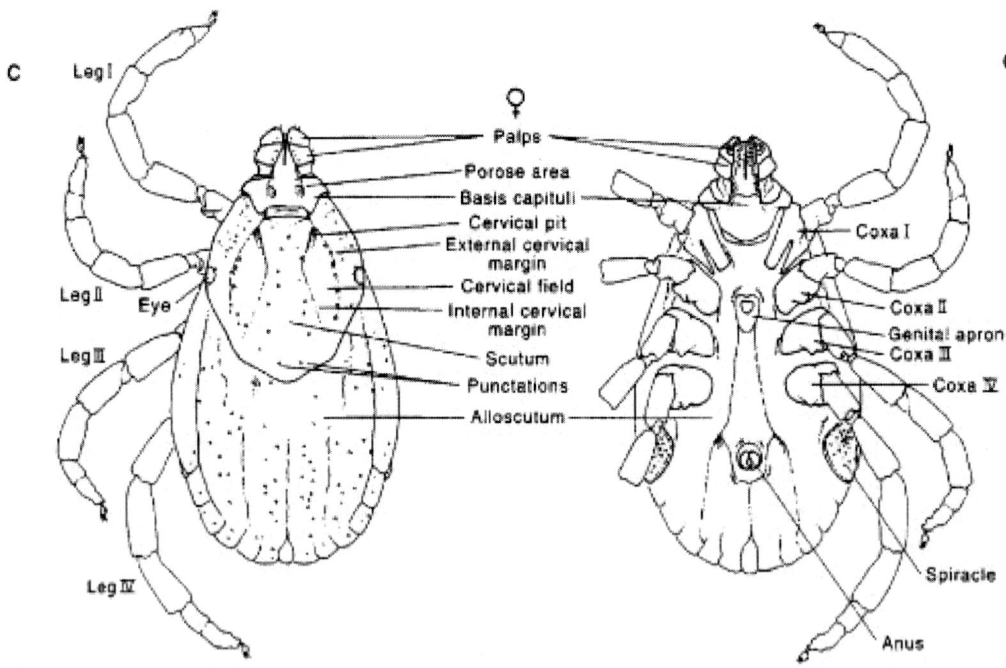

Figure 2. Morphological features of *Rhipicephalus* spp. adult's female a) Posterior b) Anterior.

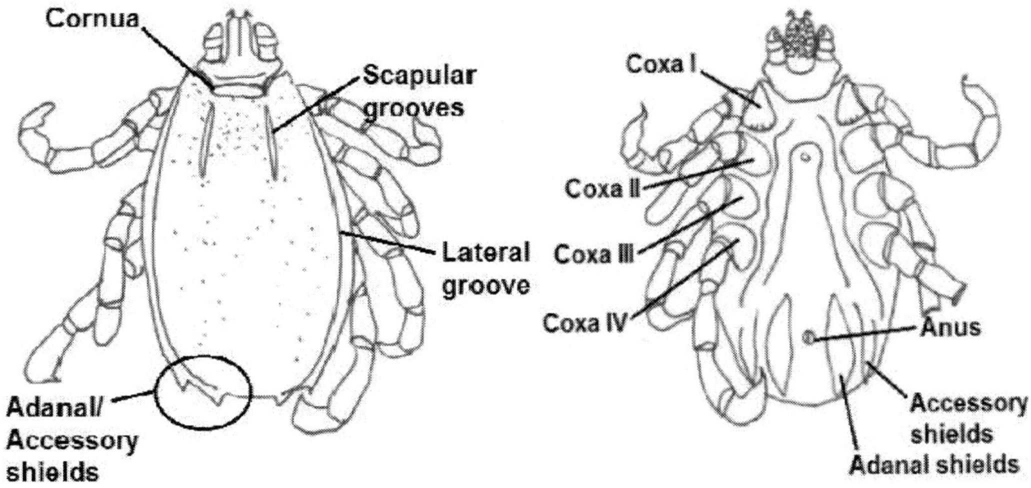

Figure 3. Morphological features of *Boophilus* spp. adult's male. (Posterior and Anterior view).

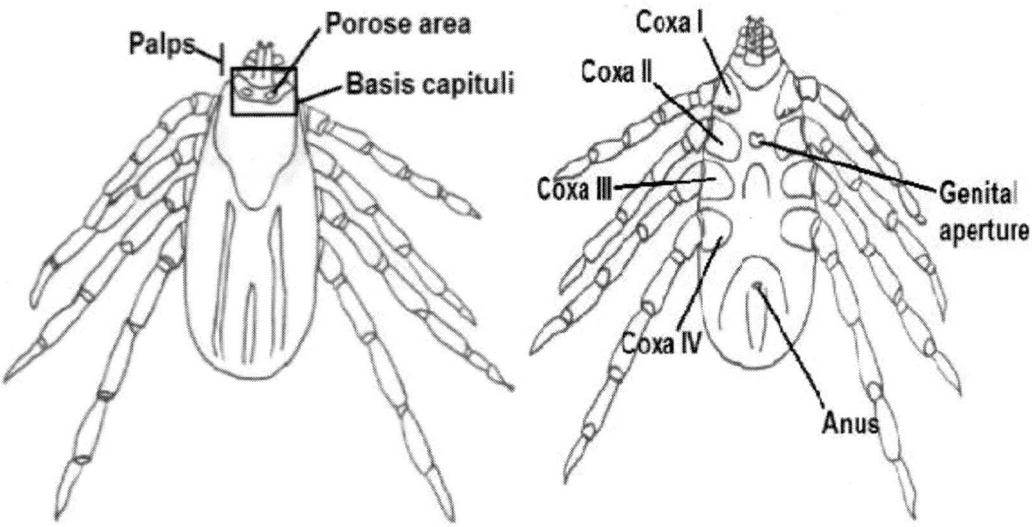

Figure 4. Morphological features of *Boophilus* spp. adult's female. (Posterior and Anterior view).

1.5. Life Cycle

The life cycle of the Ixodidae family (soft ticks) consists of three active developmental stages e.g., larva, nymph, and adult. These ticks have only one nymphal instar as compared to the argasid ticks (hard ticks) have 3 nymphal instars. As ticks are obligate parasites so they are blood feeders during some or all stages in their life cycle. Larvae attack hosts, after feeding it detached off from the host and developed in a sheltered microenvironment and molt to the nymphal stage. Subsequently, the nymph seeks a host and may attach to the same or different host for feeding, then it drops and molts to the adults. Adult ticks seek their hosts, do mating and feeding. In case of engorged females drop off the host to lay their eggs and die. Females of this family have only one gonotrophic cycle. (Gravid female takes more than a period of several weeks to lay all of their eggs). Male consumes less amount of blood than female and body mass of male increases by only 2 times. Male without the requirement of blood meal can mate with the female either it is on the host or not (Sonenshine et al., 2013).

All the genus and subgenus of *Rhipicephalus* e.g., *R. sanguineus*, *R. appendiculatus* (three-host ticks), *R. evertsi* (two host tick) *R. annulatus*, *R. microplus*, and *R. decoloratus* (one host ticks) locate the host after hatching from the eggs in the environment where they complete their other subsequent stages. After laying eggs in the environment female leave the host. *R. singuineus* uses dogs as a host and feeds on it for 5-21 days. Assembly of 3 basic life cycles of family Ixodidae is shown in figure 5.

1.6. Behavior (Attachment, Feeding and Mating)

As ticks are flood feeder, for that purpose ticks must seek a suitable host followed by attachment and piercing host skin with its specific mouthparts to gain access to the blood veins for getting the protein-rich blood. If environmental conditions are favorable Ixodidae ticks start to hunt the host with the help of several kinds of stimuli like hosts' body temperature, air-borne vibrations, and different chemical stimuli (e.g., CO_2 and ammonia). These host-seeking strategies may vary from species to species in different geographical situations. Ticks grab the hosts when the animal brushes against the vegetation, some ticks start to move towards their host when they emit from their habitat. While some species hide in the host's nest and borrows and wait for the host. These are different strategies adopted by the different tick species. For example, *R. sanguineus* is a hunter (host-seeking behavior) and three-host tick. *R. appendiculatus* is also a three-host tick. During blood-feeding ticks exhibit two types of strategies first is slow feeding in which there is continuous blood digestion in the midgut. Another one is the posterior period involved in rapid engorgement. (Dantas-Torres, F., 2008). Regurgitation occurs during the alternative process of blood-sucking and salivation. This process has great importance for the transmission of the pathogen. After engorgement tick detaches and drops off the host. Blood feeding process duration depends upon the tick species, development stage, and environmental conditions (Parola P and Raoult D, 2001). Ticks mate on the hosts and fertilized female drop of the host and oviposit in soil.

1.7. Tick Sampling

- Flagging-dragging
- Carbon dioxide traps
- Collection from captured hosts
- Collection from ambulatory human hosts
- Extraction from soil and litter samples

three-host tick. *R. appendiculatus* is also a three-host tick. During blood-feeding ticks exhibit two types of strategies first is slow feeding in which there is continuous blood digestion in the midgut. Another one is the posterior period involved in rapid engorgement. (Dantas-Torres, F., 2008). Regurgitation occurs during the alternative process of blood-sucking and salivation. This process has great importance for the transmission of the pathogen. After engorgement tick detaches and drops off the host. Blood feeding process duration depends upon the tick species, development stage, and environmental conditions (Parola P and Raoult D, 2001). Ticks mate on the hosts and fertilized female drop of the host and oviposit in soil.

1.7. Tick Sampling

- Flagging-dragging
- Carbon dioxide traps
- Collection from captured hosts
- Collection from ambulatory human hosts
- Extraction from soil and litter samples

1.7.1. The Flagging-Dragging Technique

In this method a flannel cloth is dragged over the area or vegetation area where ticks are expected to be present. Ticks present in that area got attached to the flannel cloth being dragged. So then carefully dip that cloth in soap water or alcohol to kill those ticks however, the person dragging that cloth is at risk of tick bite.in the flagging technique a cloth is attached to a pole or stick and is dragged over the area suspected for ticks its even useful in bushy areas (Newman. et al., 2019)

1.7.2. Carbon Dioxide Traps

Is an effective technique for ticks control and ticks sampling. Ticks are attracted to the carbon dioxide being emitted from their hosts. So for this purpose, we can use dry ice to attract the ticks present in any area by placing the dry ice in a trap. Holes are made at the bottom of the trap and flannel cloth or tape with sticky end up is placed to capture the ticks being attracted to carbon dioxide. Place the ticks on tape or flannel cloth carefully in alcohol or soap water to kill them (Gherman et al., 2012).

1.7.3. Collection from Captured Hosts

In this technique, animals are captured and ticks are then removed carefully from their boy using forceps. This can be done normally or by either tranquilizing the animals or carefully collecting the ticks from its body. Tick is held using forceps and then gently twisted in a clockwise manner to remove it from the host body with intact mouthparts. If abruptly removed the mouthparts of ticks got broken from its body and remain inside anchored into the host body.

1.7.4. Collection from Ambulatory Human Host's

Human dealing with places infested with ticks or with host animals sometimes act as ambulatory host and transfer ticks from one place to another. Ticks are also collected by using forceps.

1.7.5. Collection from Soil and Litter

Different techniques are used for this purpose. Tullgren Funnels (Berlese Funnels) is the easiest and common technique to capture ticks living in litter or soil. It consists of a large funnel on which a wire mesh is placed with a bulb hanging above it. A collection chamber is placed under the funnel. Sample containing ticks is placed on mesh wire heat and light is provided by the bulb. Sample dries and ticks fall into the collection chamber placed below. A sampling of ticks is also done by opening the cracks and crevices by chiseling soil sample is collected and ticks are collected from that sample manually or by the method described above. (Macfadyen et al., 1961).

1.8. Ticks Associated Pathology/Clinical Importance

They are involved in *disease transmission, blood loss* (which may lead to severe anemia or death), *wound production* (produce an invasive site for the penetration of blowflies or screwworms), *tick worry* (causes weight loss, and decreased grazing), *ticks paralysis, and damage to hides.*

Ticks are excellent vectors because they are persistent feeder/ slow feeder e.g *Rhipicephalus* (*B.*) *annulatus* remains on the host for 6-8 days. This means they *have sufficient time* for the transmission of the pathogens. Ticks have a *low host specificity*, encourages them for potential exposure to various pathogens due to having a variety of host sources. They have *longevity* due to having much time to acquire and transmit a pathogen throughout their life. There is the following *mode of transmissions* which plays an important role to prove them as an excellent vector. Like *Transovarian* (transmission to the next generation), *Transstadial* (retain pathogen from one life stage to the next), and high fecundity (18000 eggs per female). (Marchiondo et al., 2013).

R. sanguineus is one of the most studied and widespread ticks in the world due to its veterinary and public health relevance. It's the vector of many important diseases like *Ehrlichia canis*, *Coxiella burnetii*, *Rickettsia ricketsii* and *Rickettsia conorii*. Being of zoonotic concern. (Dantas-Torres et al., 2010). *R. appendiculatus* transmits the *Theileria* (*T*) *parva* which is the cause of most economically important tick-borne disease East coast fever. *R. evertsii*, *R. microplus* is the vector of *T. ovis*, *T. equi* and *Babesia caballi* which causes the disease ovine theileriosis, equine piroplasm and also transmit anaplasmosis. *R. annulatus* is the primary vector of Nairobi sheep disease in livestock.

1.9. Treatment /Control

1.9.1. Chemical Control

Usage of dips and dip vat and tropical application of aqueous formulation like topicals, whole body sprays, dust, insecticides, and systematic application like acaricide boluses and injectant. Usage of insecticides impregnated devices (ear tags and collars) etc.

There are some acaricides that can be used for the control of tick on the body of the animal-like organophosphates, carbamates, chlorinated hydrocarbons, arsenic, and synthetic pyrethroids, etc. Among all of these arsenic considered the most effective, a cheap, and soluble agent that was the first acaricide used for the control of ticks in 1893 in South Africa, before the resistance become a problem (George et al., 2000, Bekker et al., 1960). When the resistance has been reported against arsenic there was the development of some synthetic acaricides like chlorinated hydrocarbons, they were used extensively (Angus 1996). Later organophosphates took the place of chlorinated hydrocarbons after the development of resistance against them. But Wharton and Roulston 1970 reported the developed resistance against organophosphates. To overcome all of these resistance problems Regassa (2000) declared that natural compounds (herbals) have some efficacy against ticks.

1.9.2. Botanical/Non-Chemical Control

To overcome these drawbacks produced by using chemical acaricides, specific attention has been given towards the production of botanical acaricides which has great potential as tick repellent and control. Various plants e.g., neem tree (*Azadiratcha indica*), African spider (Gynandropsis gynandra), tamarind tree (*Tamarindus Indica*), Jamaica pepper (*Pimenta dioica*), chenopodium (*Chenopodium album*), and wormwood (*Artemisia brevifolia*) have been evaluated for their anti-tick properties. (Nwanade et al., 2020). The use of herbal products gaining more importance for folded benefits e.g., cost-effective and relatively safer. So, there are more opportunities for testing and applications of these plants as anti-tick. It's better to apply these non-chemical control strategies with chemical control. Usage of brush for vegetation removal but it requires much time and labour, its cost is also very high. Development of resistant cattle breed like Brahman, Sahiwal cattle, etc. Habitat change is a very necessary step to control the ticks. Sealing of cracks and cervices, cutting of shrubs and weeds, indoor application of residual acaricides if you saw the ticks crawling in the animal shed and near the water and feed troughs. The animal owner should be instructed to examine and remove the ticks manually from the animal body periodically. Special precaution is required while removing the ticks from the animal's body because there are chances for *Rickettsia ricketsii* transmission if you got exposure to tick fluid while ticks removing from the animal body. So it's important to use gloves. Keeping in view the side effects of chemical acaricides much study has been directed towards the Biological control methods. One of them is the application of entomopathogenic fungi (*Beauveria bassiana* and *Metarhizium anisopliae*) which can be helpful to control these natural enemies (Kirkland et al., 2004). Production strategies should be needed while the selection of appropriate entomopathogenic fungi strains and to bring them for practical uses is the dire need of time should be implemented. Unfortunately, there is no vaccine available for all types of tick-borne diseases. If there is the presence of any vaccine against tick (e.g., Australian tick vaccine). There is no evidence that it is feasible. Moreover, the cost-benefit relationship is also not evaluated. Predators and parasites provide the naturally protective against tick but they have the little impact. The futuristic approach involves the development of a combination of chemical and vaccine strategies, directed towards both tick eradication and transmission of the tick-borne pathogens. New emerging technologies should be used for gene discovery, a combination of key protective antigens would help in target identification, cocktail vaccines formulation and delivery. Not only tick vaccine but the usage of endosymbiotic approach (necessary for the survival of arthropod hosts, seems to be an unexplored and potential target for tick eradication), the role played by *Buphagus erythrorhynchus* for tick control and fungal biocontrol agents (formulation of fungal spores) will be used in near future research for the control of ticks.

REFERENCES

[1] Ali, A. N. Suzan, D. Al-azizz, A. A. and Afrah, A. (2014). Detection of ectoparasites between domesticated animals in Basrah City/Southern Iraq. *Journal of International Academic Research for Multidisciplinary*, 2(7): 2320-5083.

[2] Angus, B. M. (1996). The history of the cattle tick *Boophilus microptus* in Australia and achievements in its control. *International journal for parasitology*, 26(12): 1341-1355.

[3] Barton, T. R. (1995). A modified technique for extracting live ticks from small soil and litter samples. *Experimental & applied acarology*, 19(6): 357-360.

[4] Bekker, P. M. (1960). *The history of dipping.* Veld, 20: 1-5.

[5] Dantas-Torres, F. (2008). The brown dog tick, *Rhipicephalus sanguineus* (Latreille, 1806) (Acari: Ixodidae): from taxonomy to control. *Veterinary parasitology*, 152(3-4): 173-185.

[6] Dantas-Torres, F. (2010). Biology and ecology of the brown dog tick, *Rhipicephalus sanguineus. Parasites & vectors*, 3(1): 1-11.

[7] George, J. E. (2000). Present and future technologies for tick control. *Annals of the New York Academy of Sciences,* 916(1): 583-588.

[8] Gherman, C. M., Mihalca, A. D., Dumitrache, M. O., Györke, A., Oroian, I., Sandor, M. and Cozma, V. (2012). CO 2 flagging-An improved method for the collection of questing ticks. *Parasites & Vectors*, 5(1): 1-7.

[9] Horak, I. G. Apanaskevich, D. A. and Kariuki, E. K. (2013). A new species of *Rhipicephalus* (Acari: Ixodidae), a parasite of giraffes in Kenya. *Journal of medical entomology*, 50(4): 685-690.

[10] Kim, H. C. Han, S. H. Chong, S. T. Klein, T. A. Choi, C. Y. Nam, H. Y. Chae, H. Y. Lee, H. Ko, S. Kang, J. G. and Chae, J. S. (2011). Ticks collected from selected mammalian hosts surveyed in the Republic of Korea during 2008-2009. *The Korean Journal of Parasitology*, 49(3): 331.

[11] Kirkland, B. H. Westwood, G. S. and Keyhani, N. O. (2004). Pathogenicity of entomopathogenic fungi Beauveria bassiana and Metarhizium anisopliae to Ixodidae tick species Dermacentor variabilis, *Rhipicephalus* sanguineus, and Ixodes scapularis. *Journal of medical entomology*, 41(4): 705-711.

[12] Lernout, T. De Regge, N. Tersago, K. Fonville, M. Suin, V. and Sprong, H. (2019). Prevalence of pathogens in ticks collected from humans through citizen science in Belgium. *Parasites & vectors*, 12(1): 550.

[13] Macfadyen, A. (1961). Improved funnel-type extractors for soil arthropods. *The Journal of Animal Ecology*: 171-184.

[14] Marchiondo, A. A. Holdsworth, P. A. Fourie, L. J. Rugg, D. Hellmann, K. Snyder, D. E. and Dryden, M. W. (2013). World Association for the Advancement of Veterinary Parasitology (WAAVP): guidelines for evaluating the efficacy of parasiticides for the treatment, prevention and control of flea and tick infestations on dogs and cats. *Veterinary Parasitology*, 194(1): 84-97.

[15] Nava, S. Beati, L. Venzal, J. M. Labruna, M. B. Szabó, M. P. Petney, T. Saracho-Bottero, M. N. Tarragona, E. L. Dantas-Torres, F. Silva, M. M. S. and Mangold, A. J. (2018). *Rhipicephalus sanguineus* (Latreille, 1806): neotype designation,

[19] Oliver, Jr. J. H. Al-Ahmadi, Z. and Osburn, R. L. (1974). Reproduction in ticks (Acari: Ixodoidea). 3. Copulation in *Dermacentor occidentalis* Marx and Haemaphysalis leporispalustris (Packard)(Ixodidae). *The Journal of parasitology*, 499-506.

[20] Paikade, S. and Chavan, R. (2019). Taxonomy of parasitic ticks of genus *Rhipicephalus* (*Boophilus*) (Ixodida: Ixodidae) from Aurangabad district of Maharashtra India. *Indian Journal of Applied Research*, 9(9).

[21] Parola, P. and Raoult, D. (2001). Ticks and tickborne bacterial diseases in humans: an emerging infectious threat. *Clinical infectious diseases*, 32(6): .897-928.

[22] Regassa, A. (2000). The use of herbal preparations for tick control in western Ethiopia. *Journal of the South African Veterinary Association*, 71(4): 240-243.

[23] Schachat, S. R., Robbins, R. G. and Goddard, J. (2018). Color patterning in hard ticks (Acari: Ixodidae). *Journal of Medical entomology*, 55(1): 1-13.

[24] Sonenshine, D. E. and Roe, R. M. eds. (2013). *Biology of ticks volume 2*. Oxford University Press.

[25] Zajac, A. M. and Conboy, G. A. eds. (2012). *Veterinary clinical parasitology*. John Wiley & Sons.

In: The Entomological Guide to *Rhipicephalus*
Editors: Sachin Kumar, Raquel Cossio Bayugar et al.
ISBN: 978-1-53619-619-1

Chapter 6

LIVESTOCK BREEDING INTERVENTION TO CONTROL TICK- AND TICK-BORNE DISEASES IN TROPICS

Jay Prakash Gupta * ***and K. P. Shyma***[2]
[1]Department ofAnimal Genetics and Breeding,
[2]Department of Veterinary Parasitology,
College of Veterinary Science and Animal Husbandry,
Sardarkrushinagar Dantiwada Agricultural University,
Sardarkrushinagar, Gujarat, India

ABSTRACT

Ticks exert a huge effect on animal health through blood-sucking and also, they are capable of transmitting a wide range of pathogens. The study revealed that about 80% of the world cattle population is at risk of ticks associated diseases. *Rhipicephalus microplus* is the most prevalent tick in most of the Indian states. Losses caused by bovine tick burdens in tropical countries have a tremendous economic impact on production systems. The control of *R. microplus* is mostly based on the use of conventional chemical acaricides. However, such a control measure has been found to be ineffective and also involving large costs. Therefore, need for the development of alternative approaches, management of animal husbandry practices, manual removal of ticks, selection for host resistance, nutritional management, selection of cattle to increase the frequency of alleles associated with tick resistance, differences in immune response profile, and associated genes, etc. have come to center stage. The phenotype(s) could also be used to improve tick control by using cattle management. Differences in the abilityof cattle to become resistant to ticks are well documented and known by the livestock keepers. Indigenous cattle those are native to India are known tick resistant individuals. Resistance to ticks is a heritable character and several workers have estimated the heritability for resistance to ticks.

Individual selection for tick resistance can be done in three possible ways; selection under normal environmental conditions, selection under challenging environment, and an indirect approach with the use of indicators of resistance can be done. Resistant animals can then be multiplied through natural breeding as well as the use of reproductive bio-

* Corresponding Author's Email: jp.prakash01@gmail.com.

techniques. After multiplying for a number of generations a resistant population can be obtained. These measures can be adopted as they are cost-effective and reliably measures the cattle tick resistance in thousands of individual animals in commercial and smallholder farmer herds in tropical and subtropical areas globally. The selection of animals for tick resistance has gained momentum as studies have indicated a good heritability estimate for tick burden in cattle, which may result in the success of the breeding program for tick resistance. This measure will yield cattle that are permanently resistant to ticks and also it may facilitate select animals at a younger age, to reduce generation interval and increase genetic gain, etc.

Keywords: breeding, heritability, host resistance, markers, selection

INTRODUCTION

A sustainable livestock production system has many dimensions and it depends on the co-sustenance of environmental conservation of natural resources available locally as well as the preservation of cultural practices prevailing among the livestock keepers. The increased human population during the recent past in tropics has created undue pressure on the livestock production system to address the issue of food and energy insecurity (Prajapatiet al., 2020) among the poor and marginalized. Animal diseases in general and tick and tick-borne diseases (TTBD) in particular are among the many factors which directly and indirectly hamper the growth of the livestock sector due to the multiple roles of livestock as a source of food and income through, generation of employment, delivery of energy (dung, biogas), fertilizer, weed control, use of marginal lands, investment and savings as well as transport (Sansoucy, 1995). Ticks are the major ectoparasites that severely impact the livestock production system in tropical countries. It is common observation (supported by various studies) that the breed of animal is an important parameter while determining its susceptibility to ticks burden as well as TTBDs (Prayaga and Henshall, 2005; Kumar et al., 2017). Over the last 2-3 centuries many of the European cattle breed either as pure breed or crossbreed has been introduced in the tropical countries and exhibited a considerable degree of susceptibility to ticks and tick-borne diseases. Since time immemorial, the animals under domestication are undergoing the process of artificial selection mainly for production traits (Sathwaraet al., 2020). Incorporation of health status in any animal selection programme, particularly in tropics is a relatively newer concept. Partly, this is because livestock keepers as well other stakeholders fail to assess the direct losses caused by disease in comparison to the losses incurred by the lower milk production. Ease of assessment of variation in the production traits and also difficulty in quantifying diseases are other factors. Initially, selection against disease susceptibility was not much appealing as it was difficult to quantify the traits associated with it but, in the wake of the report of treatment failure and drug resistance across the country, hue-cry about drug residue in animal products and concerns for animal welfare, it seems to be the long-term viable strategy to select for disease resistance. Under the long-term natural selection, accompanied by traditional selection with human intervention has led to the development of few locally adapted cattle breeds which are resistant to tick infestation. Identification of naturally resistant breeds is the best solution to counter the problem of acaricide resistance (Hayward, 1981). Animal performance should be seen holistically and now dairy producers appreciate this fact. Animal selection for the traits which were not considered till now, is duly considered like, fertility, health, disease resistance, and overall fitness, etc. (Haskell et al., 2014). Indigenous animals, however, during the course of evolution have developed better disease resistance along with the superior draft ability and heat tolerance. Alternative or complementary control strategies are required and breeding for increased host resistance to infection or infestation is one such approach. The ability of the animal to allow the ticks to survive till maturity is defined as tick resistance (Utechet al., 1978). Selection for disease resistance particularly tick resistance is a work under progress and has huge potential, cue may be taken from poultry where selection for disease resistance has been applied successfully and up to a limited extent in the pigs and sheep in some countries (Shymaet al., 2015).

Concept of Resistance and Tolerance

Resistance is actually an ecological consideration of the interaction between the host and the pathogen species (Grenfell and Dobson, 1995), and it is the ability of the host to get control over the pathogen life cycle up to a certain degree (Bishop, 2012). Therefore, resistance is usually relative rather than absolute. Tolerance is rather different from resistance and may be defined as the net impact on the performance of a given level of infection, i.e., the regression of performance on (a function of) pathogen load. Another similar term is resilience which is related to the productivity of an animal in the face of infection.

Breeding for resistance or tolerance would enhance animal health but have a potentially different effect on the prevalence of infection. The former would reduce disease transmission but possibly impose selection pressure upon the pathogen, whereas the latter would not necessarily affect the pathogen. Breeding animals tolerant to disease would be disadvantageous when the aim is to reduce the transmission of infection.

Breeding for Resistance to Tick Infestation

- Record the incidence of tick infestation in the progeny and select those parents that produce the progeny with the lowest incidence of tick infestations.
- MHC antigens are supposed to be involved in immune response regulation as well as certain immune function traits which are having high heritability. Such animals may be incorporated in to the breeding stock and will be an integral part of a breeding programme for tick resistance.
- It is always imperative to assess the heritability few immune functions related traits as mentioned above, to plan the breeding plan for tick resistance accordingly.
- Data recording on the infestation of ticks for generations is quite important as it could be used to study variation among animals with respect to tolerance and resistance to ticks. Further, genetic selection of animals for the high immune response against ticks using some kind of index that combines estimated breeding values may be utilized. Estimates of genetic parameters based on a large amount of data and pedigree records that needs to be maintained.

Appropriate Selection Strategies

The ideal selection strategy will aim for the direct selection for detection of genetic variation in resistance by applying some of the basic criteria of selection *viz.* individual or family selection in normal as well as in the challenged environment. In the case of large animals, this approach has been utilized in case of tick resistance (Shymaet al., 2015), etc. This selection is based on the evaluation of sires based on their progenies or sib performance. However, this approach incurs a high cost, large infrastructure as well as time.

Finding direct markers for tick resistance is a cumbersome and thereby alternate way is to look for the markers that themselves are non-pathological, easy to detect, and are associated with the disease. Such markers are,

Observable or Phenotypic Markers

Observation of skin, hair, mucous membranes, secretions like tears, urine, odor, skin secretions, etc. may act as such markers. The variation between the hosts in such secretions may be associated with certain pathological conditions. Such marker is tick count on the body. Hernandez et al., (2003) suggested that immune responsiveness would be a useful indicator of disease resistance in cattle.

Genetic Markers

Sequencing of the human genomes and few livestock species has led to the discovery of several genetic markers and even genes related to the immune system. Various genetic markers such as microsatellites, SSCP, RFLPs, RAPDs, AFLPs have been utilized to select animals for resistance against specific infectious diseases. These molecular techniques were able to find out the association of genes with specific disease conditions. The resistant animals are selected based on the marker allele linked to the trait. Genes within or closely linked to the immune histocompatibility system have been shown to regulate resistance to several retroviruses. Association between BoLA antigens and disease susceptibility has also been reported.

Hereditary Architecture of Host Resistance

Differences in the ability of cattle to become resistant to ticks, whether *Bos indicus* or *Bos taurus* or within the *B. taurus* breeds have long been recognized, as has the fact that the ability to acquire resistance is heritable (Utechet al., 1978). In *Bos indicus*-cross cattle, for example, the heritability estimate for the burden of *R. microplus* is moderate ($h^2 = 0.34$, Mackinnon et al., 1991). It has also been shown that *Bos indicus* or their crossbreeds are more able to survive babesiosis (a tick-borne disease transmitted by *B. bovis* and *B. bigemina*) than *B. taurus* animals (Bock et al., 1997). The potential reduction in acaricide requirement arising from concerted selection and breeding of cattle for increased host resistance is very substantial. Indicine cattle carry between 10 and 20% of the number of ticks that taurine cattle would carry, given the same level of larval tick challenge (Jonssonet al., 2014). Whereas the most rapid gains in host resistance can be made by replacing taurine cattle with indicine breeds or crossbreeds, molecular genetic markers of host resistance have been identified and with further development hold promise of more rapid selection for host resistance within breeds (Porto Netoet al., 2011).

The manifestation of host resistance to certain pathogen or disease mainly occurs through the large part to the variability in host immune responses to infection (Bishop, 2012). Host resistance of cattle to ticks was pointed out by Wikel (1996) where, he associates it with reduced number of ticks on host from feeding to engorgement, reduced egg production and reduced egg viability. How the host interacts with invading ticks is very complex though, some of its components are inherited. Heritability estimated by various workers is highly variable though ranging from low to moderate. Animals exhibit varying degree of resistance or susceptibility to disease with the display of variation in responses and these varying responses make it possible to identify regions of the genome which offer resistance. Therefore, in principle, it may be possible to improve genetic resistance to most diseases, although ascertaining resistance phenotypes under field conditions can be challenging. It may be both feasible to measure resistance traits on sufficient animals to determine genotypes for resistance and economically worthwhile to incorporate such traits into breeding goals. Breeding for disease resistance requires to measuring resistance phenotypes which can be costly and difficult thereby, disease resistance traits attract genomic studies.

Phenotypic Basis of Tick-Resistance

Phenotypic physical barriers such as skin or coat thickness, skin or coat colour, tongue papillae, odour of animals, the ability of animal to drive away from the ticks by the movement of body parts (self-grooming), etc. are important phenotypic markers for tick resistance (Martinez et al., 2006; Ibelliet al., 2012; Marufuet al., 2013; Veríssimoet al., 2015). It has also been observed that the tongue papillae with smaller spacing in between are more effective in removing ticks (*R. microplus*) larvae from the skin (Veríssimoet al., 2015). However, there are few studies which have shown that tick infestation is not significantly influenced by skin morphology (Wagland, 1978; Doube and Wharton, 1980). Physical features associated with tick resistance are though scarce but will prove to be a simple tool for individual and mass selection particularly for smallholder's production system in the tropics.

Immunological Basis of Tick-Resistance

Immunological indicators of tick resistance are supposed to be acquired, changes subsequent to exposure recorded by O'Kelly et al., (1971) include the change in leukocytes (increased lymphocytes and eosinophils and decreased neutrophils), increase in serum globulin, etc. It is important to note that in animals particularly in bovines, an immune factor, considered to be the mediator of resistance is histamine and its associated molecules (Robbertseet al., 2017). Engracia Filho et al., (2006) in their study on artificially tick-infested F_2 Holstein x Gir crossbred cattle, concluded that the tick infestation may lead to an environmental modification in the dermis of the host due to the massive migration of mast cells or their local proliferation which was parasitized. The host's ability to produce histamine to counter the histamine binding proteins of several tick species may be used for selection for resistance. The hypersensitivity reaction induced in the host by injecting the extract of diluted tick larva can be used as a measure of tick resistance and indicate that antigen can be used for inducing the hypersensitivity reaction (Riek, 1962; Becharaet al., 2000; Marufuet al., 2013; Burrow et al., 2019). Gordon and Allen (1991) in an experiment with rabbits reported that salivary glands of tick possess anticoagulant activities against factor V and VII and it may alter the coagulation time in the host. Later, Carvalhoet al., (2010) in cattle demonstrated that cattle relatively susceptible to tick infestations (with reference to coat thickness) have delayed clotting, reverse was also true with relatively resistant animals. The possibility that resistant hosts may manipulate tick expression patterns is an exciting observation with regard to the selection of resistant breeds. This phenotype (downregulation of tick expression) may also be considered as a potential criterion for genetic improvement of tick resistance (Franzinet al., 2017; Maruyama et al., 2017).

Genetic Approach

Though phenotypic and immunological markers are very much useful at field level and to practicing livestock keepers for identification of animals with resistance against tick infection, the exploration of genetic component that provides resistance to host is highly desirable so that marker-assisted selection may be performed in order to get animals with pretty good resistance. Tick resistance is a very complex trait that is considered to be under the control of many genes, with each one of them having a small contribution to the overall trait. Examples cited in the above sections suggest that there are both innate and adaptive immune mechanisms against TTBDs. Peixotoet al., (2008) in their experiment with an F_2 segregating population of 332 animals from a Gyr x Holstein cross, identified two new Quantitative Trait Loci (QTL) that influence tick resistance. Major Histocompatibility Complex (MHC) is the most highly polymorphic in the entire mammalian genome, enabling the immune system to detect a universal group of foreign epitopes, although the genes within the MHC system perform a crucial role in recognizing and initiating adaptive immune responses. There are many genes other than MHCs that also contribute to protective immune responses to genetic selection for disease resistance. The Dairy industry in certain countries continues to emphasize selection on health traits (Parker Gaddis et al., 2014). Since very beginning, the bovine major histocompatibility complex (MHC) or bovine leukocyte antigen (BoLA) system has been studied for association with tick infestations (Stearet al., 1984, 1989). Later on, many workers have reported a significant association of different alleles (Particularly the microsatellite alleles) of BoLA with tick resistance (Acosta-Rodríguez et al., 2005; Untalanet al., 2007). Three genome-wide significant regions on chromosomes 7, 10, and 19 were identified for tick counts by Mapholiet al., (2016). These chromosomal regions were found to harbour loci associated with tick count. Netoet al., (2010) in their study on the association of SNPs with tick resistance identified the SNP in the intronic region of gene ELTD1 and it was significantly associated with tick burden. The Gene expression profiles of the pro-inflammatory genes, such as TLR-5, chemokine ligand-2, and chemokine receptor-1, are involved in resistance mechanisms. These genes were downregulated at attachment sites in Brahman cattle than in Holstein-Friesian cattle (Piper et al., 2008).

GENOME-WIDE ASSOCIATION STUDIES FOR TICK-RESISTANCE

Genome-wide association study, using single nucleotide polymorphisms (SNP), is another approach that was used to detect evidence of association with tick resistance throughout the genome. Machado et al., (2010) in the experiments with Gyr and Holstein cross have identified six QTL regions strongly associated with tick resistance in cattle on the chromosomes BTA 2, 5, 10, 11, 23, and 27 out of these QTLs, those present on chromosome number 23 may be associated with MHC. Otto et al., (2018) in a GWAS study identified candidate genes TREM1, TREM2, TREML1, CD83, MYO5A, and PRSS16 for tick resistance in an experimental F2 (Gir × Holstein) population for cattle tick resistance. GWAS studies have been utilized to test whether any of them had direct effects on tick resistance. Turner et al., (2010) in such a study confirmed that genes ABCG2, DGAT1, GHR, and PRLR, showed no statistically significant effect on tick burden. Now, the advancement and availability of information regarding genomics, identification of genes granting resistance can further be explored (Soller and Andersson, 1998). Further, detection of genes responsible for resistance would provide a means of getting an environmentally sound and sustainable approach. There is also a need to explore the interaction of ticks with the host, for identification of the genes or any molecules involved. Genome-wide association studies revealed that antibody-mediated immune responses, as well as cell-mediated immune responses are the two key components of the adaptive immune system (Thompson Crispi et al., 2014). It is estimated that 2,000 to 3,000 genes control host defense, providing mammals with a large range of immune responses to fight against foreign organisms (Breuer et al., 2013). The benefit of the genomic approach is the ability to select animals using DNA-based selection without the need to expose them to infection. Identification of Quantitative Trait Loci (QTL) or SNP-chip-based genomic predictors for resistance is needed (Meuwissenet al., 2001).

GENETIC RESISTANCE TO TICK-BORNE DISEASES

There is observed clarity on the MHC-I restricted cytotoxic T-lymphocyte response in the tick-borne infection of *Theileria parva* and *Theileria annulata* in cattle (Prajapatiet al., 2017). In an animal, macrophages get activated by IFN-γ which was produced by CD4+ and NK cells. The activated macrophages then will help in phagocytosis as well as cytokine and nitric oxide production which will ultimately either kill or inhibit the growth of the parasites. Synthesis of opsonizing anti-parasite immunoglobulin (Ig) G2 antibodies gets assistance from memory T-helper cells. Overproduction of cytokines, such as TNF-α, leads to an exaggeration of the S52 (types of IgE) clinical symptoms and pathological reactions associated with TTBDs (Ahmed, 2002).

Kumar et al., (2017) in their experiment on two groups of animals (Tharparkar and crossbred cattle), reported that the modulation of the host gene expression in response to *T. annulata* infection was evident from a large number of differentials expressed host-genes between these two. In their study on the global gene expression profile of genes though they concluded that the breed-specific differences in mRNA of crossbred and Tharparkar cattle may be responsible for resistance against tropical theileriosis at an early stage of infection but they further observed that multiple mechanisms may be involved. Their study reflects the difference in response of Tharparkar and crossbred cattle to *T. annulata* infection and provides an ideal model to explore the mechanism behind tolerance to tropical theileriosis.

The best way to control tick-borne diseases may include, utilisation of relatively resistant breeds such as Kankrej (Joshi and Phillips, 1953; Mason, 1996) Kenana from Sudan (Bakheit and Latif, 2002) and Sahiwal (Maryam et al., 2012) which have been reported to be relatively resistant to TTBDs.

Ticks and Tick-Borne Disease Resistant Breeds of India

The cattle breeds of tropical countries like India are considered to be evolved from continued selection against the infectious agent prevalent in those countries. In India, native nondescript cattle population established cattle breeds like Kankrej, Tharparkar, Sahiwal, Ongole, etc., and buffaloes are considered as resistant to tick infestation (Gaur et al., 2002; Kumar et al., 2017; Debbermaet al., 2018). It is assumed that local native animals may have a different genetic makeup that makes them resistant to parasites which may also be aided by specific climatic conditions of the region which do not favour the disease in these animals.

It is believed that the skin coat of Ongole cattle makes it difficult for ticks to adhere (Joshi and Phillips, 1953). The presence of a large number of sweat glands per unit area of skin is a significant deterrent for ticks (Ghosh et al., 2006). Another indigenous cattle breed has been utilised extensively in dairy production, as it is resistant to tick-borne diseases. The breed has the ability to limit the pathology of tick-borne disease (Theileriosis) by preventing the over-stimulation of pathways involving the cytokines (Glass et al., 2005). The cattle maintained in the hill region are also considered to be resistant as they are exposed to the continuous natural challenge during grazing in the forest land. The Kankrej cattle breed, a native of the North Gujarat region of India is also reported to be resistant to the diseases caused by ticks (Mason, 1996). One of the few systematic studies conducted in this regard is available for the Tharparkar breed of cattle which reveals that there is a difference in the expression of genes involved in the defense against tick-borne disease in comparison to crossbred cattle (Kumar et al., 2017). Similar reports of resistance have been reported in buffaloes too. Batista et al., (2018) in their study on the Brazilian buffalo population come up with the conclusion that the buffaloes are known to have some kind of resistance to tick infestations. In India also, buffaloes are considered to be more resistant to the TTBDs, one of the recently recognised buffalo breeds, Gojari is specifically mentioned for its notable resistance to TTBDs (Tribune India, 2017).

Conclusion

The main ticks controlling agent at the field is still the chemical acaricides. The development of resistance in the tick population against almost all the common acaricide is alarming. Many of the indigenous *B. indicus* breeds of cattle are low producer but significantly resistant to many diseases including those caused by TTBDs. Concerted efforts need to be directed to identify the gene(s) for tick resistance in the promising indigenous breeds which can later be incorporated into the strategic breeding programme to develop animal stocks with high genetic resistance to TTBDs. Breeding for resistance against tick infestation with appropriate selection strategies may provide an alternative solution. An Appropriate selection strategy needs some markers associated with the variations. Phenotypic markers though easily measurable, will not be of much use because of their low inheritance. Genetic markers will be highly useful and future research should be directed towards the identification of such markers and associate them with the incidence of tick infestation as well as tick-borne diseases.Indigenous cattle breed may be the source of such marker as they are evolved under such challenges through Natural Selection. The high producing animals having low resistance to such infestation may be incorporated with resistance-conferring genes through introgression.

REFERENCES

Acosta-Rodriguez, R., Alonso-Morales, R., Balladares, S., Flores-Aguilar, H., Garcia-Vazquez, Z. and Gorodezky, C. (2005). Analysis of BoLA class II microsatellites in cattle infested with *Boophilus microplus* ticks: class II is probably associated with susceptibility. *Veterinary Parasitology*, 127(3-4): 313-321.

Ahmed, J., Yin, H., Schnittger, L. and Jongejan, F. (2002). Ticks and tick-borne diseases in Asia with special emphasis on China. *Parasitology Research*, 88(1): S51-S55.

Bakheit, M. A. and Latif, A. A. (2002). The innate resistance of Kenana cattle to tropical theileriosis (*Theileria annulata* infection) in the Sudan. *Annals of the New York Academy of Sciences*, 969(1): 159-163.

Batista, H. R., Sarturi, C., Stelmachtchuk, F. N., Oliveira, D. R., Morini, A. C., Gennari, S. M., Marcili, A., Bastos, F. A. N., Barata, L. E. S. and Minervino, A. H. H. (2018). Prevalence and risk factors associated with ectoparasite infestation of buffaloes in an Amazonian ecosystem. *Parasites & vectors*, 11(1): 335.

Bechara, G. H., Morelli, J. and Szabó, M. P. J. (2000). Skin test and tick immune status in susceptible and resistant cattle in Brazil. *Annals-New York Academy of Sciences*, 916: 570-575.

Bishop, S. (2012). A consideration of resistance and tolerance for ruminant nematode infections. *Frontiers in Genetics*, 3: 168.

Bock, R. E., De Vos, A. J., Kingston, T. G. and McLellan, D. J. (1997). Effect of breed of cattle on innate resistance to infection with *Babesiabovis*, *B. bigemina* and *Anaplasmamarginale*. *Australian veterinary journal*, 75(5): 337-340.

Breuer, K., Foroushani, A. K., Laird, M. R., Chen, C., Sribnaia, A., Lo, R., Winsor, G. L., Hancock, R. E., Brinkman, F. S. and Lynn, D. J. (2013). InnateDB: systems biology of innate immunity and beyond—recent updates and continuing curation. *Nucleic acids research*, 41(D1): D1228-D1233.

Burrow, H. M., Mans, B. J., Cardoso, F. F., Birkett, M. A., Kotze, A. C., Hayes, B. J., Mapholi, N., Dzama, K., Marufu, M. C., Githaka, N. W. and Djikeng, A. (2019).

Towards a new phenotype for tick resistance in beef and dairy cattle: a review. *Animal Production Science*, 59(8): 1401-1427.

Carvalho, W. A., Franzin, A. M., Abatepaulo, A. R. R., de Oliveira, C. J. F., Moré, D. D., da Silva, J. S., Ferreira, B. R. and de Miranda Santos, I. K. F. (2010). Modulation of cutaneous inflammation induced by ticks in contrasting phenotypes of infestation in bovines. *Veterinary parasitology*, 167(2-4): 260-273.

Debbarma, A., Pandit, S., Jas, R., Baidya, S., Mandal, S. C. and Jana, P. S. (2018). Prevalence of hard tick infestations in cattle of West Bengal, India. *Biological Rhythm Research*, 49(5): 655-662.

Doube, B. M. and Wharton, R. H. (1980). The effect of locality, breed and previous tick experience on seasonal changes in the resistance of cattle to *Boophilus microplus* (Ixodoidea: Ixodidae). *Experientia*, 36(10): 1178-1179.

Engracia Filho, J. R., Bechara, G. H. and Teodoro, R. L. (2006). Dermal mast cell counts in F_2 Holstein x Gir crossbred cattle artificially infested with the tick *Boophilus microplus* (Acari: Ixodidae). *Annals of the New York Academy of Sciences*, 1081(1): 476-478.

Franzin, A. M., Maruyama, S. R., Garcia, G. R., Oliveira, R. P., Ribeiro, J. M. C., Bishop, R., Maia, A. A. M., Moré, D. D., Ferreira, B. R. and de Miranda Santos, I. K. F. (2017). Immune and biochemical responses in skin differ between bovine hosts genetically susceptible and resistant to the cattle tick *Rhipicephalus microplus*. *Parasites & vectors*, 10(1): 51.

Gaddis, K. P., Cole, J. B., Clay, J. S. and Maltecca, C. (2014). Genomic selection for producer-recorded health event data in US dairy cattle. *Journal of Dairy Science*, 97(5): 3190-3199.

Gaur, G. K., Kaushik, S. N. and Garg, R. C. (2002). Ongole cattle status in India. *Animal Genetic Resources/Resources génétiquesanimales/Recursosgenéticosanimales*, 32: 27-34.

Gaur, G. K., Singh, A., Singh, P. K. and Pundir, R. K. (2004). Morphometric characteristics and present status of Ponwar cattle breed in India. *Animal Genetic Resources/Resources génétiquesanimales/Recursosgenéticosanimales*, 34: 17-25.

Ghosh, S., Azhahianambi, P. and de la Fuente, J. (2006). Control of ticks of ruminants, with special emphasis on livestock farming systems in India: present and future possibilities for integrated control—a review. *Experimental & applied acarology*, 40(1): 49-66.

Glass, E. J., Preston, P. M., Springbett, A., Craigmile, S., Kirvar, E., Wilkie, G. and Brown, C. D. (2005). *Bos taurus* and *Bos indicus* (Sahiwal) calves respond differently to infection with *Theileria annulata* and produce markedly different levels of acute phase proteins. *International Journal for Parasitology*, 35(3): 337-347.

Gordon, J. R. and Allen, J. R. (1991). Factors V and VII anticoagulant activities in the salivary glands of feeding *Dermacentorandersoni*ticks. *The Journal of parasitology*, 77: 167-170.

Grenfell, Bryan T., Andrew P. Dobson, and H. K. Moffatt, eds. *Ecology of infectious diseases in natural populations*. Vol. 7. Cambridge University Press, 1995.

Haskell, M. J., Simm, G., and Turner, S. P. (2014). Genetic selection for temperament traits in dairy and beef cattle. *Frontiers in Genetics*, 5: 368.

Hayward, S. (1981). Opening address. In: Whitehead, G.B. & Gibson, J.D (eds), *Tick biology and control.* Tick Research Unit, Rhodes University, Grahamstown South Africa. pp xiii –xiv.

Hernández, A., Karrow, N. and Mallard, B. A., (2003). Evaluation of immune responses of cattle as a means to identify high or low responders and use of a human microarray to differentiate gene expression. *Genetics selection evolution*, 35(1): 1-15.

Ibelli, A. M. G., Ribeiro, A. R. B., Giglioti, R., Regitano, L. C. A., Alencar, M. M., Chagas, A. C. S., Paço, A. L., Oliveira, H. N. D., Duarte, J. M. S. and Oliveira, M. C. S. (2012). Resistance of cattle of various genetic groups to the tick *Rhipicephalus microplus* and the relationship with coat traits. *Veterinary Parasitology*, 186(3-4): 425-430.

Jonsson, N. N., Piper, E. K. and Constantinoiu, C. C. (2014). Host resistance in cattle to infestation with the cattle tick *Rhipicephalus microplus*. *Parasite Immunology*, 36(11): 553-559.

Joshi, N. R. and Phillips, R. W. (1953). Zebu cattle of India and Pakistan. *Zebu cattle of India and Pakistan.*, 19: 255.

Kumar, A., Gaur, G. K., Gandham, R. K., Panigrahi, M., Ghosh, S., Saravanan, B. C., Bhushan, B., Tiwari, A. K., Sulabh, S., Priya, B. and Gupta, J.P., Wani, S. A., Sahu, A. R. and Sahoo, A. P. (2017). Global gene expression profile of peripheral blood mononuclear cells challenged with *Theileria annulata* in crossbred and indigenous cattle. *Infection, Genetics and Evolution*, 47: 9-18.

Machado, M. A., Azevedo, A. L. S., Teodoro, R. L., Pires, M. A., Peixoto, M. G. C., de Freitas, C., Prata, M. C. A., Furlong, J., da Silva, M. V. G., Guimarães, S. E. and Regitano, L. C. (2010). Genome wide scan for quantitative trait loci affecting tick resistance in cattle (*Bos taurus*× *Bos indicus*). *BMC genomics*, 11(1): 280.

Mackinnon, M. J., Meyer, K. and Hetzel, D. J. S. (1991). Genetic variation and covariation for growth, parasite resistance and heat tolerance in tropical cattle. *Livestock Production Science*, 27(2-3): 105-122.

Mapholi, N. O., Maiwashe, A., Matika, O., Riggio, V., Bishop, S. C., MacNeil, M. D., Banga, C., Taylor, J. F. and Dzama, K. (2016). Genome-wide association study of tick resistance in South African Nguni cattle. *Ticks and Tick-Borne Diseases*, 7(3): 487-497.

Martinez, M. L., Machado, M. A., Nascimento, C. S., Silva, M. V. G. B., Teodoro, R. L., Furlong, J., Prata, M. C. A., Campos, A. L., Guimarães, M. F. M., Azevedo, A. L. S. and Pires, M. F. A. (2006). Association of BoLA-DRB3. 2 alleles with tick (*Boophilus microplus*) resistance in cattle. *Genetics and Molecular Res*earch 5(3): 513-524.

Marufu, M. C., Chimonyo, M., Mans, B. J. and Dzama, K. (2013). Cutaneous hypersensitivity responses to *Rhipicephalus* tick larval antigens in pre-sensitized cattle. *Ticks and tick-borne diseases*, 4(4): 311-316.

Maruyama, S. R., Garcia, G. R., Teixeira, F. R., Brandão, L. G., Anderson, J. M., Ribeiro, J. M., Valenzuela, J. G., Horackova, J., Veríssimo, C. J., Katiki, L. M. and Banin, T. M. (2017). Mining a differential sialotranscriptome of *Rhipicephalus microplus* guides antigen discovery to formulate a vaccine that reduces tick infestations. *Parasites & vectors*, 10(1): 206.

Maryam, J., Babar, M. E., Nadeem, A. and Hussain, T. (2012). Genetic variants in interferon gamma (IFN-γ) gene are associated with resistance against ticks in *Bos taurus* and *Bos indicus*. *Molecular biology reports*, 39(4): 4565-4570.

Mason, I. L. (1996). *A world dictionary of livestock breeds, types and varieties, C.A.B International,* fourth edition, 273.

Meuwissen, T. H. and Goddard, M. E. (2001). Prediction of identity by descent probabilities from marker-haplotypes. *Genetics Selection Evolution*, 33(6): 605.

Neto, L. R. P., Jonsson, N. N., Michael, J. D. and Barendse, W. (2011). Molecular genetic approaches for identifying the basis of variation in resistance to tick infestation in cattle. *Veterinary parasitology*, 180(3-4): 165-172.

O'kelly, J. C., Seebeck, R. M. and Springell, P. H. (1971). Alterations in host metabolism by the specific and anorectic effects of the cattle tick (*Boophilus microplus*) II. Changes in blood composition. *Australian journal of biological sciences*, 24(2): 381-390.

Otto, P. I., Guimarães, S. E., Verardo, L. L., Azevedo, A. L. S., Vandenplas, J., Soares, A. C., Sevillano, C. A., Veroneze, R., Maria de Fatima, A. P., de Freitas, C. and Prata, M. C. A. (2018). Genome-wide association studies for tick resistance in *Bos taurus*× *Bos indicus* crossbred cattle: A deeper look into this intricate mechanism. *Journal of dairy science*, 101(12): 11020-11032.

Peixoto, M. G. C. D., Azevedo, A. L. S., Teodoro, R. L., Pires, M. F. A., Verneque, R. S., Prata, M. C. A., Furlong, J., Regitano, L. C. A. and Machado, M. A. (2008). Identification of QTL for tick resistance using a bovine F_2 population in tropical area. In *Proceedings of the livestock and global climate change conference* pp. 100-102.

Piper, E. K., Jackson, L. A., Bagnall, N. H., Kongsuwan, K. K., Lew, A. E. and Jonsson, N. N. (2008). Gene expression in the skin of *Bos taurus* and *Bos indicus* cattle infested with the cattle tick, *Rhipicephalus (Boophilus) microplus*. *Veterinary immunology and immunopathology*, 126(1-2): 110-119.

Porto Neto, L. R., Bunch, R. J., Harrison, B. E. and Barendse, W. (2011). DNA variation in the gene ELTD1 is associated with tick burden in cattle. *Animal genetics*, 42(1): 50-55.

Prajapati, B. M., Gupta, J. P., Pandey, D. P., Parmar, G. A. and Chaudhari, J. D. (2017). Molecular markers for resistance against infectious diseases of economic importance. *Veterinary world*, 10(1): 112-120.

Prajapati, B. M., J. P. Gupta, J. D. Chaudhari, G. A. Parmar, R. N. Sathwara, H. H. Panchasara, P. A. Patel, and M. N. Prajapati. (2020). Utility of first lactation fat energy corrected milk yield as a trait for genetic evaluation of Mehsana buffalo bulls using various sire evaluation methods. *Indian Journal of Animal Sciences* 90(2): 259-263.

Prayaga, K. C., and J. M. Henshall. (2005). Adaptability in tropical beef cattle: genetic parameters of growth, adaptive and temperament traits in a crossbred population. *Australian Journal of Experimental Agriculture* 45(8): 971-983.

Riek, R. F. (1962). Studies on the reactions of animals to infestation with ticks. VI. Resistance of cattle to infestation with the tick *Boophilus microplus (Canestrini)*. *Australian Journal of Agricultural Research*, 13(3): 532-550.

Robbertse, L., Richards, S. A. and Maritz-Olivier, C. (2017). Bovine immune factors underlying tick resistance: integration and future directions. *Frontiers in Cellular and Infection Microbiology*, 7: 522.

Sansoucy, R. (1995). Livestock-a driving force for food security and sustainable development. *World* 3074 (5389): 1035.

Sathwara, R. N., Gupta, J. P., Chaudhari, J. D., Parmar, G. A., Prajapati, B. M., Srivastava, A. K., Chauhan, H. D., Patel, P. A. and Prajapati, M. N. (2020). Analysis of association between various fertility indicators and production traits in Mehsana buffaloes. *Tropical Animal Health and Production*, 52: 2585-2592.

Shyma, K. P., Gupta, J. P. and Singh, V. (2015). Breeding strategies for tick resistance in tropical cattle: a sustainable approach for tick control. *Journal of Parasitic Diseases*, 39(1): 1-6.

Soller, M. and Andersson, L. (1998). Genomic approaches to the improvement of disease resistance in farm animals. *Revue Scientifiqueet Technique-Office International des Epizooties*, 17(1): 329-345.

Stear, M. J., Newman, M. J., Nicholas, F. W., Brown, S. C. and Holroyd, R. G. (1984). Tick resistance and the major histocompatibility system. *Australian Journal of Experimental Biology and Medical Science*, 62(1): 47-52.

Stear, M. J., Nicholas, F. W., Brown, S. C. and Holroyd, R. G. (1989). Class I antigens of the bovine major histocompatibility system and resistance to the cattle tick (*Boophilus microplus*) assessed in three different seasons. *Veterinary Parasitology*, 31(3-4): 303-315.

The Tribune (2017, December 17). *Tick-resistant Gojri adds to buffalo breed.* The New Indian Express. Retrieved from https://www.tribuneindia.com/news/archive/features/tick-resistant-gojri-adds-to-buffalo-breed-514431.

Thompson-Crispi, K. A., Sargolzaei, M., Ventura, R., Abo-Ismail, M., Miglior, F., Schenkel, F. and Mallard, B. A. (2014). A genome-wide association study of immune response traits in Canadian Holstein cattle. *BMC genomics*, 15(1): 1-10.

Turner, L. B., Harrison, B. E., Bunch, R. J., Neto, L. R. P., Li, Y. and Barendse, W. (2010). A genome-wide association study of tick burden and milk composition in cattle. *Animal Production Science*, 50(4): pp.235-245.

Untalan, P. M., Pruett, J. H. and Steelman, C. D. (2007). Association of the bovine leukocyte antigen major histocompatibility complex class II DRB3* 4401 allele with host resistance to the Lone Star tick, *Amblyomma americanum*. *Veterinary Parasitology*, 145(1-2): 190-195.

Utech, K. B. W., Wharton, R. H. and Kerr, J. D. (1978). Resistance to *Boophilus microplus* (Canestrini) in different breeds of cattle. *Australian Journal of Agricultural Research*, 29(4): 885-895.

Veríssimo, C. J., D'Agostino, S. M., Pessoa, F. F., de Toledo, L. M. and de Miranda Santos, I. K. F. (2015). Length and density of filiform tongue papillae: differences between tick-susceptible and resistant cattle may affect tick loads. *Parasites & vectors*, 8(1): 1-5.

Wagland, B. M. (1978). Host resistance to cattle tick (*Boophilus microplus*) in Brahman (*Bos indicus*) cattle. II.* The dynamics of resistance in previously unexposed and exposed cattle. *Australian Journal of Agricultural Research*, 29(2): 395-400.

Wikel, S. K. (1996). Host immunity to ticks. *Annual review of entomology*, 41(1): 1-22.

In: The Entomological Guide to *Rhipicephalus*
Editors: Sachin Kumar, Raquel Cossio Bayugar et al.
ISBN: 978-1-53619-619-1

Chapter 7

STRATEGIES FOR THE CONTROL OF *RHIPICEPHALUS MICROPLUS* (FOCUS ONLY ON CHEMICAL ACARICIDES)

F. Alba-Hurtado* and M. A. Muñoz-Guzmán
Departamento de Ciencias Biológicas, Facultad de Estudios Superiores Cuautitlán, Universidad Nacional Autónoma de México, Mexico

ABSTRACT

Currently, chemical acaricide use remains the most common method of *Rhipicephalus microplus* control; however, its excessive use has led to the appearance of tick strains resistant to the main classes of conventional acaricides. When there is a lack of response to tickticide treatment, the farmers' reaction is to increase the frequency of treatments and the dose of acaricides, which increases the cost and selective pressure for resistance. Due to the above, there has been a need to implement strategies to intelligently use acaricides and thus increase their useful life and delay the appearance of resistance. This chapter describes the main classes of conventional acaricides, growth inhibitors, and macrocyclic lactones used in the control of *R. microplus* ticks. Additionally, the most common strategies that involve the use of chemical acaricides for tick control are described, such as the use of synergized acaricides, acaricide mixture formulations, use of growth inhibitors and development of new acaricides with mechanisms of action that are different from conventional ones. It is clear that the use of chemical acaricides as the only form of tick control is a mistake. Therefore, we must consider that tick control should include the use of chemical acaricides together with a systematized combination of strategies and technologies for integrated tick control.

Keywords: *Rhipicephalus microplus*, acaricides, tick control, chemical acaricides

* Corresponding Autor's Email: Email: fealba@hotmail.com.

Abbreviations

DDT: dichlorodiphenyltrichlorethane
BHC: benzenehexachloride
HCH: hexachlorocyclohexane
ppm: parts per million
mg/kg b.w.: milligrams per kilogram of body weight
g/L: grams per litter
mg/L: milligrams per litter
GABA: γ-aminobutyric acid
GluCl: glutamate-gated chloride

Introduction

Tick infestation is one of the most important limiting factors for cattle production in tropical and subtropical areas around the world. Infestations of cattle produce anemia, decrease in growth and reproductive parameters, reduction of meat and milk production, reduction in quality of pelts and increased transmission of diseases, such as babesiosis and anaplasmosis (Rodríguez-Vivas et al., 2005). The economic losses produced by the presence of ticks in livestock are associated with low productivity, control costs, treatment of other diseases, and difficulty in the mobility and export of livestock. Calculating the total economic impact of tick infestation in the world, even in a single country, is extremely complicated, and there are no data that include the cost of all of the possible factors involved, so the available calculations are only approximate. It has been calculated that the feeding of each engorged female produces a loss of 1.18 to 1.37 g of body weight in cattle and 0.26 kg/engorged female/year (Rodriguez- Vivas et al., 2005; Jonsson, 2006). Table 1 presents some data on the annual economic losses cited in the literature and their approximate value in 2020.

Table 1. Annual economic losses caused by ticks in livestock

Country	Year	Annual cost in millions of dollars	Cost calculated to 2020	References
Australia	1974	62	345	Springell, 1974
South America	1987	1000	2330	Horn, 1987
World	1984	8000	20320	Brown and Askenase, 1984
Mexico	2013	573.61	642.4	Rodríguez- Vivas et al., 2017
Brazil	2014	3240	3564	Grisi et al., 2014
Mexico	2014	942	1036	Domínguez et al., 2016
Colombia	2018	168	174.7	Betancourt-Hurtado and Giraldo-Ríos, 2018

The most important tick in cattle in many parts of the world is *Rhipicephalus microplus,* and although there are other forms of control, chemical acaricide use remains the most common method; however, its excessive use has led to the appearance of tick strains resistant to the main classes of conventional acaricides. Therefore, intelligent use of the molecules available on the market is required to increase the useful life of most of them. Therefore, the first step is to know the history, mechanism of action, resistance, synergies and strategic use for each available molecule.

BRIEF HISTORICAL REVIEW

For a long time and until the end of the 19th century, various substances were used to control ticks, including cottonseed oil, fish oil, crude oil, kerosene, creosote, tobacco extract, soap and sulfur, which are administered two or three times a week with the help of sponges and brushes (Shaw, 1969). The first news of the use of arsenic as an acaricide was in South Africa in 1893. The first record of its use in "dipping vats" for the control of *R.* (*Boophilus*) *microplus* in cattle was in Australia in 1896, and this preparation was known as "Queensland Dip" or "Australian Dip" (Angus, 1996; George, 2000). The use of arsenic quickly replaced other treatments for tick control in many regions of the world. The first records of arsenic resistance date from 1935 in Australia and 1940 in Argentina (Angus, 1996). Due to its high toxicity, the accumulation of this compound in animal tissues and the appearance of resistance, arsenic was gradually replaced by synthetic acaricides (Botana, 2002; George et al., 2004).

In the 1940s, a group of synthetic chemical compounds known as organochlorines was introduced commercially and widely used for the control of ticks and other arthropods in regions with resistance to arsenic. These compounds include dichlorodiphenyltrichlorethane (DDT), hexachlorocyclohexane (HCH), benzenehexachloride (BHC), toxaphene, lindane, and dieldrin (Cobbett, 1947). In the 1950s, the first cases of organochlorine-resistant ticks were detected. Subsequently, organochlorines were no longer used for the control of ticks, and some were prohibited due to their high toxicity, their accumulation in the adipose tissue of animals and their persistence in the environment (George et al., 2004).

In the 1950s, organophosphate compounds such as diazinon, dioxathion, coumaphos and chlorpyriphos were introduced, which gradually replaced the organochlorines. (Angus, 1996; Botana et al., 2002). In the following years, they were widely used due to their high efficiency; however, due to their high toxicity and the appearance of resistant tick strains, their use has gradually decreased worldwide (George et al., 2004). In this same decade, a first group of carbamates for tick control (carbaryl) was developed and commercialized, although they quickly presented cross-resistance with organophosphates, so their use was very limited in cattle (George, 2000; Eddleston and Clark, 2011). Currently, a group of ethyl carbamates has been developed that do not show cross-resistance with organophosphates, but they are still in the experimental phase (Pérez-González et al., 2014).

In the 1970s, pyrethroid cypermethrin (pyrethroid type 1) was developed, and in the 1980s, permethrin (pyrethroid type II) was developed. Due to their high efficacy and low environmental impact, they are widely used; however, resistance to these compounds quickly appears, so currently, their use has decreased (Ujihara, 2019).

In the same years, formamidines and cycloamidines were synthesized, such as chlordimeform, cholomethiuron, clenpyrin and amitraz (George et al., 2004; Botana et al., 2002). Chlordimeform was used in Australia as an additive in immersion baths to treat animals with organophosphate-resistant strains. However, in 1976, it was withdrawn from the market due to its carcinogenic effect. Amitraz is the most widely used in the world because it is a highly effective compound, although resistant tick strains have also been reported (George et al., 2004).

Ivermectin was the first macrocyclic lactone launched on the market in the early 1980s for the control of parasitic helminths and arthropods of domestic animals; in the following years, new molecules of this group were developed. Later, they were used successfully in programs for the control and eradication of human filariasis worldwide. Initially, due to their high cost, they were not commonly used in cattle; in the late 1990s, they began to be used massively for the control of ticks, especially *R. microplus,* and although resistant tick strains have appeared, they remain some of the most used compounds for tick control (Fernandez-Salas et al., 2012; Laing et al., 2017).

Fipronil is a member of a relatively small class of pesticides, phenyl pyrazoles or fiproles. Fipronil was introduced to the market in 1993 and is currently widely used for tick control. Some resistant strains of *R. microplus* have been reported, and concerns have been voiced about their environmental and human health effects (Tingle et al., 2003; Miller et al., 2013). In the same decade, fluazuron (Acatak®), a chitin synthesis inhibitor, was launched on the market. Due to its high efficacy and selective toxicity, as well as its low toxicity to mammals, it is widely used for the control of ticks in cattle worldwide. However, strains of ticks resistant to fluazuron have also been detected (Gaudêncio et al., 2017; Reck et al., 2014).

Chemical Acaricides Currently Used in Tick Control

Organophosphates

Organophosphates are organic esters derived from phosphoric acid, and they are characterized by having a radical S=P or O=P in their molecule. Due to their high lipid solubility, they can be absorbed through the skin, digestive and respiratory tracts. Most of these molecules are eliminated in urine and occasionally in milk, and small amounts are fixed in fatty tissue; however, continuous exposure causes the accumulation of larger amounts. The mechanism of action of organophosphates is the inhibition of acetylcholinesterase by affinity with the stearic site of the enzyme, which produces an increase in acetylcholine in nerve synapses and the consequent sustained contraction of muscle fibers. These compounds have greater affinity and stability to bind to arthropod acetylcholinesterase than mammals, which explains their high toxicity to arthropods and their lower toxicity to mammals (including humans). Organophosphates have been used to control lice, fleas, flies, fly larvae, scabies-producing mites, and ticks. Some of the most commonly used organophosphates for tick control are chlorfenvinphos, chlorpyrifos, coumaphos, diazinon, dichlorvos, malathion and trichlorfon (Pundir and Malik, 2019).

Carbamates

Carbamates are esters of carbamic acid, initially derived from physostigmine; however, these compounds are currently of synthetic origin (Botana et al., 2002; Gupta, 2006). Most of the carbamates described so far are arthropod and mammalian acetylcholinesterase inhibitors; therefore, they present cross-resistance with organophosphates and high toxic potential in mammals. Despite their proven efficacy against ticks, their use in cattle was discouraged due to instability in immersion baths and the appearance of strains with cross-resistance between organophosphates and carbamates. Currently, they are used mainly as domestic insecticides and ectoparasitides in small animals (Núñez et al., 1985). The most commonly used carbamates for arthropod control in domestic animals are propoxur and carbaryl (Mullens et al., 2017; Nogueira et al., 2020).

Recently, a series of ethyl carbamates have been developed with a mechanism of action different from that of conventional carbamates. These ethyl carbamates induce apoptosis in the ovary and other organs of *R. microplus*, behaving in a similar way to growth regulators (Prado-Ochoa et al., 2014c). Their toxicity in mammals is low, and their efficacy on multiresistant tick strains is high, making them promising candidates for use in tick control (Pérez-González et al., 2014; Prado-Ochoa et al., 2014 a, b).

Pyrethroids

Pyrethroids are synthetic analogs of natural pyrethrins obtained from chrysanthemum. Natural pyrethrins are esters of cyclopropanecarboxylic acid and a cyclopentenolone alcohol, which were synthetically modified in the pyrethroids to increase the insecticidal effect and prolong their longevity in the presence of humidity and sunlight (Elliott 1995).

The main mechanism of action of pyrethroids is the prolonged opening of sodium channels in the nervous system (Narahashi, 2001). Depending on the target site of action in the arthropod, there are two types of pyrethroids. Type I types are devoid of a cyano group in their molecule (allethrin, permethrin, tetramethrin, and cismethrin) and act mainly on the sensory and motor nerves of the central nervous system. Type II types have a cyano group at the α-benzylic position (cypermethrin, deltamethrin, fenvalerate, fenpropanate) and act by depolarizing the potential of the axon membranes (George et al., 2004; Kamita et al, 2017).

These compounds are more toxic to arthropods than to mammals because pyrethroids have a higher affinity for sodium channels of arthropods compared to mammals. In addition, the binding of pyrethroids to sodium channels in arthropods is irreversible, while in mammals, it is generally reversible. The effect of pyrethroids is negatively dependent on temperature, so pyrethroids exert a higher biological activity on ticks that have a lower body temperature than their host. Another factor is the greater capacity of mammals to detoxify these compounds (Song and Narahashi, 1996; Narahashi, 2001). Although synthetic pyrethroids are more stable than natural pyrethrins, they are denatured by sunlight and do not accumulate in the environment, which is why they are considered the most environmentally friendly pesticides. The route of administration in bovines is epicutaneous (pour-on) and topical through spray or immersion baths. Pyrethroids have also been used to control lice, fleas, flies, fly larvae, scabies-producing mites, and ticks.

Amidines

Amidines (formamidines) are organic compounds derived from oxacids that have a functional group C (= NH) NH_2, equivalent to an amide where the oxygen with a double bond is substituted by an NH group. Currently, the only amidine used for tick control is amitraz. Octopamine is an adrenergic amine similar to norepinephrine that acts as a neurohormone, neurotransmitter, or modulator of neuromuscular activity in arthropods. Amitraz blocks adenylate cyclase-dependent octopaminergic receptors in arthropods, delaying axonal depolarization, which causes an increase in excitability and motor activity, a decrease in feeding behavior and finally causes death (Costa, 2020). Amitraz acts on α1, α2 and histamine receptors and inhibits monoamine oxidase and prostaglandin synthetase, for which it is considered to have high toxic potential in mammals. The route of administration in bovines is topical by spray or immersion baths. In immersion baths, it is necessary to add 0.5% CaOH to the amitraz emulsion to raise the pH to 12 and keep the active principle stable for long periods.

Phenylpyrazoles

Fipronil is an N-phenylpyrazole with a trifluoromethylsulfinyl substituent that is used for the control of ticks, lice and flies. The most common route of administration is epicutaneous (pour-on). Fipronil acts on γ-aminobutyric acid (GABA) receptors to block glutamate-gated chloride (GluCl) channels, which causes hyperexcitation of the muscular and nervous systems of arthropods.

Benzoylphenyl Ureas

These compounds act as growth regulators in arthropods. The main components of the cuticle of ticks are proteins and chitin. The latter is an amino polysaccharide homopolymer composed of N-acetyl-D-glucosamine units that is continuously synthesized for tick growth and development. The benzoylphenyl ureas inhibit chitin synthesis in tick larvae and nymphs, which prevents ecdysis and decreases the outer chitin layer, so the tick loses fluids (mainly hemolymph) and dies from dehydration. Some molecules in this group are lufenuron, flufenoxuron, diflubenzuron, methoprene, novaluron, and fluazuron. The most commonly used in the control of ticks in cattle is fluazuron (pour-on), which, in addition to inhibiting ecdysis of larvae and nymphs, inhibits the oviposition of engorged females and the hatching of larvae (Gaudêncio et al., 2017). Novaluron is widely used in agriculture, and there are some commercial presentations against ticks in some countries; however, its efficacy is limited to the larval stages (Lohmeyer et al., 2012). In general, benzoylphenyl ureas are considered low-toxicity products to mammals and have low environmental impact (Alves et al., 2019; Kryger et al., 2007).

Macrocyclic Lactones

These are semisynthetic compounds derived from the natural molecules avermectin and milbemycin produced during fermentation carried out by Streptomyces avermitilis and Streptomyces cyaneogriseus, respectively. They are efficient endectocides for the control of nematodes and arthropods that affect domestic animals. In general, macrocyclic lactones are molecules of high molecular weight (600-900 kDa); chemically, they are made up of a macrocyclic lactone ring composed of 16 elements to which different radicals are attached. Avermectins have a carbohydrate radical at carbon 13, which milbemycins lack (Burg et al., 1979; Díaz-Carrasco et al., 1997). Although the chemical differences are relatively small, they are responsible for the observed differences in pharmacokinetics between the different macrocyclic lactones.

Macrocyclic lactones act on GluCl channels present exclusively in the neurons and muscle cells of invertebrates. The binding of these compounds to GluCl channels produces an increase in the permeability of the cell membrane to chlorine ions, and the hyperpolarization produced in the cells produces flaccid paralysis and death of the parasite (Lifschitz et al., 2002: Rodríguez-Vivas et al., 2010; Atif et al., 2019).

Macrocyclic lactones are hydrophobic and fat-soluble compounds, so they have good intestinal and skin absorption. The most commonly used routes of administration are oral, subcutaneous and epicutaneous. The main avermectins used in tick control are ivermectin, abamectin, doramectin, selamectin, and eprinomectin. The most important milbemycin is moxidectin.

Methods of Administration of Acaricides

How ticks come into contact with the chemical acaricide is important to ensure their effectiveness. The most common forms of application of chemical acaricides are spray baths, dip immersion, epicutaneous application (pour-on) and injectables. The doses/concentrations and routes of administration recommended by the manufacturers of the most common acaricides used for the chemical control of ticks are presented in Table 2.

Spray baths are based on the formation of an aerosol from an emulsion of acaricide with a manual spray pump, motorized spray pump or spray race dipping. It is necessary to completely soak the animal, including places difficult to access such as the axillary area, inguinal region, tail folds and pinna, so that the chemical compounds come into contact with the surface of all the ticks present on the bovine. Depending on the size of the bovine, 3-5 liters of emulsion per animal are required. It is essential to use the dosage and conditions of use recommended by the manufacturer of each commercial product. The main advantage is its low cost, and the main disadvantage is the continuous risk of acaricide underdosing due to poor application techniques, inadequate equipment, little planning of the workload of the operators and inadequate preparation of the emulsion. It is recommended to use a manual spray pump in farms with up to 100 cattle and a motorized spray pump in farms with up to 300 cattle and in farms with a greater number of cattle; the cost of spray race dipping is justified. In addition, the risk to operators from being in constant contact with pesticides for long periods must be considered.

Table 2. Dose/concentration recommended by the manufacturers of the main compounds used for the control of *R. microplus* in cattle

Chemical acaricide	Administration routes	Dose/concentration
Coumaphos	Spray or immersion baths	200-500 ppm = 0.2-0.5 g/L
Trichlorfon	Spray baths	1500 ppm = 1.5 g/L
Clorpyrifos	Spray or immersion baths	240-500 ppm
	Epicutaneous (pour-on)	5-10 mg/kg b.w.
Chlorphenvinphos	Spray or immersion baths	200-400 ppm = 0.2-0.4 g/L
Dichlorvos	Spray or immersion baths	400-1125 ppm = 0.4-1.12 g/L
	Epicutaneous (pour-on)	0.5-4 mg/kg b.w.
Deltamethrin	Spray or immersion baths	25 ppm = 0.025 g/L
	Epicutaneous (pour-on)	1.5 mg/kg b.w.
Permethrin	Spray or immersion baths	50-400 ppm = 0.05-0.4 g/L
	Epicutaneous (pour-on)	2-10 mg/kg b.w.
Cypermethrin	Spray or immersion baths	200 ppm = 0.2 g/L
Flumethrin	Spray or immersion baths	30 ppm = 0.03 g/L
	Epicutaneous (pour-on)	1 mg/kg b.w.
Alpha-methrin	Spray or immersion baths	50 ppm = 0.05 g/L
Fipronil	Epicutaneous (pour-on)	1 mg/kg b.w.
Amitraz	Spray baths	250 ppm = 0.25 g/L
	Immersion baths	200 ppm = 0.20 g/L
Fluazuron	Epicutaneous (pour-on)	2.5 mg/kg b.w.
Ivermectin, doramectin, abamectin	Subcutaneous injection	0.2 mg/kg b.w.
	Epicutaneous (pour-on)	

The immersion bath is based on completely submerging the cattle in a dipping tub. It is the most effective, fast and economical method in the long term when it is necessary to bathe a large number of cattle (> 300). The turbulence of the emulsion generated in the tub by the movement of the animals produces a constant homogenization of the emulsion and good penetration of the acaricide between the hair of the bovines and the arrival of the product to the anatomical regions that are difficult to access. For the design and dosage of an immersion bath, the advice of a specialized technician and technical guides for the design and official standards of each country or region are required (Bavera, 2005; FAO/UN, 1984). The main disadvantages of immersion baths are the initial cost of construction and of the immersion fluid, stress caused in cattle, especially in pregnant females, in which abortion can be caused, and the environmental contamination generated when the emulsion is discarded from the tub.

Epicutaneous or pouring application consists of spilling the acaricide product on the dorsal midline of cattle from the base of the neck to the base of the tail. The formulations used in these treatments have been designed so that the acaricide adheres to the surface of the bovine's body, with the secretions of the sweat and sebaceous glands, forming an emulsion that acts as a vehicle to distribute the acaricide throughout the skin. The formulations act on the parasites by contact and through the vapors that emanate, forming an atmosphere that surrounds the body of the treated animal. This route of administration is recommended in the case of pregnant cows or with calves and in places where there are no facilities for the handling of cattle. Due to its direct application, the amount of pesticides discharged into the environment is less than that with baths.

Macrocyclic lactones are the main endecticides administered by subcutaneous injections. The main form of contact with ticks is when they feed on blood and tissue fluids. Macrocyclic lactones are highly effective, but the cost of treatment per bovine is higher.

RESISTANCE TO ACARICIDES

Resistance has been defined as an inherited characteristic that imparts an increased tolerance to a pesticide, or group of pesticides, such that the resistant individuals survive a concentration of the compound(s) that would normally be lethal to the species (WHO, 1992). Resistance is the result of the indiscriminate exposure of a population of ticks (*R. microplus*) to an acaricide without a planned strategy, resulting in the selection of individuals who possess resistance genes. A comprehensive review of the subject is found in chapters 7 and 8 of this book and in the review by Rodríguez-Vivas et al. (2018).

STRATEGIES TO REDUCE AND CONTROL THE APPEARANCE OF RESISTANCE TO ACARICIDES

The proper use of current chemical compounds for tick control allows us to be more efficient in delaying the appearance of resistance, extending the useful life of acaricides and reducing costs per treatment. Reducing the use of acaricides through manual removal of ticks, the use of natural predators, vaccines and the use of phytostracts allows reduction of costs and selection pressure produced by chemical acaricides (FAO, 2004). These forms of control are described and discussed in Chapters 10, 11, and 14 of this book.

Not all cattle are equally susceptible to tick infestation. Resistance of cattle to tick infestation is associated with a reduced number of engorged females, reduced egg production and reduced egg viability. Breeds of African origin (Zimbabwean Sanga and Nguni breeds) have been shown to be more resistant than breeds of Asian origin and their crosses (Brahman, Indobrazil and Santa Gertrudis breeds), and the latter are more resistant than breeds of European origin (Holstein and Simental breeds). Resistance is a phenotypic characteristic with high heritability (h^2 = 0.42-0.64) in resistant breeds (Rechav and Kostrzewski, 1991; Jonsson et al., 2014). The crosses of *Bos indicus* breeds with *Bos taurus* breeds show medium resistance above the most susceptible progenitor and moderate heritability (h^2 = 0.34). The use of resistant breeds reduces the number of treatments with chemical acaricides and therefore reduces costs and the selective pressure of resistance to acaricides by ticks.

The rotation of acaricides is the alternation of application between two or more chemical acaricides with different mechanisms of action, which reduces the selective pressure produced by a single acaricide. In laboratory experiments, it has been clearly shown that the appearance of resistance to pyrethroids is delayed in strains of *R. microplus* treated alternately with deltamethrin and coumaphos (Thullner et al., 2007).

The mixture of chemical acaricides, preferably with a different mechanism of action, has been used in some countries to control ticks; in some of these mixtures, a synergistic effect has been reported between acaricides. However, the use of mixtures is controversial, and in some countries, it is prohibited. While some researchers promote their use arguing that these mixtures increase the time of use of some acaricides in places where resistant strains of ticks have appeared, others warn that the use of these mixtures favors the appearance of multidrug resistance. Table 3 shows some of the acaricide mixtures that have shown a positive effect on the control of *R. microplus*. Some of these mixtures are experimental, and others are commercially available in some countries.

Table 3. Mixtures of chemical acaricides with positive effects on the control of *R. microplus*

Acaricide mix	Strain origin	Reference
Cymiazole 0.035% cypermethrin 0.005% "Ectogan®" (adult immersion test)	Mexico	Céspedes et al., 2002
Permethrin 0.005 + Chlorpyrifos 0.024% "Garraban®" (adult immersion test)		
Deltamethrin 0.0005% + amitraz 0.025% (spray dip)	New Caledonia	Barré et al., 2008.
Permethrin 3294 μg/cm^{-2} + amitraz 11.0 μg/cm^{-2} (larval packet test)	Mexico and Brazil	Li et al., 2007
Coumaphos + piperonyl butoxide	Mexico	Miller et al., 2008
Cypermethrin 0.2% + amitraz 0.00025 (spray dip)	Mexico	Rodríguez-Vivas et al., 2013
Cypermethrin 0.2% + piperonyl butoxide 0.005% (spray dip)		
Cypermethrin 0.2% + amitraz 0.005% + piperonyl butoxide 0.005% (spray dip)		
Tetrachlorvinphos 0.069% + dichlorvos 0.015% (spray dip)	Mexico	Davey el al., 2013
Coumaphos + verbutin (larval packet test)	Mexico	Li et al., 2010
Permethrin + verbutin (larval packet test)		
Amitraz + verbutin (larval packet test)		
Fipronil 1.25 mg/kg b.w. + fluazuron 2.5 mg/kg b.w. (pour-on)	Brazil	Lopes et al., 2017.
Cypermethrin + chlorpyrifos + fenthion (adult immersion test and larval packet test).	Brazil	da Silva Rodrigues et al., 2018
Cypermethrin + chlorpyrifos + piperonyl butoxide (adult immersion test)	Brazil	Higa et al., 2019
Cypermethrin + chlorpyrifos (adult immersion test)		
Cypermethrin 15 g + chlorpyrifos 30 g + fenthion 15 g (adult immersion test)		
Fluazuron 3 mg/kg b.w. + abamectin 0.5 mg/kg de b.w. (pour-on)	Brazil	Mendonça, 2010.
Fluazuron 3 mg/kg b.w. + abamectin 0.5 mg/kg de b.w. (pour-on)	Brazil	Cruz et al., 2014

Growth Regulators

Growth regulators are characterized by not immediately killing ticks but instead blocking their biological cycle, which significantly decreases the number of individuals in the next generation. Benzoylphenyl ureas are the main group of molecules used for the control of insects and mites. Several molecules from this group have been used to control insects, but fluazuron has been shown to be highly effective and specific for regulating tick development. The main effect of fluazuron in ticks is to block chitin synthesis and thus prevent the molting of larvae to nymphs and nymphs to adults. Fluazuron-treated engorged females produce normal numbers of eggs, but the larvae within them are not strong enough to break the egg and hatch. In addition to the effects on development, fluazuron affects the digestive tract and salivary glands, interfering with the digestion and nutrition of the tick. This is related to changes in color and shape, as well as a greater fragility of adult ticks (Kemp et al., 1990).

After treatment with fluazuron, adult *R. microplus* ticks can remain on cattle for up to two weeks before shedding, much longer than after treatment with other ixodicides. The above approach, applied many times from the perspective of the farmer, means that it does not kill the ticks and therefore the treatment shows little effectiveness. However, when cattle are treated with fluazuron early in the tick season (preventative treatment) before the cattle are infested with the first active larvae on pasture, these larvae will die after ingesting blood. None or only a few preadult females will develop into engorged females and shed to produce eggs and infest grasses. However, the number of larvae that develop in grasses is considerably lower. The residual effect is 8-12 weeks, and fluazuron is eliminated in milk, so lactating calves of treated females are protected against ticks. It has been shown that treatment with fluazuron does not interfere with the transmission of *Babesia* and *Anaplasma*; therefore, it does not interfere with the premunity conferred by tick infestation against these hemoparasites (Junquera et al., 2019).

DEVELOPMENT OF NEW ACARICIDES

The appearance of resistance to acaricides for commercial use has generated the need for the development of new ixodicides with mechanisms of action different from conventional ones to avoid cross-resistance. Many of these new acaricides with effects on ticks are of natural origin extracted from plants. A review of these was carried out by Banumathi et al. (2017) and in chapter 10 of this book; therefore, in this chapter, we will only describe new chemical acaricides of synthetic origin.

Nanotechnology has provided new materials for the formulation of nanoparticles with toxic effects for ticks and with high potential for their control. Chemically synthesized and/or green-synthesized metal nanoparticles have been used against ticks. Using chemical synthesis, nanoparticles of zinc oxide (Kirthi et al., 2011), copper acetate (Ramyadevi et al., 2011) and nickel-hydrazine (Rajakumar et al., 2013) have been produced. These copper, nickel and zinc oxide nanoparticles conferred 100% mortality in *R. microplus* larvae at 8, 25 and 10 mg/L, respectively.

The global trend has been to reduce the ecological impact of chemical synthesis (green synthesis) by inventing novel reactions that can maximize the desired products and minimize byproducts, designing new synthetic schemes and apparati that can simplify operations in chemical production, and seeking greener solvents (Li and Trost, 2008). For green synthesis, nanoparticles of zinc oxide (Banumathi et al., 2016), silver chloride (Marimuthu et al., 2011; Rajakumar and Rahuman, 2012) and titanium dioxide (Marimuthu et al., 2013; Rajakumar et al., 2015) are applied using plant extracts as reducing and capping agents for nanosynthesis. These nanoparticles have shown high toxicity (mortality between 89-100%) *in vitro* against *R. microplus*.

Nanotechnology has also developed nanoparticle formulations that include some conventional acaricides, for example, deltamethrin and zinc oxide, which showed greater efficacy (95%) than conventional deltamethrin (15%) on a pyrethroid-resistant *R microplus* strain (Benelli et al., 2017). This shows that the nanopaticularization of some chemical acaricides could in part reverse the resistance of ticks and increase their useful life.

Another way to obtain new acaricidal molecules is computational design and the prediction of their activity *in silico* through studies of the structure-activity relationship. Using these tools, a series of new carbamates with activity against strains susceptible and resistant to conventional acaricides have been developed. The new carbamates, ethyl-4-bromophenyl-carbamate (LQM 919) and ethyl-4-chlorophenyl-carbamate (LQM 996), inhibit ovogenesis and damage ovary cells and therefore affect the reproduction of *R. microplus* (effective doses LQM 919 = 0.687 mg/mL and LQM 996 = 0.279 mg/mL), and although they are carbamates such as carbaryl and carbofuran, these new ethylcarbamates are not inhibitors of acetylcholinesterase. These effects occur in both strains that are susceptible and strains that are resistant to amidines, pirethroids and organophosphates (Pérez-González et al., 2014; Prado-Ochoa et al., 2013; Prado-Ochoa et al., 2014c). Furthermore, these ethyl carbamates are of low toxicity to mammals and of low environmental impact. Therefore, they are strong prospects to be used as growth regulators, but with a different mechanism than fluazuron (Prado-Ochoa et al. 2014 a, b; Prado-Ochoa et al., 2016; Iturbe-Requena et al., 2019; Iturbe-Requena et al., 2020).

Conclusion

The economic impact of ticks on bovine production promoted the development of various groups of chemical acaricides for their control. However, the appearance of strains of ticks resistant to most of them has unleashed a constant struggle between the development of new acaricides and the appearance of new resistances. As a consequence of this war, some groups, such as arsenicals and organochlorines, have disappeared as ixodicides, and others, such as organophosphates, carbamates, amidines and pyrethroids, have considerably reduced their use. Although the development of phenylpyrazoles, benzoylphenyl ureas and macrocyclic lactones has given a break to the problem of resistance, the presence of resistant strains has already been demonstrated, and it is a matter of time before these spread. Newly synthesized nanoparticles and acaricides are currently under development and show promise as ixodicides, but those that are not used properly will have the same fate. Therefore, trying to control ticks only with chemical acaricides is a mistake. All tick control must intelligently integrate a systematic combination of strategies and technologies, such as the use of vaccines, pasture management, rotation of acaricides, use of entomopathogenic organisms, and resistant bovine breeds, among others, that allow reduction of the use of chemical acaricides and tick populations without affecting cattle productivity, human health and the environment.

References

Alves, P.R.L., Bandeira, F.O., Giraldi, M., Presotto, R., Segat, J.C., Cardoso, E.J.B.N., and Baretta, D. (2019). Ecotoxicological assessment of Fluazuron: effects on *Folsomia candida* and *Eisenia andrei*. *Environmental Science and Pollution Research,* 26(6): 5842-5850.

Angus, B.M. 1996. The history of the cattle tick *Boophilus microplus* in Australia and achievements in its control. *Australian Society Parasitology,* 26(12): 1341-1355.

Atif, M., Smith, J.J., Estrada-Mondragon, A., Xiao, X., Salim, A.A., Capon, R.J., Lynch, J.W. and Keramidas, A. (2019). GluClR-mediated inhibitory postsynaptic currents reveal targets for ivermectin and potential mechanisms of ivermectin resistance. *PLoS Pathogens,* 15(1): e1007570.

Banumathi, B., Malaikozhundan, B., Vaseeharan, B., (2016). *In vitro* acaricidal activity of ethnoveterinary plants and green synthesis of zinc oxide nanoparticles against *Rhipicephalus* (*Boophilus*) *microplus*. *Veterinary Parasitology,* 216: 93-100.

Banumathi, B., Vaseeharan, B., Rajasekar, P., Prabhu, N. M., Ramasamy, P., Murugan, K., Canale, A. and Benelli, G. (2017). Exploitation of chemical, herbal and nanoformulated acaricides to control the cattle tick, *Rhipicephalus* (*Boophilus*) *microplus*–a review. *Veterinary Parasitology,* 244: 102-110.

Barré, N., Li, A.Y., Miller, R.J., Gaïa, H., Delathiere, J.M., Davey, R.B., and George, J. E. (2008). *In vitro* and *in vivo* evaluation of deltamethrin and amitraz mixtures for the control of *Rhipicephalus* (*Boophilus*) *microplus* (Acari: Ixodidae) in New Caledonia. *Veterinary Parasitology,* 155(1-2): 110-119.

Bavera, G.A. (2005). *Cursos parcial de Producción Bovina de Carne [Partial courses of Bovine Meat Production], FAV UNRC.* http://www.produccion-animal.com.ar/informacion_tecnica/instalaciones/14-banaderos.pdf

Benelli, G., Maggi, F., Romano, D., Stefanini, C., Vaseeharan, B., Kumar, S., Higuchig, A., Alarfajh, A.A., Mehlhorn, H., and Canale, A. (2017). Nanoparticles as effective acaricides against ticks - a review. *Ticks and tick-borne diseases*, 8(6): 821-826.

Betancourt-Hurtado, O.J., Giraldo-Ríos, C. (2018). Economic and health impact of the ticks in production animals. In: Abubakar M. (Ed.), *Ticks and Tick-Borne Pathogens,* IntechOpen.

Botana, L. (2002). *Farmacología y Terapéutica Veterinaria* [*Veterinary Pharmacology and Therapeutics*]. McGraw-Hill, España.

Brown, S.J. and Askenase, P.W. (1984). Analysis of host components mediating immune resistance to ticks. In: Griffiths, D.A. and Bowman, C.E. (Eds.), *Acarology VI,* Vol. 2, Ellis Horwood, Chichester, pp. 1040-1050.

Burg, R.W., Miller, B.M., Baker, E.E., Birnbaum, J., Currie, S.A., Hartman, R., Kong, Y.L., Monaghan, R.L., Olson, G., Putter, I., Tunac, J.B., Wallick, H., Stapley, E.O., Oiwa, R. and Ōmura, S. (1979). Avermectins, new family of potent anthelmintic agents: producing organism and fermentation. *Antimicrobial agents and Chemotherapy,* 15(3): 361-367.

Céspedes, N.S., Vargas, M.S., Sánchez, H.F., and Vázquez, Z.G. (2002). Primer caso de resistencia al amitraz en la garrapata del ganado *Boophilus microplus* en México [First case of resistance to amitraz in the cattle tick Boophilus microplus in Mexico]. *Técnica Pecuaria en México*, 40(1): 81-92.

Cobbett, N.G. (1947). Preliminary tests in Mexico with DDT, cube, hexachlorocyclohexane (benzene hexachloride), and combinations thereof, for the control of the cattle fever tick, *B. annulatus*. *American Journal of Veterinary Reseach,* 8(28): 280-283.

Costa, L.G. (2020). Neurotoxicity of amitraz, a formamidine pesticide. In. Aschner M. and Costa, L.G. (Eds.), *Advances in Neurotoxicology* Vol. 4, Academic Press, London, pp. 255-276.

Cruz, B.C., Teixeira, W.F.P., Maciel, W.G., Felippelli, G., Fávero, F.C., Cruz, A.C., Buzzulini, C., Soares, V.E., Gomes, L.V.C., Lopes, W.D.U., De Oliveira, G.P. and da Costa A.J. (2014). Effects of fluazuron (2.5 mg/kg) and a combination of fluazuron (3.0 mg/kg) + abamectin (0.5 mg/kg) on the reproductive parameters of a field population of *Rhipicephalus* (*Boophilus*) *microplus* on experimentally infested cattle. *Research in Veterinary Science,* 97(1): 80-84.

da Silva Rodrigues, V., Bonatte, P., Garcia, M.V., Higa, L.D.O.S., Piña, F.T.B., Zimmermann, N.P., Oliveira-Duartea, P., Cavalcante-Barros, J., and Andreotti, R. (2018). Efficacy profile of Cypermethrin and Chlorpyrifos based acaricides on *Rhipicephalus microplus* control on cattle in the rearing phase, naturally infested and exposed to tick fever agents in central Brazil. *Veterinary Parasitology: Regional Studies and Reports,* 12: 43-48.

Davey, R.B., Thomas, D.B., Pound, J.M., Lohmeyer, K.H., and Miller, R.J. (2013). Efficacy of an organophosphate mixture against an organophosphate-resistant strain of *Rhipicephalus* (*Boophilus*) *microplus* (Acari: Ixodidae). *Journal of Entomological Science,* 48(4): 306-316.

Díaz-Carrasco, M.D., Espuny, A., Escudero, E., and Cárceles, Y.C.M. (1997). Farmacologia de los endectocidas: aplicaciones terapéuticas [Farmacologia de los endectocidas: aplicaciones terapéuticas]. *Anales de Veterinaria de Murcia, 13*(14), 3-22.

Domínguez, D., Torres, F., Rosario-Cruz, R. (2016). Evaluación económica del control de garrapatas *Rhipicephalus microplus* en México [Economic evaluation of the control of *Rhipicephalus* microplus ticks in Mexico]. *Revista Iberoamericana de las Ciencias Biológicas y Agropecuarias,* 5(9): 43-52.

Eddleston, M. and Clark, R.F. (2011). Insecticides: Organic Phosphorus Compounds and Carbamates. In: Robert, S. Hoffman, R.S., Howland, M.A., Lewin, N.A., Nelson, L.S., Goldfrank, L.R. (Eds.), *Goldfrank's Toxicologic Emergencies.* Tenth Edition. McGraw-Hill, New York, pp. 1450-1476.

Elliott, M. (1995). Chemicals in insect control. In: Casida, J.E. and Quistad G.B. (Eds.), *Pyrethrum Flowers: Production Chemistry, Toxicology, and Uses*, Oxford University Press, New York, pp. 3-26.

FAO, (1984). *Ticks and tick borne disease control. A practical field manual.* Volume I. Tick control. Rome.

FAO, (2004). Module 1. Ticks: Acaricide resistance: Diagnosis, management and prevention. In: Rome (Italy): FAO Animal Production and Health Division (Ed.), *Guidelines Resistance Management and Integrated Parasite Control in Ruminants,* pp. 25-77.

Fernández-Salas, A., Rodríguez-Vivas, R.I., and Alonso-Díaz, M.A. (2012). First report of a *Rhipicephalus microplus* tick population multi-resistant to acaricides and ivermectin in the Mexican tropics. *Veterinary Parasitology,* 183(3-4): 338-342.

Gaudêncio, F.N., Cordeiro, M.D., Rodrigues, J.D.A. V., Baêta, B.A., Bittencourt, V.R.E. P., da Fonseca, A.H., Angelo, I.C., and Pinheiro, J. (2017). Effects of fluazuron on the biological parameters of engorged females of *Rhipicephalus microplus*. *Brazilian Journal of Veterinary Medicine,* 39(4): 231-238.

George, J. (2000). Present and future technologies for tick control. *Annals of the New York Academy of Science,* 916: 583-588.

George, J.E., Pound, J.M. and Davey, R.B. (2004). Chemical control of ticks on cattle and the resistance of these parasites to acaricides. *Parasitology,* 129: S353-S366.

Grisi, L., Leite, R.C., Martins, J.R.D.S., Barros, A.T.M.D., Andreotti, R., Cançado, P.H.D., Leon, A.A., Pereira, J.B., and Villela, H.S. (2014). Reassessment of the potential economic impact of cattle parasites in Brazil. *Revista Brasileira de Parasitologia Veterinária,* 23(2): 150-156.

Gupta, R.C. (2006). Classification and uses of organophosphates and carbamates. In Gupta, R.C. (Ed.), *Toxicology of Organophosphate & Carbamate Compounds.* Academic Press. pp. 5-24.

Higa, L.D.O.S., Garcia, M.V., Rodrigues, V.D.S., Junior, P.B., Barradas Pina, F.T., Barros, J.C., and Andreotti, R. (2019). Effects of cypermethrin, chlorpyrifos and piperonyl butoxide-based pour-on and spray acaricides on controlling the tick *Rhipicephalus microplus*. *Systematic and Applied Acarology,* 24(2): 278-286.

Horn, S.C., Arteche, C.C.P., and Gómez, G. (1987). *Ticks in Brazil programs of tick control in the countries of the Southern region* (No. IICA-PM A4/UY No. 87-001). IICA, Montevideo (Uruguay).

Iturbe-Requena, S.L., Prado-Ochoa, M.G., Muñoz-Guzmán, M.A., Velázquez-Sánchez, A. M., Angeles, E., and Alba-Hurtado, F. (2019). Toxic effects of new ethyl-carbamates on the morphology, mortality and acetylcholinesterase activity of *Eisenia foetida. Ecotoxicology and Environmental Safety,* 176: 219-225.

Iturbe-Requena, S.L., Prado-Ochoa, M.G., Velázquez-Sánchez, A.M., García-Hernández, F., Cossío-Bayúgar, R., Muñoz-Guzmán, M.A., Angeles, E., and Alba-Hurtado, F. (2020). Oogenesis and embryogenesis inhibition induced by two new ethyl-carbamates in the cattle tick *Rhipicephalus microplus*. *Ticks and Tick-borne Diseases*, 101326.

Jonsson, N.N. (2006). The productivity effects of cattle tick (*Boophilus microplus*) infestation on cattle, with particular reference to *Bos indicus* cattle and their crosses. *Veterinary Parasitology,* 137(1-2): 1-10.

Jonsson, N.N., Piper, E.K., and Constantinoiu, C.C. (2014). Host resistance in cattle to infestation with the cattle tick *Rhipicephalus microplus*. *Parasite Immunology,* 36(11): 553-559.

Junquera, P., Hosking, B., Gameiro, M., and Macdonald, A. (2019). Benzoylphenyl ureas as veterinary antiparasitics. An overview and outlook with emphasis on efficacy, usage and resistance. *Parasite,* 26: 26.

Kamita, S.G., Kang, K.D., Hammock, B.D., and Inceoglu, A.B. (2010). 10 Genetically modified Baculomviruses for pest insect control. In: Gilbert, L.I and Gill, S.S. (Eds.), *Insect Control: Biological and Synthetic Agents*. Elsevier, Academic Press, pp. 331-381.

Kemp, D.H., Hughes, S., Binnington, K.C., Bird, P., and Nolan, J. (1990). Mode of action of CGA 157419 on the cattle-tick *Boophilus microplus*. *Bulletin de la Société Française de Arasitologie, 8*(Suppl. 2): 1048.

Kirthi, A.V., Rahuman, A.A., Rajakumar, G., Marimuthu, S., Santhoshkumar, T., Jayaseelan, C., and Velayutham, K. (2011). Acaricidal, pediculocidal and larvicidal activity of synthesized ZnO nanoparticles using wet chemical route against blood feeding parasites. *Parasitology research,* 109(2): 461-472.

Kryger, U., Deschodt, C., Davis, A.L.V., and Scholtz, C.H. (2007). Effects of cattle treatment with a fluazuron pour-on on survival and reproduction of the dung beetle species *Onthophagus gazella* (Fabricius). *Veterinary parasitology,* 143(3-4): 380-384.

Laing, R., Gillan, V., and Devaney, E. (2017). Ivermectin old drug, new tricks?. *Trends in Parasitology, 33*(6): 463-472.

Li, A.Y., Chen, A.C., Miller, R.J., Davey, R.B., and George, J.E. (2007). Acaricide resistance and synergism between permethrin and amitraz against susceptible and resistant strains of *Boophilus microplus* (Acari: Ixodidae). *Pest Management Science: Formerly Pesticide Science,* 63(9): 882-889.

Li, A.Y., Davey, R.B., and Miller, R.J. (2010). Laboratory evaluation of Verbutin as a synergist of acaricides against larvae of *Rhipicephalus* (*Boophilus*) *microplus* (Acari: Ixodidae). *Journal of Economic Entomology,* 103(4): 1360-1364.

Li, C.J., and Trost, B.M. (2008). Green chemistry for chemical synthesis. *Proceedings of the National Academy of Sciences,* 105(36): 13197-13202.

Lifschitz, A, Virkel, G., Imperiale, F., Pis, A., and Lanusse, C. (2017). Fármacos endectocidas: avermectinas y milbemicinas [Endectocide drugs: avermectins and milbemycins]. En: Botana, L.M., Landoni, F., Matín-Jiménez, T. (Eds.), *Farmacología y Terapéutica Veterinaria.* McGraw-Hill-Interamericana, Madrid, pp. 545-558.

Lohmeyer, K.H., Davey, R.B., and Pound, J.M. (2012). Therapeutic and residual efficacy of a pour-on formulation of novaluron against *Rhipicephalus* (*Boophilus*) *microplus* (Acari: Ixodidae) on infested cattle. *Journal of Entomological Science,* 47(3): 238-246.

Lopes, W.D.Z., Chiummo, R.M., Vettorato, L.F., de Castro Rodrigues, D., and Sonada, R.B. (2017). The effectiveness of a fixed-dose combination pour-on formulation of 1.25% fipronil and 2.5% fluazuron against economically important ectoparasites and associated pharmacokinetics in cattle. *Parasitology International,* 66(5): 627-634.

Marimuthu, S., Rahuman, A.A., Jayaseelan, C., Kirthi, A.V., Santhoshkumar, T., Velayutham, K., Bagavan, A., Kamaraj, C., Elango, G., Iyappan, M., Siva, C., Karthik, L., and Rao, K.V., (2013). Acaricidal activity of synthesized titanium dioxide nanoparticles using *Calotropis gigantea* against *Rhipicephalus microplus* and *Haemaphysalis bispinosa. Asian Pacific Journal of Tropical Medicine.* 6(9), 682-688.

Marimuthu, S., Rahuman, A.A., Rajakumar, G., Santhoshkumar, T., Kirthi, A.V., Jayaseelan, C., Bagavan, A., Zahir, A.A., Elango, G., and Kamaraj, C., (2011). Evaluation of green synthesized silver nanoparticles against parasites. *Parasitology Reseach,* 108: 1541-1549.

Mendonça, R.P. (2010). Atividade *endectocida, segurança clinica e farmacocinetica de residuos de uma nova alternativa terapeutica* (*fluazuron + abamectina*) *em bovinos* [*Endectocidal activity, clinical safety and pharmacokinetic residues of a new therapeutic alternative* (*fluazuron + abamectin*) *in cattle*]. Ph.D. Thesis, Universidade Estadual Paulista, Faculdade de Ciências Agrárias e Veterinárias, Unesp, Campus de Jaboticabal. Brasil.

Miller, R.J., Almazán, C., Ortíz-Estrada, M., Davey, R.B., George, J. E., and De León, A. P. (2013). First report of fipronil resistance in *Rhipicephalus* (*Boophilus*) *microplus* of Mexico. *Veterinary Parasitology,* 191(1-2): 97-101.

Miller, R.J., Li, A.Y., Tijerina, M., Davey, R.B., and George, J.E. (2008). Differential response to diazinon and coumaphos in a strain of *Boophilus microplus* (Acari: Ixodidae) collected in Mexico. *Journal of Medical Entomology,* 45(5): 905-911.

Mullens, B.A., Murillo, A.C., Zoller, H., Heckeroth, A.R., Jirjis, F., and Flochlay-Sigognault, A. (2017). Comparative *in vitro* evaluation of contact activity of fluralaner, spinosad, phoxim, propoxur, permethrin and deltamethrin against the northern fowl mite, *Ornithonyssus sylviarum*. *Parasites and Vectors,* 10(1): 1-7.

Narahashi, T. (2001). Recent progress in the mechanism of action of insecticides: pyrethroids, fipronil and indoxacarb. *Journal of Pesticide Science,* 26(3): 277-285.

Nogueira, S.N.L., da Silva, M.F., de Andrade Furtado, R., Júnior, D.P., Soares, M.C., Andrade, M.M.A., Nascimento, E.G., Lopes-Ferreira, L., Soares, V.D., Welber, D.Z.L., and de Mendonça, R.P. (2020). *In vitro* test for the evaluation of the efficacy of topical products for the control of *Cochliomyia hominivorax* larvae. *Parasitology,* 147(7): 816-821.

Nuñez, J.L., Muñoz-Cobeñas, M.E., and Moltedo, H.L. (1985). Tick Control off the Host. In: *Boophilus microplus* Springer, Berlin, Heidelberg, pp. 153-160.

Pérez-González, I.E., Prado-Ochoa, M.G., Muñoz-Guzmán, M.A., Vázquez-Valadez, V. H., Velázquez-Sánchez, A.M., Avila-Suárez, B.L., Cuenca-Verde, C., Angeles, E., and Alba-Hurtado, F. (2014). Effect of new ethyl and methyl carbamates on *Rhipicephalus microplus* larvae and adult ticks resistant to conventional ixodicides. *Veterinary Parasitology,* 199(3-4): 235-241.

Prado-Ochoa, M.G., Abrego-Reyes, V.H., Velázquez-Sánchez, A.M., Muñoz-Guzmán, M. A., Ramírez-Noguera, P., Angeles, E., and Alba-Hurtado, F. (2014a). Subchronic toxicity study in rats of two new ethyl-carbamates with ixodicidal activity. *BioMed research international,* 2014.

Prado-Ochoa, M.G., Gutiérrez-Amézquita, R.A., Abrego-Reyes, V.H., Velázquez-Sánchez, A.M., Muñoz-Guzmán, M.A., Ramírez-Noguera, P., Angeles, E., and Alba-Hurtado, F. (2014b). Assessment of acute oral and dermal toxicity of 2 ethyl-carbamates with activity against *Rhipicephalus microplus* in rats. *BioMed Research International*, 2014.

Prado-Ochoa, M.G., Muñoz-Guzmán, M.A., Abrego-Reyes, V.H., Velázquez-Sánchez, A.M., Lara-Rocha, M., Cuenca-Verde, C., Angeles, E., and Alba-Hurtado, F. (2013). Effect of new ethyl and methyl carbamates on biological parameters and reproduction of the cattle tick *Rhipicephalus microplus*. *Veterinary Parasitology,* 194(1): 49-57.

Prado-Ochoa, M.G., Muñoz-Guzmán, M.A., Vázquez-Valadez, V.H., Velázquez-Sánchez, A.M., Salazar, A.M., Ramírez-Noguera, P., Angeles, E., and Alba-Hurtado F. (2016). Genotoxicity and cytotoxicity assessment of new ethyl-carbamates with ixodicidal activity. *Mutation Research/Genetic Toxicology and Environmental Mutagenesis,* 807: 8-14.

Prado-Ochoa, M.G., Ramírez-Noguera, P., Díaz-Torres, R., Garrido-Fariña, G.I., Vázquez-Valadez, V.H., Velázquez-Sánchez, A.M., Muñoz-Guzmán, M.A., Angeles, E., and Alba-Hurtado, F. (2014c). The action of two ethyl carbamates on acetylcholinesterase and reproductive organs of *Rhipicephalus microplus*. *Veterinary Parasitology,* 199(3-4): 215-224.

Pundir, C.S., and Malik, A. (2019). Bio-sensing of organophosphorus pesticides: A review. *Biosensors and Bioelectronics,* 140: 111348.

Rajakumar, G., and Rahuman, A.A., (2012). Acaricidal activity of aqueous extract and synthesized silver nanoparticles from *Manilkara zapota* against *Rhipicephalus* (*Boophilus*) *microplus*. *Research in Veterinary Science,* 93: 303-309.

Rajakumar, G., Rahuman, A.A., Roopan, S.M., Chung, I.M., Anbarasan, K., and Karthikeyan, V., (2015). Efficacy of larvicidal activity of green synthesized titanium dioxide nanoparticles using *Mangifera indica* extract against blood-feeding parasites. *Parasitology Research,* 14: 571-581.

Rajakumar, G., Rahuman, A.A., Velayutham, K., Ramyadevi, J., Jeyasubramanian, K., Marikani, A., Elango, G., Kamaraj, C., Santhoshkumar, T., Marimuthu, S., Zahir, A.A., Bagavan, A., Jayaseelan, C., Kirthi, A.V., Iyappan, M., Siva, C. (2013). Novel and simple approach using synthesized nickel nanoparticles to control blood-sucking parasites. *Veterinary Parasitology,* 191(3-4): 332-339.

Ramyadevi, J., Jeyasubramanian, K., Marikani, A., Rajakumar, G., Rahuman, A.A., Santhoshkumar, T., Kirthi, A.V., Jayaseelan, C., and Marimuthu, S. (2011). Copper nanoparticles synthesized by polyol process used to control hematophagous parasites. *Parasitology Research,* 109(5): 1403-1415.

Rechav, Y., and Kostrzewski, M. W. (1991). The relative resistance of six cattle breeds to the tick *Boophilus decoloratus* in South Africa. *Onderstepoort Journal of Veterinary Research,* 58: 181-186.

Reck, J., Klafke, G.M., Webster, A., Dall'Agnol, B., Scheffer, R., Souza, U.A., Corassini, V.B., Vargas, R., dos Santos, J.S., and de Souza Martins, J.R. (2014). First report of fluazuron resistance in *Rhipicephalus microplus*: a field tick population resistant to six classes of acaricides. *Veterinary Parasitology,* 201(1-2): 128-136.

Rodríguez-Vivas, R.I., Arieta-Román, R.J., Pérez-Cogollo, L.C., Rosado-Aguilar, J.A., Ramírez-Cruz, G.T., and Basto-Estrella, G. (2010). Uso de lactonas macrocíclicas para el control de la garrapata *Rhipicephalus* (*Boophilus*) *microplus* en el ganado bovino [Use of macrocyclic lactones for the control of the *Rhipicephalus* (*Boophilus*) *microplus* tick in cattle]. *Archivos de Medicina Veterinaria,* 42(3): 115-123.

Rodríguez-Vivas, R.I., Grisi, L., Pérez de León, A.A., Silva-Villela, H., Torres-Acosta, J. F., Fragoso-Sánchez, H., Romero-Salas, D., Rosario-Cruz, R., Saldiema, F. García-Carrasco, D. (2017). Potential economic impact assessment for cattle parasites in Mexico. Review. *Revista Mexicana de Ciencias Pecuarias,* 8(1): 61-74.

Rodríguez-Vivas, R.I., Jonsson, N.N., and Bhushan, C. (2018). Strategies for the control of *Rhipicephalus microplus* ticks in a world of conventional acaricide and macrocyclic lactone resistance. *Parasitology Research,* 117(1): 3-29.

Rodríguez-Vivas, R.I., Li, A.Y., Ojeda-Chi, M.M., Trinidad-Martínez, I., Rosado-Aguilar, J.A., Miller, R.J., and de León, A.P. (2013). *In vitro* and *in vivo* evaluation of cypermethrin, amitraz, and piperonyl butoxide mixtures for the control of resistant *Rhipicephalus* (*Boophilus*) *microplus* (Acari: Ixodidae) in the Mexican tropics. *Veterinary Parasitology,* 197(1-2): 288-296.

Rodríguez-Vivas, R.I., Quiñones, A.F., and Fragoso, S.H. (2005). Epidemiología y control de la garrapata *Boophilus* en México [Epidemiology and control of the Boophilus tick in

Mexico]. En: Rodríguez-Vivas, R.I. (Ed.), *Enfermedades de Importancia Económica en Producción Animal*. McGraw-Hill, México DF, pp. 571-592.

Shaw, R.D. (1969). Tick control on domestic animals (i) A brief history of the economic significance of tick infestations. *Tropical Science,* 11: 113-119.

Song, J.H., and Narahashi, T. (1996). Modulation of sodium channels of rat cerebellar Purkinje neurons by the pyrethroid tetramethrin. *Journal of Pharmacology and Experimental Therapeutics,* 277(1): 445-453.

Springell, P.H. (1974). The cattle tick in relation to animal production in Australia. *World Animal Review,* 10: 19-23.

Thullner, F., Willadsen, P., and Kemp, D. (2007). Acaricide rotation strategy for managing resistance in the tick *Rhipicephalus* (*Boophilus*) *microplus* (Acarina: Ixodidae): laboratory experiment with a field strain from Costa Rica. *Journal of Medical Entomology,* 44(5): 817-821.

Tingle, C.C., Rother, J.A., Dewhurst, C.F., Lauer, S., and King, W.J. (2003). Fipronil: environmental fate, ecotoxicology, and human health concerns. In: Ware, G.W. (Ed.), *Reviews of Environmental Contamination and Toxicology.* Springer, New York, pp. 1-66.

Ujihara, K. (2019). The history of extensive structural modifications of pyrethroids. *Journal of Pesticide Science,* 44(4): 215-224.

WHO Expert Committee on Vector Biology and Control, & World Health Organization. (1992). *Vector resistance to pesticides: fifteenth report of the WHO Expert Committee on Vector Biology and Control* [*meeting held in Geneva from 5 to 12 March 1991*]. World Health Organization.

In: The Entomological Guide to *Rhipicephalus* ISBN: 978-1-53619-619-1
Editors: Sachin Kumar, Raquel Cossio Bayugar et al.

Chapter 8

TYPES OF ACARICIDE RESISTANCE

Raquel Cossio-Bayugar[1,*], Estefhan Miranda-Miranda[1], Hugo Aguilar Diaz[1] and Enrique Reynaud[2]

[1]Centro Nacional de Investigaciones Disciplinarias en Salud Animal e Inocuidad, Instituto Nacional de Investigaciones Forestales Agrícolas y Pecuarias (INIFAP), Jiutepec, Morelos, México

[2]Departamento de Genética del Desarrollo y Fisiología Molecular, Instituto de Biotecnología Universidad Nacional Autónoma de México, Cuernavaca, Morelos, México

ABSTRACT

The cattle tick infestations and the bovine infectious diseases transmitted by them, are controlled mainly by the massive use of acaricides on the livestock, however, this strategy has led to the emergence of tick resistance to most acaricide formulations, and the lack of acaricide effectiveness has become a severe threat to the strategies to control of ticks and tick borne diseases, consequently, it is a necessity to monitor the spread of acaricide resistancein order to design preventive and alternative measures of tick control in endemic areas, for a long time this has been done by toxicological bioassays using living field-isolated ticks of various development stages exposed to the different acaricide formulations, these procedures are time-consuming and expensive because they require abundant specialized labor and facilities, therefore, the use of fast molecular test is required to overcome the disadvantages of outdated bioassays. The design of a molecular test requires in the first place, the identification of the molecular mechanism of acaricide resistance in the cattle tick to the different acaricide chemical formulations, acaricide resistance is a genetically inherited phenomenon that involves the expression products of one or several genes. Two major genetic-driven mechanisms are recognized nowadays, namely: a) enhanced metabolic detoxification involving an activity increase of one or several xenobiotic metabolizing enzymes, b) altered target sites involving DNA mutations of the genes that code for the proteins that are targeted by the acaricide, this chapter describes themolecular and biochemical mechanisms that cause resistance to the most common acaricides used for tick control.

[*]Corresponding Author's Email: cossio.raquel@inifap.gob.mx.

Keywords: tick control, acaricide resistance, resistance mechanisms, metabolic resistance and target-site insensitivity

INTRODUCTION

Ixodid ticks are medically important vectors of serious diseases in humans and livestock, this blood-sucking ectoparasites are responsible for zoonotic illnesses such as Lime disease, that plague wildlife and humans in the northern hemisphere, and bovine diseases like theileriosis, babesiosis, anaplasmosis and cowdriosis, also known as heartwater, affecting the production of cattle and small ruminants in Latin America, Australia and Africa (Domingos et al., 2013). Ixodid ticks are classified among different genera such as: *Hyalomma* spp, *Amblyomma* spp, *Haemaphysalis* spp, and *Rhipicephalus* spp, have a major role as vectors of pathogens (Domingos et al., 2013). *Theileria parva, T. annulate, T orientalis/buffeli* are the causative agents of the disease known as Theileriosis in bovines (Domingos et al., 2013; Cossio-Bayugar et al., 2002; Hammer et al., 2015) while *Theileria ovis, T. separata, T. uilenbergi, T. recondita, T. lestoquardi, T. luwenshuni* and *T annulata*are the causative agent of theileriosis in sheep and goat (Li et al., 2009; Li et al., 2014; Schnittger et al., 2000; Zhang et al., 2014; Chae et al., 1999); the protozoan parasite *Babesia bovis* and *bigemina* causes babesiosis in cattle while *B. motasi, B. crassa,* and *B. ovis* causes the disease in small ruminants (Dehkordiet al., 2010), as well as *Cowdriasis* caused by the rickettsias *Ehrlichia ruminantium* and *Ehrlichia minasensis* (Domingos et al., 2013) and anaplasmosis by the rickettsias *Anaplasma marginale, A. centrale, A. ovis, A. bovis, A. phagocytophilum, A. capra* (Rennekeret al., 2013; Zhan et al., 2010; Liu et al., 2012; Kocan et al., 2004; Li et al., 2015). *Rhipicephalus* (*Boophilus*) *microplus* is the most important livestock tick worldwide and is the major vectors of Babesiosis and Anaplasmosis (Domingos et al., 2013; Abbas et al., 2014). This specie and other of the subgenus *Boophilus* produce massive infestations in cattle and other ungulates. (Nicholson et al., 2019).

Rhipicephalus microplus ticks are obligated blood sucking bovine ectoparasites that have an adverse impact on the cattle industry, which is reflected on reduced meat and milk production, animal weight losses, depression of immune functions and direct harm on the skin of the animals while feeding, reducing the quality of the leather derivatives. The cattle tick is controlled by massive use of pesticides, this practice has become the major strategy to prevent tick damages and tick-borne diseases (Abbas et al., 2014). Pesticides usedto prevent tick infestation of the cattle are known as acaricides and are the sole or the major component of integrated ticks control methods worldwide (Domingos et al., 2013), the reasons for this chemicals to be the major strategy to control tick and tick-borne diseases (TTBD), is that they are relatively nontoxic to mammal and because they offer a fairly quick and cost-effective method for the suppression of tick populations (Abbas et al., 2014).

However, the high intensity and prolonged use of acaricides for tick control, has a major disadvantage, as large populations of ticks are exposed to the different acaricide formulations, a Darwinian selection of acaricide-tolerant and acaricide-resistant ticks occurs and resistant phenotypes tend to emerge, eventually producing homogenous resistant tick populations to practically all major classes of acaricides (Rodriguez-Vivas et al., 2018), once the acaricide resistance is present in a cattle grazing area, tick control is compromised adversely affecting cattle production and the capacity to control this blood sucking parasite, it is therefore a major challenge to identify and avoid the spread of acaricide resistance, when tick control relies only on cattle pesticide treatment. Evidently, the key to avoid acaricide resistance is the monitoring of the presence or absence of resistant ticks to specific acaricides, making the selection of pesticide formulation for a specific area a scientific choice, thus avoiding the guessing that normally plagues the selection and use of pesticides. In this chapter we explain the mode of action of the main acaricides used for tick control and the molecular type of resistance that have being associatedwith resistance to the most common used acaricides.

TYPES OF RESISTANCE

Pesticides are poisonous substances widely used in agriculture, livestock industry and medical disciplines, when pesticides are used to kill insects, they are called insecticides, when these chemicals are used to kill mites and ticks, they are called acaricides, and ixodicides when used specifically to kill ixodid ticks. Several formulations of ixodicides have been designed by the specialized industry, they represent a large segment of the market for the pesticide brands and the major acaricides formulations used for tick control are depicted on Table 1. The classification used on the table is based on the Insecticide Resistance Action Committee (IRAC) classification (IRAC, 2020; Sparks and Nauen, 2015). The first generation of Ixodicides was constituted by several compounds of arsenic salts, witch were soon discontinued due to their obvious poisonous effects on the cattle and the environment (George et al., 2004). The second generation of acaricides group synthetic organochlorines, close related to Dichlorodiphenyltrichloroethane, better known as DDT, as well as carbamates and organophosphates, the latter related to the nervous agent phosgene, used as weapon during WWI. The third generation contains synthetic pyrethroids and amidines. The most recent generation of ixodicides are less toxic for animals and environment, constituted by phenypyrazoles, insect growth regulators (IGR) and macrocyclic lactones, this compounds nowadays are the major class of acaricides used in ticks control programs worldwide (Rodriguez-Vivas et al., 2018; Abbas et al., 2014), most of the before mentioned ixodicides affect the nervous system of the ticks, for example the target site for organophosphate and carbamates are the tick's acetylcholinesterase, for organochlorides and fipronilare theγ-aminobutyric acid (GABA) gated chloride channels (Bloomquist, 1993; Bloomquist, 2003), for pyrethroids the target site is the voltage-gated para-sodium channel of the axonal membranes (Dong et al., 2014),and for macrocyclic lactoneare the glutamate-gated chloride channel (Glu-GCC) (Lumaret et al., 2012;Omura, 2008) and the γ-aminobutyric acid-gated chloride channel (GABA-GCC) (Bloomquist, 2003).

The IRAC define resistance as "*a heritable change in the sensitivity of a pest population that is reflected in the repeated failure of a product to achieve the expected level of control when used according to the label recommendation for that pest species*" (IRAC, 2020). The cattle tick is consider between the 12 top species of arthropods to generate resistance to different pesticide products, having developed resistance to more than 44 chemical formulations and by 2014 were reported 167 cases of resistance to one or more acaricide compounds (Sparks and Nauen, 2015). It is a conventional procedure when reporting results of acaricide resistance bioassays, to mention two types of resistance; the phenotypic resistance, that are the toxicologic profiles of the tick strains or field isolates to a specific acaricide formulation, this is used for field surveillance reports to detect those acaricides that fail to efficiently control the ticks under laboratory conditions, this results are reported as mortality rate of tick larvae under various acaricide concentrations and compared to reference sensitive and resistant ticks strains. The second type of resistance describes the molecular mechanism involved on this process and is related to the expression level or structural changes of the gene products involved in the resistance process as determined by a molecular detection assay such as gene sequencing, polymerase chain reaction (PCR) or proteomic assay. This second type of resistance is a major scientific challenge, since it requires the identification and description of the molecular mechanisms underlying the observed acaricide resistance detected by the phenotypic resistance to each type of acaricides. The reasoning behind the search of the molecular resistance is that we may be able to pinpoint the molecule or molecules responsible for each specific type of acaricide resistance so we can eventually have a faster and cheaper diagnostic of resistance by mean of a molecular test, in theory, the knowledge of the precise mechanism that make a resistant tick, should help us to develop strategies to use alternative acaricides or integrated methods for the control of acaricide-resistant ticks, this strategy should also give us the ability to monitor the emergence of resistance alleles and predict which acaricides will stop functioning for tick control in a region and have a more efficient use of acaricides.

Table 1. The IRAC classification of the major classes of acaricides used to control ticks

IRAC Main group Primary site of action MOA Group descriptors	Chemical group	Commercial Acaricides Brands
Acetylcholinesterase (AChE) inhibitors Nerve action Inhibit AChE, causing muscle hyperexcitation or tetanization. AChE is the enzyme that terminates the action of the excitatory neurotransmitter acetylcholine at nerve synapses.	Carbamates	Carbaryl, propoxur
	Organophosphates	Ethion, Chlorpyrifos, Chlorfenvinphos, Coumaphos, Diazinon
GABA gated chloride channel blockers (antagonist) Nerve action Block the GABA-Gated Chloride Channel (GABA-GCC), GABA is the major inhibitory neurotransmitter in insects, causing hyperexcitation and convulsions.	Cyclodiene Organochlorines	Chlordane, Lindane, Dieldrin
	Phenylpyrazoles	Fipronil, Pyripol
Voltage gated sodium channel modulator Nerve action Keep sodium channels open, causing hyperexcitation and, in some cases, nerve block. Sodium channels are involved in the propagation of action potentials along nerve axons.	Pyrethroids and Pyrethrins	Cypermethrin, Deltamethrin, Cyhalothrin, Flumethrin, Permethrin, Cyfluthrin
Glutamate Chloride Channel Activators Nerve and Muscle Targets Allosterically activate glutamate-gated chloride channels (Glu-GCC), Glutamate is an important inhibitory neurotransmitter in insect causing paralysis.	Avermectins Milbemycins	Ivermectin Doramectin, Selamectin, Abamectin, and Eprinomectin Milbemectin Moxidectin, Milbemycin Oxime
Octopamine Receptor Nerve Action Activate octopamine receptors, leading to hyperexcitation. Octopamine is the insect equivalent of adrenaline, the fight-or-flight neurohormone.	Formamidines	Amitraz
Inhibitors of chitin biosynthesis affecting CHS1 (chitin synthases 1)	Benzoylureas	Fuazuron

MOLECULAR MECHANISMS OF ACARICIDE RESISTANCE

Acaricide resistance is inherited, this means that that it is genetic in nature and there are genetic changes that cause resistant phenotypes through changes in the expression levels or the structure of a protein or a functional RNA, importantly, it is reasonable to assume that acaricide resistance involvesthe expression products of one or several genesacting at the same time. It is evident when reviewing the scientific literature on the subject that there are two major types of inheritable or genetic resistance mechanisms identified and characterized in ixodid ticks (Table 2) which are:

- *Enhanced metabolic detoxification resistance,* this type of resistance is mediated by xenobiotic modifying enzymes (XMEs), by biotransformation or sequestration of acaricides, in this type of resistance, the target of the acaricide it is not involved, instead, the culprit may be an increase in the activity of metabolic detoxification pathways were the chemical structure of the ixodicide is modified by the action of altering its structure and toxic effect (Dzul et al., 2007). Increased enzyme activity can be caused by gene amplification, increased upregulation, coding sequence mutations or by a combination of these mechanisms (Karunaratne et al., 2018). Several studies suggest that the arthropod pesticide resistance is based principally on the action of XMEs (Parkinson, 1996). It is because of this phenomenon that most acaricide resistance studies focus intensively on the expression of XMEs such as cytochrome P450, glutathione-S-transferase, carboxylesterase and cholinesterase (Parkinson, 1996). The XMEs biochemically transform the endogenous and exogenous xenobiotics like the ixodicides permethrin, rendering phenoxyphenyl-methanol and dichloroethenyl-dimethylcyclopropanecarboxylate which are less toxic and easily eliminated, this metabolic process is known as biotransformation and usually are described as a process were a first line of XMEs modify the pesticide that now are eliminated and secreted.
- *Target site modification,* this implicates a change in the structure of the target site by means of a DNA mutation that leads to an amino acid substitution that modifies the pesticide affinity to the target and therefore its toxicity (Bull and Ahrens, 1988), several reports on altered target sites are available for acaricides based on carbamates, organophosphates and pyrethroid formulations.
- *Reduced penetration,* this type of resistance requires changes in chemical composition or thickening of the cuticle, it is mediated by enzymes that are responsible for the synthesis of the cuticle like chitin Synthase causing reduced penetration of acaricides to the hemolymph, chitin Synthase is the target molecule of benzoylureas acaricides (Douris et al., 2016; Sun et al., 2015), and it should be considered a target modification-type of resistance and the presence of carboxylesterases type B in the integument (Villarino et al., 2001; Villarino et al., 2003), which should be considered an enhanced metabolic detoxification-type of resistance, however it is the less studied acaricide mechanism and more studies are required before making a assessment of acaricide-resistance classification.

Table 2. Types of resistance in ticks

Resistance Mechanism	Description	Chemical group that presented the type of resistance	Molecules involved in resistance
Metabolic detoxification (enzymatic)	Resistance mediated by xenobiotic modifying enzymes (XMEs), working in enhanced detoxification by biotransformation or sequestration of acaricides	Organophosphate Carbamates	Esterases, GST[a]
		Pyrethroids	Esterase, P450, GST [b]
		Formamidines	Monoamine oxidases, esterases, GST. ATP binding cassette transporter activity [c]
		Macrolytic lactones	esterase, glutathione-S-transferases, cytochrome P450 and ABC transporters [d]
Reduced sensitivity at the site of action	Alteration of the structure of the target site by mean of a DNA single point mutation or/and amino acid substitution modifying the pesticide affinity to the target and the toxicity of it	Organophosphate Carbamates	Modified AChE[e]
		Pyrethroid	Modified VGSC[f]
		Cyclodiene Organochlorines	Modified GABA-GCC [g]
		Phenylpyrazoles	Modified GABA-GCC[h]
		Formamidines	Modified OR [i] Modified O/TyrR[j]
		Macrolytic Lactones	Modified Glu-GCC (No yet reported for ticks)
Reduced penetration of acaricide	Changes in chemical composition or thickening of cuticle causing reduced penetration of acaricides	Organophosphate	Synthase increased activity, Esterase expression in integument[k]

AChE Acetylcholinesterase, VGSC Voltage gated sodium channel modulator, GABA-GCC GABA gated chloride channel, OR Octopamine receptor, O/TyrR Octopamine/Tyramine receptorGlu-GCCGlutamate gated chloride channel, [a](Fular et al., 2020;Baxter and Barker, 2002; Miranda et al., 1995; Baffi et al., 2008), [b](Cossío-Bayúgar et al., 2018; Cossío-Bayúgar et al., 2009; De Jersey et al., 1985; Baffi et al., 2008),[C] (Jonsson et al., 2018;Lara et al., 2015)[d](Le Gall et al., 2018), [e](Temeyer et al., 2019; Temeyer et al., 2007), [f] (Stone et al., 2014;Morgan et al., 2009;He et al., 1999;Cossío-Bayúgar et al., 2020), [g](Tian et al., 2019),[h] (Tian et al., 2019),[i] (Baron et al., 2015), [j] (Corley et al., 2013;Jonsson et al., 2018),[k] (Sun et al., 2015;Douris et al., 2016;Villarino et al., 2003).

Acaricide Cross Resistance

The research on acaricide resistance highlights the importance of knowing the acaricide mode of action (AMOA) at the molecular level, it is only then; when the reasons for the occurrence of acaricide cross-resistance becomes evident, and it also allows to explain why different acaricides with a similar mode of action exhibit cross-resistance, and the reasons for acaricides structurally related are ineffective to show cross-resistance. Acaricides with similar structure usually share a common target site within the tick (same AMOA), and is common that when the interaction of an acaricide with is target site is altered, will confer cross-resistance to molecules with the same target site (IRAC, 2020; Abbas et al., 2014; Janadaree Bandara and Parakrama Karunaratne, 2017). This pattern has being observed among organophosphate and carbamate in strains of *R. microplus* and field samples (Li et al., 2005; Madder et al., 2011; Abbas et al., 2014; Janadaree Bandara and Parakrama Karunaratne, 2017). It is important to consider to the use rotation or alternation of different groups of acaricides that have no cross resistance in order to reduce the selection pressure for resistance of any particular acaricide group (Abbas et al., 2014).

Acaricide Resistance to Organophosphate and Carbamates

Acaricides of the organophospate and carbamate families target acetylcholinesterase (AChE). The AChE is present in a great variety of organism including vertebrates and arthropods making its inhibitors highly toxic for both groups of animals, this enzyme regulates cholinergic nervous transmission by hydrolyzing the neurotransmitter acetylcholine into choline and acetate in the vertebrate neuromusclular synaptic cleft and in the arthropod nervous system (NS) (Boublik et al., 2002). The organophosphates (OP) and carbamates, reacts with a serine inside the active site of the AChE, producing a quasi-irreversible phosphorylation and inactivating the enzyme (Figure 2A; Wright and Ahrens, 1988). The canonical example of target modification of this type of resistance is a modified AChE that is capable of resist phosphorylation of the active site. This mechanism had being identified in several organisms: *Musca domestica* (Kim et al., 2003), *Culex pipiens* (Weillet al., 2002), *Anopheles albimanus* (Dzulet al., 2007) *Drosophila* sp. (Boublik et al., 2002) and in *R. microplus* (Wright and Ahrens, 1988; Temeyer et al., 2007). An example of an enhanced metabolic detoxification is an overexpression of a putative AChE that may be a carboxyl esterase the only purpose of this overexpression is to sequester and detoxify the excess Organophosphate or carbamate in the hemolymph (Miranda et al., 1995; Baxter and Barker, 2002; Cossío-Bayúgar et al., 2009). The carboxylesterases (CE) are enzymes close related with the AChE and part of the XMEs may also produce increased acaricide detoxification thru the ester hydrolysis organophosphate (Hemingway and Karunaratne, 1998). The AChE and CE share a close amino acid identity of up to 70% and are very ubiquitous in arthropods, however, unlike the AChE, the CE may hydrolyze the phospho-ester bonds within the acaricide and both enzymes share affinity for the same synthetic subtracts (Manchenko, 2003). The CE are involved on the metabolism of xenobiotic compounds either by sequestering or hydrolyzing the acaricides (Scott, 1999; Karunaratne and Hemingway, 2001). Likewise CEs have been reported in several arthropods species in pesticide resistance, like the sheep ectoparasite *Lucilia cuprina* (Newcomb et al., 1997; Hartley et al., 2006), the hematophagous mosquitoes *Culex* sp., *Aedes* sp. and *Anopheles* sp. (Hemingway and Karunaratne, 1998; Paton et al., 2000; Karunaratne and Hemingway, 2001); and in the peach-potato aphid *Myzus persicae* (Field and Devonshire, 1998;Field et al., 1999). In the cattle tick has being reported an increased esterase activity in OP resistant strains (Miranda et al., 1995; Rosario-Cruz et al., 1997; Brito et al., 2017; Fular et al., 2020; Baxter and Barker, 2002; Nagar et al., 2016; Baffi et al., 2008).

In the cattle tick, it was first reported that OP and carbamate resistance was partially due to altered AChE insensitivity to common OP in the resistant strains, this was determined by enzymatic activity and inhibition studies with OP (Wright and Ahrens, 1988; Baxter et al., 1999; Pruett, 2002; Li et al., 2003; Li et al., 2005). In *R microplus* has being reported at least three transcripts encoding AChE; dubbed *BmAChE1* (Baxter and Barker, 2002), *BmAChE2* (Hernandez et al., 1999) and *BmAChE3* (Temeyer et al., 2004). These enzymes were expressed in the baculovirus system and biochemically characterized confirming their role as

AChE with synthetic substrates preference for acetylcholine with different sensitivity and affinity constants and sensitivity and inhibition by AChE specific inhibitors like eserin and the OP acaricides Maloxon and Paroxon (Temeyeret al., 2006; Temeyer et al., 2010) and the three are expressed on the synganglion, with different expression levels suggesting different functions (Temeyer et al., 2012). The *BmAChE1* has been demonstrated to also be expressed on ovaries (Baxter and Barker, 2002) and on the salivary glands (Baxter and Barker, 2002), with a probable function on tick feeding and/or pathogen transmission (Temeyer and Tuckow, 2016). Several mutations have been associated with phenotypic OP resistance but rarely a biochemical confirmation has been done. BmAChE2 mutation has been previously reported in OP-resistant field-isolates and an OP resistant reference strain, but the contribution to acaricide OP-resistance of these mutation has yet to be determined (Ghosh et al., 2015). A tenfold increase in the expression of *BmAChE2* has been reported in field isolated OP acaricide-resistant Mexican ticks (Cossío-Bayúgar et al., 2009), suggesting that a target modification and an enhanced detoxification may be working at the same time for OP-resistance in the cattle tick. Expressed mutant and wildtype BmAChE1 in the baculovirus system followed by inhibition assays to determine the sensitivity to Coroxon, Paraxon and Maloxon were analyzed to determine if these mutation results in an altered inhibition to OP (Temeyer et al., 2019). Many of the mutation analyzed varied its sensibility depending on the OP inhibitor, most of them do not achieved substantial reduced sensitivity resistance by themselves with the exemption of one AChE dubbed D188G it shows an amino acid (AA) substitution at the position 188 from Aspartic Acid (D) to Glycine (G), that exhibited a significant reduced sensitivity to OP inhibition and 181 fold resistance to Coroxon, but little or no resistance to Maloxon or Paroxon, the other AChE mutations that have proved a statistically significant resistance to Coroxon were, dubbed S282G showing an AA substitution at position 282 from Serine (S) to Glycine (G), this mutation also produced Malaoxon cross-resistance, F390S, showing an AA substitution at position 390 from Phenylalanine (F) to Serine (S),and G142S, showing an AA substitution at position 142 from Glycine (G) to Serine (S) this mutation also produced Malaoxon cross-resistance (Temeyeret al., 2019). The first mutation found in *BmAChE3* gene associated with OP resistance as reduced AChE insensitive to OP mechanisms was the R86Q showing an AA substitution at position 286 from Arginine (R) to Glutamine (Q), with an apparent 20 fold reduction in paraoxon sensitivity, this mutation that was not in the active site of the AChE, instead, it was found in the peripheral AA chain of the enzyme, which it is not sufficient to explain OP resistance by itself suggesting other mechanism may be involved (Temeyer et al., 2007). In an survey in India several mutation originating AA substitution in the AChE were found in OP-resistant isolates of the cattle tick, three of this previously reported: I48L, I54V. R86Q was also found in Mexican strains, with more frequency on resistant strains (Temeyer et al., 2007) and three new substitution found in India isolates V71A, I77M and S79P (Jyoti et al., 2016).

Besides Esterases, other XMEs show an active role in acaricide resistance of the cattle tick, Glutathion S Transferase (GST) activity, has also been associated with OP resistance as key metabolizing enzyme and in the house fly (Wang et al., 1991; Wei et al., 2001), but its low participation on diazinon resistance was reported when analyzed by the used of synergist (Li et al., 2003), GST was found to upregulate upon Coumaphos exposure on tick strains indicating a probable acaricide metabolizing activity of GST (Saldivaret al., 2008; Fular et al., 2020). This enzyme shows an increased expression in salivary glands and midgut during *R, microplus* blood feeding (Rosa de Lima et al., 2002), same kind of metabolism exhibited by the hard tick *Haemophysalis longicornis,* a probably indication of protective role during oxidative stress while blood feeding (Hernandez et al., 2018).

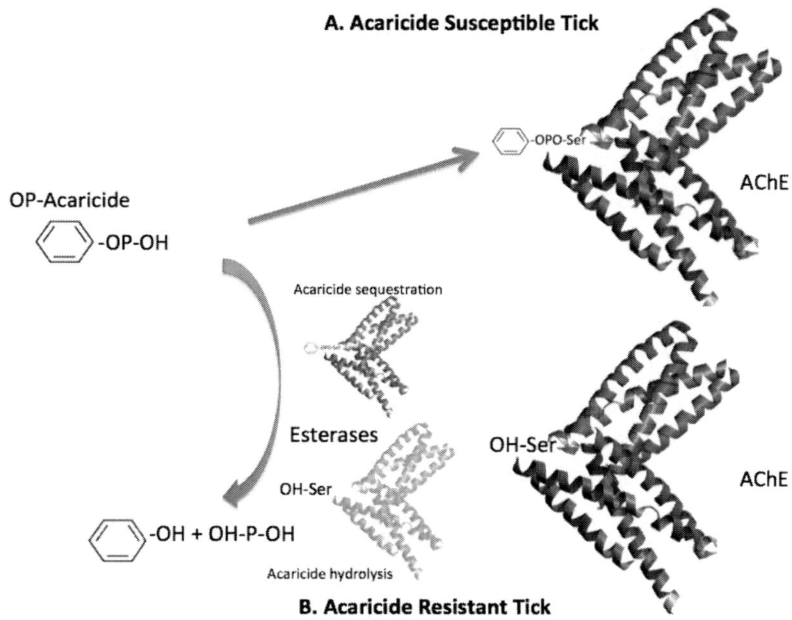

Figure 1. The organophosphate resistance involving esterase is characterized by the overexpression of these enzymes. When Esterases are involved in Organophosphate-acaricide products, a mimetism of the target molecule is produced, so the organophosphate-acaricide covalently bounds to the esterase instead of the target molecule. A) In susceptible ticks, the acaricides react covalently with active site of the AChE. B) In the case OP-acaricide resistance involving these overexpresed enzymes, interact with the acaricide by two possible mechanisms; 1 Sequestration of the acaricides by a covalent reaction with its active site, preventing the acaricide from reaching the target molecule, allowing the survival of the resistant tick. 2. Hydrolytic degradation of the acaricide into free radicals reducing its toxicity.

ACARICIDE RESISTANCE TO SYNTHETIC PYRETHROIDS

Synthetic pyrethroids are molecules derived from six natural pyrethrins that are the insecticidal constituents of extracts of *Chrysanthemum cinerariaefolium* and *C. coccineum*: Pyrethrin I and II, Cinerin I and II and Jasmolin I and II. These molecules are esters of a cyclopropanecarboxylic acid and a cyclopentenolone alcohol. Synthetic pyrethroids were developed by sequentially replacing structural elements of the original natural pyrethrins using novel chemical substitutions that conserved the molecular shape and physical properties of the original template structure. Natural pyrethrins are very efficient insecticides, however they are easily degraded by light. Commercial synthetic pyrethroids retain the rapid insecticidal activity and relatively low acute mammalian toxicity of the natural pyrethrins having increased photostability (Soderlund, 2012; Soderlund et al., 2002). Pyrethroids are quite structurally divergent and there is no specific structural motive that can be recognized as the reactive entity, or molecular moiety that confers the toxophore insecticidal activity characteristic of this family of molecules (Soderlund et al., 2002). The ability of these molecules to bind to their target apparently results from the appropriate fit of the entire molecule at its site of action, and it has been demonstrated that a wide variety of chemical structures satisfy the requirements to fit at that site. On the other hand the fact that the whole surface of the pyrethroid interacts with its target site explains the profound stereospecific effect on the insecticidal activities that these molecules have (Soderlund, 2010a). Pyrethroids are neurotoxins whose primary toxic action is thru the interference of the gating kinetics of voltage-sensitive sodium channels slowing their inactivation and therefore prolonging the sodium current and enhancing membrane depolarization (Soderlund and Bloomquist, 1989). Sodium influx and membrane depolarization induces hyperexcitability, bursts of action potentials and nerve blockage, the effects varying according to the species, nerve preparation and pyrethroid.

Pyrethroid insecticides fall into two groups, designated types I and II. These two types have different poisoning symptoms and neural actions. Examples of type I pyrethroids are: tetramethrin, allethrin, resmethrin, phenothrin, and permethrin. Type I compounds induce repetitive firing caused thru the modification of sodium and potassium channel gating kinetics causing symptoms of hyperactivity and uncoordinated movement in poisoned insects (Gammon et al., 1981). DDT, which is not a pyrethroid, is considered to have similar action to the type I pyrethroids. However, toxicity of type I pyrethroids is not completely explained by these repetitive firings suggesting that other targets may exist for these toxins, and they appear to be either intracellular or membrane bound. Type II pyrethroids have an α-cyano substituent, examples of these molecules include cypermethrin, deltamethrin and fenvalerate and their mechanism of action is different and better characterized than type I compounds. Type II pyrethroids induce a very potent depolarization of nerve axons and terminals do not cause extensive repetitive firing. The behavioral effect of type II pyrethroids is also very different, instead of displaying hyperactivity and uncoordinated movements, insects poisoned by type II pyrethroids briefly convulse and rapidly present flaccid paralysis, that resemble the advanced stages of type I poisoning (Gammon et al., 1981). The main target for type I and II pyrethroids is the voltage-gated sodium channels of the *para* family; which codes for the core alfa subunit of the channel, and it is highly conserved among animals. The most direct interpretation for these differential physiological and behavioral effects is that type II pyrethroids have more affinity and are more specific to the voltage-gated sodium channels than type I. Pyrethroids disrupt nerve function by inhibiting channel deactivation, and stabilizing its open configuration; the prolongation of the open channel configuration promotes axon depolarization that rapidly leads to axonal conduction block that stops neuromuscular transmission causing flaccid paralysis (Osborne and Hart, 1979; Adams and Miller, 1980; Salgado et al., 1983a; b; Seabrook et al., 1988; Soderlund and Bloomquist, 1989; Seabrook et al., 1989; Pepper and Osborne, 1993; Dong, 2007; Soderlund, 2010; Cossío-Bayúgar et al., 2012; Soderlund, 2012; Silver et al., 2014; Cossío-Bayúgar et al., 2015, 2020). It has been reported in many arthropods the existenceof several single nucleotide polymorphisms (SNPs) in the gene that codes for the para-sodium channel that confer pyrethroid resistance; these mutations result in the substitution of single aminoacids in its protein sequence. Phylogenetical analysis of arthropod orthologues of these channels allowed the identification of homologous mutations in the subclass *Acari* that have been shown to also correlate with pyrethroid resistance but until recent, no measurable physiological effect that correlates with the presence of these mutations and resistance was reported (Cossío-Bayúgar et al., 2020).

As mentioned before, the arthropod para-sodium gene codes for the core of the voltage gated sodium channel and even if the absence of β subunits or other channel modulating proteins it is capable forming channels that conduct Na^+in a voltage-gated way. The core channel is structured into four homologous repeat domains (I–IV), each consisting of six α-helical transmembrane segments (S1–S6) that form the pore itself, which is also the molecular binding site for DDT and pyrethroids (Davies et al., 2008; Lovis et al., 2012; Stone et al., 2014; Du et al., 2016). The reported SNPs in *Acari* including *Rhipicephalus microplus* and other arthropods that cause amino acid substitutions that correlate with resistance are always referred as homologous to the ones found in *Musca domestica* (aminoacid substitutions are listed in Table 3, *Rhicephalus microplus*; GenBank accession number: AF134216.2) although their relative amino acid position may vary depending on the species. The overall sequence and structural identity of the voltage gated sodium channel and the homologous position of the SNPs that confer pyrethroid resistance mechanistically explain the independent emergence of convergent resistant strains (He et al., 1999; Davies et al., 2008; Morgan et al., 2009; O'Reilly et al., 2014; Stone et al., 2014; Kumar et al., 2020).

Table 3. Aminoacid substitutions that confer pyrethroid resistance in *Rhipicephalus microplus*. GenBankaccession number: AF134216.2

Location in sodium channel	Nucleotide position and substitution	Aminoacid position and substitution	First detection
Domain II	C148T	L50P	(Stone et al., 2014)
	T170C	M57T	
	G184C	G62R	
Domain II S4-5 linker	C190A	L64I	(Morgan et al., 2009)
	C190G	L64V	(Stone et al., 2014)
	G215T	G72V	(Jonsson et al., 2010)
Domain III S6	T2134A	F712I	(He et al., 1999)
	T2134C	F712L	(Villar et al., 2020)

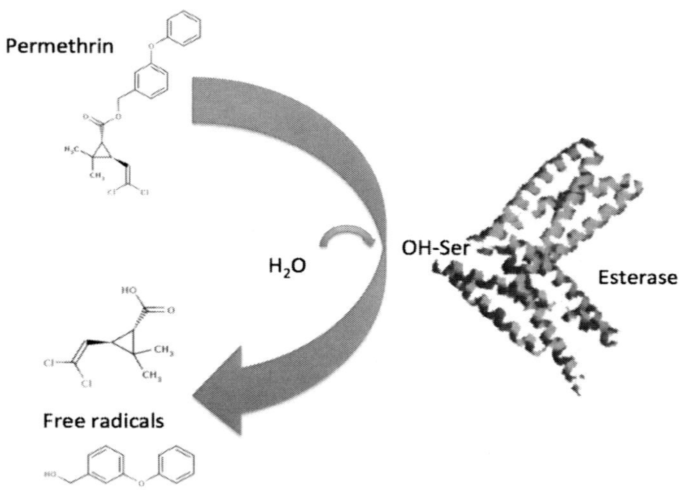

Figure 2. Pyrethroid Hydrolytic Esterases are responsible for Pyr-acaricide resistance in some strains of *R. microplus*. Some synthetic pyrethoids widely used as acaricides like Permetrin have been reported to by degraded via hydrolysis by some isoforms of carboxyl esterases in *R. microplus* (De Jersey *et al.*, 1985).

Pyrethroids are still widely used for cattle tick control although widespread Pyrethoid-resistance is reported around the world. Target site insensitivity and modification is the primary mechanism found on the literature but, enhanced metabolic detoxification is increasingly been involved as pyrethroid resistance including XME-CE, has also being demonstrated to participate on detoxification of synthetic Pyrethroid by esther bond hydrolysis of the acaricide (De Jersey et al., 1985a), increased CE enzymatic activity and gene expression has been reported in *R. microplus* pyrethroid resistant strains (Jamroz et al., 2000; Hernandez et al., 2000; Pruett et al., 2002; Cossío-Bayúgar et al., 2009; Gupta et al., 2016) and in a pyrethroid resistant field samples (Cossío-Bayúgar et al., 2018; Singh and Rath, 2014; Gaur et al., 2017; Baffi et al., 2008; Nagar et al., 2016) (Figure 2).

Monooxygenase in Organophosphate and PyrethroidResistance

The monooxygenase like the cytochrome P450 (CYP), belong to a big family of ubiquitous enzymes that regulate bioactive endogenous molecules like hormones. These enzymes are part of EMX that are involved in metabolism regulation and in the elimination of harmful chemical for the cells like plant toxins, drugs and pesticides in a large number of arthropods (Singh and Rath, 2014). Several arthropods species have naturally increased levels of CYP, for these reason are naturally insensible to pyrethroids, and have demonstrated that these chemicals are useless for their control (Scott, 1999). Several studies with agriculture pests with high tolerance or/and resistance to pyrethroid, has shown that the metabolism mediated CYP is involved on the deactivation of the pyrethroids in several resistant arthropods, such as *Culex* spp. mosquitoes (Kasai, 2004), the crop pest *Helicoverpa armiguera* (Chen et al., 2005), and house fly *Musca* spp. (Kasai and Scott, 2000; Scott and Tomita, 1995). The indirect role of CYP on organophosphate and pyrethroid resistant *R. microplus* ticks has been demonstrated by the used of the pesticide synergist piperonyl butoxide, that inhibit the mixed function oxidase enzymes and alter the detoxification metabolism with an increased on the toxicity of pesticides (OP and pyrethroids) in resistant ticks (Li et al., 2003; Miller et al., 1999). High levels of monooxygenase expresión like the CYP has been reported in pyrethroid resistant ticks, where as low levels of this enzyme was found on multi resistant ticks, suggesting that ticks modulate de the expression of this gene to counteract the effect of diverse pesticides (Guerrero et al., 2007; Cossío-Bayúgar et al., 2008; Cossío-Bayúgar et al., 2018; Graham et al., 2016). On the other hand organophosphate resistance, requires different expression levels of CYP, because this kind of acaricides are required to be enzymatically metabolized by CYP to bioactivate the pesticide to a more toxic molecule (Parkinson, 1996) this is the case for the OP coumphos and diazinon, where the tolerance or resistance in *R. microplus* has been related with a reduced genetic expression and/or enzymatic activity of CYP to avoid the biotransformation to a more toxic OP (Li et al., 2003; Cossío-Bayúgar et al., 2008; Cossío-Bayúgar et al., 2018).

RESISTANCE TO AMITRAZ

One of the most used acaricides when resistance to OP-acaricides and pyrethroids-acaricides emerge is Amitraz (George et al., 2004), a formamidine that interacts with the octopamine (OA) receptors located on the arthropods neurons (Bloomquist, 1996; Baxter and Barker, 2002). OA intervenes in many energy-demanding processes and behaviors in arthropods, such as egg laying, aggressive behavior, feeding rates and defensive jumping it is considered to be a neurotransmitter, neurohormone and neuromodulator. The Octopamine/Tyramine system appears to be analogous of the vertebrate epinephrine-norepinephrine system, where OA is analogous to epinephrine and tyramine (TA) to norepinephrine; TA can function as is either an agonist or an antagonist of OA depending on of its concentration and the receptor onto which it is binding (Roeder, 2005; Roeder et al., 2003). Amitraz blocks the oviposition, has an larvicide effect and produces tick shedding (George et al., 2004). It has been reported that the most potent molecules to block oviposition and ovary contraction are TA and other alfa adrenergic compounds (Cossío-Bayúgar et al., 2012; Cossío-Bayúgar et al., 2015). The gene reported originally as anoctopamine receptor (Baxter and Barker, 2002) was further analyzed and demonstrated to be a type tyramine/octopamine receptor (Cossío-Bayúgar et al., 2012; Gross et al., 2015).

This receptor was heterologously expressed in Chinese hámster ovary (CHO-K1) and exhibited higher affinity toward tyramine than to octopamine, amitraz has a modulatory activity on this receptor suggesting that is a potential target for other posible form amidine acaricides (Gross et al., 2015). Two single nucleotide polymorphism (SNP) were reported on this gene that were present on a resistant strain but not in a susceptible strain (Chen et al., 2007), these SNP produce two amino acid substitutions on the tyramine/octopamine receptor, one in AA 8 (threonine to proline), and the other on AA 22 (leucine toserine). A positive correlation was reported between Amitraz resistance and the presence of these SNP, in South African field population of *R microplus* (Baron et al., 2015). A beta adrenergic-like octopamine receptor (βAOR) was identified (Corley et al., 2012) and a SNP denominated *I61F* (substitution of a isoleucine to phenylalanine in AA 61) was also reported, when the 419 AA protein is aligned against the susceptible sequence, this AA substitution is located within the first transmembrane domain and also has a positive correlation with Amitraz resistance (Corley et al., 2013), further studies also confirmed the presence of this SNP in South American Amitraz resistant tick isolates was (Jonsson et al., 2018).

Both receptors: the tyramine/octopamine, and the beta adrenergic like octopamine receptor (βAOR) are likely targets of Amitraz, and the SNPs found on these genes could be responsible of Amitraz producing insensitive target site resistance. Indirect involvement of several XMEs like Monoamine oxidases and esterases has also been proposed as alternative mechanisms of Amitraz resistance (Jonsson and Hope, 2007). Using inhibitors of P450 Piperonyl Butoxide (PB) and Carboxylesterases triphenyl phosphate (TPP), induced enhanced Amitraz toxicity in the cattle tick strains analyzed, suggesting a probable role of this enzymes in Amitraz detoxification and resistance (Li et al., 2004). Up regulation of GST in an Amitraz resistant strain identified using gene microarray analysis strongly suggested that this enzyme is involved in the Amitraz resistance mechanism (Saldivar et al., 2008).

The ABC transporters are part of integral membrane glycoproteins involved in the elimination of endogenous or xenobiotic lipophilic substances from cells after prior chemical modification and conjugation to an anionic moiety and transport across the plasma membrane into the extracellular space (Homolya et al., 2003; Jonsson et al., 2018). Lara et al., (2015) also reported ATP-binding (ABC) cassette transporter which is a member of a group of this group of membrane transport proteins which is involved in Amitraz and heme detoxification in the tick midgut cells.

THE GABA-GATED CHLORIDE CHANNEL AND ACARICIDE RESISTANCE

The mode of action of acaricides such as chlorinated hydrocarbon cyclodienes (Dieldrin, Aldrin and Endrin) and Fipronil is blocking the γ-aminobutyric acid (GABA)-gated chloride channel (GABA-GCC) (Bloomquist, 1993; Cole et al., 1993) and the avermectins (ivermectin) acts as agonists of this channel (Bloomquist, 1993; Campbell et al., 1983; Bloomquist, 2003). The GABA receptor is a pentameric transmembrane chloride channel (Bloomquist, 1993), that mediates synapsis on the insects central nervous system. Each of the five subunits that form the pore has three distinctive domains: the extracellular domain, the transmembrane domain (TM1-4) that is formed by four segments of α-helices and the cytoplasmic loop between TM3 and 4 α-helices (Tian et al., 2019). The ion channel is formed by the five TM2 domains, in typical physiological conditions, the chlorine ions move from the outside of the cell into the cytoplasm lowering the neuronal resting potential and reducing the probability of induction of an action potential (Tian et al., 2019). The binding site of the noncompetitive antagonists such as Cyclodienes and Fipronil is believed to be the TM2 domains (Tian et al., 2019). Binding of these type of acaricides to the GABA channels blocks the entrance of chloride ions into the cytoplasm thus depolarizing the cell and leading to neuronal hyperexcitability which in turn induces arthropod over excitement, convulsions and death (Tian et al., 2019). Resistance to one molecule correlates with cross resistant to other pesticide molecules that have the same target site (Bloomquist, 2003). In contrast to the GABA antagonist the Avermectins cause a different poisoning syndrome that cause an ataxia or sedation effect. The Avermectines can block the action of GABA at low concentrations but cause an irreversible activation of the GABA channels at higher concentrations inducing nervous system depression. Thus, GABA antagonists produce tremors and convulsions in exposed insects while GABA agonists induce irreversible channel activation causing ataxia and sedation (Bloomquist, 2003).

RESISTANCE TO CYCLODIENE ORGANOCHLORINES

In the case of *Rhipicephalus microplus* a mutation that confers resistance to dieldrin has been reported to be in the TM2 domain, but its effect has not yet been studied (Hope et al., 2010). Changes that cause mutations on the GABA receptor TM2 have being shown to alter the binding site of dieldrin. However these changes also alter the channel electrophysiological properties in *Drosophila* (ffrench-Constant et al., 1993; ffrench-Constant et al., 2004).

RESISTANCE TO FIPRONIL

Fipronil resistance in another example of target site modification, it has been referred as a substitution of 5 different amino acids found on the GABA-GCC receptor of the *R. microplus* populations from Uruguay an Brazil, this ticks had been previously demonstrated to be resistant to fipronil and lindane (Castro Janer et al., 2019), one amino acid changeA286S/L, this is unusual since changes Alanine at position 286 to either Serine or Leucine, this substitution found in the TM2 domain is believed that is mechanistically homologous to the (A302S) change found in the same receptor's aminoacid sequence of *Drosophila* flies that are resistant to cyclodienes and phenylpyrazoles (Ffrench-Constant et al., 2004; Remnant et al., 2014). The other 4 mutations found in *R. microplus* resistant strains were (S281T, V317I, T328 A and A329S) but always found with the A286S mutation (Castro Janer et al., 2019). It has being reported that fipronil is a potent open channel blocker of glutamate-activated chloride channels in cockroach neurons (Zhao et al., 2004) but has not being reported its effect on ticks on this channel, however, this issue has to be addressed in order to found out if these type of modifications in this channel has a role in fipronil tick resistance.

RESISTANCE TO MACROCYCLIC LACTONES

Macrocyclic lactones (ML) are a group of antiparasitic molecules extensively used in veterinary medicine. They are endectocides which means that have astrong activity against endoparasites and ectoparasites (Lumaret et al., 2012; Campbell et al., 1983). They are two groups of ML commonly used for control of nematodes (roundworms) and arthropods (insects, ticks, and mites); the Avermectins derivatives known as Ivermectin, Doramectin, Eprinomectin, Abamectin and Milbemycin also known as Moxidectin (Rodriguez-Vivas et al., 2010). The avermectins and Milbemycins are structurally related and are products of fermentation of soil derived *Streptomyces avermilitis* for avermectins and *Streptomyces cyanogriseus* for milbemycins (Campbell et al., 1983). Ivermectin and abamectin are commonly used to control ecto and endoparasites of livestock and ivermectin is the most widely used in cattle (Lumaret et al., 2012).

The Avermectins and Milbemycins inhibit the Gamma-Aminobutyric Acid (GABA) neurotransmission between nervous and muscular cells in arthropods, they also bind to the the Glutamate-Gated Chloride Channel (Glu-GCC) receptors which are found only in invertebrates, Ivermectin enhances their Chloride currents thereby paralyzing pharyngeal and somatic muscles of most arthropods (Lumaret et al., 2012; Omura, 2008) and have a minor effect on the γ-aminobutyric acid (GABA)-gated chloride channel prolonging the duration of $GABA_{A}RS$ currents without altering its conductivity (Atif et al., 2017; Bloomquist, 2003). The glutamate-gated binding sites apparently occur in close proximity to GABA-gated chloride channels and the macrolide endectocides may increase GABA-gated sites as well (Lumaret et al., 2012). The avermectins and Milbemycins bind to receptors on neuronal membranes of nematodes and myoneural junctions of arthropods (Lumaret et al., 2012). The chloride ion influx lowers cell membrane resistance and produces a hyperpolarization of the post-synaptic cells. This in turn alter the neurotransmission leading to flaccid paralysis, death and/or expulsion of the parasite (Lumaret et al., 2012).

A target-site mutation was described in a *Drosophila* strain with resistance to nodulisporic acid, a natural insecticidal product isolated from an endophytic fungus and also presented and weak (threefold) cross- resistance to ivermectin (Kane et al., 2000; ffrench-Constant et al., 2016). Mutations on the glutamate-gated chloride (GluCl) channel receptors has also being reported in *T. urticae* with resistance to theavermectin miticide abamectin (Dermauw et al., 2012; Kwon et al., 2010), but no mutations has yet being reported for ticks.

Exposure of *R. annulatus* to ivermectin resulted with a decrease GST activity (Batiha et al., 2019), Le Gall et al. (2018) reported that the esterase, glutathione-S-transferases, and cytochrome-oxidases played and a role in ivermectin detoxification, but detoxification mediated by ABC transporters was probably the most important of all detoxification mechanisms analyzed.

REDUCED PENETRATION OF ACARICIDE

Cuticular resistance is frequently describe as a minor resistance mechanism and refers to the reduced uptake of the acaricide (Surbhi et al., 2018) and it is likely due to a thickening or change in chemical composition of the cuticle (Karunaratne et al., 2018), however, this process depends on an increased activity of the enzyme chitin synthase and could be interpreted as an additional type of resistance mediated by XMEs as only have been found in mites (Sun et al., 2015; Douris et al., 2016). Very few studies have been done to analyze this mechanism Schnitzerling et al. (1974), however, when reduced penetration was probed as a probable mechanism of organophosphate resistance in the cattle tick, no relevant differences between resistance and susceptible strains were found but the presence of a thicker cuticle, Villarino et al., (2003) showed increased esterase activity in the inner layer of the integument of organophosphates resistant adult female *R. microplus* ticks compared with the OP susceptible ticks suggesting that an enhanced metabolic detoxification occurs at the intertegument.

CONCLUSION

Ticks have developed resistances to acaricides based on Arsenic, Cyclodiene-Pesticides, Chlorinated hydrocarbons, Organophosphorus compounds, Pyrethroids, Ivermectin and Formamidines. Most cases of Acaricide-resistance in the cattle tick have been related to either Enhanced Metabolic Detoxification or Target site Modification, and most of the efforts are directed to identify genetic markers of resistance on the nucleotide sequence of the gene that codes for the target of the acaricides or the genes and or levels of RNA transcripts of XMEs that may correlate with acaricide neutralization by metabolism. It is important to develop monitoring assays required to provide the tools to rapidly identify resistant populations, so that this knowledge can be used to implement better tick control strategies. The discovery of new acaricide compounds is expensive, time consuming and complex, therefore, it is very important to make a more rational use of the current acaricides by the use of reliable molecular detection of acaricide-resistances in order to safeguard their long-term efficacy and utility. Simultaneously, a plan for integral tick control is also required to delay the emergence of acaricide resistances, research should be intensified to find novel and ecologically sound tick control practicesand methods. Above all, acaricide management is of fundamental importance and should consider acaricide rotation, acaricide mixtures and its possible use in combination with anti-tick vaccines, biological control of ticks, and the exploitation of tick resistant hosts. Vector control programs nowadays consider important factors like the presence of different vector species with varying vectoral capacities, tick genomic diversity, up regulation and post-transcriptional editing of resistance conferring genes and different binding specificities of the acaricide to its target site; this information will be invaluable in order to make a reliable assessment and develop a control strategy of acaricide resistance emergencein the *Rhipicephalus* ticks (Karunaratneet al., 2018; Sparks and Nauen, 2015).

REFERENCES

Abbas, R. Z., Zaman, M. A., Colwell, D. D., Gilleard, J. and Iqbal, Z. (2014). Acaricide resistance in cattle ticks and approaches to its management: the state of play. *Veterinary Parasitology,* 203: 6–20. doi: 10.1016/j.vetpar.2014.03.006.

Adams, M. E. and Miller, T. A. (1980). Neural and behavioral correlates of pyrethroid and DDT-type poisoning in the house fly, *Musca domestica* L. *Pesticide Biochemistry and Physiology,* 13: 137–147. doi: 10.1016/0048-3575(80)90065-6.

Atif, M., Estrada-Mondragon, A., Nguyen, B., Lynch, J. W. and Keramidas, A. (2017). Effects of glutamate and ivermectin on single glutamate-gated chloride channels of the parasitic nematode *H. contortus. PLOS Pathogens,* 13: e1006663. doi: 10.1371/journal. ppat.1006663.

Baffi, M. A., de Souza, G. R. L., de Sousa, C. S., Ceron, C. R. and Bonetti, A. M. (2008). Esterase enzymes involved in pyrethroid and organophosphate resistance in a Brazilian population of *Riphicephallus* (*Boophilus*) *microplus* (Acari, Ixodidae). *Molecular and Biochemical Parasitology,* 160: 70–73. doi: 10.1016/j.molbiopara.2008.03.009.

Baron, S., van der Merwe, N. A., Madder, M. and Maritz-Olivier, C. (2015). SNP analysis infers that recombination is involved in the evolution of amitraz resistance in *Rhipicephalus microplus. PLoS ONE,* 10. doi: 10.1371/journal.pone.0131341.

Batiha, G. E., El-Far, A. H., El-Mleeh, A. A., Alsenosy, A. A., Abdelsamei, E. K., Abdel-Daim, M. M., El-Sayed, Y. S. and Shaheen, H. M. (2019). *In vitro* study of ivermectin efficiency against the cattle tick, *Rhipicephalus* (*Boophilus*) *annulatus*, among cattle herds in El-Beheira, Egypt. *Veterinary World,* 12: 1319–1326. doi: 10.14202/vetworld. 2019.1319-1326.

Baxter, G. D. and Barker, S. C. (2002). Analysis of the sequence and expression of a second putative acetylcholinesterase cDNA from organophosphate-susceptible and organophosphate-resistant cattle ticks. *Insect Biochemistry and Molecular Biology,* 32: 815–820. doi: 10.1016/s0965-1748(01)00168-0.

Baxter, G. D., Green, P., Stuttgen, M. and Barker, S. C. (1999). Detecting resistance to organophosphates and carbamates in the cattle tick *Boophilus microplus*, with a propoxur-based biochemical test. *Experimental & Applied Acarology,* 23: 907–914. doi: 10.1023/a:1006364816302.

Bloomquist, J. R. (1993). Toxicology, mode of action and target site-mediated resistance to insecticides acting on chloride channels. *Comparative Biochemistry and Physiology Part C: Pharmacology, Toxicology and Endocrinology,* 106: 301–314. doi: 10.1016/0742-8413(93)90138-B.

Bloomquist, J. R. (1996). Ion channels as targets for insecticides. *Annual Review of Entomology,* 41: 163–190. doi: 10.1146/annurev.en.41.010196.001115.

Bloomquist, J. R. (2003). Chloride channels as tools for developing selective insecticides. *Archives of Insect Biochemistry and Physiology,* 54: 145–156. doi: 10.1002/arch.10112.

Boublik, Y., Saint-Aguet, P., Lougarre, A., Arnaud, M., Villatte, F., Estrada-Mondaca, S. and Fournier, D. (2002). Acetylcholinesterase engineering for detection of insecticide residues. *Protein Engineering,* 15: 43–50. doi: 10.1093/protein/15.1.43.

Brito, L. G., de Oliveira Nery, L., da Silva Barbieri, F., Huacca, M. E. F., dos Santos Pereira, S., da Silva, R. R., de Freitas Fernanades, C. C. and de Sena Oliveira, M. C. (2017). Molecular quantitative assay for esterase-mediated organophosphate resistance in *Rhipicephalus microplus. Ticks and Tick-borne Diseases,* 8: 725–732. doi: 10.1016/j. ttbdis.2017.05.006.

Bull, D. L. and Ahrens, E. H. (1988). Metabolism of coumaphos in susceptible and resistant strains of *Boophilus microplus* (Acari: Ixodidae). *Journal of Medical Entomology,* 25: 94–98. doi: 10.1093/jmedent/25.2.94.

Campbell, W. C., Fisher, M. H., Stapley, E. O., Albers-Schönberg, G. and Jacob, T. A. (1983). Ivermectin: a potent new antiparasitic agent. *Science* (New York, N.Y.), 221: 823–828. doi: 10.1126/science.6308762.

Castro Janer, E., Klafke, G. M., Fontes, F., Capurro, M. L. and Schumaker, T. S. S. (2019). Mutations in *Rhipicephalus microplus* GABA gated chloride channel gene associated with fipronil resistance. *Ticks and Tick-Borne Diseases,* 10: 761–765. doi: 10.1016/ j.ttbdis.2019.03.009.

Chae, J. S., Allsopp, B. A., Waghela, S. D., Park, J. H., Kakuda, T., Sugimoto, C., Allsopp, M. T., Wagner, G. G. and Holman, P. J. (1999). A study of the systematics of *Theileria* spp. based upon small-subunit ribosomal RNA gene sequences. *Parasitology Research,* 85: 877–883. doi: 10.1007/s004360050651.

Chen, S., Yang, Y. and Wu, Y. (2005). Correlation between fenvalerate resistance and Cytochrome P450-mediated O-Demethylation activity in *Helicoverpa armigera* (Lepidoptera: Noctuidae). *Journal of Economic Entomology,* 98: 943–946. doi: 10.1603/ 0022-0493-98.3.943.

Chen, A. C., He, H. and Davey, R. B. (2007). Mutations in a putative octopamine receptor gene in amitraz-resistant cattle ticks. *Veterinary Parasitology,* 148: 379–383. doi: 10. 1016/j.vetpar.2007.06.026.

Cole, L. M., Nicholson, R. A. and Casida, J. E. (1993). Action of phenylpyrazole insecticides at the GABA-gated chloride channel. *Pesticide Biochemistry and Physiology,* 46: 47–54. doi: 10.1006/pest.1993.1035.

Corley, S. W., Piper, E. K. and Jonsson, N. N. (2012). Generation of full-length cDNAs for eight putative GPCnR from the cattle tick, *R. microplus* using a targeted degenerate PCR and sequencing strategy. *PLoS ONE,* 7. doi: 10.1371/journal.pone.0032480.

Corley, S. W., Jonsson, N. N., Piper, E. K., Cutullé, C., Stear, M. J. and Seddon, J. M. (2013). Mutation in the RmβAOR gene is associated with amitraz resistance in the cattle tick

Rhipicephalus microplus. *Proceedings of the National Academy of Sciences,* 110: 16772–16777. doi: 10.1073/pnas.1309072110.

Cossio-Bayugar, R., Pillars, R., Schlater, J. and Holman, P. J. (2002). *Theileria buffeli* infection of a Michigan cow confirmed by small subunit ribosomal RNA gene analysis. *Veterinary Parasitology,* 105: 105–110. doi: 10.1016/S0304-4017(02)00003-1.

Cossío-Bayúgar, R., Miranda-Miranda, E., Ortiz-Najera, A. and Neri-Orantes, S. (2008). *Boophilus microplus* pyrethroid resistance associated to increased levels of monooxygenase enzymatic activity in field isolated Mexican ticks. *Journal of Biological Sciences,* 8: 404–409. doi: 10.3923/jbs.2008.404.409.

Cossío-Bayúgar, R., Miranda-Miranda, E., Portilla-Salgado, D. and Osorio-Miranda, J. (2009). Quantitative PCR detection of cholinesterase and carboxylesterase expression levels in acaricide resistant *Rhipicephalus* (*Boophilus*) *microplus*. *Journal of Entomology,* 6: 117–123. doi: 10.3923/je.2009.117.123.

Cossío-Bayúgar, R., Miranda-Miranda, E., Narváez Padilla, V., Olvera-Valencia, F. and Reynaud, E. (2012). Perturbation of tyraminergic/octopaminergic function inhibits oviposition in the cattle tick *Rhipicephalus* (*Boophilus*) *microplus*. *Journal of Insect Physiology,* 58: 628–633. doi: 10.1016/j.jinsphys.2012.01.006.

Cossío-Bayúgar, R., Miranda-Miranda, E., Fernández-Rubalcaba, M., Narváez Padilla, V. and Reynaud, E. (2015). Adrenergic ligands that block oviposition in the cattle tick *Rhipicephalus microplus* affect ovary contraction. *Scientific Reports,* 5. doi: 10.1038/srep15109.

Cossío-Bayúgar, R., Martínez-Ibañez, F., Aguilar-Díaz, H. and Miranda-Miranda, E. (2018). Pyrethroid acaricide resistance is proportional to P-450 cytochrome oxidase expression in the cattle tick *Rhipicephalus microplus*. *BioMed Research International,* 2018: 8292465. doi: 10.1155/2018/8292465.

Cossío-Bayúgar, R., Miranda-Miranda, E., Martínez-Ibañez, F., Narváez-Padilla, V. and Reynaud, E. (2020). Physiological evidence that three known mutations in the para-sodium channel gene confer cypermethrin knockdown resistance in *Rhipicephalus microplus*. *Parasites & Vectors,* 13: 370. doi: 10.1186/s13071-020-04227-7.

Davies, T. E., O'Reilly, A. O., Field, L. M., Wallace, B. and Williamson, M. S. (2008). Knockdown resistance to DDT and pyrethroids: from target-site mutations to molecular modelling. *Pest Management Science,* 64: 1126–1130. doi: 10.1002/ps.1617.

De Jersey, J., Nolan, J., Davey, P. A. and Riddles, P. W. (1985). Separation and characterization of the pyrethroid-hydrolyzing esterases of the cattle tick, *Boophilus microplus*. *Pesticide Biochemistry and Physiology,* 23: 349–357. doi: 10.1016/0048-3575(85)90096-3.

Dehkordi, Z. S., Zakeri, S., Nabian, S., Bahonar, A., Ghasemi, F., Noorollahi, F. and Rahbari, S. (2010). Molecular and biomorphometrical identification of ovine babesiosis in iran. *Iranian Journal of Parasitology,* 5: 21–30.

Dermauw, W., Ilias, A., Riga, M., Tsagkarakou, A., Grbić, M., Tirry, L., Van Leeuwen, T. and Vontas, J. (2012). The cys-loop ligand-gated ion channel gene family of *Tetranychus urticae*: implications for acaricide toxicology and a novel mutation associated with abamectin resistance. *Insect Biochemistry and Molecular Biology,* 42: 455–465. doi: 10.1016/j.ibmb.2012.03.002.

Domingos, A., Antunes, S., Borges, L. and Rosário, V. E. do (2013). Approaches towards tick and tick-borne diseases control. *Revista Da Sociedade Brasileira De Medicina Tropical,* 46: 265–269. doi: 10.1590/0037-8682-0014-2012.

Dong, K. (2007). Insect sodium channels and insecticide resistance. In *Invertebrate Neuroscience*, pp. 17–30. NIH Public Access doi: 10.1007/s10158-006-0036-9.

Dong, K., Du, Y., Rinkevich, F., Nomura, Y., Xu, P., Wang, L., Silver, K. and Zhorov, B. S. (2014). Molecular biology of insect sodium channels and pyrethroid resistance. *Insect Biochemistry and Molecular Biology,* 50: 1–17. doi: 10.1016/j.ibmb.2014.03.012.

Douris, V., Steinbach, D., Panteleri, R., Livadaras, I., Pickett, J. A., Leeuwen, T. V., Nauen, R. and Vontas, J. (2016). Resistance mutation conserved between insects and mites unravels the benzoylurea insecticide mode of action on chitin biosynthesis. *Proceedings of the National Academy of Sciences,* 113: 14692–14697. doi: 10.1073/pnas.1618258113.

Du, Y., Nomura, Y., Zhorov, B. S. and Dong, K. (2016). Sodium channel mutations and pyrethroid resistance in *Aedes aegypti. Insects,* 7: 1–11. doi: 10.3390/insects7040060.

Dzul, F., Penilla Navarro, R. and Rodríguez, A. (2007). Susceptibility and inseticide resistance mechanisms in *Anopheles albimanus* from the southern Yucatan Peninsula, Mexico. *Salud pública de México,* 49: 302–11.

Ffrench-Constant, R. H., Steichen, J. C., Rocheleau, T. A., Aronstein, K. and Roush, R. T. (1993). A single-amino acid substitution in a gamma-aminobutyric acid subtype A receptor locus is associated with cyclodiene insecticide resistance in *Drosophila* populations. *Proceedings of the National Academy of Sciences of the United States of America,* 90: 1957–1961. doi: 10.1073/pnas.90.5.1957.

Ffrench-Constant, R. H., Daborn, P. J. and Goff, G. L. (2004). The genetics and genomics of insecticide resistance. *Trends in Genetics,* 20: 163–170. doi: 10.1016/j.tig.2004.01.003.

Ffrench-Constant, R. H., Williamson, M. S., Davies, T. G. E. and Bass, C. (2016). Ion channels as insecticide targets. *Journal of Neurogenetics,* 30: 163–177. doi: 10.1080/01677063.2016.1229781.

Field, L. M. and Devonshire, A. L. (1998). Evidence that the E4 and FE4 esterase genes responsible for insecticide resistance in the aphid *Myzus persicae* (Sulzer) are part of a gene family. *Biochemical Journal,* 330 (Pt 1): 169–173. doi: 10.1042/bj3300169.

Field, L. M., Blackman, R. L., Tyler-Smith, C. and Devonshire, A. L. (1999). Relationship between amount of esterase and gene copy number in insecticide-resistant *Myzus persicae* (Sulzer). *Biochemical Journal,* 339 (Pt 3): 737–742.

Fular, A., Gupta, S., Sharma, A. K., Kumar, S., Upadhaya, D., Shakya, M., Nagar, G. and Ghosh, S. (2020). Standardization of tick specific biochemical tools for estimation of esterases, monooxygenases and glutathione S-transferase for characterization of acaricide resistance. *Pesticide Biochemistry and Physiology,* 164: 130–139. doi: 10.1016/j.pestbp.2020.01.008.

Gammon, D. W., Brown, M. A. and Casida, J. E. (1981). Two classes of pyrethroid action in the cockroach. *Pesticide Biochemistry and Physiology,* 15: 181–191. doi: 10.1016/0048-3575(81)90084-5.

Gaur, R. S., Sangwan, A. K., Sangwan, N., Ghosh, M. and Kumar, S. (2017). Comparative study of esterases in deltamethrin and diazinon resistant *Rhipicephalus microplus* and *Hyalomma anatolicum* ticks collected from the Trans-Gangetic plains of India. *Experimental and Applied Acarology,* 73: 115–127. doi: 10.1007/s10493-017-0175-1.

George, J. E., Pound, J. M. and Davey, R. B. (2004). Chemical control of ticks on cattle and the resistance of these parasites to acaricides. *Parasitology.* 129 Suppl: S353-366. doi: 10.1017/s0031182003004682.

Ghosh, S., Kumar, R., Nagar, G., Kumar, S., Sharma, A. K., Srivastava, A., Kumar, S., Ajith Kumar, K. G. and Saravanan, B. C. (2015). Survey of acaricides resistance status of *Rhipiciphalus* (*Boophilus*)*microplus* collected from selected places of Bihar, an eastern state of India. *Ticks and Tick-Borne Diseases,* 6: 668–675. doi: 10.1016/j.ttbdis.2015.05.013.

Graham, K. M., Sparagano, O. A. E. and Finn, R. D. (2016). Isolation of the monooxygenase complex from *Rhipicephalus* (*Boophilus*) *microplus* – clues to understanding acaricide resistance. *Ticks and Tick-borne Diseases,* 7: 614–623. doi: 10.1016/j.ttbdis.2016.01.012.

Gross, A. D., Temeyer, K. B., Day, T. A., Pérez de León, A. A., Kimber, M. J. and Coats, J. R. (2015). Pharmacological characterization of a tyramine receptor from the southern cattle tick, *Rhipicephalus* (*Boophilus*) *microplus*. *Insect Biochemistry and Molecular Biology,* 63: 47–53. doi: 10.1016/j.ibmb.2015.04.008.

Guerrero, F. D., Bendele, K. G., Chen, A. C., Li, A. Y., Miller, R. J., Pleasance, E., Varhol, R., Rousseau, M.-E. and Nene, V. M. (2007). Serial analysis of gene expression in the southern cattle tick following acaricide treatment of larvae from organophosphate resistant and susceptible strains. *Insect Molecular Biology,* 16: 49–60. doi: 10.1111/j.1365-2583.2007.00699.x.

Gupta, S., Ajith Kumar, K. G., Sharma, A. K., Nagar, G., Kumar, S., Saravanan, B. C., Ravikumar, G. and Ghosh, S. (2016). Esterase mediated resistance in deltamethrin resistant reference tick colony of *Rhipicephalus* (*Boophilus*) *microplus*. *Experimental and Applied Acarology,* 69: 239–248. doi: 10.1007/s10493-016-0032-7.

Hammer, J. F., Emery, D., Bogema, D. R. and Jenkins, C. (2015). Detection of *Theileria orientalis* genotypes in *Haemaphysalis longicornis* ticks from southern Australia. *Parasites & Vectors,* 8. doi: 10.1186/s13071-015-0839-9.

Hartley, C. J., Newcomb, R. D., Russell, R. J., Yong, C. G., Stevens, J. R., Yeates, D. K., La Salle, J. and Oakeshott, J. G. (2006). Amplification of DNA from preserved specimens shows blowflies were preadapted for the rapid evolution of insecticide resistance. *Proceedings of the National Academy of Sciences of the United States of America,* 103: 8757–8762. doi: 10.1073/pnas.0509590103.

He, H., Chen, A. C., Davey, R. B., Ivie, G. W. and George, J. E. (1999). Identification of a point mutation in the para-type sodium channel gene from a pyrethroid-resistant cattle tick. *Biochemical and Biophysical Research Communications,* 261: 558–561. doi: 10.1006/bbrc.1999.1076.

Hemingway, J. and Karunaratne, S. H. (1998). Mosquito carboxylesterases: a review of the molecular biology and biochemistry of a major insecticide resistance mechanism. *Medical and Veterinary Entomology,* 12: 1–12. doi: 10.1046/j.1365-2915.1998.00082.x.

Hernandez, R., He, H., Chen, A. C., Ivie, G. W., George, J. E. and Wagner, G. G. (1999). Cloning and Sequencing of a Putative Acetylcholinesterase cDNA from *Boophilus microplus* (Acari: Ixodidae). *Journal of Medical Entomology,* 36: 764–770. doi: 10.1093/jmedent/36.6.764.

Hernandez, R., He, H., Chen, A. C., Waghela, S. D., Wayne Ivie, G., George, J. E. and Gale Wagner, G. (2000). Identification of a point mutation in an esterase gene in different populations of the southern cattle tick, *Boophilus microplus. Insect Biochemistry and Molecular Biology,* 30: 969–977. doi: 10.1016/S0965-1748(00)00069-2.

Hernandez, E. P., Kusakisako, K., Talactac, M. R., Galay, R. L., Hatta, T., Matsuo, T., Fujisaki, K., Tsuji, N. and Tanaka, T. (2018). Characterization and expression analysis of a newly identified glutathione S-transferase of the hard tick *Haemaphysalis longicornis* during blood-feeding. *Parasites & Vectors,* 11: 91. doi: 10.1186/s13071-018-2667-1.

Homolya, L., Váradi, A. and Sarkadi, B. (2003). Multidrug resistance-associated proteins: Export pumps for conjugates with glutathione, glucuronate or sulfate. *BioFactors,* 17: 103–114. doi: 10.1002/biof.5520170111.

Hope, M., Menzies, M. and Kemp, D. (2010). Identification of a dieldrin resistance-associated mutation in *Rhipicephalus* (*Boophilus*) *microplus* (Acari: Ixodidae). *Journal of Economic Entomology,* 103: 1355–1359. doi: 10.1603/EC09267.

IRAC (2020). *Insecticide Resistance Action Committee.* IRAC Insecticide Resistance Action Committee. www.irac-online.org or enquiries@irac-online.org.

Jamroz,R. C., Guerrero,F. D., Pruett, J. H., Oehler,D. D. and Miller, R. J. (2000). Molecular and biochemical survey of acaricide resistance mechanisms in larvae from Mexican strains of the southern cattle tick, *Boophilus microplus. Journal of Insect Physiology,* 46: 685–695. doi: 10.1016/s0022-1910(99)00157-2.

Janadaree Bandara, K. M. U. and Parakrama Karunaratne, S. H. P. (2017). Mechanisms of acaricide resistance in the cattle tick *Rhipicephalus* (*Boophilus)microplus* in Sri Lanka. *Pesticide Biochemistry and Physiology,* 139: 68–72. doi: 10.1016/j.pestbp.2017.05.002.

Jonsson, N. N. and Hope, M. (2007). Progress in the epidemiology and diagnosis of amitraz resistance in the cattle tick *Boophilus microplus. Veterinary Parasitology,* 146: 193–198. doi: 10.1016/j.vetpar.2007.03.006.

Jonsson, N. N., Cutullè, C., Corley, S. W. and Seddon, J. M. (2010). Identification of a mutation in the para-sodium channel gene of the cattle tick *Rhipicephalus microplus* associated with resistance to flumethrin but not to cypermethrin. *International Journal for Parasitology,* 40: 1659–1664. doi: 10.1016/j.ijpara.2010.07.006.

Jonsson, N. N., Klafke, G., Corley, S. W., Tidwell, J., Berry, C. M. and Koh-Tan, H. C. (2018). Molecular biology of amitraz resistance in cattle ticks of the genus *Rhipicephalus. Frontiers in Bioscience (Landmark Edition),* 23: 796–810. doi: 10.2741/4617.

Jyoti, Singh, N. K., Singh, H., Singh, N. K. and Rath, S. S. (2016). Multiple mutations in the acetylcholinesterase 3 gene associated with organophosphate resistance in *Rhipicephalus* (*Boophilus*) *microplus* ticks from Punjab, India. *Veterinary Parasitology,* 216: 108–117. doi: 10.1016/j.vetpar.2015.12.004.

Kane, N. S., Hirschberg, B., Qian, S., Hunt, D., Thomas, B., Brochu, R., Ludmerer, S. W., Zheng, Y., Smith, M., Arena, J. P., Cohen, C. J., Schmatz, D., Warmke, J. and Cully, D. F. (2000). Drug-resistant *Drosophila* indicate glutamate-gated chloride channels are targets for the antiparasitics nodulisporic acid and ivermectin. *Proceedings of the National Academy of Sciences,* 97: 13949–13954. doi: 10.1073/pnas.240464697.

Karunaratne, S. H. and Hemingway, J. (2001). Malathion resistance and prevalence of the malathion carboxylesterase mechanism in populations of mosquito vectors of disease in Sri Lanka. *Bulletin of the World Health Organization,* 79: 1060–1064.

Karunaratne, S. H. P. P., De Silva, W. A. P. P., Weeraratne, T. C. and Surendran, S. N. (2018). Insecticide resistance in mosquitoes: Development, mechanisms and monitoring. *Ceylon Journal of Science* 47:299–309. doi: http://doi.org/10.4038/cjs.v47i4.7547.

Kasai, S. (2004). Role of cytochrome P450 in mechanism of pyrethroid resistance. *Journal of Pesticide Science,* 29: 220–221. doi: 10.1584/jpestics.29.220.

Kasai, S. and Scott, J. G. (2000). Overexpression of cytochrome P450 CYP6D1 is associated with monooxygenase-mediated pyrethroid resistance in house flies from Georgia. *Pesticide Biochemistry and Physiology,* 68: 34–41. doi: 10.1006/pest.2000.2492.

Kim, C. S., Kim, W. T., Boo, K. S. and Kim, S. I. (2003). Cloning, mutagenesis, and expression of the acetylcholinesterase gene from a strain of *Musca domestica*; the change from a drug-resistant to a sensitive enzyme. *Molecules and Cells,* 15: 208–215.

Kocan, K. M., de la Fuente, J., Blouin, E. F. and Garcia-Garcia, J. C. (2004). Anaplasma marginale (Rickettsiales: Anaplasmataceae): recent advances in defining host-pathogen adaptations of a tick-borne rickettsia. *Parasitology,* 129 Suppl: S285-300. doi: 10.1017/s0031182003004700.

Kumar, R., Klafke, G. M. and Miller, R. J. (2020). Voltage-gated sodium channel gene mutations and pyrethroid resistance in *Rhipicephalus microplus*. *Ticks and Tick-borne Diseases*, 101404. doi: 10.1016/J.TTBDIS.2020.101404.

Kwon, D. H., Yoon, K. S., Clark, J. M. and Lee, S. H. (2010). A point mutation in a glutamate-gated chloride channel confers abamectin resistance in the two-spotted spider mite, *Tetranychus urticae* Koch. *Insect Molecular Biology,* 19: 583–591. doi: 10.1111/j.1365-2583.2010.01017.x.

Lara, F. A., Pohl, P. C., Gandara, A. C., Ferreira, J. da S., Nascimento-Silva, M. C., Bechara, G. H., Sorgine, M. H. F., Almeida, I. C., Vaz, I. da S. and Oliveira, P. L. (2015). ATP binding cassette transporter mediates both heme and pesticide detoxification in tick midgut cells. *PLoS ONE,* 10. doi: 10.1371/journal.pone.0134779.

Le Gall, V. L., Klafke, G. M. and Torres, T. T. (2018). Detoxification mechanisms involved in ivermectin resistance in the cattle tick, *Rhipicephalus* (*Boophilus*) microplus. *Scientific Reports,* 8: 12401. doi: 10.1038/s41598-018-30907-7.

Li, A. Y., Davey, R. B., Miller, R. J. and George, J. E. (2003). Resistance to coumaphos and diazinon in *Boophilus microplus* (Acari: Ixodidae) and evidence for the involvement of an oxidative detoxification mechanism. *Journal of Medical Entomology.* 40: 482–490. doi: 10.1603/0022-2585-40.4.482.

Li, A. Y., Davey, R. B., Miller, R. J. and George, J. E. (2004). Detection and characterization of amitraz resistance in the southern cattle tick, *Boophilus microplus* (Acari: Ixodidae). *Journal of Medical Entomology*, 41: 193–200. doi: 10.1603/0022-2585-41.2.193.

Li, A. Y., Davey, R. B. and George, J. E. (2005). Carbaryl resistance in Mexican strains of the southern cattle tick (Acari: Ixodidae). *Journal of Economic Entomology,* 98: 552–556. doi: 10.1093/jee/98.2.552.

Li, Y., Luo, J., Guan, G., Ma, M., Liu, A., Liu, J., Ren, Q., Niu, Q., Lu, B., Gao, J., Liu, Z., Dang, Z., Tian, Z., Zhang, B., He, Z., Bai, Q. and Yin, H. (2009). Experimental transmission of *Theileria uilenbergi* infective for small ruminants by *Haemaphysalis longicornis* and *Haemaphysalis qinghaiensis*. *Parasitology Research,* 104: 1227–1231. doi: 10.1007/s00436-009-1347-7.

Li, Y., Chen, Z., Liu, Z., Liu, J., Yang, J., Li, Q., Li, Y., Cen, S., Guan, G., Ren, Q., Luo, J. and Yin, H. (2014). Molecular identification of *Theileria* parasites of northwestern Chinese Cervidae. *Parasites & Vectors,* 7: 225. doi: 10.1186/1756-3305-7-225.

Li, H., Zheng, Y.-C., Ma, L., Jia, N., Jiang, B.-G., Jiang, R.-R., Huo, Q.-B., Wang, Y.-W., Liu, H.-B., Chu, Y.-L., Song, Y.-D., Yao, N.-N., Sun, T., Zeng, F.-Y., Dumler, J. S., Jiang, J.-F. and Cao, W.-C. (2015). Human infection with a novel tick-borne *Anaplasma* species in China: a surveillance study. *Lancet Infectious Diseases,* 15: 663–670. doi: 10. 1016/S1473-3099(15)70051-4.

Liu, Z., Ma, M., Wang, Z., Wang, J., Peng, Y., Li, Y., Guan, G., Luo, J. and Yin, H. (2012). Molecular survey and genetic identification of *Anaplasma* species in goats from central and southern China. *Applied and Environmental Microbiology,* 78: 464–470. doi: 10. 1128/AEM.06848-11.

Lovis, L., Guerrero, F. D., Miller, R. J., Bodine, D. M., Betschart, B. and Sager, H. (2012). Distribution patterns of three sodium channel mutations associated with pyrethroid resistance in *Rhipicephalus* (*Boophilus*) *microplus* populations from North and South America, South Africa and Australia. *International Journal for Parasitology: Drugs and Drug Resistance,2*: 216–224. doi: 10.1016/J.IJPDDR.2012.08.001.

Lumaret, J.-P., Errouissi, F., Floate, K., Römbke, J. and Wardhaugh, K. (2012). A review on the toxicity and non-target effects of macrocyclic lactones in terrestrial and aquatic environments. *Current Pharmaceutical Biotechnology,* 13: 1004–1060. doi: 10.2174/ 138920112800399257.

Madder, M., Thys, E., Achi, L., Touré, A. and De Deken, R. (2011). *Rhipicephalus* (*Boophilus*) *microplus*: a most successful invasive tick species in West-Africa. *Experimental and Applied Acarology,* 53: 139–145. doi: 10.1007/s10493-010-9390-8.

Manchenko, G. P. (2003). *Handbook of Detection of Enzymes on Electrophoretic Gels. Second Edition.*, 2nd Edn. CRC Press LLC, Boca Raton.

Miller, R. J., Davey, R. B. and George, J. E. (1999). Characterization of pyrethroid resistance and susceptibility to coumaphos in Mexican *Boophilus microplus* (Acari: Ixodidae). *Journal of Medical Entomology,* 36: 533–538. doi: 10.1093/jmedent/36.5.533.

Miranda, E., Cossio-Bayugar, R., Tellez-Alaniz, M., Garcia-Vazquez, Z., Rosario-Cruz, R. and Ortiz-Estrada, M. (1995). An enzymatic marker for ixodicide resistance detection in the cattle tick *Boophilus microplus*. *Advances in Agricultural Research,* 4: 1–8.

Morgan, J. A. T., Corley, S. W., Jackson, L. A., Lew-Tabor, A. E., Moolhuijzen, P. M. and Jonsson, N. N. (2009). Identification of a mutation in the para-sodium channel geneof the cattle tick *Rhipicephalus* (*Boophilus*) *microplus* associated with resistanceto synthetic pyrethroid acaricides. *International Journal for Parasitology,* 39: 775–779. doi: 10.1016/ j.ijpara.2008.12.006.

Nagar, G., Anjali, K., Gupta, S., Kumar, S., Saravanan, B. C., Rai, A. and Ghosh, S. (2016). Over-expression of esterases in acaricide resistant isolates of *Rhipicephalus* (*Boophilus*) *microplus*. *International Journal of Science, Environment and Technology,* 5: 3457–3465.

Newcomb, R. D., Campbell, P. M., Ollis, D. L., Cheah, E., Russell, R. J. and Oakeshott, J. G. (1997). A single amino acid substitution converts a carboxylesterase to an organophosphorus hydrolase and confers insecticide resistance on a blowfly. *Proceedings of the National Academy of Sciences of the United States of America,* 94: 7464–7468.

Nicholson, W. L., Sonenshine, D. E., Noden, B. H. and Brown, R. N. (2019). Chapter 27 - Ticks (Ixodida). In *Medical and Veterinary Entomology (Third Edition)* (ed. Mullen, G. R. and Durden, L. A.), pp. 603–672. Academic Press doi: 10.1016/B978-0-12-814043-7.00027-3.

Omura, S. (2008). Ivermectin: 25 years and still going strong. *International Journal of Science, Environment and Technology,* 31: 91–98. doi: 10.1016/j.ijantimicag.2007.08.023.

O'Reilly, A. O., Williamson, M. S., González-Cabrera, J., Turberg, A., Field, L. M., Wallace, B. A. and Davies, T. G. E. (2014). Predictive 3D modelling of the interactions of pyrethroids with the voltage-gated sodium channels of ticks and mites. *Pest Management Science,* 70: 369–377. doi: 10.1002/ps.3561.

Osborne, M. P. and Hart, R. J. (1979). Neurophysiological studies of the effects of permethrin upon pyrethroid resistant (kdr) and susceptible strains of dipteran larvae. *Pesticide Science* 10: 407–413. doi: 10.1002/ps.2780100507.

Parkinson, A. (1996). Biotransformation of xenobiotics, In *Cassarett and Doulls Toxicology: The Basic Science of Poisons*, pp. 113–196. McGraw-Hill Medical, New York, NY.

Paton, M. G., Karunaratne, S. H., Giakoumaki, E., Roberts, N. and Hemingway, J. (2000). Quantitative analysis of gene amplification in insecticide-resistant Culex mosquitoes. *Biochemical Journal,* 346: 17–24.

Pepper, D. R. and Osborne, M. P. (1993). Electrophysiological identification of site-insensitive mechanisms in knockdown-resistant strains (kdr, super-kdr) of the housefly larva (*Musca domestica*). *Pesticide Science,* 39: 279–286. doi: 10.1002/ps.2780390405.

Pruett, J. H. (2002). Comparative inhibition kinetics for acetylcholinesterases extracted from organophosphate resistant and susceptible strains of *Boophilus microplus* (Acari: Ixodidae). *Journal of Economic Entomology,* 95: 1239–1244. doi: 10.1603/0022-0493-95.6.1239.

Pruett, J. H., Guerrero, F. D. and Hernandez, R. (2002). Isolation and identification of an esterase from a Mexican strain of *Boophilus microplus* (Acari: Ixodidae). *Journal of Economic Entomology,* 95: 1001–1007. doi: 10.1603/0022-0493-95.5.1001.

Remnant, E. J., Morton, C. J., Daborn, P. J., Lumb, C., Yang, Y. T., Ng, H. L., Parker, M. W. and Batterham, P. (2014). The role of Rdl in resistance to phenylpyrazoles in *Drosophila melanogaster*. *Insect Biochemistry and Molecular Biology,* 54: 11–21. doi: 10.1016/j.ibmb.2014.08.008.

Renneker, S., Abdo, J., Salih, D. E. A., Karagenç, T., Bilgiç, H., Torina, A., Oliva, A. G., Campos, J., Kullmann, B., Ahmed, J. and Seitzer, U. (2013). Can *Anaplasma ovis* in small ruminants be neglected any longer? *Transboundary and Emerging Diseases,* 60 (Suppl 2): 105–112. doi: 10.1111/tbed.12149.

Rodriguez-Vivas, R. I., Arieta-Román, R., Pérez-Cogollo, L., Rosado-Aguilar, J., Ramírez-Cruz, G. and Basto-Estrella, G. (2010). So de lactonas macrocíclicas para el control de la garrapata *Rhipicephalus* (*Boophilus) microplus* en el ganado bovino [Macrocyclic lactones for the control of the tick *Rhipicephalus* (*Boophilus*) *microplus* in cattle]. *Archivos de Medicina Veterinaria,* 42: 115–123.

Rodriguez-Vivas, R. I., Jonsson, N. N. and Bhushan, C. (2018). Strategies for the control of *Rhipicephalus microplus* ticks in a world of conventional acaricide and macrocyclic lactone resistance. *Parasitology Research*, 117: 3–29. doi: 10.1007/s00436-017-5677-6.

Roeder, T. (2005). Tyramine and octopamine: ruling behavior and metabolism. *Annual Review of Entomology,* 50: 447–477. doi: 10.1146/annurev.ento.50.071803.130404.

Roeder, T., Seifert, M., Kähler, C. and Gewecke, M. (2003). Tyramine and octopamine: antagonistic modulators of behavior and metabolism. *Archives of Insect Biochemistry and Physiology,* 54: 1–13. doi: 10.1002/arch.10102.

Rosa de Lima, M. F., Sanchez Ferreira, C. A., Joaquim de Freitas, D. R., Valenzuela, J. G. and Masuda, A. (2002). Cloning and partial characterization of a *Boophilus microplus* (Acari: Ixodidae) glutathione S-transferase. *Insect Biochemistry and Molecular Biology,* 32: 747–754. doi: 10.1016/S0965-1748(01)00157-6.

Rosario-Cruz, R., Miranda-Miranda, E., Garcia-Vasquez, Z. and Ortiz-Estrada, M. (1997). Detection of esterase activity in susceptible and organophosphate resistant strains of the cattle tick *Boophilus microplus* (Acari: Ixodidae). *Bulletin of Entomological Research,* 87: 197–202. doi: 10.1017/S0007485300027358.

Saldivar, L., Guerrero, F. D., Miller, R. J., Bendele, K. G., Gondro, C. and Brayton, K. A. (2008). Microarray analysis of acaricide-inducible gene expression in the southern cattle tick, *Rhipicephalus* (*Boophilus*) *microplus*. *Insect Molecular Biology,* 17: 597–606. doi: 10.1111/j.1365-2583.2008.00831.x.

Salgado, V. L., Irving, S. N. and Miller, T. A. (1983a). Depolarization of motor nerve terminals by pyrethroids in susceptible and kdr-resistant house flies. *Pesticide Biochemistry and Physiology,* 20: 100–114. doi: 10.1016/0048-3575(83)90126-8.

Salgado, V. L., Irving, S. N. and Miller, T. A. (1983b). The importance of nerve terminal depolarization in pyrethroid poisoning of insects. *Pesticide Biochemistry and Physiology,* 20: 169–182. doi: 10.1016/0048-3575(83)90021-4.

Schnittger, L., Yin, H., Jianxun, L., Ludwig, W., Shayan, P., Rahbari, S., Voss-Holtmann, A. and Ahmed, J. S. (2000). Ribosomal small-subunit RNA gene-sequence analysis of *Theileria lestoquardi* and a *Theileria* species highly pathogenic for small ruminants in China. *Parasitology Research,* 86: 352–358. doi: 10.1007/s004360050680.

Schnitzerling, H. J., Schuntner, C. A., Roulston, W. J. and Wilson, J. T. (1974). Characterization of the organophosphoms-resistant Mt Alford, Gracemere and Silkwood strains of the cattle tick, *Boophilus microplus*. *Australian Journal of Biological Sciences,* 27: 397–408. doi: 10.1071/bi9740397.

Scott, J. G. (1999). Cytochromes P450 and insecticide resistance. *Insect Biochemistry and Molecular Biology,* 29: 757–777. doi: 10.1016/s0965-1748(99)00038-7.

Scott, J. G. and Tomita, T. (1995). Characterization of a cytochrome P450 responsible for pyrethroid resistance in the house fly. *Pesticide Science,* 43: 244–245. doi: 10.1002/ps.2780430312.

Seabrook, G. R., Duce, I. R. and Irving, S. N. (1988). Effects of the pyrethroid cypermethrin on l-glutamate-induced changes in the input conductance of the ventrolateral muscles of the larval house fly, *Musca domestica*. *Pesticide Biochemistry and Physiology,* 32: 232–239. doi: 10.1016/0048-3575(88)90106-X.

Seabrook, G. R., Duce, I. R. and Irving, S. N. (1989). Spontaneous and evoked quantal neurotransmitter release at the neuromuscular junction of the larval housefly, *Musca domestica*. *Pflügers Archiv European Journal of Physiology,* 414: 44–51. doi: 10.1007/BF00585625.

Silver, K. S., Du, Y., Nomura, Y., Oliveira, E. E., Salgado, V. L., Zhorov, B. S. and Dong, K. (2014). Voltage-Gated Sodium Channels as Insecticide Targets. *Advances in insect physiology,* 46: 389–433. doi: 10.1016/B978-0-12-417010-0.00005-7.

Singh, N. K. and Rath, S. S. (2014). Esterase mediated resistance against synthetic pyrethroids in field populations of *Rhipicephalus* (*Boophilus*) *microplus* (Acari: Ixodidae) in Punjab districts of India. *Veterinary Parasitology,* 204: 330–338. doi: 10.1016/j.vetpar.2014.05.035.

Soderlund, D. M. (2010a). Toxicology and Mode of Action of Pyrethroid Insecticides. In *Hayes' Handbook of Pesticide Toxicology*, pp. 1665–1686. Academic Press doi: 10.1016/B978-0-12-374367-1.00077-X.

Soderlund, D. M. (2010b). State-dependent modification of voltage-gated sodium channels by pyrethroids. *Pesticide Biochemistry and Physiology,* 97: 78–86. doi: 10.1016/j.pestbp.2009.06.010.

Soderlund, D. M. (2012a). Molecular mechanisms of pyrethroid insecticide neurotoxicity: recent advances. *Archives of Toxicology,* 86: 165–181. doi: 10.1007/s00204-011-0726-x.

Soderlund, D. M. and Bloomquist, J. R. (1989). Neurotoxic actions of pyrethroid insecticides. *Annual Review of Entomology,* 34: 77–96. doi: 10.1146/annurev.en.34.010189.000453.

Soderlund, D. M., Clark, J. M., Sheets, L. P., Mullin, L. S., Piccirillo, V. J., Sargent, D., Stevens, J. T. and Weiner, M. L. (2002). Mechanisms of pyrethroid neurotoxicity: implications for cumulative risk assessment. *Toxicology,* 171: 3–59. doi: 10.1016/s0300-483x(01)00569-8.

Sparks, T. C. and Nauen, R. (2015). IRAC: Mode of action classification and insecticide resistance management. *Pesticide Biochemistry and Physiology,* 121: 122–128. doi: 10.1016/j.pestbp.2014.11.014.

Stone, N. E., Olafson, P. U., Davey, R. B., Buckmeier, G., Bodine, D., Sidak-Loftis, L. C., Giles, J. R., Duhaime, R., Miller, R. J., Mosqueda, J., Scoles, G. A., Wagner, D. M. and Busch, J. D. (2014). Multiple mutations in the para-sodium channel gene are associated with pyrethroid resistance in *Rhipicephalus microplus* from the United States and Mexico. *Parasites & Vectors,* 7. doi: 10.1186/s13071-014-0456-z.

Sun, R., Liu, C., Zhang, H. and Wang, Q. (2015). Benzoylurea chitin synthesis inhibitors. *Journal of Agricultural and Food Chemistry,* 63: 6847–6865. doi: 10.1021/acs.jafc.5b02460.

Surbhi, Gupta, S. and Gupta, S. K. (2018). Acaricide resistance mechanisms and monitoring tools available for *Rhipicephalus* (*Boophilus*) *microplus*. *Pharma Innovation Journal,* 7: 398–405.

Temeyer, K. B. and Tuckow, A. P. (2016). Tick Salivary Cholinesterase: A probable immunomodulator of host-parasite interactions. *Journal of Medical Entomology,* 53: 500–504. doi: 10.1093/jme/tjv252.

Temeyer, K. B., Davey, R. B. and Chen, A. C. (2004). Identification of a third *Boophilus microplus* (Acari: Ixodidae) cDNA presumptively encoding an acetylcholinesterase. *Journal of Medical Entomology,* 41: 259–268. doi: 10.1603/0022-2585-41.3.259.

Temeyer, K. B., Pruett, J. H., Untalan, P. M. and Chen, A. C. (2006). Baculovirus expression of BmAChE3, a cDNA encoding an acetylcholinesterase of *Boophilus microplus* (Acari: Ixodidae). *Journal of Medical Entomology,* 43: 707–712. doi: 10.1093/jmedent/43.4.707.

Temeyer, K. B., Pruett, J. H., Olafson, P. U. and Chen, A. C. (2007). R86Q, a mutation in BmAChE3 yielding a *Rhipicephalus microplus* organophosphate-insensitive acetylcholinesterase. *Journal of Medical Entomology,* 44: 1013–1018. doi: 10.1603/0022-2585(2007)44[1013:ramiby]2.0.co;2.

Temeyer, K. B., Pruett, J. H. and Olafson, P. U. (2010). Baculovirus expression, biochemical characterization and organophosphate sensitivity of rBmAChE1, rBmAChE2, and rBmAChE3 of *Rhipicephalus* (*Boophilus*) *microplus*. *Veterinary Parasitology,* 172: 114–121. doi: 10.1016/j.vetpar.2010.04.016.

Temeyer, K. B., Chen, A. C., Davey, R. B., Guerrero, F. D., Howell, J. M., Kammlah, D. M., Li, A. Y., Lohmeyer, K. H., Olafson, P. U., Perez de Leon, A. A., Phillips, P. L., Pound, J. M. and Welch, J. B. (2012). Novel Approaches for control of *Rhipicephalus (Boophilus) microplus*. *Revista mexicana de ciencias pecuarias,* 3: 25–40.

Temeyer, K. B., Schlechte, K. G. and McDonough, W. P. (2019). Baculoviral Expression of presumptive OP-resistance mutations in BmAChE1 of *Rhipicephalus* (*Boophilus*) *microplus* (Ixodida: Ixodidae) and biochemical resistance to OP Inhibition. *Journal of Medical Entomology,* 56: 1318–1323. doi: 10.1093/jme/tjz062.

Tian, Y., Gao, Y., Chen, Y., Liu, G. and Ju, X. (2019). Identification of the fipronil resistance associated mutations in *Nilaparvata lugens* GABA receptors by molecular modeling. *Molecules,* 24: 4116. doi: 10.3390/molecules24224116.

Villar, D., Klafke, G. M., Rodríguez-Durán, A., Bossio, F., Miller, R., Pérez de León, A. A., Cortés-Vecino, J. A. and Chaparro-Gutiérrez, J. J. (2020). Resistance profile and molecular characterization of pyrethroid resistance in a *Rhipicephalus microplus* strain from Colombia. *Medical and Veterinary Entomology,* 34: 105–115. doi: 10.1111/mve.12418.

Villarino, M. A., Waghela, S. D. and Wagner, G. G. (2001). Histochemical localization of esterases in the integument of the female *Boophilus microplus* (Acari: Ixodidae) tick. *Journal of Medical Entomology,* 38: 780–782. doi: 10.1603/0022-2585-38.6.780.

Villarino, M. A., Waghela, S. D. and Wagner, G. G. (2003). Biochemical Detection of Esterases in the adult female integument of organophosphate-resistant *Boophilus microplus* (Acari: Ixodidae). *Journal of Medical Entomology,* 40: 52–57. doi: 10.1603/0022-2585-40.1.52.

Wang, J., McCommas, S. and Syvanen, M. (1991). Molecular cloning of a glutathione S-transferase overproduced in an insecticide-resistant strain of the housefly (*Musca domestica*). *Molecular and General Genetics,* 227: 260–266. doi: 10.1007/BF00259679.

Wei, S. H., Clark, A. G. and Syvanen, M. (2001). Identification and cloning of a key insecticide-metabolizing glutathione S-transferase (MdGST-6A) from a hyper insecticide-resistant strain of the housefly *Musca domestica*. *Insect Biochemistry and Molecular Biology,* 31: 1145–1153. doi: 10.1016/s0965-1748(01)00059-5.

Weill, M., Fort, P., Berthomieu, A., Dubois, M. P., Pasteur, N. and Raymond, M. (2002). A novel acetylcholinesterase gene in mosquitoes codes for the insecticide target and is non-homologous to the ace gene in Drosophila. *Proceedings. Biological Sciences,* 269: 2007–2016. doi: 10.1098/rspb.2002.2122.

Wright, F. C. and Ahrens, E. H. (1988). Cholinesterase insensitivity: a mechanism of resistance in Mexican strains of *Boophilus microplus* (Acari: Ixodidae) against coumaphos. *Journal of Medical Entomology,* 25: 234–239. doi: 10.1093/jmedent/25.4.234.

Zhan, L., Cao, W.-C., Jiang, J.-F., Zhang, X.-A., Wu, X.-M., Zhang, W.-Y., Liu, W., Zuo, S.-Q., Cao, Z.-W., Yang, H., Richardus, J. H. and Habbema, J. D. (2010). *Anaplasma phagocytophilum* in livestock and small rodents. *Veterinary Microbiology,* 144: 405–408. doi: 10.1016/j.vetmic.2010.02.018.

Zhang, X., Liu, Z., Yang, J., Chen, Z., Guan, G., Ren, Q., Liu, A., Luo, J., Yin, H. and Li, Y. (2014). Multiplex PCR for diagnosis of *Theileria uilenbergi*, *Theileria luwenshuni*, and *Theileria ovis* in small ruminants. *Parasitology Research,* 113: 527–531. doi: 10.1007/s00436-013-3684-9.

Zhao, X., Yeh, J. Z., Salgado, V. L. and Narahashi, T. (2004). Fipronil is a potent open channel blocker of glutamate-activated chloride channels in cockroach neurons. *The Journal of Pharmacology and Experimental Therapeutics,* 310: 192–201. doi: 10.1124/jpet.104.065516.

In: The Entomological Guide to *Rhipicephalus*
Editors: Sachin Kumar, Raquel Cossio Bayugar et al.
ISBN: 978-1-53619-619-1

Chapter 9

REFERENCE TICK STRAINS AS AN IMPORTANT BIOLOGICAL MATERIAL FOR ACARICIDE-RESISTANCE CHARACTERIZATION

Francisco Martinez-Ibañez[1], Estefan Miranda-Miranda[2], Carlos Enrique Jasso-Villazul[1] and Raquel Cossio-Bayugar[2,*]

[1]Servicio Nacional de Sanidad, Inocuidad y Calidad Agroalimentaria, Secretaria de Agricultura y Desarrollo Rural (SADER), Mexico

[2] Centro Nacional de Investigaciones Disciplinarias en Salud Animal e Inocuidad, Instituto Nacional de Investigaciones Forestales Agrícolas y Pecuarias (INIFAP), Mexico

ABSTRACT

We describe the selection, culture, maintenance and rearing of the cattle tick *Rhipicephalus* (*Boophilus) microplus* at CENAPA-SENASICA-SADER, Mexico, where reference susceptible and acaricide-resistant tick strains are reared and toxicologically tested. The purpose of maintaining these tick strains is to evaluate the efficacy of diverse acaricides prior to requesting commercial permits to the Mexican Agriculture Ministry. Ticks are also used for assessments of acaricide resistance and compared to field-isolated ticks in acaricide bioassays, such as the larval package test and engorged female test. Five reference cattle tick strains are maintained for these purposes, and all of them are constantly toxicologically characterized. A complete description of their susceptibility and resistance to acaricide formulations, such as organophosphates, pyrethroids, amitraz, fipronil and ivermectin, is provided along with detailed procedures of acaricide bioassays and strain provenance.

Keywords: tick strains, acaricide resistance, toxicological analyses

[*] Corresponding Author's Email: bayugar@yahoo.es.

Abbreviations

OP	Organophosphate
OC	Organochloride
Pyr	Pyrethroids
DD	Discriminating dose
LD	Lethal dose or concentration
RF	Resistance factor
DF	Dilution factor
IC	Initial concentration
CFT	Final working concentration
MR	Mortality rate.

Introduction

The first and most important step of bioassays for acaricide resistance relies on the selection, culture-rearing and maintenance of reference tick strains that have been exhaustively tested for acaricide resistance under laboratory conditions. These strains are denominated reference strains, and ideally, the toxicological bioassay requires an acaricide-susceptible tick strain to compare the susceptibility of field isolates to the different acaricide formulations. A second strain with demonstrated resistance to one or several acaricide compounds is also required to compare with possible acaricide-resistant field isolates. Here, we describe our experience in implementing bioassays of field isolates against comparative reference strains of the cattle tick *Rhipicephalus* (*Boophilus) microplus*. Bioassays for monitoring acaricide resistance were routinely performed on laboratory-reared ticks at the Incubation Laboratory of the Ecotoparasites and Dipteran Department at Centro Nacional de Servicios de Constatación en Salud Animal (National Animal Health Verification Services Center CENAPA-SENASICA-SADER). CENAPA maintains reference susceptible and acaricide-resistant tick strains of *R. (Boophilus) microplus,* which were selected and toxicologically characterized by the staff of technicians. The purpose of maintaining these tick strains is to use those ticks for evaluations of the efficacy of diverse commercial acaricides presented to the Mexican Secretaría de Agricultura y Desarrollo Rural (SADER, Ministry of Agriculture and Rural Development) for evaluation and Mexican commercial registration, as well as commercial permits according to the current Mexican government regulation of commerce. Reference tick strains are also used to compare resistance patterns in the diagnosis of field ticks during surveys and testing for acaricide resistance. Five reference *R. microplus* tick strains are routinely maintained in this laboratory, all of which belong to the species *R. microplus*. All of these strains have been toxicologically characterized using bioassays under laboratory conditions against different acaricides and named as follows: "Susceptible," which exhibits a 100% mortality rate when exposed to all acaricides; "Mora," which is organophosphate (OP)- and pyrethroid (Pyr)–resistant; "San Alfonso," which is organophosphate-, pyrethroid- and amitraz-resistant; and "Rancho Nuevo- and Vargas," which are organophosphate-, pyrethroid-, amitraz- and fipronil-resistant. The toxicological profiles and provenances for all of these strains are described and summarized in Table 1.

During the nonparasitic stages of their life cycle (egg-laying, egg masses and larvae), ticks are maintained in incubators at 28° C ± 2° C with 80 to 90% relative humidity, and for the parasitic stages, 15- to 50-day-old larvae are used to infest Aberdeen Angus cattle. These reference ticks exhibit a parasitic stage of 20- to 23-days and a nonparasitic stage of 40- to 60- days. Fifteen-day-old larvae were used for the packaged larval test for tick toxicology bioassay profiling according to previously reported methodologies (Stone and Haydock, 1962; Drummond et al., 1967; FAO, 1971; FA0, 1984), and 19- to 20-day postinfestation engorged females were used for the immersion test, as described in a previous report (Drummond et al., 1967; FAO, 1999) (Table 2). These tests are used for the verification or control of acaricide efficacy and to determine their toxicological characteristics, as well as to evaluate products on the proper strain that is required for a particular purpose. This analysis is requested by acaricide manufacturers or agriculture authorities, and bovine infestations with the proper tick strains are programmed.

Table 1. Reference strains and resistance profiles. Ticks maintained at our facilities are routinely tested under laboratory conditions by performing larval package bioassays (Stone and Haydock, 1962; FAO, 1987) and engorged female immersion tests (Drummond et al., 1973; FAO, 1999)

Strain (abbreviation)	Pyrethroid Resistance	Organophosphate Resistance	Amidine Resistance	Fipronil Resistance	Provenance
Susceptible (Su)	-	-	-	-	Zacatecas
Mora (Mo)	+	+	-	-	Tabasco
San Alfonso (SA)	+	+	+	-	Tabasco
Rancho Nuevo (RN)	+	+	+	+	Tamaulipas
Vargas (Va)	+	+	+	+	Tamaulipas

Table 2. Acaricide bioassay used to evaluate acaricide resistance and products

Type of Acaricide Bioassay	Development stage	Interpretation Parameters	Type of acaricide test Suitability
Engorged tick immersion[a]	20-day postinfestation engorged females	Adult mortality rate Oviposition rate Eggs mass weight Eclosion rate	OP Pyr Fipronil
Larvae Package Test[b]	15-day-old larvae	Larvae mortality rate	All acaricides
Larvae Immersion Test[c]	15-day-old larvae	Larvae mortality rate	Amitraz Ivermectin[d]

[a] (Drummond et al., 1967; FAO, 1999).
[b] (Stone and Haydock, 1962; FAO, 1987).
[c] (Shaw Rd, 1966; FA0, 1984); [d] (Klafke et al., 2006).

Figure 1. Provenance of acaricide-resistant reference strains. The geographical sites at which all acaricide-resistant strains were isolated are located in Mexican states on the shores of the Gulf of Mexico, and the cattle grazing areas within this region are dedicated to the substantial exploitation of livestock with excessive use of various formulations of acaricides. The red circles indicate the location at which the strains were isolated, and the abbreviations of the strains are described in Tables 1 and 5.

Reference Strains in Mexico

Rhipicephalus microplus reference strains maintained at CENAPA are used for several *in situ* biological evaluations of acaricides and other assessments, *e.g.,* atypical behavior of tick populations, such as tolerance and/or resistance to commonly used brands of acaricides in Mexico. Reference strains are also used for constant evaluations of the effectiveness of acaricides sold in Mexico against different developmental stages of ticks. The acaricide susceptibility and provenance of ticks maintained at the reference tick CENAPA depository are described in Table 1.

Rearing, management and maintenance of ticks exhibiting medical and veterinary importance, such as those that belong to the genus *Rhipicephalus sp.,* are reproduced, selected and characterized by the CENAPA staff of technicians for acaricide resistance procedures by discriminant doses and Probit according to procedures recommended by the FAO (FAO, 2004), and the results of the resistance phenotypes for each strain and their comparative behaviors are reported in CENAPA public records (CENAPA-PD-404 0/2, 2019).

Since *R. microplus* is a cattle ectoparasite, rearing procedures require the availability of Aberdeen Angus cattle for propagation of the parasitic stages of the ticks. For this experimental infestation, approximately 10,000 15-day-old larvae are required and deposited over the back of the animal host with the help of a brush as a tick-spreading tool. One bovine is required for the propagation of each strain of ticks. Additional precautions to consider preventing crossbreeding between tick strains include restricting the movement of the cattle to avoid self-grooming that may detach the ticks and prevent tick-infested bovines from contacting each other and thus preventing cross-transfer of ticks from different strains and strain contamination. After 24 hours, larvae attach themselves to the host skin and feed on blood for 4- to 6-days to become meta-larvae that differentiate into nymphs after 7-9 days. The first meta-nymphs appear on days 10 to 12, with the males differentiating first and female meta-nymphs appearing at days 13 to15. Young adult molting occurs on days 16-18, and engorgement and mating occur from days 16-18. After 20- to 21-days, the samples are ready to be collected, or natural detachment occurs, depending on the strain, ambient temperature and relative humidity (Figure 2) (Núñez et al., 1982).

Once collected, the engorged females are cleaned, and 40 to 50 of them are placed in 90 X 15 mm Petri dishes labeled with the following data: collection date, genus and species, and number of ticks. An average of 4 to 5 Petri dishes are collected. Engorged females are collected from infested bovines for 4- to 5-days, depending on the programmed acaricide products to be evaluated, the type of tick strain used and the requirement of the laboratory to verify acaricide resistance and susceptibility levels of each strain of ticks (CENAPA-PD-404 0/2, 2019).

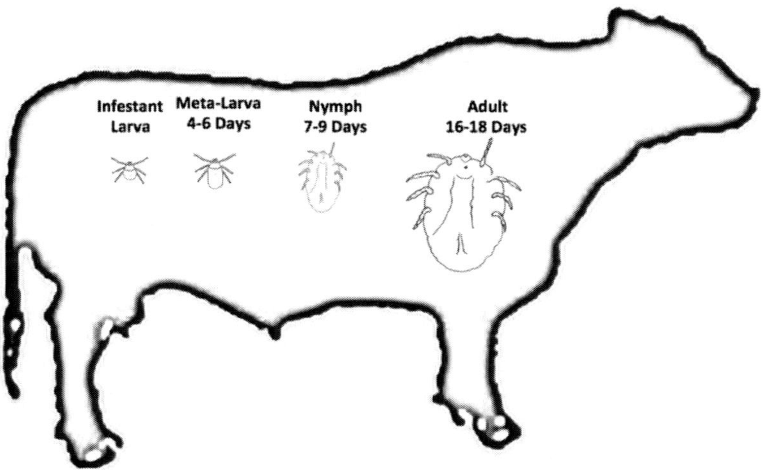

Figure 2. Parasitic stages of *R. microplus* over experimentally infested bovines. Larvae obtained from incubated egg masses are spread over the back of the host. Once attached, they feed on blood for 4-6 days before molting into a meta-larva. After 7-9 days, meta-larvae become nymphs and finally reach adulthood after 16-18 days when they mate and detach from the host (Figure 2).

Egg-laying engorged females are placed in Petri dishes, and their data are registered during the 2- to 3-day long preoviposition period. The next 10- to 12-days are allowed for egg laying to be completed. This egg-laying period may vary according to the different strains, but regardless, a standard procedure in our laboratory is to collect the egg masses at day 14. One gram of eggs containing approximately 20000 eggs is placed in 10 ml glass vials plugged with cotton lids, and each vial is labeled with the egg mass in grams, collection date, genus, species and weight of the eggs. Each vial is incubated until larvae hatch, starting at 7- to 10-days and culminating at 14-days, when larvae climb the vial and accumulate on the cotton plug. From this point, larvae are incubated for 7 more days to ensure that the larvae are ready for use in laboratory procedures such as larvae package tests and larvae immersion tests dedicated to acaricide-resistance diagnostics or Probit analyses (Robertson et al., 2007) to determine resistance indexes at lethal concentrations or doses of LD_{50}, LD_{90} and LD_{99}, as well as the discriminating dose (DD) test (FAO, 2004) (CENAPA-PD-402 3/1, 2019).

In 2008, the CENAPA staff characterized the susceptible reference strain by performing a Probit analysis of the larval package test. The LD_{50} and LD_{90} values for the toxicological assessment are summarized in Table 3 with the average Probit results (2009-2012). Thirteen toxicological verifications were performed in subsequent years to screen the susceptibility phenotype to OP and Pyr acaricides (Martínez et al., 2008).

Table 3. Historical Probit results for the susceptible strain over time

Ixodicide		LD_{50}	LD_{90}	Year
Organophosphorous	Coumaphos	0.0084*	0.072*	2008*
	Chlorfenvinphos	0.0288	0.0493	Average Probit results for 2009-2012
		0.017	0.062	2008*
		0.0154	0.0648	Average Probit results for 2009-2012
	Diazinon	0.0103	0.0280	2008*
		0.0094	0.0334	Average Probit results for 2009-2012
Pyrethroids	Flumethrin	0.0005	0.0017	2008*
		0.0032	0.0104	Average Probit results for 2009-2012
	Deltamethrin	0.005	0.022	2008*
		0.0038	0.0158	Average Probit results for 2009-2012
	Cypermethrin	0.026	0.18	2008*
		0.0194	0.0738	Average Probit results for 2009-2012

* (Martínez et al., 2008). LD_{50} = lethal dose 50; LD_{90} = lethal dose 90.

Procedures for Acaricide Susceptibility-Resistance Diagnostics on Larvae and Adult Stages of the Cattle Tick *Rhipicephalus microplus*

Larvae Package Test

The larvae package test was developed by Australian researchers Stone and Haydock in 1962 and has been widely used for toxicological assessments in tick strains and the detection of acaricide resistance in ixodid ticks. This procedure has received international recognition for its practical usefulness; therefore, the FAO has made this procedure the gold standard for acaricide-resistance diagnostics in ticks worldwide (FAO, 1987). This test basically consists of the exposure of tick larvae to filter papers impregnated with different concentrations of an acaricide. After a certain time, the mortality rate of tick larvae is determined, and the results may be analyzed by performing a discriminating dose (DD) test of the acaricide (the number of ticks that are not killed by a double concentration of acaricide that kills 99% (LD_{99}) ticks of the susceptible strain) or the ratio of the dose of acaricide required to kill a particular proportion of a test population (i.e., 50, 90 or 99%) compared with a susceptible strain and analyzed using Probit (Robertson et al., 2007), which is a logarithmic procedure that matches the logarithmic dilutions of acaricides. The discriminating dose test is a laboratory test that is used to evaluate field isolates, the results indicate acaricide resistance in natural populations of ticks and are useful for the application of alternative control methods when acaricide resistance is detected (FAO, 1987; Stone and Haydock, 1962). An objective of this method is to use the larvae package test for acaricide-resistance diagnostics from field-isolated ticks during surveillance programs by testing tick larvae against OP, organochloride, Pyr and fipronil acaricides; for this purpose, 7- to 15-day-old larvae are used (CENAPA-PD-404 0/2, 2019). The difference between a simple package test analysis and a Probit analysis is basically that the results of the first test are reported as the mortality rate as a percentage when the larvae are exposed to acaricide dilutions that are not necessary are exponential or logarithmic, whereas the latter is always an assessment of logarithmic dilutions compatible with a logarithmic plotting scale, and the results are always expressed as LD_{50} or LD_{90}. Additionally, the Probit analysis facilitates the calculation of the resistance factor (RF) (FAO, 1987). Probit analysis is also used to determine the biological efficacy and to compare the biological effects of different chemical compounds. The procedure of the larval package test used is described in the next section.

Initial Acaricide Stock Dilution

The initial concentration in terms of the percentage of the active formula for every commercial acaricide presentation is important to know in the analytical presentation (95, 98 or 100%). This procedure allows to calculate the weight in grams of the liquid required for the first dilution using the following formula (Figure 3):

$$\frac{(\%\ \text{C.I.})}{(\text{f x } 3)}\ \frac{(100\%)}{(\%\ \text{CFT.})} = \text{grams of principle active component of the acaricide}$$

where:

- % C.I. = percentage of the desired concentration.
- f = fractional value assigned to the volume used to prepare the solution based on 100 ml as the whole.

e.g.: 50 ml (50%) is ½: 25 ml (25%) is ¼; 20 ml (20%) is 1/5, etc., where 3 is a number used as a constant. This constant expresses the fractional value of the next element of the equation, which is the fixer-diluter mixture. Olive oil is used as the fixer of the dilution, and trichloroethylene is used as the diluter in a proportion of 1:2; hence, the constant value of 3 becomes obvious.

The acaricide active principle components are mixed with the dissolvent, as stated in the equation, used to impregnate the filter paper and dried for one hour for 2/3 of the volume of the trichloroethylene to evaporate, which is why the concentration of the acaricide as a function of the olive oil alone (1/3 of the mix) must be calculated (Figures 4-6).

% CFT. Concentration in percentage of the analytical presentation of the acaricide to test; this parameter is displayed on the label of the analytical grade acaricide container.

The 100% relationship with the percent CFT expresses the percent purity of the analytical presentation of the acaricide, and this value must be known to accurately report the percentage of active principle components in the first dilution necessary for the bioassay.

Example:

An initial dilution of 50 ml must be prepared with a concentration of 1%, and the label of the analytical grade acaricide shows a CFT of 58.78%. How many grams are required for this dilution?

$$\frac{1\,\%}{(2)(3)} \times \frac{100\,\%}{58.78\%} = (0.1666)\,(1.7012) = 0.2834\ \text{g}^{*}$$

* = Liquids and solids of the analytical presentation of acaricides are measured in grams.

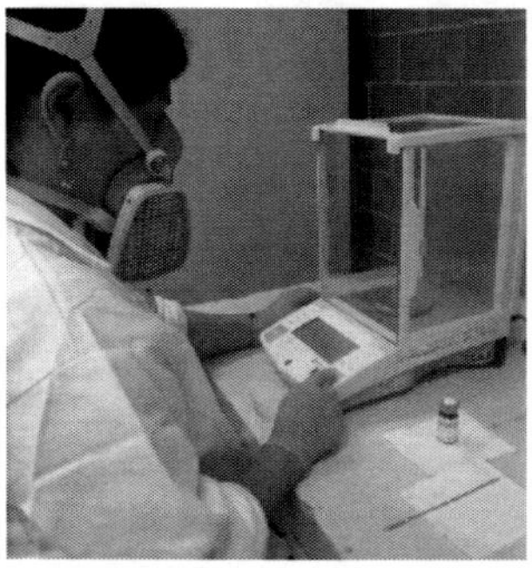

Figure 3. Weighing of the principle active component of the acaricide on an analytical scale.

Figure 4. Fixer-diluent mix of olive oil and trichloroethylene.

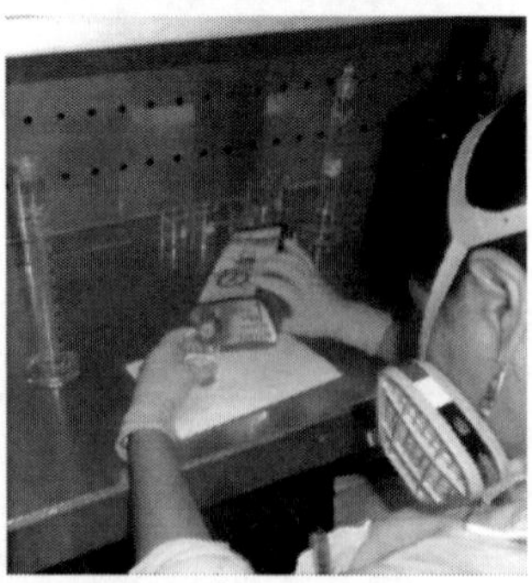

Figure 5. Olive oil measurement for solvent preparation.

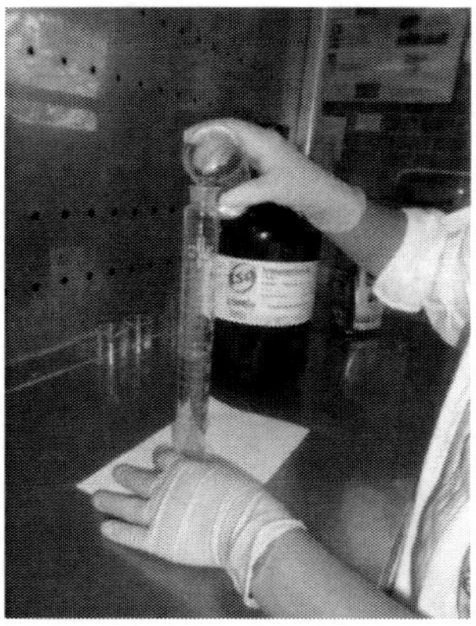

Figure 6. Active principle component mixed with the solvent.

Preparation of Discriminant Doses (DDs)

The DD is the acaricide concentration capable of discriminating between susceptibility and resistance in ticks; it is capable of killing 100% of susceptible ticks but allows the survival of ticks exhibiting acaricide resistance (Figure 7).

Table 4. Discriminant doses obtained for the "Susceptible" reference strain during the assessment of acaricide resistance in samples of *R (Boophilus) microplus* (CENAPA-PD-404 0/2, 2019)

Chemical compound	Ixodicide	Discriminant Doses (DD)
Organophosphates (OPs)	Coumaphos	0.2%
	Diazinon	0.08%
	Chlorfenvinphos	0.2%
	Chlorpyrifos	0.3%
Synthetic Pyrethroids (SPs)	Flumetrine	0.01%
	Deltamethrin	0.09%
	Cypermethrin	0.5%
Organochlorades (Ocls)	Lindane	0.1%
Phenyl pyrazolones	Fipronil	0.05%

From the initial dilution, a DD is prepared using the following equation:

$$Ci \quad Vi = Cf \ Vf$$

where:

Ci = Initial concentration of the acaricide
Vi = Initial volume of the initial solution to be diluted
Cf = Final concentration of the desired dilution
Vf = Desired volume to prepare

Example:

For a flumethrin DD of 0.01% in a volume of 10 ml of dissolvent from a starting stock dilution of 1%.

What is the volume of the initial stock solution needed?

Ci Vi Cf Vf

(1 %) (X) = (0.01 %) (10 ml)

$$\text{where: } (X) = \frac{(0.01\ \%)\ (10\ \text{ml})}{(1\%)} \qquad X = 0.1\ \text{ml}$$

A volume of 0.1 ml of the initial stock solution of 1% plus 99.9 ml of diluter is necessary to obtain a DD of 0.01% for flumethrin.

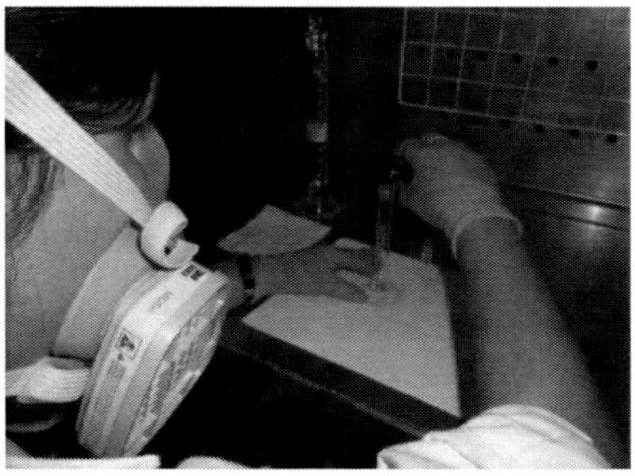

Figure 7. DD preparation.

Procedure for Preparing Dilutions

Six to eight DD dilutions are required for a typical assay on field-isolated ticks, and to obtain LD50, 90 and 99 values, the Dilution Factor (DF) must be included in the laboratory procedure. For instance, a double serial dilution is a DF of 0.5 where a series of graduated cylinders of 10 ml is filled with 10 ml of initial dilution and 5 ml is transferred to the next cylinder containing 5 ml of diluter for eight dilutions. Other DFs are 0.6, 0.7 or higher, depending on the experience and preliminary results.

Paper Impregnation

Once prepared, diluted samples are applied on rectangular filter papers with dimensions of 75 X 85 mm in duplicate experiments, used to contain the larvae and labeled with the following data using a graphite pencil: sample number, strain, acaricide formula and concentration.

Example:

M-004	(Sample number or strain name)
Flumethrin	(Acaricide formula)
R. (Boophilus) microplus (Bm)	(Genus and species)
0.01%	(concentration)

Every sample should be analyzed in at least two replicates for each prepared dilution, including controls. Impregnation is performed using a 1000 µl micropipette by applying 670 µl (0.67 ml) on each filter paper. This step should be performed in an extraction hood away from the incubators. Impregnated papers are placed over aluminum rails prepared for this purpose with the labeled side facing down, the indicated volume of diluted acaricide is applied on the clean side of the paper and allowed to dry while pressed with paper clips, avoiding contact with other filters and any other surface, and hanging from a universal lab support (Figure 8).

For a Probit analysis, this procedure is followed for every dilution, starting with the controls, which are only impregnated with the dilution solution followed by the lowest acaricide solution for evaluation. After drying, the packages are formed using 3 paper clips starting with the controls followed by the lowest acaricide solution to evaluate. For Probit assays, each filter paper is folded in the middle using 96% alcohol-cleaned gloves, with the impregnated side inside and avoiding contact with any surface (Figure 9).

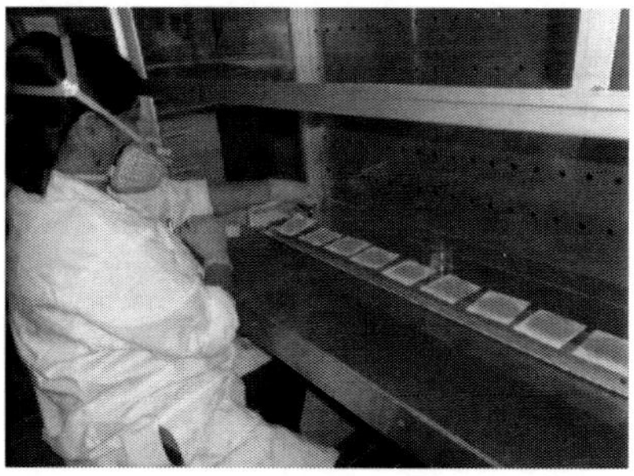

Figure 8. Filter paper impregnation.

Figure 9. Preparation of impregnated filter papers.

Larvae Filling of Packages

Empty packages are placed on a metal tray awaiting the larvae filling procedure. A vial containing the larvae is unplugged and placed on a Petri dish containing soap diluted in water to avoid the dispersal of larvae, along with the brush and dissection needle used to insert the larvae inside the packages (Figure 10). This procedure is conducted within a large tray containing soapy water with a Petri dish upside down at the center where the package to be filled will be supported and kept dry. The larvae filling process starts with the controls, followed by the lowest acaricide solution to evaluate. A brush and dissection needle are used to insert the larvae, removing the upper paper clip and placing approximately 100 larvae while avoiding touching the paper edges with the brush or the needle. Once this procedure is complete, the package is closed with the upper paper clip, and the process proceeds until the remaining packages are filled. When the larvae are filled in all packages, they are placed on a clean tray in an incubator at 28 ± 2° C with 80-90% relative humidity for 24 hours.

Figure 10. Larvae filling in packages.

Interpretation of the Results

After 24 hours, the results are interpreted by counting the larvae that survived the test starting with the controls, followed by the lowest acaricide solution to evaluate. Counts are recorded by capturing all mobile larvae with adherent tape in one hand, while counting them with a mechanical counter with the other hand. The parameters to evaluate are either mobility or paralysis, and larvae that can move by themselves are considered alive. Only living larvae on the filter paper are captured by the tape to be counted to facilitate the process. The results are carefully handwritten on a report form to be processed for calculations of death rate at every LD and further Probit analyses (Figure 11).

The Following Equation Is Used to Estimate the Mortality Rate (MR)

$$\frac{(\text{Dead Larvae})\ (100)}{\text{Total Larvae}} = \text{MR.}$$

When a mortality rate > 10% is observed in the controls, the test is repeated.

Figure 11. Interpretation of the larvae package test.

Larvae Immersion Test (Shaw Immersion Sandwich Technique)

Some laboratory challenges associated with amitraz acaricide-resistance diagnostics elude the experience gained with the larvae package test that is widely used for other acaricide formulas; instead, a new procedure of larval immersion was designed to overcome the difficulties (Shaw, 1966) and modified by FAO (1984). A commercial amitraz acaricide was used (Taktic®) instead of the analytical grade. DDs are depicted in Table 4, and results were interpreted after 72 hours. Probit analyses require 8 to 10 dilutions and 3 replicates for each dilution. The objective of this procedure is to establish amitraz acaricide-resistance diagnostics for *R. microplus* field-isolated ticks or toxicological characteristics of reference strains according to FAO guidelines (FAO, 1987).

Table 5. DD obtained for the susceptible strain used for amitraz acaricide-resistance diagnostics for field-isolated *R. microplus* ticks

Chemical Compound	Ixodicide	D. D.
Amidines	Amitraz	0.0002%

Figure 12. Placement of larvae on amitraz-water-impregnated filter paper.

Figure 13. Treatment of larvae with amitraz-water DD solutions.

Figure 14. Image of the larvae immersed in the DD solutions.

Figure 15. Package filled with larvae from the control and amitraz-submerged larvae test.

For the DD determination, the same equation is used as that for the larvae package test with analytical grade acaricide, and the commercial product of amidines is provided at a concentration of 12.5% (CENAPA-PD-404 0/2, 2015). Seven- to 15-day-old larvae are placed on 150 mm Petri dishes containing 125 mm Whatman No. 1 filter papers impregnated with the solutions starting with the control solution (water) followed by the lowest DD. Next, 300-500 larvae are placed on the paper with the aid of a brush, and a second filter paper is placed on top of the larvae. The immersion time is 10 min, starting with larval placement on the paper (Figures 12-14).

When the immersion period concludes, filter papers are extracted from the solutions with the aid of dissection pliers and placed over paper towels to eliminate the excess acaricide solution. The filter paper is removed, and approximately 100 larvae are removed and placed in a 75 X 85 mm envelope made of Whatman No. 1 filter paper held by paper clips that has been previously labeled with the type of acaricide, concentration and number of replicates. When the larvae filling of packages is concluded, the packages are placed on a clean tray in an incubator at 28 ± 2° C with 80-90% relative humidity (Figure 15).

After 72 hours, the results are interpreted by counting those larvae that survived the test starting with the controls, followed by the lowest acaricide solution to evaluate. This procedure is performed by capturing all mobile larvae with adherent tape in one hand, while counting them with a mechanical counter using the other hand. The parameters to evaluate are either mobility or paralysis, and all larvae that can move by themselves are considered alive. Only living larvae on the filter paper are captured with the tape living death to be counted and to facilitate the process; the results are carefully handwritten on a report form to be processed for calculations of death rate at every LD and further Probit analyses.

The following equation is used to estimate the mortality rate:

$$\text{\% OF MORTALITY} = \frac{\text{(Dead Larvae)}}{\text{Total Larvae}}(100)$$

When a Probit analysis is performed using this procedure, the results of every concentration and the mean values for MR for every concentration are plotted on semilogarithmic paper, drawing a regression line to establish the LD_{50} of the resistant strain by searching for the value in the graph and comparing the values between acaricide-resistant and acaricide-susceptible strains. When the LD_{50} of the resistant strain is divided by the LD_{50} of the susceptible strain, the resistance factor (RF) is obtained (Figures 16-17).

For the ivermectin resistance test, a modification of the larval immersion test was designed (Shaw, 1966) and modified by Klafke et al. (2006). Fourteen- to 21-day-old larvae were immersed in 1.5 ml vials with diluted ivermectin solutions for 10 min, placed on Whatman No. 1 filter papers, introduced into an incubator at 28 ± 2° C and 80-90% relative humidity for 24 hours, analyzed to obtain the percent mortality at different dilutions, and analyzed by performing a Probit assessment of the LD_{50} and LD_{99} of the tested sample relative to the reference strain to obtain the resistance index (RF).

Figure 16. Larvae package interpretation.

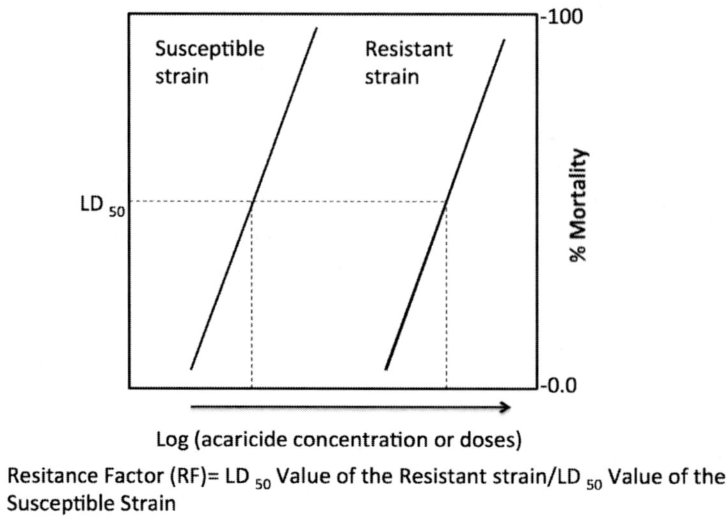

Figure 17. Determination of the resistance factor using a Probit analysis. The larval package test may be used for the Probit analysis by providing the proper logarithmic dilutions and plotting the results over a logarithmic scale, where the Y axis represents the mortality rate as a percentage, and the X axis represents the logarithm of the acaricide dilutions. The LD_{50} value of each strain is determined by extrapolating the plot at a 50% mortality rate. The division of the LD_{50} value of the resistant strain by the LD_{50} value of the susceptible strain represents the resistance factor (RF).

ADULT IMMERSION TEST

This test was described by Drummond et al. (1967) to test the adult developmental stage of ticks, which is the stage most likely to be treated by an acaricide over the bovine host (FAO, 1999). It uses handpicked engorged female ticks submerged in different acaricide dilutions for a period that ranges from 1 to 30 min. The test is suitable for testing the efficacy of commercial products at the manufacturer's recommended concentrations, but it is also useful for bioassays of acaricide resistance using Probit analysis (FAO, 1999).

Clean engorged females that have been collected no more than 24 hours prior to the bioassay procedure are required. The most commonly used acaricide for this type of testing is amitraz due to the difficulties in testing this acaricide using the other bioassay procedures. Dilutions are prepared in water using 25 ml beakers, double serial dilutions are prepared, and ten ticks are submerged in each dilution. Treated ticks are removed from the solution with a sieve and allowed to dry on a paper towel. Once dried, the ticks are placed in Petri dishes and incubated at 27° C and 80-90% relative humidity for seven days. Ticks are observed daily, egg-laying and mortality rates are recorded using the following equations:

Number of egg-laying treated ticks/Number of egg-laying control ticks (100) =Mortality rate.

Another parameter to measure from submerged engorged females is the egg laying inhibition rate using the following equation:

Weight of ticks in the treated group/Weight of ticks in the control group - Weight of eggs in the treated group/Weight of eggs in the control group (100) = Egg-laying inhibition rate

Acaricide Resistance in Mexico

R. microplus is endemic to 53% of the Mexican territory (1,043,772.4 Km^2), an area where 70% of the cattle is found (Solis, 1991; SENASICA-CENAPA, 2020). Every acaricide designed for tick control in cattle has been used in Mexico, including organochlorides, organophosphorus, carbamates, pyrethroids, amidines, macrocyclic lactones, phenyl pyrazolones and development inhibitors (IGR), as well as an anti-cattle tick vaccine. The most commonly used acaricides are those formulated as emulsion concentrates suitable for aspersion or immersion, followed by cutaneous pour-on applications and injectables. Currently, in Mexico, acaricides, repellents and synergists are sold as mixtures of several compounds.

Organophosphorus (OP) Acaricide Resistance

In 1976, a national cattle tick eradication campaign was officially established. OP acaricides were the only compounds authorized for use at the national level, and after extensive use during the next seven years, OP acaricide resistance was documented in 1982 in the cattle-grazing areas around Tuxpan city, which is located in the Gulf of Mexico state of Veracruz (Camino et al., 1984). Ticks isolated from that area were brought to our laboratories, and an acaricide-resistant strain was purified by acaricide pressure selection using the larvae package test. As a result, the first OP acaricide-resistant reference strain was obtained and named Tuxpan, after its region of origin, with resistant factors (RFs) of 9.55, 14.39 and 11.79 for coumaphos, chlorpyriphos and ethion, respectively (Aguirre et al., 1986) (Table 5). In 1984, another acaricide-resistant strain exhibiting double organochloride (OC) and OP acaricide resistance was isolated from the Tempoal region adjacent to Tuxpan city, which was also characterized using the larvae package test (CENAPA-PD-404 0/2, 2019). Since those first efforts to implement an *in vitro* assessment of toxicological acaricide resistance in cattle ticks, the adopted protocols have constantly incorporated refinements to include recently detached engorged female immersion tests suitable for assessing acaricide products during the adult stage of tick development, which has been overlooked until this point (Drummond et al., 1967).

Pyrethroid Acaricide Resistance

In 1985, Pyr acaricides were granted permission for sale in Mexico, Pyr acaricide resistance was detected in 1993 in the same region of the Gulf of Mexico, which was confirmed by the DD calculated using the larvae package test (Stone and Haydock, 1962). New Pyr-resistant reference strains named San Jorge and La Mora were isolated, exhibiting resistance to flumethrin, deltamethrin, cypermethrin and residual OP resistance to coumaphos and chlorfenvinphos (Ortiz et al., 1995). In 1994, a cattle tick acaricide resistance survey program was initiated in high-risk areas in nine Mexican states, and widespread acaricide resistance was observed in cattle ticks isolated from all surveyed areas (Fragoso *et al.*, 1996; Ortiz et al., 1995). Two new Pyr-resistant reference strains named Coatzacoalcos and Aldama were obtained during this survey (Fragoso and Soberanes, 2001). The tick reference strains and their acaricide resistance phenotypes are depicted in Table 5.

Table 6. Chronology of the isolation and characterization of acaricide-resistant reference strains maintained at CENAPA. Strains are described by the date of isolation and their Resistance Factors[a], as described in the text, showing the evolution of acaricide resistance in Mexico

	Strains							
Name (Abbreviation)	Tuxpan (Tx)	Tempoal (Te)	Mora (Mo)	Aldama (Al)	San Jorge (SJ)	Coatzacoalcos (Co)	San Alfonso (SA)	Vargas (Va)
Year [b]	1982[c]	1984[d]	1994[d]	1994[d]	1994[d]	1994[d]	2001[e]	1991[f]
Provenance	Tuxpan, Veracruz	Tempoal, Veracruz	Emiliano Zapata, Tabasco	Aldama, Tamaulipas	Veracruz	Coatzacoalcos Veracruz	Tabasco	Aldama Tamaulipas
	Acaricide Resistance Factor							
Coumaphos	4	11	5.05	2.9	11	-	2	1.95
Chlorfenvinphos	10	4	0.0	3.5	4	-	1.5	32.77*
Diazinon	11	10	8.6	2.9	10	-	6.9	40.67
Lindane	2	535	-	-	535	-	-	-
Flumethrin	-	-	352.9	65.95	450	-	+	2.62
Deltamethrin	-	-	104	3.4	166.1	-	+	7.20
Cypermethrin	-	-	118.75	2.83	63.1	16.19	+	3.91
Amitraz	-	-	-	-	-	-	41.9	49
Fipronil	-	+	-	-	-	-	-	8.29

[a] RF, resistance factor= LD_{50} of the test strain divided by the LD_{50} of the susceptible reference strain.
[b] Year of collection, [c] (Aguirre et al., 1986), [d] (Ortiz et al., 1995), [e] (Fragoso and Soberanes, 2001; Soberanes-Céspedes et al., 2002), + the authors could not determine the resistance factor, [f] (Osorio et al., 2012).

The reference tick strains Tuxpan and Coatzacoalcos were shared with the USDA CFTRL in the Mission as a collaboration between countries. This strain was established between 1994 and 2001. The Coatzacoalcos strain was challenged with permethrin, and Tuxpan strains were challenged with coumaphos to increase and maintain resistance to pyrethroid and OP acaricides, respectively (Davey and George, 1998; Miller et al., 1999; Li et al., 2003), to elucidate the resistance mechanism to these acaricides.

Amitraz Resistance

Since 1993 and after the detection of Pyr and OP acaricide resistance in cattle ticks, a survey program was implemented to identify resistance to the amidine acaricide, since sales of this product had increased by 1000%. It was detected in 2002 in the Mexican state of Tabasco (Soberanes-Céspedes et al., 2002) and named San Alfonso, a multiresistant strain that also exhibited acaricide resistance to Pyr and OP in addition to amitraz, a similar phenotype to the Australian strain Ultimo and the Brazilian strain Cavalcaty (Kemp et al., 1998). The San Alfonso strain exhibited resistance to multiple acaricides, including Pyr, OP and amidines (Table 5), and has been maintained without selective pressure during this period. As a result, it became susceptible to OP acaricides, but maintained resistance to amitraz and pyrethroids, similar to the resistance exhibited when it was isolated (Cossío-Bayúgar et al., 2020).

Fipronil and Ivermectin Acaricide Resistance

Table 7. Reference strains maintained at CENAPA and used in collaboration with other institutions to study mechanisms of acaricide resistance

Strain used	Reference	Study
Susceptible, Tuxpan, Mora	Miranda et al., 1995	Esterase detection in SDS gels and organophosphate-resistant ticks
Susceptible, Tuxpan	Rosario-Cruz et al., 1997	Detection of esterase activity in susceptible and organophosphate-resistant ticks
Tempoal, Aldama, Tuxpan, Mora, Susceptible	Soberanes-Céspedes et al., 2005	Changes in general esterase activity and its relationship with organophosphate resistance
San Alfonso	Fragoso-Sanchez et al., 2011	Amitraz resistance and genetic analysis
Mora, San Alfonso, Tuxpan	Cossío-Bayúgar et al., 2005	Increase phospholipid-hydroperoxide glutathione peroxidase (PHGPx) activity in resistant strains
Susceptible and CENAPA acaricide-resistant field isolates of ticks	(Cossio-Bayugar et al., 2008a)	Pyrethroid resistance associated with increased levels of the monooxygenase enzyme detected using zymograms
Susceptible, Mora, San Alfonso, Tuxpan	(Cossio-Bayugar et al., 2008b)	Pyrethroid resistance associated with increased cytochrome P450 expression
Mora, San Alfonso and susceptible	Cossío-Bayúgar et al., 2009	Cholinesterase and esterase expression levels in acaricide-resistant ticks
Mora, susceptible	Miranda-Miranda et al., 2009	Age induced carboxylesterase expression
Susceptible	Miranda-Miranda et al., 2010	Detection of *Staphylococcus saprophyticus* as a tick pathogen
Susceptible	Cossío-Bayúgar et al., 2012	Tyramine, octopamine, epinastine and 83 adrenergic compounds were screened for their ability to block oviposition in *Rhipicephalus microplus*
San Alfonso, Mora, susceptible	Pérez-González et al., 2014	Effects of new ethyl and methyl carbamates on *Rhipicephalus microplus* larvae and adult ticks
San Alfonso	Ibarra-Velarde et al., 2015	Evaluation of permethrin, cypermethrin or zeta-cypermethrin mixed with plant extracts against susceptible and resistant (San Alfonso) *Rhipicephalus* (*Boophilus*) *microplus* (Acari: Ixodidae) strains
Susceptible and CENAPA acaricide-resistant tick field isolates	Cossío-Bayúgar et al., 2018	P450 cytochrome oxidase expression and pyrethroid resistance
San Alfonso, susceptible	Cossío-Bayúgar et al., 2018	Hemolymph proteomic and enzymatic analyses
San Alfonso	Iturbe-Requena et al., 2020	Oogenesis and embryogenesis inhibition induced by two new ethyl-carbamates in the cattle tick *Rhipicephalus microplus*
San Alfonso, susceptible	Cossío-Bayúgar et al., 2020	Pyrethroid resistant, para-sodium channel gene

When *R. microplus* resistant to multiple acaricides was confirmed in Mexico, it was detected as a widely dispersed phenomenon in cattle grazing areas of the country. As a result, CENAPA was commissioned to determine the DDs of fipronil and ivermectin as alternative to cattle tick control, finding an LD_{50} of 0.05 and 2.09, respectively (Osorio et al. 2008). Once these tests were implemented, a number of samples were analyzed until one from Aldama Tamaulipas was found to be fipronil acaricide-resistant (Table 5). The technical staff performed a complete toxicological assessment and dubbed Aldama the new fipronil acaricide-resistant strain (Table 5).

Conclusion

In Mexico, CENAPA is the FAO National Reference Center for antiparasitic compounds. It is one of the few countries allowed to maintain reference strains of different parasites for product evaluation using bioassays applying the norm NOM-006-ZOO-1993 and constantly applies these assessments either for strain characterization or acaricide evaluations. Additionally, CENAPA supports acaricide resistance monitoring surveys and provides biological samples dedicated to scientific research related to the biology of ticks, molecular approaches to acaricide resistance diagnostics, etc. A sample of scientific publications is depicted in Table 6. Acaricide resistance surveys have identified zones with problems of resistance to multiple acaricides, such as OP, Pyr, OC, and fipronil, and only ivermectin acaricide resistance remains elusive to date.

References

Aguilar-Díaz, H., Esquivel-Velázquez, M., Quiroz-Castañeda, R. E., Miranda-Miranda, E., Conde-Baeye, R. J. P., Cobaxín-Cárdenas, M., Ostoa-Saloma, P. and Cossío-Bayúgar, R. (2018). Comparative hemolymph proteomic and enzymatic analyses of two strains of *Rhipicephalus* (*Boophilus*) *microplus* ticks resistant and susceptible to ixodicides. *BioMed Research International*: e9451547. doi: https://doi.org/10.1155/2018/9451547.

Aguirre, E. J., Sobrino, A., Santamaria, V. M., Aburto, S. A., Roman, S. E., Hernandez, C. M., Ortiz, E. M. and Ortiz, N. A. (1986). Resistencia de garrapatas en Mexico [Tick resistance in Mexico]. In *Memorias del Seminario Internacional de Parasitologia Animal*, pp. 281–306. AMPAVE-SARH, Cuernavaca, Morelos, Mexico.

Camino, L. M., Butler, J. S., Rios, R. G. and Quintero, M. T. (1984). The development of an integrated pest managment system of the cattle tick, *Boophilus microplus* (Canestrini) in Morelos State México. In *Acarology VI*, pp. 1220–1231. Ellis Horwood Ltd, London.

CENAPA-PD-402 3/1 (2019). *Procedimiento para el cultivo y mantenimiento de cepas de garrapatas (CENAPA-PD-402 3/1)* [*Procedure for the cultivation and maintenance of tick strains (CENAPA-PD-402 3/1)*]. Departamento de Ectoparásitos y Dípteros, CENAPA-SENASICA, SADER (Mexican Ministry of Agriculture and Rural Development) Dirección General de Salud Animal, available upon request. Morelos Mexico.

CENAPA-PD-404 0/2 (2015). Procedimientos de técnicas para el diagnóstico de susceptibilidad a acaricidas en larvas y adultas de la garrapata R. (B) microplus. (CENAPA-PD-404 0/2) [Techniques procedures for the diagnosis of susceptibility to acaricides in larvae and adults of the tick R. (B) microplus. (CENAPA-PD-404 0/2)]. Departamento de Ectoparásitos y Dípteros. *Técnica de inmersión Larval (Shaw`s Immersion Sandwich Technique, -LIT).*, Official Laboratory techniques, CENAPA-SENASICA, SADER (Mexican Ministry of Agriculture and Rural Development) Dirección General de Salud Animal, available upon request.

CENAPA-PD-404 0/2 (2019). Procedimientos de técnicas para el diagnóstico de susceptibilidad a acaricidas en larvas y adultas de la garrapata R. (B) microplus. (CENAPA-PD-404 0/2) [Techniques procedures for the diagnosis of susceptibility to acaricides in larvae and adults of the tick R. (B) microplus. (CENAPA-PD-404 0/2)]. Departamento de Ectoparásitos y Dípteros. Prueba de Paquete de larvas *Stone y Haydock, Larval Packet Test-LPT)*, Tick Resistance analysis, CENAPA-SENASICA, SADER (Mexican Ministry of Agriculture and Rural Development) Dirección General de Salud Animal, available upon request, Morelos Mexico.

Cossío-Bayúgar, R., Miranda, E. and Holman, P. J. (2005). Molecular cloning of a phospholipid-hydroperoxide glutathione peroxidase gene from the tick, *Boophilus microplus* (Acari: Ixodidae). *Insect Biochemistry and Molecular Biology,* 35:1378–1387. doi: 10.1016/j.ibmb.2005.08.008.

Cossio-Bayugar, R., Miranda-Miranda, E., Ortiz-Najera, A. and Neri-Orantes, S. (2008a). *Boophilus microplus* pyrethroid resistance associated to increased levels of monooxygenase enzymatic activity in field isolated Mexican ticks. *Journal of Biological Sciences,* 8: 404-409. doi: 10.3923/jbs.2008.404.409.

Cossio-Bayugar, R., Miranda-Miranda, E., Ortiz-Najera, A., Neri-Orantes, S. and Olvera-Valencai, F. (2008b). Cytochrome P-450 monooxygenase gene expression supports a multifactorial origin for acaricide resistance in *Ripicephalus microplus*. *Research Journal of Parasitology,* 3: 59–66. doi: 10.3923/jp.2008.59.66.

Cossío-Bayúgar, R., Miranda-Miranda, E., Portilla-Salgado, D. and Osorio-Miranda, J. (2009). Quantitative PCR detection of cholinesterase and carboxylesterase expression levels in acaricide resistant *Rhipicephalus* (*Boophilus*) *microplus*. *Journal of Entomology,* 6: 117–123. doi: 10.3923/je.2009.117.123.

Cossío-Bayúgar, R., Miranda-Miranda, E., Narváez Padilla, V., Olvera-Valencia, F. and Reynaud, E. (2012). Perturbation of tyraminergic/octopaminergic function inhibits oviposition in the cattle tick *Rhipicephalus* (*Boophilus*) *microplus*. *Journal of Insect Physiology,* 58: 628–633. doi: 10.1016/j.jinsphys.2012.01.006.

Cossío-Bayúgar, R., Martínez-Ibañez, F., Aguilar-Díaz, H. and Miranda-Miranda, E. (2018). Pyrethroid acaricide resistance is proportional to P-450 cytochrome oxidase expression in the cattle tick *Rhipicephalus microplus*. *BioMed Research International*, 2018: e8292465. doi: https://doi.org/10.1155/2018/8292465.

Cossío-Bayúgar, R., Miranda-Miranda, E., Martínez-Ibañez, F., Narváez-Padilla, V. and Reynaud, E. (2020). Physiological evidence that three known mutations in the para-sodium channel gene confer cypermethrin knockdown resistance in *Rhipicephalus microplus*. *Parasites & Vectors,* 13: 370. doi: 10.1186/s13071-020-04227-7.

Davey, R. B. and George, J. E. (1998). *In vitro* and *in vivo* evaluations of a strain of *Boophilus microplus* (Acari: Ixodidae) selected for resistance to permethrin. *Journal of Medical Entomology,* 35: 1013–1019. doi: 10.1093/jmedent/35.6.1013.

Drummond, R. O., Graham, O. H. and Ernest, S. E. (1967). Evaluation of insecticides for the control of *B. annulatus* (Say) and *B. microplus* (Canestrini) (Acarina: Ixodidae) on cattle. In *2nd International Congress of Acarology*, pp. 493–498.

Drummond, R. O., Ernst, S. E., Trevino, J. L., Gladney, W. J. and Graham, O. H. (1973). *Boophilus annulatus* and *B. microplus*: laboratory tests of insecticides. *Journal of Economic Entomology*, *66*: 130–133. doi: 10.1093/jee/66.1.130.

FA0 (1984). *Ticks and tick-borne disease control: a practical field manual. v. 1: Tick control.- v. 2: Tick-borne disease control.* Food and Agriculture Organization of the United Nations, Rome.

FAO (1971). Recommended methods for the detection and measurement of resistance of agricultural pests to pesticides. Tentative method for larvae of cattle ticks, *Boophilus* spp.-FAO Method no. 7. *Plant Protection Bulletin, F.A.O.* 19,.

FAO (1987). *Control de garrapatas y las enfermedades que transmiten. Manual práctico de campo* [*Control of ticks and the diseases they transmit. Field practical manual*]. Food and Agriculture Organization of the United Nations, Rome.

FAO (1999). *Standardization of diagnostic test. Sub-group 2. Adult Inmersion Test (AIT). The FAO working group and FAO/industry contact group on parasite resistance.* Embrapa-Gado De Leite. Food and Agriculture Organization of the United Nations, Juis de Fora M.G. Brazil.

FAO (2004). *Guidelines, resistance management and integrated parasite control in ruminants.* Animal Production and Health Division, Agriculture Dept., Food and Agriculture Organization of the United Nations, Rome.

Fragoso, S. H. and Soberanes, C. N. (2001). Control de la resistencia a los ixodicidas a la luz de los conocimientos actuales [Control of resistance to ixodicides in light of current knowledge]. In *Memorias*, pp. 40–48. Asociación Mexicana de Médicos Especialistas en Bovinos, A. C, Veracruz, Mexico.

Fragoso, S. H., Soberanes, C. N., Ortiz, E. M., Santamaria, V. M. and Ortiz, N. A. (1995). Epidemiologia de la resistencia a ixodicidas piretroides en garrapatas Boophilus microplus, en la República Mexicana [Epidemiology of resistance to pyrethroid ixodicides in Boophilus microplus ticks, in the Mexican Republic]. In *III Seminario Internacional de Parasitología VeterinariaResistencia y control en garrapatas y moscas de importancia veterinaria*, p. 45+57. SAGAR-CANIFARMA-FAO-IICA-INIFAP.

Fragoso-Sanchez, H., Garcia-Vazquez, Z., Tapia-Perez, G., Ortiz-Najera, A., Rosario-Cruz, R. and Rodriguez-Vivas, I. (2011). Response of Mexican *Rhipicephalus* (*Boophilus*) *microplus* ticks to selection by amitraz and genetic analysis of attained resistance. *Journal of Entomology,* 8: 218–228. doi: 10.3923/je.2011.218.228.

Ibarra-Velarde, F., Alcala-Canto, Y. and Vera-Montenegro, Y. (2015). *In vivo* and *in vitro* evaluation of permethrin, cypermethrin or zeta-cypermethrin mixed with plant extracts against susceptible and resistant (San Alfonso) *Rhipicephalus* (*Boophilus*) *microplus* (Acari: Ixodidae) strains. *Pharmacology & Pharmacy,* 6:34–40. doi: 10.4236/pp.2015.61005.

Iturbe-Requena, S. L., Prado-Ochoa, M. G., Velázquez-Sánchez, A. M., García-Hernández, F., Cossío-Bayúgar, R., Muñoz-Guzmán, M. A., Ángeles, E. and Alba-Hurtado, F.

(2020). Oogenesis and embryogenesis inhibition induced by two new ethyl-carbamates in the cattle tick *Rhipicephalus microplus*. *Ticks and Tick-borne Diseases,* 11: 101326. doi: 10.1016/j.ttbdis.2019.101326.

Kemp, D. H., Thulner, F., Nari, A. and Sabatini, G. A. (1998). *Acaricide resistance in the cattle ticks Boophilus microplus and Boophilus annulatus*. Animal Health Services. FAO, 1998.

Klafke, G. M., Sabatini, G. A., de Albuquerque, T. A., Martins, J. R., Kemp, D. H., Miller, R. J. and Schumaker, T. T. S. (2006). Larval immersion tests with ivermectin in populations of the cattle tick *Rhipicephalus* (*Boophilus*) *microplus* (Acari: Ixodidae) from State of Sao Paulo, Brazil. *Veterinary Parasitology,* 142: 386–390. doi: 10.1016/j.vetpar.2006. 07.001.

Li, A. Y., Davey, R. B., Miller, R. J. and George, J. E. (2003). Resistance to oumaphos and Diazinon in *Boophilus microplus* (Acari: Ixodidae) and evidence for the involvement of an oxidative detoxification mechanism. *Journal of Medical Entomology,* 40: 482–490. doi: 10.1603/0022-2585-40.4.482.

Martínez, I. F., Ortiz, N. A. and Osorio, M. J. (2008). Comportamiento toxicológico a organofosforados y piretroides de *Boophilus microplus* cepa Susceptible (CENAPA) del año 2002 al 2005 posterior a su caracterización [Toxicological behavior to organophosphates and pyrethroids of Boophilus microplus Susceptible strain (CENAPA) from 2002 to 2005 after their characterization]. In *VI Seminario Internacional de Parasitología Animal*, p. SAGARPA- AMPAVE- CANIFARMA- INIFAP, Boca del Rio, Veracruz, Mexico.

Miller, R. J., Davey, R. B. and George, J. E. (1999). Characterization of pyrethroid resistance and susceptibility to coumaphos in Mexican *Boophilus microplus* (Acari: Ixodidae). *Journal of Medical Entomology,* 36:533–538. doi: 10.1093/jmedent/36.5.533.

Miranda, E., Cossio-Bayugar, R., Tellez-Alaniz, M., Garcia-Vazquez, Z., Rosario-Cruz, R. and Ortiz-Estrada, M. (1995). An enzymatic marker for ixodicide resistance detection in the cattle tick *Boophilus microplus*. *Advances in Agricultural Research,* 4:1–8.

Miranda-Miranda, E., Cossio-Bayugar, R., quezada Delgado, R., Olvera-Valencia, F. and Neri-Orantes, S. (2009). Age-Induced carboxylesterase expression in acaricide-resistant *Rhipicephalus microplus*. *Research Journal of Parasitology,* 4: 70–78. doi: 10.3923/jp. 2009.70.78.

Miranda-Miranda, E., Cossio-Bayugar, R., Rosario Quezada-Delgado, M. del, Sachman-Ruiz, B. and Reynaud, E. (2010). *Staphylococcus saprophyticus* is a pathogen of the cattle tick *Rhipicephalus* (*Boophilus*) *microplus*. *Biocontrol Science and Technology,* 20:1055–1067. doi: 10.1080/09583157.2010.505325.

NOM-006-ZOO-1993 (1994). Norma oficial mexicana NOM-006-ZOO-1993, requisitos de efectividad biológica para los ixodicidas de uso en bovinos y método de prueba [Official Mexican standard NOM-006-ZOO-1993, biological effectiveness requirements for ixodicides for use in cattle and test method]. *Servicio Nacional de Sanidad, Inocuidad y Calidadgob.mx*.

Núñez, J. L., Muñoz-Cobeñas, M. E. and Moltedo, H. L. (1982). *Boophilus microplus: la garrapata común del ganado vacuno* [*Boophilus microplus: the common cattle tick*]. Hemisferio Sur. S. A., Buenos Aires, Argentina.

Ortiz, E. M., Santamaria, V. M., Ortiz, N. A., Soberanes, C. N., Osorio, M. J., Franco, B. R., Martínez, I. F., Quezada, D. R. and Fragoso, S. H. (1995). Caracterización de la resistencia de *Boophilus microplus* a ixodicidas en México [Characterization of the resistance of Boophilus microplus to ixodicides in Mexico]. In *Memorias del III Seminario Internacional de Parasitología Animal.*, pp. 58–66. SAGAR-CANIFARMA-FAO-IICA-INIFAP, Acapulco, Guerrero, Mexico.

Osorio, M. J., Martínez, I. F., Delabra, V. G. and Pelaez, F. A. (2012). Caracterización toxicológica de una cepa de garrapatas *Boophilus microplus* resistente a fenilpyrazolonas [Toxicological characterization of a Boophilus microplus strain resistant to phenylpyrazolones]. In *Memorias del VII Seminario Internacional de Parasitología Animal y IX Congreso Nacional de Parasitología Animal*, p. SAGARPA- AMPAVE-CANIFARMA- INIFAP, Querétaro, Qro, México.

Pérez-González, I. E., Prado-Ochoa, M. G., Muñoz-Guzmán, M. A., Vázquez-Valadez, V. H., Velázquez-Sánchez, A. M., Avila-Suárez, B. L., Cuenca-Verde, C., Angeles, E. and Alba-Hurtado, F. (2014). Effect of new ethyl and methyl carbamates on *Rhipicephalus microplus* larvae and adult ticks resistant to conventional ixodicides. *Veterinary Parasitology,* 199: 235–241. doi: 10.1016/j.vetpar.2013.07.042.

Robertson, J. L., Savin, N. E., Savin, N. E., Preisler, H. K., Savin, N. E., Savin, N. E. and Preisler, H. K. (2007). *Bioassays with Arthropods*. CRC Press doi: 10.1201/978142 0004045.

Rosario-Cruz, R., Miranda-Miranda, E., Garcia-Vasquez, Z. and Ortiz-Estrada, M. (1997). Detection of esterase activity in susceptible and organophosphate resistant strains of the cattle tick *Boophilus microplus* (Acari: Ixodidae). *Bulletin of Entomological Research* 87*:* 197–202. doi: 10.1017/S0007485300027358.

SENASICA-CENAPA (2020). Situación actual campaña nacional para el control de la garrapata *Boophilus* spp [Current situation national campaign to control the tick Boophilus spp]. http://www.gob.mx/senasica/documentos/situacion-actual-del-control-de-la-garrapata-boophilus-spp.

Shaw, R. D. (1966). Culture of an organophosphorus-resistant strain of *Boophilus microplus* (Can.) and an assessment of its resistance spectrum. *Bulletin of entomological research,* 53: 389–405. doi: 10.1017/s0007485300056480.

Soberanes-Céspedes, N., Santamaria Vargas, M., Fragoso Sánchez, H. and Garcia Vázquez, Z. (2002). Primer caso de resistencia al amitraz en la garrapata del ganado *Boophilus microplus* en México [First case of resistance to amitraz in the cattle tick Boophilus microplus in Mexico]. *Revista Mexicana de Ciencias Pecuarias,* 40: 81–92.

Soberanes-Céspedes, N., Rosario-Cruz, R., Santamaria Vargas, M. and Garcia Vázquez, Z. (2005). General esterase activity variation in the cattle tick *Boophilus microplus* and its relationship with organophosphate resistance. *Tecnica Pecuaria en Mexico,* 43: 239–246.

Solis, S. S. (1991). Epidemiología de la garrapata *Boophilus* spp y *Amblyomma* spp [*Epidemiology of the tick Boophilus spp and Amblyomma spp*]. In *II Seminario de Parasitología Animal*, pp. 19–30. SARH-IICA.

Stone, B. F. and Haydock, K. P. (1962). A method for measuring the acaricide-susceptibility of the cattle tick *Boophilus microplus* (Can.). *Bulletin of Entomological Research,* 53: 563–578. doi: 10.1017/S000748530004832X.

In: The Entomological Guide to *Rhipicephalus* ISBN: 978-1-53619-619-1
Editors: Sachin Kumar, Raquel Cossio Bayugar et al.

Chapter 10

ACARICIDE RESISTANCE IN INDIA: PROBLEM AND MITIGATION

Anil Kumar Sharma[1,*], Srikanta Ghosh[1], Sachin Kumar[1] and Gajanan M. Chigure[2]

[1]Division of Parasitology, ICAR-Indian Veterinary Research Institute, Izatnagar, Bareilly, UP, India

[2]Department of Veterinary Parasitology, College of Veterinary & Animal Sciences, MAFSU, Parbhani, Maharashtra, India

ABSTRACT

In India, *Rhipicephalus microplus* is the most economically important tick species, widely prevalent and infesting animals. The infestation causes many problems in animals including skin damage, blood loss, fatal diseases like babesiosis and anaplasmosis, reduced growth and milk production. Frequent application of chemical acaricides on hosts is the most common tick management method adopted by the animal owners. Indiscriminate use of different generations of chemical acaricides during the last several decades has led to the development of many problem linked with environment, human, animal health and development of resistant tick populations. During the last ten years, the resistance to many commonly used chemical acaricides in tick populations collected from different states of India has been reported. Along with organophosphates, synthetic pyrethroids and amidines, the tick population in India is developing resistance to ivermectin. The rapid and sensitive molecular and biochemical techniques have paved ways for early detection and understanding the possible mechanisms of resistance. The possible mechanisms involving metabolic detoxification mediated by esterases, glutathione S-transferase and monooxygenase as well as point mutations at target sites were reported in Indian ticks. The chapter will provide the status of acaricide resistance in ticks, the possible mechanisms of resistance operating in the tick population, factors contributing to the development of resistance and the managemental strategies.

Keywords: acaricides, contributing factors, *Rhipicephalus microplus*, resistance, mechanisms

INTRODUCTION

The livestock sector is significantly contributing to the Indian economy by providing milk, meat, bones and hides. Apart from this, livestock plays a major role in maintaining soil fertility, energy production and rural transportation. The bovine population of 299.9 million which includes 190.90 million cattle and 108.7 million buffaloes (DAHD, 2015) accounts for 16.2 and 56.9% of the world bovine population, respectively. Despite very low productivity per animal, India stands first in the world milk production, accounting for 20% of world production (DAHD, 2018). The overall contribution of the livestock sector in total GDP is nearly 4.11% at current prices during 2012-13 (DAHD, 2015). However, this sector is suffering from many health-related problems posed by bacteria, viruses, fungi, parasites etc. Of the different problems, management of the multidimensional problems created by tick infestations is very challenging. Due to ideal environmental conditions, almost all the dairy animals are suffering from ticks and tick-borne diseases (TTBDs) (Ghosh et al., 2007). The control cost of TTBDs in India has been estimated at the tune of 498.7 million US$/annum through a questionnaire survey (Minjauw and McLeod, 2003).

In India, the problem of TTBDs is particularly relevant because large scale maintenance of susceptible crossbred animals to improve and sustain the production of milk and dairyproducts. Despite the effortsmade to develop other non-chemical methods to manage ticks, the use of chemical acaricides is still the predominant approach implemented in the field (Ghosh and Nagar, 2014). Amongst 106 valid tick species documented from India, *Rhipicephalus microplus* is widely prevalent and is the most damaging species infesting livestock, zoo and wild animals (Ghosh et al., 2007). Due to short and single host life cycle with high fecundity rate, the *R. microplus* has been subjected to maximum exposure of acaricides and thus developed a remarkable ability to become resistant to the toxic effect of the chemicals used to control them. Globally, *R. microplus* has become resistant to almost all the classes of chemicals that are registered for the management of ticks, for example; against organophosphates and carbamates (Li et al., 2005; Kumar et al., 2011); pyrethroids (Ghosh et al., 2015; Klafke et al., 2017a); amidines (Miller et al., 2002, Klafke et al., 2017b); macrocyclic lactones (Martins and Furlong, 2001; Perez-Cogollo et al., 2010a,Klafke et al., 2017a; Nandi et al., 2018); fipronil (Castro-Janer et al., 2010; Klafke et al., 2017a) and against fluazuron (Reck et al., 2014).

The tick management programme is focused on indiscriminate application of synthetic chemicals coming under organophosphates, pyrethroids, formamidines and macrocyclic lactones groups of insecticides. Uses of amitraz, a formamidines compound and ivermectin, a macrocyclic lactonesare comparatively less but now it is rising due to the declining efficacy of organophosphates and pyrethroids. By a questionnaire survey in sampled population, FAO (2004) predicted the presence of widespread acaricidal resistance in India. Later, with the establishment of country specific discriminating concentration (DC) of commonly used acaricides, the reports of resistance and its possible mechanism in *R. microplus* against various types of acaricides are published from different parts of the country (Singh et al., 2010; Kumar et al., 2011; Sharma et al., 2012; Kumar Rinesh et al., 2013; Jyothimol et al., 2014; Kumar et al., 2014; Singh et al., 2015a; Kumar et al. 2017; Nagar et al., 2018; Sharma et. al., 2018; Godara et al., 2019) (Table 1). The chapter will provide overall acaricide resistance scenario and an overview of the current state of knowledge, research gaps, present and future national initiatives required to tackle the complex problem.

Table 1. Reports of acaricide resistance in tick species from different states of India

States	Synthetic pyrethroids				Organophosphate			Amidines	Macrocyclic Lactones	Phenyl pyrazoles	References
	DLM	CYP	FEN	FLU	DZN	MAL	CMP	AMZ	IVM	FIP	
Bihar	√	√	-	-	√	-	-	√	-	-	Kumar et al. (2011), Sharma et al. (2012), Kumar Rinesh et al. (2013), Kumar et al. (2014), Ghosh et al. (2015)
Rajasthan	√	√	-	-	√	-	-	√	√	-	Kumar et al. (2014), Kumar et al. (2016), Sharma et al. (2012), Kumar Rinesh et al. (2013), Gaur et al. (2016), Nandi et al. (2018)
Uttar Pradesh	√	√	-	-	√	-	√	√	√	√	Kumar et al. (2011), Sharma et al. (2012), Kumar et al. (2014), Kaur et al. (2017), Chigure et al. (2018), Nandi et al. (2018), Nagar et al. (2018)
Haryana	√	√	-	-	√	-	-	-	-	-	Shyma et al. (2013)
Kerala	√	-	-	-	-	-	-	-	-	-	Jyothimol et al. (2014), Ravindran et al. (2014)
Jammu & Kashmir	√	√	-	-	-	√	-	√	-	-	Ahanger et al. (2015), Khajuria et al. (2014), Dutta et al. (2017), Godara et al. (2019)
West Bengal	√	√	-	-	√	-	-	√	-	-	Kumar et al. (2011), Sharma et al. (2012), Kumar Rinesh et al. (2013), Kumar et al. (2014)
Tripura	√	√	-	-	√	-	-	-	-	-	Kumar et al. (2011), Sharma et al. (2012)
Punjab	√	√	√	-	√	√	√	√	√	-	Rath et al. (2006), Singh et al. (2010), Kumar et al. (2011), Sharma et al. (2012), Singh et al. (2012), Kumar Rinesh et al. (2013), Singh and Rath (2014), Jyoti et al. (2014), Jyoti et al. (2016b), Singh et al. (2014a), Singh et al. (2014b), Singh et al. (2015b), Nandi et al. (2015), Nandi et al.

Table 1. (Continued)

States	Synthetic pyrethroids				Organophosphate			Amidines	Macrocyclic Lactones	Phenyl pyrazoles	References
	DLM	CYP	FEN	FLU	DZN	MAL	CMP	AMZ	IVM	FIP	
											(2018), Sagar et al. (2019 unpublished data)
Gujarat	√	√	-	√	-	-	√	√	-	√	Singh et al. (2015a), Shyma et al. (2015), Sharma et al. (2018), Sagar et al. (2019 unpublished data)
Madhya Pradesh	√	√			√		√				Katuri et al. (2017) Sagar et al. (2019 unpublished data)
Uttarakhand	√	√	√	-	-	-	-	-	-	-	Vatsya and Yadav (2011), Abdullah et al. (2013)
Andhra Pradesh	√	√	-	-	-	-	-	√	-	-	Kumar et al. (2017)
Telangana	√	-	-	-	-	-	-	-	-	-	Lenka et al. (2016)
Maharashtra	√	-	-	-	-	-	-	-	-	-	Lenka et al. (2016)
Tamil Nadu	√	-	-	-	-	-	-	-	-	-	Lenka et al. (2016)
Karnataka	√	√	-	-	-	-	-	√	-	-	Pradeep et al. (2012), Lenka et al. (2016)

DEL-deltamethrin, CYP-cypermethrin, FEN-fenvaerate, FLU-flumethrin, DZN-diazinon, MAL-malathion, CMP-coumaphos, AMZ-amitraz, IVM-ivermectin, FIP-fipronil.

Reference Tick Strains

Due to non-availability of reference tick strains, no country specific bioassays, biochemical and molecular assays were standardized until 2011 for the assessment of resistance. The systematic work on acaricide resistance characterization in tick populations was started after national initiatives through World Bank Funded, National Agricultural Innovation Project (Project No. NAIP/Comp-4/ C2066/2008-09) in the year 2008. In the subsequent years, reference strains of *R. microplus* which is a prerequisite for resistance characterization were established and are being maintained in the entomology laboratory of the Indian Veterinary Research Institute (IVRI), Izatnagar, India (Table 2). These strains are extensively used for standardization of bioassays, tick specific biochemical assays for quantification of resistance linked metabolic enzymes, identification of mutations in the targeted genes in the resistant populations and drug development research. It was realized that there couldn't be a universal DC for each acaricide globally to monitor the level of resistance because of multiple factors contributing to the development of resistance. The possible factors are geographical isolation, geographical population differences and difference in thehistory ofacaricides usage, selection pressure and climate (He et al., 1999; Pasay et al., 2008).

Reference strains were also established in different tick research laboratories viz., CFTRL, Texas (Miller et al., 1999), Centre for Scientific and Industrial Research Organization, Australia (Kunz and Kemp, 1994; Jonsson and Hope, 2007), Department of

Primary Industries and Fisheries, Australia (Jonsson et al., 2000) and the country specific DCs of a number of chemical acaricides were determined for resistance monitoring. In India, the DCs of nine acaricides were established using the reference strains and are being adopted by different laboratories (Kumar et al., 2011, 2015; Sharma et al., 2012; Nandi et al., 2018). Subsequently, to study the possible role of over-expression of different enzymes and target site insensitivity in conferring acaricide resistance in ticks, the baseline values of esterases, monooxygenases and glutathione S-transferases (Kumar Rinesh et al., 2013; Ghosh et. al., 2017; Fular et al., 2018) and sequence data of the targeted genes (JQ693154, KX670384, KT215345, KX670386, KC493616, KC493617, HM579825, GU222462, HQ337623) were generated and are now being used by tick researchers (Shyma et al., 2015; Gaur et al., 2016, 2017; Sharma et al., 2018; Nagar et al., 2018).

Table 2. List of reference tick lines of *Rhipicephalus. microplus* and *Hyalomma anatolicum* in India

S. No.	Reference tick line	Registration No.	Characters	Reference
1.	IVRI-I *(R. microplus)*	NBAII/IVRI/BM /1/1998	Reference susceptible tick line. Susceptible to deltamethrin, cypermethrin, diazinon, malathion, coumaphos, amitraz, ivermectin and fipronil	Ghosh & Azhahianambi, 2007; Kumar Rinesh et al., 2013
2.	IVRI-II *(H. anatolicum)*	NBAII/IVRI/HA /1/1998	Reference susceptible tick line. Susceptible to deltamethrin, cypermethrin and diazinon	Ghosh & Azhahianambi, 2007; Kumar Rinesh et al., 2013
3.	IVRI-IV *(R. microplus)*	NBAIR/IVRI/B M/4/2009	Pyrethroid resistant tick line. Resistant to deltamethrin (RF = 194) and cypermethrin (RF = 26.6)	Ghosh et al., 2017
4.	IVRI-V *(R. microplus)*	----------	Multi-acaricide resistant tick line. Resistant to deltamethrin (RF = 22.9), cypermethrin (RF = 20.0), fenvalerate (RF = 25.1) and diazinon (RF = 25.4)	Fular et al., 2018

RF-Resistance Factor.

ACARICIDE RESISTANCE STATUS IN INDIA

The *R. microplus* population infesting cattle kept at the Latoli kraal farm of the Indian Veterinary Research Institute, Mukteswar, developed resistance to benzene hexachloride (BHC) and reported by Chaudhari and Naithani (1964). Later, resistance to carbaryl, a broad-spectrum carbamate pesticide, was reported by Basu and Haldar (1997). Various reports on acaricide resistance were published from time to time but in most of the experiments, the efficacy of commercially available acaricides was assessed (Khan and Srivastava, 1990; Jani et al., 1991; Ghosh and Nanda, 1997; Srivastava et al., 2001; Sinha et al., 2010). After the establishment of DCs of commonly used acaricides, several resistance reports from different states have been published (Table 1).

Resistance to Synthetic Pyrethroids (SPs)

Synthetic pyrethroids, mainly deltamethrin are the most aggressively marketed acaricide that has been used indiscriminately for the management of tick infestations on livestock animals for the last one and half decade. Analysis of the literature published by a number of tick researchers depicted a significantly high level of resistance in *R. microplus* population collected from different parts of the country (Vatsya and Yadav, 2011; Sharma et al., 2012; Kumar Rinesh et al., 2013; Khajuria et al., 2014; Singh and Rath, 2014; Ahanger et al., 2015; Gaur et al., 2016; Lenka et al., 2016; Gaur et al., 2017; Dutta et al., 2017; Godara et al., 2019; Upadhaya et al., 2020). A remarkable decrease in deltamethrin potency against tick population collected from Rajasthan, Punjab (Sharma et al., 2012), Jammu and Kashmir (Khajuria et al., 2014), Bihar (Ghosh et al., 2015) and Gujarat (Sharma et al., 2018) has been reported. It was observed that of the three compounds used for tick management, the resistance to deltamethrin was comparatively higher than cypermethrin and fenvalerate. In some of the tested populations, the resistance factors (RF) between the SP compounds were distantly different. Pradeep et al. (2012) documented 10-30% resistance to deltamethrin and cypermethrin in *R. microplus* population from Bangalore, southern India using tea bag method, filter paper impregnation and immersion method while a low level of deltamethrin resistance was reported in the tick population of *R. annulatus* collected from cattle farms of different districts of Kerala using LPT (Jyothimol et al., 2014; Ravindran et al., 2014).

Resistance to Organophosphates (OPs)

Before the establishment of DC of OP acaricides, the first report of resistance in *R. microplus* population against malathion was documented from the Punjab state (Rath et al., 2006). After the establishment of DC, the presence of widespread resistance to diazinon following adult immersion test (AIT) was reported in twenty isolates of *R. microplus* collected from six agro-climatic regions of Indiawhere intensive animal husbandry is practised and the highest RF of 26.65 (Kumar et al., 2011) was recorded. However, a low level of malathion resistance in *R. microplus* and *H. anatolicum* population from two different districts of Punjab was detected using adult ticks (Jyoti et al., 2014, 2015). Further, loss in the efficacy of diazinon in *R. microplus* collected from Bihar, Haryana and Rajasthan has been reported using both larval and adult stages (Ghosh et al., 2015; Gaur et al., 2016) while resistance against coumaphos in Assam state was also reported using larvae and adults (Upadhaya et al., 2020). The larval stage was more sensitive than adult against OP compounds and hence, the variations in DCs of OP compounds for larvae and adults were observed (Kumar et al., 2015). It is observed that the researchers have adopted both the adult immersion test (AIT) and larval packet test (LPT) to characterize the OP resistance in one and multi-host tick species. The standard bioassay promoted by FAO for resistance characterization is the LPT originally described by Stone and Haydock (1962). However, AIT has been successfully adopted in different laboratories (Sabatini et al., 2001; Jonsson et al., 2007; Mendes et al., 2007; Castro-Janer et al., 2009; Andreotti et al., 2011; Sharma et al., 2012).

Resistance to Formamidine Compounds

The development of resistance against SP and OP compounds has driven the farmers to rely on formamidines (amitraz) which was introduced in the market in 1972. Amitraz is under use in different states where SP compounds were not working. The first case of amitraz resistance in India was reported from Gujarat state (Singh et al., 2015a) and subsequently, Kumar et al. (2014) by using AIT reported resistance ranging from level I to level III in eleven isolates of north India with a maximum RF of 27.3 in Sikar isolates from Rajasthan. Similarly, Dutta et al. (2017) reported a low level of resistance using AIT in two tick isolates collected from four districts of Jammu.

Resistance to Macrocyclic Lactones (MLs)

Macrocyclic lactones have been introduced in the market in 1981 (Geary, 2005) under two categories, ivermectins and milbemycins. It was proposed that MLs are systemically active against ticks, have a longer residual activity than synthetic pyrethroids, and are active against a range of arthropods and nematodes (Taylor, 2001). Despite the positive impact of MLs in terms of tick management, the selective pressure due to widespread application of these drugs during the past few years has triggered the development of resistance in tick population and the resistance to ivermectin has been reported from Brazil (Klafke et al., 2006, 2010) and Mexico (Perez-Cogollo et al., 2010a, 2010b). In India, due to the ease of application methodand availability, the use of MLs has been adopted at a very faster speed and consequently, resistance against the drug has been reported. For example, through limited sample size resistance was reported from Punjab state (Singh et al., 2015b). Further, Nandi et al. (2018) reported different level of resistance (level I to level III) in seven isolates of *R. microplus* collected from the three Indian states (Table 1). Still, MLs are working satisfactorily but there is a need to monitor the situation so that the life span of this effective compound can be extended.

Resistance to Fipronil

Fipronil is a member of phenyl pyrazole insecticides which was introduced in the market in 1993 by Rhone-Poulenc Agro (now Bayer Crop Science). It was initially used in agriculture (Anadon and Gupta, 2012) and later in the livestock sector in the mid-1990s for the management of ticks and horn flies (*Haematobia irritans*) (Guglielmone et al., 2000; Alberti et al., 2001). This is the most recently introduced acaricide in the market, however, failure of efficacy has also been reported (Castro-Janer et al., 2010; Klafke et al., 2017a). In India, no systematic study has been conducted and thus the status of the compound is not yet clear. Reviewing the published information in Indian subcontinent, it is observed that fipronil is used comparatively less in most of the country except in some pockets of Gujarat, Haryana, Madhya Pradesh and Uttarakhand states. As Indian ticks have developed resistance to a number of acaricides, the resistance/susceptibility status of tick isolates collected from 25 districts belonging to six Indian states was screened against fipronil using discriminating concentration (DC) determined earlier (Kumar et al., 2015).

The resistance literature indicated that Indian *R. microplus* populations have developed multi-acaricide resistance and in some cases, the RF has reached beyond the manageable level. However, as in the case of publishedreports from other countries, the protocol to characterize resistance has not been properly followed and sample collection was biased in several cases. No suitable statistical sampling protocol was adopted and sufficient biological materials were not used for resistance characterization. On an average engorged *R. microplus* lays 1200-1500 eggs in laboratory condition and 60% eggs hatched into larvae (Ghosh unpublished data). Shakya et al. (2020) reported the development of fipronil resistance in tick populations collected from six states of India. Few engorged ticks are sufficient to generate the good number of larvae. Although there is a significant difference in the susceptibility status of adults and larvae of the same population to a particular acaricide, the larval stage was used as per the convenience of the researchers. There is a need to have a uniform and effective protocol for resistance characterization and concomitantly, researchersand the publication houses need to be sensitized suitably so that the data generation and reporting process is scientifically validated.

POSSIBLE MECHANISMS OF ACARICIDE RESISTANCE

Metabolic Detoxification of Enzymes

Ticks are using a detoxifying enzyme system to break down or sequester broad spectrum of acaricides due to its wide substrate specificity. Metabolic resistance mediated by detoxifying enzymes such as esterases, glutathione S-transferases (GST) and cytochrome P450 monooxygenases are reportedly involved in conferring resistance inticks (Crampton et al., 1999; Jamroz et al., 2000; Enayati et al., 2010; Temeyer et al., 2013). The involvement of the esterases in development of resistance to synthetic pyrethroids (Miller et al., 1999) and amitraz (Klafke et al., 2017a) in *R. microplus* using esterase inhibitor, triphenyl-phosphate (TPP) was reported. Further, many workers correlated high esterase activity in synthetic pyrethroid resistant ticks (Baffi et al., 2008; Lovis et al., 2012). In India, the work in this area was initiated very late due to the paucity of suitable reference population. By comparative quantitative assay Kumar Rinesh et al. (2013) reported 4.35 and 2.92 times more expression of esterases using α- and β-naphthyl acetate as substrate in field resistant population in comparison to reference susceptible population. Subsequently, Singh and Rath (2014) studied the esterase mediated resistance against synthetic pyrethroids in the field population of *R. microplus* collected from thirteen districts of Punjab state and revealed significantly elevated range of α- (3.34 ± 0.30 to 13.75 ± 1.33µmol/min/mg protein) and β- esterases (1.31 ± 0.15 to 8.09 ± 0.68 µmol/min/mg protein) in nine tick isolates resistant to deltamethrin (RF 2.4 to 38.54) and cypermethrin (RF 1.48 to 11.22). Ghosh et al. (2015) found 1.9 to 4.24 times higher level of esterase activity in seven tick isolates resistant to deltamethrin (RF 2.1 to 34.9) and cypermethrin (RF 1.31 to 5.58). Similarly, Chigure et al. (2018) demonstrated significantly elevated enzyme ratio of 7.79 in comparison to reference susceptible population in SP resistant *R. microplus* ticks collected from different districts of Uttar Pradesh state. A direct correlation between SP resistance and higher activity of esterases has also been reported by Kumar et al. (2016) and Nandi et al. (2015). Further, while characterizing and establishing a multi-acaricide resistance, Fular et al. (2018) recorded significantly higher esterase ratio of 4.78 and 5.84 in the reference multiacaricide resistant tick strain (IVRI-V) of *R. microplus*using α-and β-naphthyl acetate as substrate.

Besides esterases, the higher activity of monooxygenases has also been associated with resistance development. A significantly higher level of cytochrome P450 activity in the resistant population of *R. bursa* from Iran (Enayati et al., 2010) and pyrethroid resistant Coatzacoalcos Mexican population of *R. microplus* (Klafke et al., 2017a). Ghosh et al. (2015) found 2.28-fold increased activity of monooxygenase in seven tick isolates resistant to deltamethrin (RF 2.1 to 34.9), cypermethrin (RF 1.31 to 5.58) and diazinon (RF 1.03 to 57.5) collected from different districts of Bihar state. Similarly, Chigure et al. (2018) recorded 3.11-fold higher monooxygenase activity in resistant *R. microplus* population of Uttar Pradesh state in comparison to susceptible strain.

Along with esterases and monooxygenases, over-expression of GST in resistant population has been recorded by many workers. For example, a significantly higher GST activity (2 to 5-fold) was reported in resistant *R. microplus* collected from Bihar (Ghosh et al., 2015), Uttar Pradesh (Chigure et al., 2018) and from Punjab (Nandi et al., 2015) states. Although an association between resistance and higher enzymes activities has been observed, the quantification protocol standardized for mosquito larvae was adopted without standardizing different variables in the tick system. The optimum age when enzyme activity is at a maximum level and the optimum number of larvae required to have readable activity are the two crucial variables on which activity of the enzymes can be quantified. In addition, reaction time is another crucial factor in the biochemical reaction. These variables are to be standardized in tick system so that a comparable data can be generated.

Utilizing the knowledge, the commercial houses are launching acaricidal formulations (Butox Plus, Naash, and Amipor®) with enzyme inhibitors like piperonyl butoxide (PBO) or other molecules for increasing the efficacy of existing acaricides. In our preliminary laboratory experiments, suitable enzyme inhibitors were identified and were found effective in reducing the LC_{50} value of deltamethrin and ivermectin against resistant field populations (Ghosh unpublished data).

Target Site Insensitivity

The global research efforts have led to the identification of at least ten mutations (SNPs) in the sodium channel gene (Miller et al., 1999; He et al., 1999; Jonsson et al., 2010; Morgan et al., 2009; Stone et al., 2014; Klafke et al. 2019) (Table 3). Out of these ten SNPs, five SNPs (T2134A, C190A, G215T, T170C and C2136A) were non-synonymous and their association with resistance has been proved (He et al., 1999; Morgan et al., 2009; Stone et al., 2014; Klafke et al. 2019). Five other synonymous SNPs viz., C189A, C2130T, C148T, G184C and C190G were not associated with resistanceor their involvement in conferring resistance remains to be investigated (Stone et al., 2014). None of the Indian isolates, across multiple studies (Kumar Rinesh et al., 2013; Kumar et al., 2016; Ghosh et al., 2015; Nagar et al., 2018) has detected T2134A mutation. The absence of T2134A mutation was also reported in various pyrethroid resistant tick isolates from Australia and Brazil (Li et al., 2007; Chen et al., 2009; Andreotti et al., 2011) (Table 3).

Table 3. Summary of the single nucleotide polymorphisms (SNPs) identified in the targeted acaricide resistant gene of *R. microplus* in India and other parts of the world

Targeted gene	Nucleotide position and substitution	Amino acid position and substitution	Reference	Notes
Sodium channel (Nucleotide sequence numbering corresponding to GenBank accession number AF134316)	T2134A	F1550I	He et al., 1999	Not Detected in Indian isolates of *R. microplus*
	C190A	L64I	Morgan et al., 2009	Detected in Indian isolates of *R. microplus*
	G215T	G72V	Jonsson et al., 2010	Not Detected in Indian isolates of *R. microplus*
	T170C	M57T	Stone et al., 2014	
	C148T	Leucine/Phenylalanine		
	G184C	G62R		
	C189A	Synonymous		
	C190G	Leucine/Valine		
	C2130T	Synonymous		
	C2136A	F712L	Klafke et al., 2019	Not Detected in Indian isolates of *R. microplus*
Carboxylesterase gene (Nucleotide sequence numbering corresponding to GenBank accession number AF183383)	G1120A	G374A	Hernandez et al., 2000	Not detected in Indian isolates of *R. microplus*
Octopamine-Tyramine (Nucleotide sequence numbering corresponding to GenBank accession number EF490687)	A22C, T65C	T8P, L22S	Chen et al., 2007 Baron et al., 2015	Not detected in Indian isolates of *R. microplus*
β-adrenergic octopamine receptor (βAOR) (Nucleotide sequence numbering corresponding to GenBank accession number EF490687)	A181T, T185C, A263C, C264A, T123C*, C126T*, A225G*	I61F, I62T, Y88C, Y88C	Corley et al., 2013 Jonsson et al., 2018	Not detected in Indian isolates of *R. microplus*
AChE2 (Nucleotide sequence numbering corresponding to GenBank accession number AJ278342)	G889A, T1090A,C1234T, G1403A, G138A*	V297I, S364T, H412Y, R468K	Ghosh et al., 2015 Nagar et al., 2016	Detected in Indian isolates of *R. microplus*
AChE3 (Nucleotide sequence numbering corresponding to GenBank accession number AY267337)	A366C, A384G, G481A, G633A, C1700G, A1866G, T436C, A455G, T459C	I48L, I54V, R86Q, V137I, I492M, T548A, V71A, I77M, S79P	Temeyer et al., 2009 Jyoti et al., 2016a	

*Synonymous mutation.

Morgan et al. (2009) reported cytosine to adenine (CTC to ATC) mutation at position 190 (C190A) in the domain II S4-5 linker of the sodium channel gene of pyrethroids resistant Parkhurst isolates of *R. microplus* from Australia. Later, this mutation was also detected in pyrethroid resistant *R. microplus* population of Brazil (Mendes et al., 2010; Andreotti et al., 2011; Lovis et al., 2012), Argentina (Lovis et al., 2012), South Africa (van Wyk et al., 2016) and West Africa (Stone et al., 2014; Yessinou et al., 2018).

In India, initially Kumar Rinesh et al. (2013) reported the presence of C190A mutation in the sodium channel gene of the pyrethroid resistant *R. microplus* collected from four different agroclimatic zones. The identification of the exact location of mutation and its respective change in the sodium channel protein (L64I) in pyrethroid resistant Indian isolates of *R. microplus* has given a significant clue to the Indian researcher. Many reports were documented in subsequent years while characterizing pyrethroid resistance in ticks collected from other states (Ghosh et al., 2015; Kumar et al., 2016; Ghosh et al., 2017; Nagar et al., 2018; Fular et al., 2018). Ghosh et al. (2015) studied the pyrethroid resistant *R. microplus* from Bihar state and did not detect the flumethrin specific G215T and putative super-kdr T170C substitutions in any of the field isolates as well as in reference resistant strain. Similarly, Nagar et al. (2018) also could not able to demonstrate the G215T and T170C substitutions in sodium channel gene of several pyrethroid resistant field isolates of *R. microplus* and in reference resistant tick strains.

Till date, only C190A point mutation was recorded in pyrethroid resistant Indian *R. microplus* population. We are in the process of genotyping field population based on C190A mutation. The data generated till date indicated that a significantly higher proportion of the *R. microplus* population is maintaining RS genotypes, however, SS genotypes are at a very low level.

Studies of acaricide resistant *R. microplus* have shown that esterases particularly carboxylesterase (CaEs) are associated with resistance. Hernandez et al. (2000) detected higher transcript of carboxylesterase in pyrethroid resistant *R. microplus* population compared to susceptible ticks suggesting its role in resistance development. Interestingly, the same tick population had a novel nucleotide substitution of adenosine (A) for guanidine (G) at position 1120 (G1120A) in CaEs encoding gene resulting in the replacement at position 374 from aspartate to asparagine (D374N). It was hypothesized that the resultant mutation (D374N) has led to pyrethroid resistance by increasing the affinity of carboxylesterase or by enhancing its ability to hydrolyze pyrethroids (Hernandez et al., 2002). This mutation creates an *EcoRI* restriction site (*CTTAAG*) in the mutant individuals and thus its use as a monitoring tool was developed in the form of a polymerase chain reaction-restriction fragment length polymorphism (PCR-RFLP) assay (Baffi et al., 2007). In later studies, this carboxylesterase marker has also been associated with organophosphate resistance in *R. microplus* (Baffi et al., 2007; Faza et al., 2013). However, Nagar et al. (2016, 2018) did not detect any such mutation at position 1120 in the 372 bp CaES gene in reference resistant IVRI-IV (reference pyrethroid resistant), IVRI-V (multi acaricide resistant) strains and in any of the field isolates of *R. microplus* (Table 3).

Being the target of organophosphates acaricides, acetylcholinesterases (AChE) are the enzymes of interest to understand the molecular basis of organophosphates resistance. In *R. microplus,* three AChE genes were characterized as AChE1, AChE2 and AChE3 (Baxter and Barker, 1998; Hernandez et al., 1999; Temeyer et al., 2004). In India, the information regarding mutations in any of the AChE genes was scanty, until four amino acid substitutions in AChE2 gene of resistant isolates collected from the state of Bihar were identified by Ghosh et al. (2015). They detected five new SNPs (G138A, G889A, T1090A, C1234T and G1403A) in the AChE2 gene of reference OP resistant *R. microplus* strain and in OP resistant field populations. One of the nucleotide substitutions (G138A) was a synonymous mutation while other four substitutions were non-synonymous replacing a valine for isoleucine in position 297 (V297I), a serine for a threonine in position 364 (S364T), histidine for a tyrosine in 412 (H412Y) and a lysine for arginine in position 468 (R468K) (Table 3). They suggested that these mutations could probably be associated with organophosphates resistance but the actual contribution of these mutations in the development of insensitive AChE2 could not be ascertained yet. This can also imply that other factors exist which contribute to insensitive acetylcholinesterase other than point mutations. Later, Nagar et al. (2016) studied the role of mutations in the AChE2 gene in the development of organophosphates resistance in *R. microplus* and identified the same four amino acid substitutions (viz. V297I, S364T, H412Y and R468K) in the resistant field populations and in reference resistant strain of *R. microplus* as detected earlier by Ghosh et al. (2015). Recently, Fular et al. (2018) identified 22 SNPs in AChE2 gene in reference multi-acaricide resistant strain (IVRI-V) of *R. microplus* and amino acid substitution was detected in 11SNPs. Four of the amino acid substitutions (V297I, S364T, and H412Y and R468K) was also reported previously in OP resistant field isolates of *R. microplus* (Ghosh et al., 2015; Nagar et al., 2016). Although at this stage of research, the presence of various mutations in AChE2 gene of OP resistant Indian isolates of *R. microplus* cannot be correlated conclusively with resistance, however, it provides an insight into one of the possible mechanisms.

Jyoti et al. (2016a) reported six point mutations and their corresponding amino acid changes in AChE3 gene (I48L, I54V, R86Q, V71A, I77M and S79P). The first three amino acids were previously associated with organophosphate resistance in the Mexican San Roman strain (Temeyer et al., 2007; Temeyer et al., 2009) and the other three were reported for the first time from India. They hypothesized that these mutations in Indian ticks are possibly associated with OP resistance but these mutations were not detected in reference OP resistant strain (Nagar et al., 2016) and in any of the OP resistant isolates. As such, the definite role of these mutations in conferring resistance could not be ascertained.

Amitraz is a triazapentadiene compound and has been used for the management of ticks. The use of amitraz became more frequent when pyrethroid resistance problems started to hinder the tick management activities. It is thought that amitraz is a potential agonist of the octopaminergic system located in the tick synganglion. The target of amitraz has been proposed to be one of the biogenic amine receptors most likely the β-adrenergic-like (βAOR), or octopamine/tyramine receptors (OCT/Tyr) (Evans and Gee, 1980; Hollingworth and Lund, 1982). Several researchers have identified (Chen et al., 2007; Baron et al., 2015; Jonsson et al., 2018; Sungirai et al., 2018) the multiple mutations in OCT/Tyr and βAOR sequences gene of *R. microplus* collected from diverse locations around the world. Several mutations have been detected to date and the development of a single comprehensive diagnostic test based on DNA sequence variation alone is likely to be difficult (Jonsson et al., 2018). At present, polymorphisms in octopamine-tyramine receptor and βAOR have some potential for molecular diagnosis of amitraz resistance, however, the diversity of mutations suggests that no single polymorphism can be relied on (Rodriguez-Vivas et al., 2018). In India, as per our knowledge, no published literature is available demonstrating the role of any gene(s) in amitraz resistance.

CONTRIBUTING FACTORS IN RESISTANCE DEVELOPMENT

Based on the managemental practices and infrastructure, the farm types are classified as organized and unorganized. Sheds with poor infrastructural and ventilation facilities provide ample breeding spaces for ticks in cracks and crevices for propagation. Besides, the habit of pilling cow dung cakes, farm machinery, feed, gunny bags, etc. in and around the animal sheds further add to breeding/shelter spaces in the sheds for easy proliferation of population.

The climatic conditions of India are very conducive for the growth and survival of the tick population throughout the year. Due to over-reliance on artificial insemination, using exotic germplasm the herd composition has been changed from lowly productive but comparatively resistant indigenous animals to highly productive but susceptible cross-breed's animals. This has resulted in spreading of tick-borne diseases (Bansal, 2005).

The liberalization of the veterinary drug industries has made acaricides even more accessible to farmers. Because of limited control, cases of irrational use of acaricides have been widely observed. Very few livestock owners seek technical advice for acaricide use and adopt a personal convenient strategy to manage ticks by increasing the number of treatments or by increasing the dose of chemicals resulted in the development and spread of resistance. Over and above the application frequency is conveniently increased by the resource rich owners in comparison to resource poor owners. Therefore, improper dose, frequency and mode of application, no or limited rotational use of acaricides andnon-adherence to the principle of use of chemicals have compounded the problem of acaricide resistance.

India has twenty nine states and each state has its own animal husbandry department responsible for the implementation of various federal and state policies. Despite the emerging menace of acaricide resistance and the upsurge in TTBDs in India,no specific policy hasbeen framed to manage tick infestations in any of the state, hence, there is no scientifically framed policy. In the majority of the cases, the acaricide procurement is not based on the resistance/susceptibility status of the ticks which resulted in multifold increase in the use of ineffective acaricide fortick management. The farmers and the field veterinarians are together facing many issues to manage heterogeneously distributed multi-acaricide resistant tick population. Despite receiving complaints from farmers regarding the ineffectiveness of the aggressively marketed chemical acaricides having different trade names for the last many years, concerted efforts have not been taken by the field veterinarians to monitor the efficacy of widely used acaricides in the field condition. Moreover,very limited workforce is available to address the ever-growing problem of acaricide resistance. All the above mentioned factors are working in synergy in large-scale resistance development.

MITIGATION STRATEGY OF ACARICIDE RESISTANCE

Being a vast landform of Indiafalling under fifteen agro-climatic zones, the implementation of a specific regime for tick management for the entire country would neither be possible nor feasible. In order to design well-structured management programs, there is an emergent need for comprehensive data about the status of acaricide resistance spanning different agro-climatic zones of the country. Therefore, periodic and appropriate monitoring of resistance using established and country specific protocol against commonly used acaricides in the particular state/zone is essential. There is a need to establish the specialized referral acaricide resistance testing laboratory in each state and one at the national level for issuing state or region wise recommendation for implementation. The coordinated activities of these laboratories will significantly improve the health status of animals and the environment by minimizing the use of non-effective acaricides which will definitely reduce the economic losses associated with the treatment of TTBDs.

A policy decision regarding initiation of an all India coordinated networking research project is the need of the hour. The project will develop a networking partnership between national and various state research laboratories working on acaricide resistance. The network of these laboratories spread all over the agro-climatic zones of the country will generate data on the status of resistance against various acaricides. The data will be helpful in designing the strategies to slow down the development of resistance and will definitely improve the overall tick management scenario. The data on resistance will also help in policy decision on the type of acaricides to be used in the field situation.

Most of the data generated on acaricide resistance have emerged either from very limited research institutions or from universities and often published in international journals less accessible to government officials of the department of animal husbandry and to the field veterinarians. There is a need to strengthen the extension services through which the laboratory test data and results can be disseminated to farmers, field veterinarians, and policymakers so, that these data can finally be translated into tick management practices in the field.

The government policy of unregulated liberalization of the acaricide market should be reanalyzed. In most of the areas, farmers are in direct contact with pharmaceuticals sales representatives of different companies promoting sales of same active ingredients in different trade names leading to the development of high level of confusion within the farming community. Regulatory guidelines are to be established to protect farmers from the highly competitive marketing strategies of companies.

Non-chemical natural product based tick management protocol is shown a lot of promise. The indigenous ethno-veterinary and medicinal knowledge of Indian farmers offer a range of herbs for their insecticidal and acaricidal properties. The acaricidal, growth-inhibiting, anti-moulting and repellent activities of many plants and herbs have been scientifically established. Chhabra and Saxena (1998) provided a compilation of plants having acaricidal properties in the Indian subcontinent. Although the list of active plants has been improved significantly in the recent years (Ghosh et al., 2014), effective phyto-formulations against acaricide resistant tick population has not yet been marketed. There is a need to develop national level public-private-partnership on natural product development efforts. Once both commercial houses and market accept the potentiality of natural anti-tick products, the exclusive dependency on chemical acaricides will be reduced significantly which in turn will promote green agriculture.

The proposed integrated approaches include pasture alternation and/or rotation combined with limited applications of effective chemical acaricides, strategic use of phyto-formulation, housing management, strengthening of dairy cattle management with alternative feeding technologies such as silage and hay to reduce contact between ticks and cattle, immunization against ticks and ticks borne diseases (de la Fuente et al., 2000; Stutzer et al., 2013). Integrated tick management will substantially reduce the over dependence on acaricides and lessen selection pressure on ticks thus will preserve the efficacy of chemicals and reducing the incidence of acaricide resistance (Jongejan and Uilenberg, 2004).

CONCLUSION

Policy decision regarding uniform tick management strategy is not a realistic target for a country like India because in general, the livestock owners have small holdings, unorganized farms with least infrastructure and minimum knowledge about the gravity of the problem. There is an urgent need for continuous monitoring of acaricide resistance in the field situation through bioassay, biochemical and molecular tools for strategic application of available acaricides and for maintaining the life span of the existing products. Short-to-medium term actions to combat the problem of acaricide resistance should include extensive education of livestock owners on appropriate acaricide use, demerits of indiscriminate use of acaricides, increased farmer access to veterinary advisory services, building human and laboratory capacity for prompt detection and timely intervention against acaricide failure or resistance. One widely agreed recommendation for fighting acaricide failure is the adoption of IPM at least at large farms that are one of the main contributors in spreading resistance tick population in the country.

REFERENCES

Abdullah, S., Yadav, C.L. and Vatsya, S. (2013). Comparative efficacy of two synthetic pyrethroids against *Rhipicephalus (Boophilus) microplus*. *Acarina,* 21: 84–87.

Ahanger, R.R., Godara, R., Katoch, R., Yadav, A., Bhutyal, A.D.S., Katoch, M., Singh, N.K. and Bader, M.A. (2015). Deltamethrin resistance in field populations of *Rhipicephalus (Boophilus) microplus* (Acari: Ixodidae) in Jammu and Kashmir, India. *Experimental and Applied Acarology,* 67: 467–475.

Alberti, H., Alberti, A.L.L., Rinaldi, P.L.F., Lamberti, D.D.G., Rodríguez, L.G. and Lima, G.L. (2001). Avaliac¸ ão da eficácia do fipronil e da deltametrina, formulac¸ ão "pour-on," no control da *Haematobia irritans*, parasitandobovinosem regime de campo, naregiãooeste do Estado de São Paulo. *A-Hora-Vet,* 20:48–51.

Anadon, A. and Gupta, R.C. (2012). Fipronil. In: Gupta RC (ed.) *Veterinary toxicology*. Academic Press, Amsterdam, pp 604–608.

Andreotti, R., Guerrero, F.D., Soares, M.A., Barros, J.C., Miller, R.J. and de Léon, A.P. (2011). Acaricide resistance of *Rhipicephalus (Boophilus) microplus* in state of Mato Grosso do Sul, Brazil. *The Revista Brasileira de Parasitologia Veterinária,* 20: 127–133.

Baffi, M.A., de Souza, G.R.L., de Sousa, C.S., Ceron, C.R. and Bonetti, A.M. (2008). Esterase enzymes involved in pyrethroid and organophosphate resistance in a Brazilian population of *Rhipicephalus (Boophilus) microplus* (Acari: Ixodidae). *Molecular and Biochemical Parasitology,* 160: 70–73.

Baffi, M.A., de Souza, G.R.L., Vieira, C.U., de Souza, C.S., Goulart L. R., and Bonetti, A.M. (2007). Identification of point mutation in a putative caboxylesterase and their association with acaricide resistance in *Rhipicephalus* (*Boophilus) microplus* (Acri:Ixodidae). *Veterinary Parasitology,* 148:301-309.

Bansal, G.C. (2005). Bovine theileriosis in India: an overview. *Proceedings of National Academy of Science, India,* 75: 134–143.

Baron, S., van der Merwe, N.A., Maddar, M. and Maritz-Olivier C. (2015). SNP analysis infers that recombination is involved in the evolution of amitraz resistance in *Rhipicephalus microplus*. *PLoS ONE,* 10(7): e0131341. doi: 10.1371/journal.pone.0131341.

Basu, A.K. and Haldar, D.P. (1997). A note on the effect of continuous use of sevin 50 wp on some cattle ticks. *Journal of Veterinary Parasitology,* 11: 183–184.

Baxter, G.D. and Barker, S.C. (1998). Acetylcholinesterase cDNA of the cattle tick, *Boophilus microplus*: characterization and role in organophosphate resistance. *Insect Biochemistry and Molecular Biology,* 28: 581–589.

Castro-Janer, E., Martins, J.R., Mendes, M.C., Namindome, A., Klafke, G.M. and Schumaker, T.T.S. (2010). Diagnoses of fipronil resistance in Brazilian cattle ticks (*Rhipicephalus (Boophilus) microplus*) using *in vitro* larval bioassays. *Veterinary Parasitology,* 173: 300-306.

Castro-Janer, E., Rifran, L., Piaggio, J., Gil, A., Miller, R.J. and Schumaker, T.T.S. (2009). In vitro tests to establish LC_{50} and discriminating concentrations for fipronil against *Rhipicephalus (Boophilus) microplus* (Acari: Ixodidae) and their characterization. *Veterinary Parasitology,* 162: 120–128.

Chaudhuri, R.P. and Naithani, R.C. (1964). Resistance to BHC in the cattle tick *Boophilus microplus* (Can.) in India. *Bulletin of Entomological Research,* 55: 405–410.

Chen, A.C., He, H. and Davey, R.B. (2007). Mutations in a putative octopamine receptor gene in amitraz-resistant cattle ticks. *Veterinary Parasitology,* 148: 379–383.

Chen, A.C., He, H., Temeyer, K.B., Jones, S., Green, P. and Barker, S.C. (2009). A survey of *Rhipicephalus microplus* populations for mutations associated with pyrethroid resistance. *Journal of Economic Entomology,* 102: 373–380.

Chhabra, M.B. and Saxena, M.J. (1998). The use of phytotherapeutic agents for the control of acariasis in animals: a review. *Journal of Veterinary Parasitology,* 12: 3–8.

Chigure, G.M., Sharma, A.K., Kumar, S., Fular, A., Sagar, S.V., Nagar, G., Upadhaya, D., Saravanan, B.C., Kumar, R. and Ghosh, S. (2018). Role of metabolic enzymes in conferring resistance to synthetic pyrethroids, organophosphates, and phenylpyrazole compounds in *Rhipicephalus microplus*. *International Journal of Acarology,* 44: 28-34.

Crampton, A.L., Green, P., Baxter, G.D. and Barker, S.C. (1999). Monooxygenases play only a minor role in resistance to synthetic pyrethroids in the cattle tick, *Boophilus microplus*. *Experimental and Applied Acarology,* 23: 897–905.

DAHD. (2015). 19th Livestock Census-2012, *All India Report*. Ministry of agriculture andfarmer's welfare, Department of Animal Husbandry, Dairying & Fisheries, Government of India, New Delhi. Pp, 20.

DAHD. (2018). National action plan for dairy development, vision 2022. *All India Report*. Ministry of Agriculture Department of Animal Husbandry, Dairying & Fisheries, Krishi Bhawan, New Delhi. pp: 13.

de la Fuente, J., Rodríguez, M. and García-García, J.C. (2000). Immunological control of ticks through vaccination with *Boophilus microplus* gut antigens. *Annals of New York Academy of Science,* 916: 617–621.

Dutta, S., Godara, R., Katoch, R., Yadav, A., Katoch, M. and Singh, N.K. (2017). Detection of amitraz and malathion resistance in field populations of *Rhipicephalus (Boophilus) microplus* (Acari: Ixodidae) in Jammu region of India. *Experimental and Applied Acarology,* 71: 291–301.

Enayati, A.A., Asgarian, F., Amouei, A., Sharif, M., Mortazavi, H., Boujhmehrani, H. and Hemingway, J. (2010). Pyrethroid insecticide resistance in *Rhipicephalus bursa* (Acari, ixodidae). *Pesticide Biochemistry and Physiology,* 97*:* 243–248.

Evans, P.D. and Gee, J.D. (1980). Action of formamidine pesticides on octopamine receptors. *Nature,* 287*:* 60–62.

F.A.O. (2004). *Resistance management and integrated parasite control in ruminants-Guidelines*, Module I- Ticks: Acaricide resistance: Diagnosis, management and prevention. Food and Agriculture Organisation, Animal Production and Health Division, FAO, Rome, pp. 25–77.

Faza, A.P., Pinto, I.S.B., Fonseca, I., Antunes, G.R., de Oliveira Monteiro, C.M., Daemon, E., de Souza Muniz, M., Martins, M.F., Furlong, J. and de Azevedo Prata, M.C. (2013). A new approach to characterization of the resistance of populations of *Rhipicephalus microplus* (Acari: Ixodidae) to organophosphate and pyrethroid in the state of Minas Gerais, Brazil. *Experimental Parasitology,* 134*:* 519–523.

Fular, A., Sharma, A.K., Kumar, S., Nagar, G., Chigure, G., Ray, D.D. and Ghosh, S. (2018). Establishment of a multi acaricide resistant reference tick (IVRI-V) of *Rhipicephalus microplus. Ticks and Tick-Borne Diseases,* 9: 1184–1191.

Gaur, R.S., Sangwan, A.K., Sangwan, N., Ghosh, M. and Kumar, S. (2017). Comparative study of esterases in deltamethrin and diazinon resistant *Rhipicephalus microplus* and *Hyalomma anatolicum* ticks collected from the Trans-Gangetic plains of India. *Experimental and Applied Acarology,* 73*:* 115–127.

Gaur, R.S., Sangwan, A.K., Sangwan, N. and Kumar, S. (2016). Acaricide resistance in *Rhipicephalus (Boophilus) microplus* and *Hyalomma anatolicum* collected from Haryana and Rajasthan states of India. *Experimental and Applied Acarology,* 69*:* 487-500.

Geary, T.G. (2005). Ivermectin 20 years on: maturation of a wonder drug. *Trends of Parasitology,* 21*:* 530-532.

Ghosh, S. and Azhahianambi, P. (2007). Laboratory rearing of *Theileria annulata*-free *Hyalomma anatolicum anatolicum* ticks. *Experimental and Applied Acarology,* 43*:* 137–146.

Ghosh, S., Bansal, G.C., Gupta, S.C., Ray, D.D., Khan, M.Q., Irshad, H., Shahiduzzaman, M., Seitzer, U. and Ahmed, J.S. (2007). Status of tick distribution in Bangladesh, India and Pakistan. *Parasitology Research,* 101: 207–216.

Ghosh, S., Gupta, S., Kumar, K.A., Sharma, A.K., Kumar, S., Nagar, G., Kumar, R., Paul, S., Fular, A., Chigure, G. and Nandi, A. (2017). Characterization and establishment of a reference deltamethrin and cypermethrin resistant tick line (IVRI-IV) of *Rhipicephalus (Boophilus) microplus*. *Pesticide Biochemistry and Physiology,* 138: 66–70.

Ghosh, S., Kumar, R., Nagar, G., Kumar, S., Sharma, A.K., Srivastava, A., Kumar, S., Ajith Kumar, K.G. and Saravanan, B.C. (2015). Survey of acaricides resistance status of *Rhipiciphalus (Boophilus) microplus* collected from selected places of Bihar, an eastern state of India. *Ticks and Tick-Borne Diseases,* 6*:* 668–675.

Ghosh, S. and Nagar, G. (2014). Problem of ticks and tick-borne diseases in India with special emphasis on progress in tick control research: A review. *Journal of Vector Borne Diseases,* 51: 259–270.

Ghosh, S., Rawat, A.K.S., Ravindran, R. and Ray, D.D. (2014). *Study of herbal acaricides as means to overcome the development of resistance in ticks to conventional acaricides.* Project report of NAIP, ICAR, New Delhi.

Ghosh, S.K. and Nanda, S.K. (1997). Therapeutic efficacy of Butox (Deltamethrin) against cattle tick *Boophilus microplus* in Tripura. *Indian Veterinary Journal,* 74: 250-251.

Godara, R., Katoch, R., Rafiqi, S., Yadav, A., Nazim, K., Sharma, R., Singh, N.K. and Katoch, M. (2019). Synthetic pyrethroid resistance in *Rhipicephalus (Boophilus) microplus* ticks from north-western Himalayas, India. *Tropical and Animal Health Products,* 51: 1203-1208.

Guglielmone, A.A., Volpogni, M.M., Mangold, A.J., Anziani, O.S. and Castelli, M.C. (2000). Evaluación de unaformulacióncomercial "pour on" con fipronil al 1% para el control de *Haematob iairritans* envaquillonas Holando natural mentein festadas. *Veterinaria (Argentina),* 17:108–113.

He, H., Chen, A.C., Davey, R.B., Ivie, G.W. and George, J.E. (1999). Identification of a point mutation in the para-type sodium channel gene from a pyrethroid-resistant cattle tick. *Biochemical and Biophysical Research Communications,* 261: 558–561.

Hernandez, R., Haiqi, H., Chen, A.C., Ivie, G., Wayne, G., John, E. and Gale, W.G. (1999). Cloning and sequencing of a putative acetylcholinesterase cDNA from *Boophilus microplus* (Acari: ixodidae). *Journal of Medical Entomology,* 36: 764–770.

Hernandez, R., He, H., Chen, A.C., Waghela, S.D., Ivie, G.W., George, J.E. and Wagner, G.G. (2000). Identificationof a point mutation in an esterase gene in different populations of the southern cattle tick, *Boophilus microplus. Insect Biochemistry and Molecular Biology,* 30: 969–977.

Hollingworth, R. M. and Lund, A.E. (1982). Biological and neurotoxic effect of amidine pesticides. *Insecticide Mode of Action*, ed Coats JR (Academic, New York), 1st Ed, pp. 189–227.

Jamroz, R.C., Guerrero, F.D., Pruett, J.H., Oehler, D.D. and Miller, R.J. (2000). Molecular and biochemical survey of acaricide resistance mechanisms in larvae from Mexican strains of the southern cattle tick, *Boophilus microplus. Journal of Insect Physiology,* 46: 685–695.

Jani, B.M., Jani, R.C., Thakur, R.M. and Avasthi, B.L. (1991). A trial with Butox against canine ectoparasites. *Journal of Veterinary Parasitology,* 5:136-138.

Jongejan, F. and Uilenberg, G. (2004). The global importance of ticks. *Parasitology,* 129: S3–S14.

Jonsson, N.N. and Hope, M. (2007). Progress in the epidemiology and diagnosis of amitraz resistance in the cattle tick *Boophilus microplus. Veterinary Parasitology,* 146: 193–198.

Jonsson, N.N., Klafke, G., Corley, S.W., Tidwell, J., Berry, C.M. and Koh-Tan, H.H.C. (2018). Molecular biology of amitraz resistance in cattle ticks of the genus *Rhipicephalus. Frontiers in Bioscience, (Landmark Ed)* 23: 796–810.

Jonsson, N.N., Mayer, D.G. and Green, P.E. (2000). Possible risk factors on Queensland dairy farms for acaricide resistance in cattle tick (*Boophilus microplus*). *Veterinary Parasitology,* 88: 79–92.

Jyothimol, G., Ravindran, R., Juliet, S., Ajithkumar, K.G., Suresh, N.N., Vimalkumar, M.B., Lenka, D.R., Varghese, S. and Ghosh, S. (2014). Low level deltamethrin resistance in ticks from cattle of Kerala, a south Indian state. *Veterinary Parasitology,* 204: 433-438.

Jyoti, Nandi, A., Singh, H., Singh, N.K. and Rath, S.S. (2016a). Fenvalerate resistance status in *Rhipicephalus (Boophilus) microplus* (Acari: Ixodidae) from Punjab, India. *Journal of Parasitic Diseases,* 40: 694-698.

Jyoti, Singh, N.K., Singh, H. and Rath, S.S. (2014). Malathion resistance in *Rhipicephalus (Boophilus) microplus* from Ludhiana district, Punjab. *Journal of Parasitic Diseases,* 38: 343–346.

Jyoti, Singh, N. K., Prerna, M., Singh, H. and Rath, S.S. (2015). Detection of Malathion Resistance in *Hyalomma anatolicum anatolicum* from Bathinda District, Punjab. *Toxicology International,* 22: 125-129.

Jyoti., Singh, N. K., Singh, H., Singh, N.K. and Rath, S.S. (2016b). Multiple mutations in the acetylcholinesterase 3 gene associated with organophosphate resistance in *Rhipicephalus (Boophilus) microplus* ticks from Punjab, India. *Veterinary Parasitology,* 216: 108-117.

Katuri, R.N., Das, G., Singh, A. K., Chalhotra, S.K. and Nath, S. (2017). Comparative efficacy of deltamethrin and and chlorpyriphos in bovine ticks in and around Jabalpur. *Journal of Parasitic Diseases,* 41: 713-715.

Kaur, D., Jaiswal, K. and Mishra, S. (2017). Studies on evaluation of resistance in common ixodid tick: A preliminary study. *International Journal of Fauna and Biological Studies,* 4: 112-115.

Khajuria, V., Godara, R., Yadav, A. and Katoch, R. (2014). Deltamethrin resistance in *Rhipicephalus microplus* of Jammu region. *Indian Veterinary Journal,* 91: 21–24.

Khan, M.H. and Srivastav, S.C. (1990). Evaluation of permethrin against rhipicephalid tick. *Journal of Veterinary Parasitology,* 4: 9-12.

Klafke, G.M., de Albuquerque T.A., Miller, R.J. and Schumaker, T.T. (2010). Selection of an ivermectin-resistant strain of *Rhipicephalus microplus* (Acari: Ixodidae) in Brazil. *Veterinary Parasitology,* 168: 97-104.

Klafke, G.M., Miller, R.J., Barreto, R., Guerrero, F.D., Kaufman, P.E. and Pérez deLeón, A.A. (2017a). Mutation in the sodium channel gene corresponds with phenotypic resistance of *Rhipicephalus sanguineus sensulato* (Acari, Ixodidae) to pyrethroids. *Journal of Medical Entomology,* 54: 1639-1642.

Klafke, G.M., Miller, R.J., Tidwell, J.P., Thomas, D.B., Sanchez, D., Feria Arroyo T. P. and Pérez de León, A.A. (2019). High-resolution melt (HRM) analysis for detection of SNPs associated with pyrethroid resistance in the southern cattle fever tick, *Rhipicephalus (Boophilus) microplus* (Acari: Ixodidae). *International Journal of Parasitology- Drugs and Drug Resistance,* 9: 100-111.

Klafke, G.M., Sabatini, G.A., de Albuquerque, T.A., Martins, J.R., Kemp, D.H., Miller, R.J. and Schumaker, T.T. (2006). Larval immersion tests with ivermectin in populations of the cattle tick *Rhipicephalus (Boophilus) microplus* (Acari: Ixodidae) from State of Sao Paulo, Brazil. *Veterinary Parasitology,* 142: 386–390.

Klafke, G.M., Webster, A., Angol, B.D., Pradel, E., Silva, J., de la Canal, L.H., Becker, M., Osório, M.F., Mansson, M., Barreto, R., Scheffer, R., Souza, U.A., Corassini, V.B., Santos, J.D., Reck, J. and Martins, J.R. (2017b). Multiple resistance to acaricides in field populations of *Rhipicephalus microplus* from Rio Grande do Sul state, Southern Brazil. *Ticks and Tick-Borne Diseases,* 8: 73-80.

Kumar Rinesh, Nagar, G., Sharma, A.K., Kumar, S., Ray, D.D., Chaudhuri, P. and Ghosh, S. (2013). Survey of pyrethroids resistance in Indian isolates of *Rhipicephalus (Boophilus) microplus*: Identification of C190A mutation in the domain II of the para-sodium channel gene. *Acta Tropica,* 125: 237-245.

Kumar, S.,Paul, S.,Sharma, A.K.,Kumar, R.,Tewari, S.S.,Chaudhuri, P., Ray, D.D., Rawat, A.K.S. and Ghosh, S. (2011). Diazinon resistant status in *Rhipicephalus (Boophilus) microplus* collected from different agroclimatic zones of India. *Veterinary Parasitology,* 181*:* 274-281.

Kumar, S., Sharma, A. K., Nagar, G. and Ghosh, S. (2015). Determination and establisnment of discriminating concentrations of Malathion, Coumaphos, Fenvalerate, Fipronil for monitoring acaricide resistance in ticks infesting animals. *Ticks and Tick-Borne Diseases,* 6: 183-187.

Kumar, S., Sharma, A. K., Ray, D.D. adn Ghosh, S. (2014). Determination of discriminating dose and evaluation of amitraz resistance status in different field isolates of *Rhipicephalus (Boophilus) microplus* in India. *Experimental and Applied Acarology,* 63: 413–422.

Kumar, S., Sharma, A.K., Nagar, G., Rawat, S.S., Tiwari, S.S., Kumar, R., Dhakad, M.L., Sharma, R.K., Saxana, R.K., Mehraniya, R.S. and Singh, R.S. (2016). Characterization of acaricide resistance in tick isolates collected from Rajasthan, India. *Indian Journal of Animal Science,* 86: 14-23.

Kumar, S.S., Rayulu, V.C., Rao, K.S. and Kumar, N.V. (2017). Acaricidal resistance in *Rhipicephalus* (*Boophilus*) *Microplus* ticks infesting cattle of Andhra Pradesh. *Journal of Entomology and Zoology Studies,* 5: 580-584.

Kunz, S.E. and Kemp, D.H. (1994). Insecticides and acaricides resistance and environmental impacts. *Scientific and Technical Review of the Office International des Epizooties (Paris)* 13: 1249–1286.

Lenka, D. R., Ravindran, R., Jyothimol, G., Udaykumar, M., Reddy, P.M.T., Sathish, N., Palla, I., Chandramohan, B., Ajithkumar, K.G., Nair, S.N., Chandrasekhar, L., Priya, M.N., Deepa, C.K., Sabu, L., Juliet, S. and Ghosh, S. (2016). Deltamethrin resistance in south Indian isolates of *Rhipicephalus (Boophilus) microplus. Veterinary Parasitology: Regional Studies and Reports,* 5: 37-41.

Li, A.Y., Chen, A.C., Davey, R.B., Miller, R.J. and George, J.E. (2007). Acaricide resistance and synergism between permethrin and amitraz against susceptible and resistant strains of *Boophilus microplus* (Acari, Ixodidae). *Pesticide Management Science,* 63: 882–889.

Li, A.Y., Davey, R.B., George, J.E., 2005. Carbaryl resistance in Mexican strains of the southern cattle tick (Acari, Ixodidae). *Journal ofEconomic Entomology,* 98: 552–556.

Lovis, L., Guerrero, F.D., Miller, R.J., Bodine, D.M., Betschart, B. and Sager, H. (2012). Distribution patterns of three sodium channel mutations associated with pyrethroid resistance in *Rhipicephalus (Boophilus) microplus* populations from North and South America, South Africa and Australia. *International Journal of Parasitology- Drugs and Drug Resistance,* 2: 216–224.

Martins, J.R. and Furlong, J. (2001). Avermectin resistance of the cattle tick *Boophilus microplus* in Brazil. *Veterinary Records,* 149: 64.

Mendes, M.C., Duarte, F.C. and Ruiz, V.L.A. (2010). Identification of pyrethroid resistance by a sodium channel mutation in population of *Rhipicephalus (Boophilus) microplus*

(Acari, Ixodidae) collected from São Paulo, Brazil. In, *XIII International Congress of Acarology*. Recife, p. 152.

Mendes, M.C., Pereira, J.R. and Prado, A.P. (2007). Sensitivity of *Boophilus microplus* (acari: ixodidae) to pyrethroids and organophosphate in farms in the Vale do Paraíba region, São Paulo, Brazil. *Arquivos do Instituto Biológico São Paulo,* 74: 81–85.

Miller, R.J., Davey, R.B. and George, J.E. (1999). Characterization of pyrethroid resistance and susceptibility to coumaphos in Mexican *Boophilus microplus* (Acari: Ixodidae). *Journal of Medical Entomology,* 36*:* 533–538.

Miller, R.J., Davey, R.B. and George, J.E. (2002). Modification of the Food and Agriculture Organisation larval packet test to measure amitraz susceptibility against Ixodidae. *Journal of Medical Entomology,* 39: 645–651.

Minjauw, B. and McLeod, A. (2003). *Tick borne disease and poverty. The impact of tick and tick-borne disease on the livelihoods of small scale and marginal livestock in India and eastern and southern Africa.* Research report, DFID Animal Health Programme, Centre for Tropical Veterinary Medicine, University of Edinburgh, UK.

Morgan, J.A.T., Corley, S.W., Jackson, L.A., Lew-Tabor, A.E., Moolhuijzen, P.M. and Jonsson, N.N. (2009). Identification of a mutation in the para sodium channel gene of the cattle tick *Rhipicephalus (Boophilus) microplus* associated with resistance to synthetic pyrethroid acaricides. *International Journal of Parasitology,* 39*:* 775–779.

Nagar, G., Sharma, A.K., Chigure, G., Manjunathachar, H.V., Saravanan, B.C., Rai, A. and Ghosh, S. (2016). Identification of mutations in acetylcholinesterase 2 gene of acaricide resistant isolates of *Rhipicephalus (Boophilus) microplus*. *International Journal of Science Environment and Technology,* 5: 3440–3447.

Nagar, G., Sharma, A.K., Kumar, S., Saravanan, B.C., Kumar, R., Gupta, S., Kumar, S. and Ghosh, S. (2018). Molecular mechanism of synthetic pyrethroid and organophosphate resistance in field isolates of *Rhipicephalus microplus* tick collected from a northern state of India. *Experimental and Applied Acarology,* 75: 319-331.

Nandi, A., Sagar, S.V., Chigure, G., Fular, A., Sharma, A.K., Nagar, G., Kumar, S., Saravanan, B.C. and Ghosh, S. (2018). Determination and validation of discriminating concentration of ivermectin against *Rhipicephalus microplus*. *Veterinary Parasitology,* 250: 30-34.

Nandi, A., Jyoti, Singh, H. and Singh, N.K. (2015). Esterase and glutathione S-transferase levels associated with synthetic pyrethroid resistance in *Hyalomma anatolicum* and *Rhipicephalus microplus* ticks from Punjab, India. *Experimental and Applied Acarology,* 66: 141-157.

Pasay, C., Arlian, A., Morgam, M., Moher-Vyszenski, D., Rose, A., Holt, D., Walton, S. and McCrthy, J. (2008). High-resolution melt analysis for the detection of a mutation associated with permethrin resistance in a population of scabies mites. *Medical and Veterinary Entomology,* 22: 82-88.

Perez-Cogollo, L.C., Rodriguez-Vivas, R.I., Ramirez-Cruz, G.T. and Rosado-Aguilar, J.A. (2010b). Survey of *Rhipicephalus microplus* resistance to ivermectin at cattle farms with history of macrocyclic lactones use in Yucatan, Mexico. *Veterinary Parasitology,* 172: 109–113.

Perez-Cogollo, L.C., Rodriguez-Vivas, R.I., Ramirez-Cruz, G.T. and Miller, R.J. (2010a). First report of the cattle tick *Rhipicephalus microplus* resistant to ivermectin in Mexico. *Veterinary Parasitology,* 168: 165–169.

Pradeep, B.S., Renukaprasad, C. and Souza Placid, E.D. (2012). Evaluation of the commonly used acaricides against different stages of the cattle tick *Boophilus microplus* by using different *in vitro* tests. *Indian Journal of Animal Research,* 46: 248 – 252.

Rath, S.S., Kumar, S. and Joia, B.S. (2006). Resistance to diazinon and malathion in *Boophilus microplus* (Acari: Ixodidae) population from Punjab, India. *Journal of Insect Science,* 19: 74-81.

Ravindran, R., Ajith Kumar, K.G., Suresh, N.N., Amithamol, K.K., Sunil, A.R., Adarshkrishna, T.P., Chithra, N.D., Jyothimol, G., Ghosh, S. and Juliet, S. (2014). Acaricidal effects of fenvalerate and cypermethrin against *Rhipicephalus (Boophilus) annulatus. Tropical Biomedicine,* 31: 449–455.

Reck, J., Klafke, G.M., Webster, A., Dall'agnol, B., Scheffer, R., Souza, U.A., Corassini, V.B., Vargas, R., dos Santos, J.S. and Martins, J.R. (2014). First report of fluazuron resistance in *Rhipicephalus microplus*, a field tick population resistant to six classes of acaricides. *Veterinary Parasitology,* 201: 128–136.

Rodriguez-Vivas R.I., Jonsson, N.N. and Bhusan, C. (2018). Strategies for the control of *Rhipicephalus microplus* ticks in a world of conventional acaricide and macrocyclic lactone resistance. *Parasitology Research,* 117: 3–29.

Sabatini, G.A., Kemp, D.H., Hughes, S., Nari, A. and Hansen, J. (2001). Test to determine LC_{50} and discriminating doses for macrocyclic lactones against the cattle tick *Boophilus microplus*. *Veterinary Parasitology,* 95: 53–62.

Shakya, M., Kumar, S., Fular, A., Upadhaya, D., Sharma, A.K., Bisht, N., Nandi, A. and Ghosh, S. (2020). Emergence of fipronil resistant *Rhipicephalus microplus* populations in Indian states. *Experimental and Applied Acarology*. https://doi.org/10.1007/s10493-020-00481-7.

Sharma, A.K., Kumar, R., Kumar, S., Nagar, G., Singh, N.K., Rawat, S.S., Dhakad, M.L., Rawat, A.K.S., Ray, D.D. and Ghosh, S. (2012). Deltamethrin and cypermethrin resistance status of *Rhipicephalus (Boophilus) microplus* collected from six agro-climatic regions of India. *Veterinary Parasitology,* 188: 337–345.

Sharma, N., Singh, V., Shyma, K.P., Solanki, V. and Gupta, J.P. (2018). Comparative resistance status of *Hyalomma anatolicum* and *Rhipicephalus (Boophilus) microplus* ticks against synthetic pyrethroids (deltamethrinand cypermethrin) from Banaskantha, Gujarat, India. *International Journal of Acarology,* 44(6): 268-275.

Shyma, K.P., Gupta, J.P., Singh, V. and Patel, K.K. (2015). *In vitro* detection of acaricidal resistance status of *Rhipicephalus (Boophilus) microplus* against commercial preparation of deltamethrin, flumethrin, and fipronil from North Gujarat, India. *Journal of Parasitology Research:* 1–7.

Singh, N.K., Gelot, I.S., Singh, V. and Rath, S.S. (2015a). Detection of amitraz resistance in *Rhipicephalus (Boophilus) microplus* from north Gujarat, India. *Journal of Parasitic Diseases,* 39: 49–52.

Singh, N.K., Haque, M., Jyoti and Rath, S.S. (2012). Deltamethrin resistance in *Rhipicephalus microplus* in Ludhiana. *Indian Veterinary Journal,* 89: 23–25.

Singh, N.K., Jyoti Haque, M. and Rath, S.S. (2010). Studies on acaricide resistance in *Rhipicephalus (Boophilus) microplus* against synthetic pyrethroids by adult immersion test with a discriminating dose. *Journal of Veterinary Parasitology,* 24: 207–208.

Singh, N.K., Jyoti Haque, M., Singh, H., Rath, S.S. and Ghosh, S. (2014a). A comparative study on cypermethrin resistance in *Rhipicephalus (Boophilus) microplus* and *Hyalommaanatolicum* from Punjab (India). *Ticks and Tick-Borne Diseases,* 5: 90–94.

Singh, N.K., Nandi, A. and Rath, S.S. (2014b). Detection of amitraz resistance in *Rhipicephalus (Boophilus) microplus* from SBS nagar, Punjab, India. *Scientific World Journal,* 2014. http://dx. doi.org/10.1155/2014/594398.

Singh, N.K. and Rath, S.S. (2014). Esterase mediated resistance against synthetic pyrethroids in field populations of *Rhipicephalus (Boophilus) microplus* (Acari: Ixodidae) in Punjab districts of India. *Veterinary Parasitology,* 204: 330–338.

Singh, N.K., Singh, H., Jyoti, Prerna, M. and Rath, S.S. (2015b). First report of ivermectin resistance in field populations of *Rhipicephalus (Boophilus) microplus* (Acari: ixodidae) in Punjab districts of India. *Veterinary Parasitology,* 30: 192–194.

Sinha, S.R.P., Sinha, S., Kumar, P., Jaychandaran, C. and Mandal, K.G. (2010). Therapeutic efficacy of deltamethrin and Piperonyl Butoxide (Butox Plus) against *Boophilus microplus* and brown dog tick *Rhipicephalus sanguineus. Journal of Veterinary Parasitology,* 24: 75-77.

Srivastava, P. K., Raizada, Aneesh. and Agarwal, P. (2001). A comparative trial on the efficacy of deltamethrin, cypermethrin and fenvalerate, as acaricides in livestock animals. *Indian Veterinary and Medical Journal,* 25:293-294.

Stone, B.F. and Haydock, P. (1962). A method for measuring the acaricides susceptibility of the cattle tick *Boophilus microplus* (Can.). *Bulletin of Entomological Research,* 53: 563–578.

Stone, N.E., Olafson, P.O., Davey, R.B., Buckmeier, G., Bodine, D., Sidak-Loftis, L.C., Giles, J.R., Duhaime, R., Miller, R.J., Mosqueda, J., Scoles, G.A., Wagner, D.M. and Busch, J.D. (2014). Multiple mutations in the para sodium channel gene are associated with pyrethroid resistance in *Rhipicephalus microplus* from the United States and Mexico. *Parasites and Vectors,* 7: 456.

Stutzer, C., van Zyl, W.A., Olivier, N.A., Richards, S. and Maritz-Olivier, C. (2013). Gene expression profiling of adult female tissues in feeding *Rhipicephalus microplus* cattle ticks. *International Journal of Parasitology,* 43: 541–554.

Sungirai, M., Baron, S., Moyo, D.Z., Clercqe, P.D., Maritz-oliver, C. and Maddar, M. (2018). Genotyping acaricide resistance profiles of *Rhipicephalus microplus* tick populations from communal land areas of Zimbabwe. *Ticks and Tick-Borne Diseases,* 9: 2-9.

Taylor, M.A. (2001). Recent developments in ectoparasiticides. *The Veterinary Journal,* 161: 253-268.

Temeyer, K.B., Davey, R.B. and Chen, A.C. (2004). Identification of a third *Boophilus microplus* (Acari: ixodidae) cDNA presumptively encoding an acetylcholinesterase. *Journal of Medical Entomology,* 41: 259–268.

Temeyer, K.B., Olafson, Jr, P.U., Brake, D.K., Tuckow, A.P., Li, A.Y. and Perez de Leon, A.A. (2013). Acetylcholinesterase of *Rhipicephalus (Boophilus) microplus* and *Phlebotomuspapatasi*: gene identification, expression, and biochemical properties of recombinant proteins. *Pesticide Biochemistry and Physiology,* 106: 118–123.

Temeyer, K.B., Olafson, P.U. and Miller, R.J. (2009). Genotyping mutations in BmAChE3: a survey of organophosphate resistant and susceptible strains of *Rhipicephalus (Boophilus) microplus*. *Journal of Medical Entomology,* 46: 1355–1360.

Upadhaya, D., Kumar, B., Kumar, S., Sharma, A.K., Fular, A., Bisht, N., Srivastava, S., Boruah, R.R., Nagar, G., Shakya, M., Nath, T., Nandi, S.P., Karivaradharaja, S. and Ghosh, S. (2020). Characterization of acaricide resistance in *Rhipicephalus microplus* populations infesting cattle in north eastern India and assessment of local plant extracts for tick management. *Veterinary Parasitology*. https://doi.org/10.1016/j.vetpar.2019.109011.

van Wyk, R.D.J., Baron, S. and Maritz-Olivier, C. (2016). An integrative approach to understanding pyrethroid resistance in *Rhipicephalus microplus* and *R. decoloratus* ticks. *Ticks and Tick-Borne Diseases,* 7: 586–594.

Vatsya, S., Yadav, C.L., 2011. Evaluation of acaricide resistance mechanisms in field populations of *Rhipicephalus (Boophilus) microplus* collected from India. *International Journal of Acarology,* 37: 405–410.

Yessinoua, R.E., Akpo, Y., Sidick, A., Adoligbea, C., Karima, I.Y.A., Akogbetob, M. and Farougoua, S. (2018). Evidence of multiple mechanisms of alpha-cypermethrin and deltamethrin resistance in ticks *Rhipicephalus microplus* in Benin, West Africa. *Ticks and Tick-Borne Diseases,* 9: 665-671.

In: The Entomological Guide to *Rhipicephalus*
Editors: Sachin Kumar, Raquel Cossio Bayugar et al.
ISBN: 978-1-53619-619-1

Chapter 11

RHIPICEPHALUS (BOOPHILUS) MICROPLUS: MOLECULAR EVOLUTION OF ACARICIDE RESISTANCE

Rodrigo Rosario-Cruz* and Delia Inés Domínguez-García
Biotechnology Health and Environment Research Unit,
Natural Science College from Autonomous Guerrero State University,
Carretera Federal Chilpancingo-Acapulco,
Las Petaquillas Guerrero, Mexico

ABSTRACT

The application of ixodicides has long been the alternative of choice for tick control, however, the use of ixodicides has currently limited efficacy in reducing tick infestations, due to the selection of resistant individuals within tick populations. Acaricide resistance is a growing problem that needs to be addressed, because it is currently affecting the competitiveness of livestock and the economy of thousands of producers in Mexico. The solution to this problem requires more research focused on the use of 'omics' and reverse genetic approaches in order to identify new target genes for diagnostics or immunoprophylactic purpouses, since the improvement of diagnostic tests and a better vaccine, will thus contribute to design better strategies and better tick control programs, in order to better management of acaricide resistance. The aim of this review is to compile different current topics related to the evolution of acaricide resistance and reverse genetics research, in order to make clear how identification of new target genes or genetic traits, can lead to improve Molecular diagnostics, vaccine development and therefore the methods currently used for tick control.

Keywords: *Rhipicephalus microplus*, acaricide resistance, tick control, molecular evolution of resistance

* Corresponding Author's Email: rockdrig@yahoo.com.mx.

Abbreviations

OP	organophosphorous
SP	Synthetic pyrethroids
RR	Resistant genotype/phenotype
SS	Susceptible genotype/phenotype
RS, or SR	Heterozygous genotype/phenotype
GABA	Gamma Amino Butiric Acid
GST	Glutathion-S-Transferase
SAGARPA	Mexico's Department of Agriculture and livestock
PCR	Plymerase Chain Reaction
LPT	larval Packet test
LIT	Larval Inmersion Test
AIT	Adult Inmersion Test
CzEST9	Coatzacoalcos, Esterase 9
Kdr,	Knock Down resistance
SENASICA	National Service for Healthy, Safety and Quality Food

Introduction

The Cattle tick *Rhipicephalus (Boophilus) microplus* belongs to the Ixodidae family. It is a thelmophagous, single host ectoparasite, which affects the tropical and subtropical regions of the world causing great economic losses in bovine livestock industry. Specifically, *R. microplus* causes direct damage due to the action of bites (Buczen and Bartozic, 2006) and indirect damage caused by the transmission of three etiological agents: *Babesia bovis, Babesia bigemina* and *Anaplasma marginale* (Jonsson et al., 2008).

Figure 1. Use of chemical treatments to control the cattle tick infestations by *Rhipicephalus (Boophilus) microplus*. (Photo from Rosario-Cruz, R., archives).

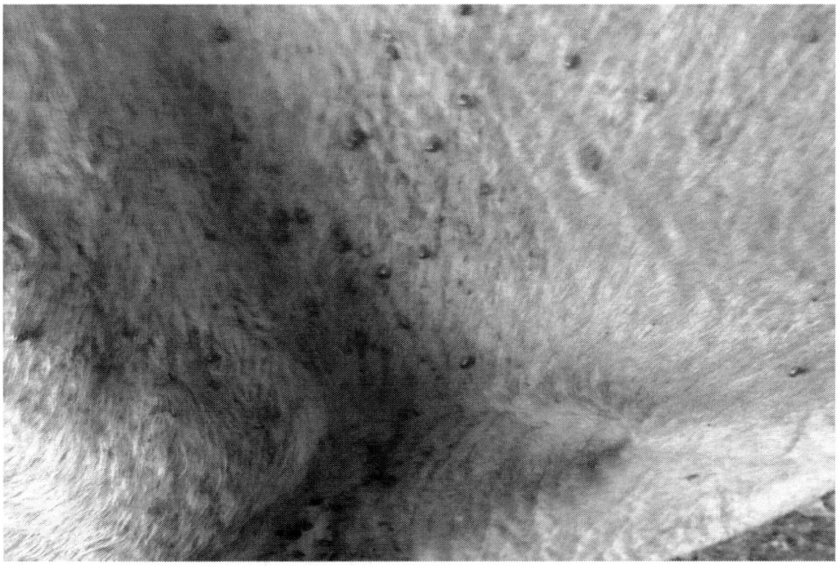

Figure 2. Cattle tick infestation by *R. microplus* resistant to Ixodicides (Photo from Rosario-Cruz, R., archives).

The frequent and continued use of acaricides has led to the selection of resistant tick populations (Nolan and Schnitzerling 1986), since the most used strategy for the treatment of tick infested animals, is the application of toxic chemicals on their body. The different types of products used for tick control, includes arsenicals, organochlorines, organophosphates, carbamates, formamidines, pyrethroids, (Aguirre, 1987), macrocyclic lactones (Pérez-Cogollo et al., 2010), fipronil (Miller et al., 2013) and fluazuron (Reck et al., 2014).

In Mexico, chemical control of the cattle tick *R. microplus* and *R. annulatus,* has been used intensively since 1975 when the National Campaign against Ticks was created and continued for approximately 10 years, in which organophosphate compounds (OP) were used, causing an OP resistance problem. The use of pyrethroids was allowed later, again presenting the problem of resistance to synthetic pyrethroids (SP); Therefore, in 1994 formamidines began to be used more frequently due to the presence of OP, SP or combined resistance of *R. microplus*, and in 2002 the first case of resistance to formamidines was reported (Soberanes et al., 2002).

Cases of multiple resistance to ixodicides is a growing problem that needs to be addressed, because at this time it is affecting the competitiveness of livestock and the economy of thousands of producers in Mexico (Figure 2). The solution to this problem requires the implementation of research programs that help to elucidate the molecular mechanisms of resistance in order to find new molecular targets for diagnosis and for the development of recombinant vaccines. The recent use of new genomic tools will accelerate the discovery of new genes for diagnostic or immunoprophylactic interest, thus contributing to the improvement of tick control strategies.

Taxonomy and Life Cycle of *Rhipicephalus (Boophilus) microplus*

Ticks belong to the phylum *Arthropoda*, subphylum *Chelicerata,* which is comprised of spiders, scorpions, mites and ticks, and the subclass Acari. The *chelicerates* include a diversified group of terrestrial and aquatic arthropods (Jeyaprakash and Hoy, 2009), and are the second largest group of arthropods after the insects. Ticks are important vectors of a significant number of pathogenic microorganisms, protozoa, *rickettsiae*, *spirochaetes* and viruses. Consequently, the tick and tick borne diseases (TTBD) complex, constitutes a global threat for livestock and human health (Jongejan and Uilenberg, 2004).

Ticks develop during the free living stage of their life, on the soil or grass (Figure 3), however, as a blood-sucking ectoparasite, it has to feed on the host animals during the parasitic stage of their biological cycle (Figure 3). They parasitize a large number of vertebrates included reptiles, birds and mammals. Ticks are considered the second most important vector of diseases to humans, domestic and wild animals, only after the mosquitos (de la Fuente et al., 2008). Ticks have been adapted to most of the terrestrial hábitats, since they can live from zero to 2600 msnm (Vecino et al., 2010). The geographical distribution of tick species is conditioned to enviromental factors such as: Temperature, humidity vegetation and host availability (Castro et al., 2007).

There is a list of 901 valid names of tick genus and species around the world (Figure 4), including two new extincted species recently identified (Peñalver, et al. 2017), clasified in four families: Ixodidae (hard ticks) (714), argasidae (Soft ticks) (185), Nuttalliellidae (one single species with characteristics from both soft and hard ticks), and Deinocrotonidae, comprised by only one single extinct species recently identified (Figure 4) (Keirans et al., 1976; Horak et al., 2002; Barker and Murrell, 2004; Peñalver et al., 2017).

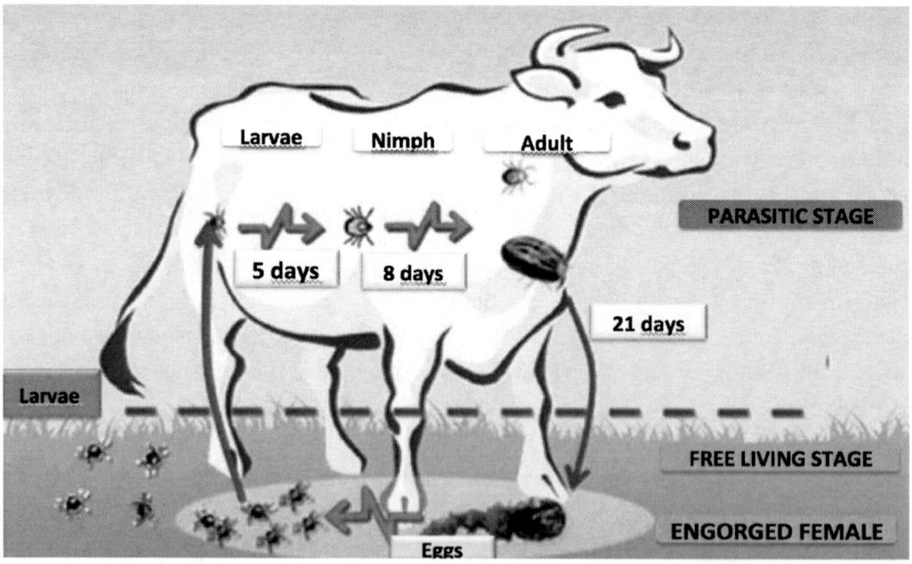

Figure 3. Life cycle of the cattle tick *Rhipicephalus (Boophilus) microplus* consists of the parasitic and the free-living stage, the former ocurring on the calf and the later on the grass (Martín Ortiz kindly provided the original ilustration).

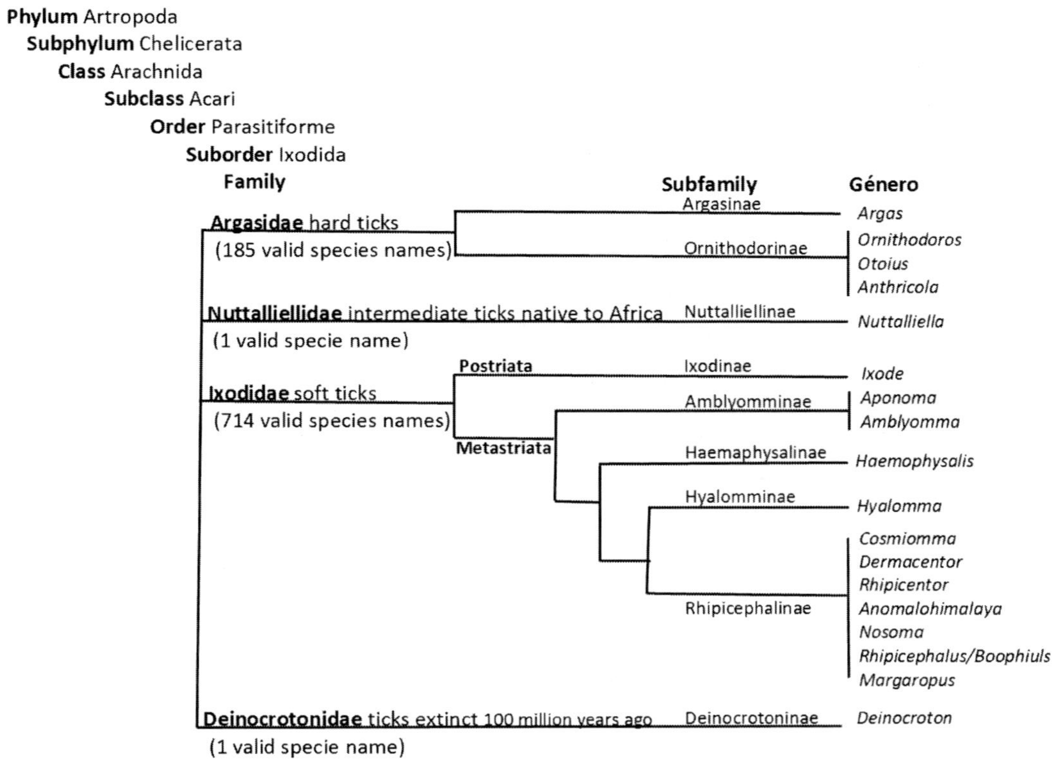

Figure 4. List of valid names of genus grouped into four families including the recent two newly classified species. Adapted from Barker and Murrell 2004.

Distribution of Ticks in Mexico

According to the National Service for Healthy, Safety and Quality Food (SENASICA), the *R. microplus* tick-free zones, include the states of Baja California, Sonora, Aguascalientes and Tlaxcala, northern Baja California Sur, and most of Chihuahua (Figure 5) (SENASICA, 2020).

The most common tick species affecting the cattle industry is *R. microplus* and its distribution has been associated with warm humid climate and high relative humidity regions (Alcala-Canto et al., 2018; Almazan et al., 2018). Its distribution in Mexico starts at the Northern state of Sinaloa to Chiapas along the pacific coast and in the northeast at the state of Tamaulipas, extending along the Gulf coastal states, to the Yucatan peninsula.

The prevalence of *R. microplus* ticks varies regionally, and in the states of the Gulf coast of Mexico, such as Tamaulipas, Veracruz, Tabasco and Yucatán, climate conditions, favors their development with cattle infestations up to 90-95% (Almazan et al., 2018) (Figure 5). The most widely distributed genera are *Rhipicephalus (Boophilus)* and *Amblyomma* within the infested regions of the country, in which prevails favorable climate factors for their development, as well as the presence of bovines, equines, dogs, white tailed and red deer (Almazan et al., 2018).

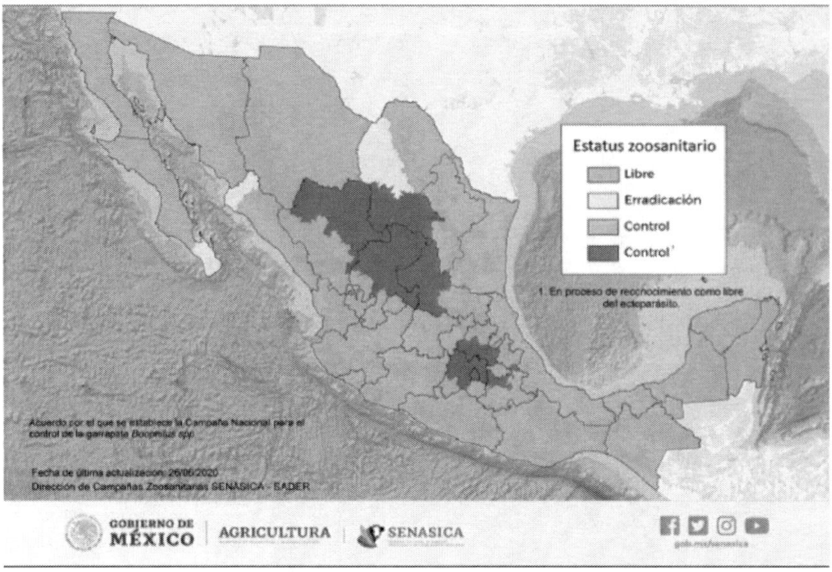

Figure 5. *Rhipicephalus* (*Boophilus*) spp. Current status of *Rhipicephalus (Boophilus) microplus* distribution in México, according to SENASICA-SADER. Retrieved in June 26th 2020.

R. annulatus is native to the Mediterranean and the Midwest, and easily adapts to climates with high temperature and low relative humidity such as North Africa (Estrada-Peña et al., 2004). In Mexico it is distributed in an area between the states of Tamaulipas, Nuevo Leon and Coahuila, along the borderline with the Rio Grande Valley, in Texas, USA. However, given the high mobilization of livestock in México, *R. annulatus* has been found in the center and south of Tamaulipas (Almazan et al., 2018), and can be found in other states of the country.

The genus *Amblyomma* is one of most diverse in terms of tick species and hosts, ranging from amphibians, reptiles, birds and mammals and it is worldwide distributed. In Mexico, have been identified 26 species, in most of the states with the exception of Aguascalientes and Tlaxcala (Guzman-Cornejo et al., 2011). In cattle, 10 species have been identified (Guzman-Cornejo et al., 2011, Almazan et al. 2016, Rodriguez-Vivas et al., 2016). However, in bovines the most important species is *A. mixtum* (Nava et al., 2014), although *A. maculatum* and *A. immitator* can also be found less frequently (Martínez-Ibáñez et al., 2006). *A. mixtum* is well adapted to those places with high temperatures such as the states of the Gulf coast of Mexico from Tamaulipas to Yucatán (Almazán et al., 2016; Rodriguez-Vivas et al., 2016).

Cattle from the Gulf coast of Mexico can be infested with *A. mixtum* throughout the year, with seasonal variations associated to high humidity, temperature, and high population density of domestic and wild animals favoring the development of this species (Almazan et al., 2016). *A. mixtum,* has been found over the pacific coast, in Baja California and Sonora, in hosts different from bovines. This tick is common in cattle from Sinaloa, Nayarit, Chiapas and some states in the center of the country (Estrada-Peña et al., 2004; Almazan et al., 2016, 2018). Elevated areas with low temperatures such as the Sierra Madre are not favorable for its development (Estrada-Peña et al., 2004).

The department of Zoosanitary Campaigns from SENASICA is responsible to recognize the zoosanitary status of the different regions of Mexico, depending on the incidence of *R. microplus* infestations. Current data indicate that 30.6% of the Mexican territory is free of

Rhipicephalus spp. currently the only officially recognized states with free status are: Sonora, Tlaxcala and Aguascalientes. 3.4% of the national territory is in the eradication phase, and includes the municipalities of Los Cabos and La Paz in Baja California Sur, Morelos and Guadalupe y Calvo in Chihuahua, Ahome, El Fuerte and Choix in Sinaloa and Cuatrocienegas, Ocampo and Sierra mojada in Coahuila. 65.9% is in the control phase and includes the central highlands and the cold mountainous region (SENASICA, 2020) (Figure 5).

Biological Aspects of Acaricide Resistance

Resistance to ixodicides is a genetic condition that confers to a tick population the capability to get successfully adapted to a toxic environment, by a selection process naturally and/or artificially promoted (Rosario-Cruz et al., 2009).

Derived from this definition, the resistance phenomenon can be understood as an important example of "recreation" of natural selection (Ffrench-Constan et al., 2004), which serves as an excellent biological model for the study of evolutionary adaptations in individuals belonging to arthropod populations subjected to constant selection pressure due to the constant use of pesticides (Figure 1) (Palumbi, 2001).

In nature, this coevolution process, driven by secondary metabolites (allelochemical interactions) produced by plants against both plants and arthropods has taken millions of years (Wheat et al., 2007) and as a result of the reciprocal interaction between these two groups of organisms (Rausher, 2001).

The arthropods have kept within their genomes, all those characteristics that have an evolutionary advantage to guarantee the survival of the species in nature. These molecules, such as toxins and repellants produced by plants functioning as defense mechanisms against herbivorous insects (Kessler and Baldwin, 2002; Wittstock and Gershenzon, 2002), have selected complex detoxification mechanisms in ordr to inhibit the toxicity of metabolites, produced by the plant on which they feed (Ranson et al., 2002). A type of resistance due to the natural exposure to secondary metabolites produced by plants.

Insects affect the reproduction and survival of plants, which in response, produce several types of secondary chemicals (phenolic acids, flavonoids, terpenoids, steroids, alkaloids and organic cyanides) starting a plant-insect coevolution in which the mutual selection pressure, has led to the processes of speciation, initiating an evolutionary game tending to preserve the natural balance.

Plants are natural factories of a vast arsenal of chemicals. Some of these defensive mechanisms, including repellants, antifeedant, and poisons are well known and we have benefited immensely from using some of these plant defense strategies to protect our plants from agricultural pests and to control insects vectors of diseases.

Allelochemical interactions are mechanisms by which various species defend themselves from others or compete for food. In addition, the most common allelopathic substances belong to some important groups of compounds; Secondary substances in the plant include phenolic acids, flavonoids and other aromatic compounds, terpenoid substances, steroids, alkaloids and organic cyanides.

Adaptation to these natural environmental conditions must lead to the development of a defense machinery naturally selected by nature. These coadaptation mechanisms exist within a wide variety of insects and their food plants. A significant portion of the entire insect and the microbial community associated with plants of the cruciferous family owe their organization to a single set of chemicals, mustard oil and its glycosides. The herbivores that respond to these compounds are the white cabbage butterflies *Pieris rapae* and *P. brassicae* the diamondback moth, *Plutella maculipennis*; the plant weevil *Lidtroderes costirostris* the mustard beetle, *Phaedib cochleariae*; the flea beetle, *Phyllotreta cruciferae* and *P. striolata* and the cabbage aphid, *Brevicoryne brassicae* (Whittaker and Feeney, 1971).

Most herbivorous species attack few among many species of food plants available. However, most plants are consumed by certain species. Herbivores have developed the means to metabolize most of the toxic material produced by plants, but not all of them are metabolized by any species. A specialized insect gains an advantage by restricting its detoxifying energy to one or a few potentially harmful substrates, whereas a polyphagous species must spend most of its energy and nutrients maintaining detoxification mechanisms for a wide range of chemically defined plants, which means that the insect carries a "higher metabolic load" (Figure 6).

The activity of multiple function oxidases (P-450) in the gut of moth and butterfly larvae is considerably higher in polyphagous species than in species restricted to a family of food plants. Pyrethrins are among the many insecticides metabolized by these enzymes (Whittaker and Feeney, 1971). Vertebrate herbivores may be less specialized than insects, due to the low cost of general detoxification ability in response to specific genetically programmed chemical signals.

Metabolic Detoxification and Evolution of Acaricide Resistance

Because of this coevolution process, the species have been diversified while a set of characteristics genetically selected have allowed each one of them to survive against the entire range of secondary compounds produced by plants. These characteristics are families of highly specialized multi-genes in the detoxification processes of both biotic and xenobiotic compounds, such as the superfamilies of glutathione transferases, multiple function oxidases and carboxylesterases (Devonshire and Field 1991), each of them capable of hydrolyzing a large number of related compounds (Figure 6).

Three groups of detoxifying enzymes are responsibles of the resistance phenomenon, glutathione-S-transferase (GST), mixed-function oxidases, and carboxylesterases (Figure 6) (Devonshire and Field 1991). Although each of these is involved with resistance, the elevation of esterases is mainly associated with organophosphate compounds as the most important resistance mechanism in the *Culex* mosquito (Hemingway and Karunaratne 1998); however, oxidases also play an important role. In the detoxification of organophosphate compounds and pyrethroids (Bisset et al., 1998). High concentrations of oxidases and esterases are present in permethrin-resistant mosquitoes (Vulule et al., 1999). On the other hand, in cockroaches the possible mechanisms of resistance to pyrethroids include detoxification processes mediated by cytochrome p450, esterases and glutathione-S-transferases (Figure 6), in addition to nerve insensitivity (*Kdr* type), detoxification by esterases and oxidases has been identified as the most frequent resistance mechanisms for this insect (Hemingway et al., 1993).

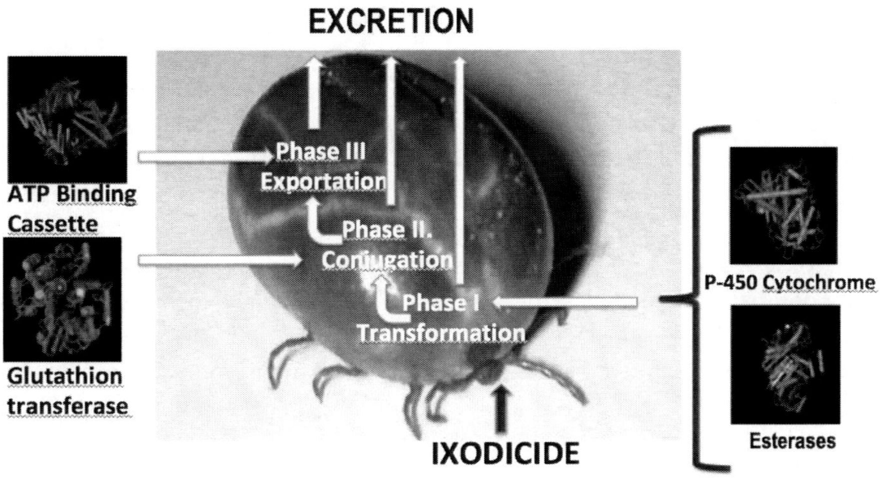

Figure 6. Molecular mchanisms of detoxification are responsible for ixodicide resistance and have been functionally conferred to multigenic family of detoxifying enzimes, participating in the metabolic process. Esterases, Multiple function oxidases (P-450) and Glutathion Transferases (GST).

Modifications in target molecules of insecticides, produced by specific mutations in the gene sequence, function as mechanisms responsible for resistance. Thus, point mutations are present in different arthropods in GABA receptor genes (Zhang et al., 1994), in genes encoding acetylcholinesterase (Mutero et al., 1994), of other detoxifying esterases (Campbell et al., 1998; Hernández et al., 2000) or in sodium channels (He et al., 1999). The latter has been associated to resistance type *Kdr* (knokdown resistance); modifying the site of action of the insecticide, in this case the target of pyrethroid insecticides (Williamson et al., 1996, Guerrero et al., 1997).

In *R. microplus*, several gene sequences associated with resistance have been cloned, recently. These alterations in the sequence of genes such as the substitution # 2134 in the sodium channel sequence reported by He et al., (1999), in which, Guerrero et al., (2001), found in the susceptible strain a thymine (T) in the susceptible strain that is substituted by an adenine (A) in the Corrales strain that is resistant to pyrethroids. This substitution results in a change of a phenylalanine (Phe) for an Isoleucine (Ile), in the sequence of the transmembrane segment S6 of domain III of the para-type sodium channel. Recent studies have demonstrated the presence of point mutations in at least two genes of the esterase family, one of them being associated with resistance to pyrethroids (Hernandez, et al., 2000) and the other mutation has been obtained from an OP resistant strain of *R. microplus* although its association with OP resistance has not been fully proven.

The mutation related to resistance to pyrethroids have been consistently found in ticks resistant to pyrethroids, which shows that, at least in pyrethroid resistance, the mechanisms may be mediated by alterations in the gene sequences. Not only of sodium channels, but also due to alterations in the sequence of enzymes such as esterases, which makes the full understanding of the molecular mechanisms of resistance more complex.

Resistance Mechanisms in the Cattle Tick *R. microplus*

Pesticide resistance in arthropods is one of the most important problems in the world, since many arthropods are natural parasites of plants and animals that are the main source of food for humans. Pesticide resistance is being studied in numerous species, for example: house flies, fruit flies, cockroaches, aphids, potato beetles, mosquitoes, and cattle ticks.

Resistance to ixodicides can occur by metabolic detoxification and/or modification of the target site (Rosario-Cruz et al., 2009). Metabolic resistance is predominantly caused by the high activity of enzymes grouped in multigene families such as cytochrome P450, glutathione transferases and carboxylesterases (Miller et al., 2008; Baffi et al., 2008, Itoh et al., 2018), which degrade and/or sequester pesticide molecules depending on which enzyme system or family of pesticides is involved. In general, the participation of mechanisms of overexpression of esterases involved in resistance to organophosphates is recognized; however, at least one esterase from the Coatzacoalcos strain has been expressed, involved in resistance to synthetic pyrethroids, especially cypermethrin. The Coatzacoalcos strain has been toxicologically characterized by the larval packet bioassay (Ortiz et al., 1995) and its molecular mechanism, elucidated. This strain possesses a significant increase in the general activity of esterases, as well as hydrolysis of permethrin, which suggests the participation of metabolic mechanisms mediated by esterases associated with resistance to permethrin (Pruet et al., 2002). This is the result of duplication events of the CzEST9 gene of the Coatzacoalcos strain, which leads to the overexpression of this particular esterase (Jamroz et al., 2000). Currently, the gene that codes for this esterase has been sequenced from the genetic material of the Coatzacoalcos strain and the 62.8 Kd protein encoded by this gene is suggested to be CzEST9 (Pruet et al., 2002), associated with resistance to permethrin. It is interesting to note that the Coatzacoalcos strain does not have the *Kdr*-type mutation found in Mexican strains, which suggests that there are two mechanisms of resistance to pyrethroids acting independently, although the most common mechanism is associated with presence of the *Kdr* mutation (Rosario-Cruz et al., 2005).

This mechanism consists of the presence of a mutation in the sodium channel gene that gives the channel the inability to bind with the pesticide, thus inhibiting its toxic effect, due to the stereochemical modification of the sodium channel structure. The sodium channel is the target site of pyrethroids, for this reason, this mechanism is known as modification of the target site or resistance type *Kdr* by its name in English (Knock down resistance) and is one of the best documented mechanisms in insects. In *R. microplus*, the substitution of a phenylalanine for an isoleucine (Phe-Ile) has been documented in the sodium channel gene of Mexican strains, located in domain III of segment 6 (III-S6), which confers resistance to the synthetic pyrethroid family (Figure 7) (He et al., 1999). Resistant phenotypes (R) are inherited as recesive traits to the progeny (RS anf SR) when homozygous resistant genotypes (RR) are crosssed with the homozygous susceptible homologs (SS), therefore the progeny inherit a susceptible (heterozygous) phenotype for Flumethrine and Deltamethrine (Figure 7) (Aguilar, et al., 2008) . Recently, in the same gene a new mutation has been identified in the sodium channel gene producing a substitution of a leucine for an isoleucine (Le-Ile). This substitution is located at position 190, at the bridge that joins segments 4 and 5 of the domain II (II-S4-5) of the sodium channel of Australian strains, similar to that found in *Bemicia tabaci*, but never before identified in ticks (Morgan et al., 2009).

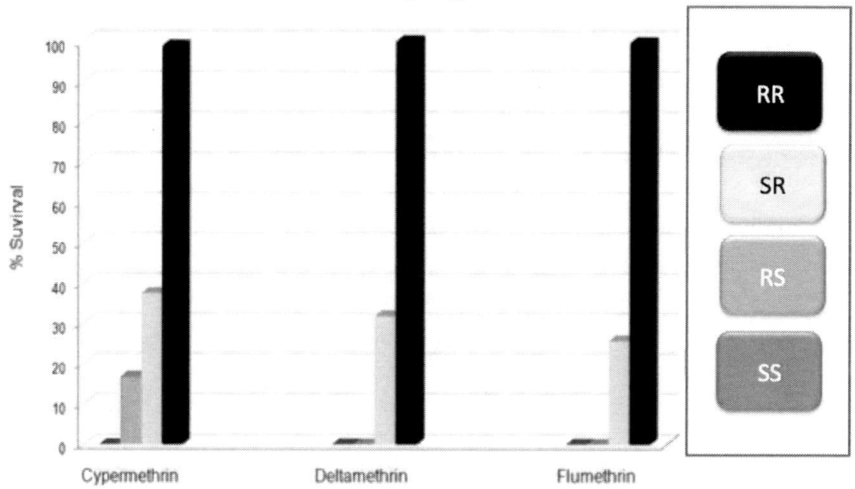

Figure 7. Survival percentage of four different tick strain genotypes, using the Larval Packet Test discriminant dose to Cypermethrine, Deltamethrine and Flumethrine, derived from resistant and susceptible diallelic crosses (RRxSS). First position in each genotype (RR or SS), indicates Male and second position indicates Female. R = resistant allele; S = susceptible allele.

The target site modification mechanism (*Kdr* Type) is the most documented in insects and other arthropod species, and is also present in *R. microplus*, where its role and contribution to SP resistance has been demonstrated (Rosario-Cruz et al., 2005). These findings have been confirmed by experiments based on the differences in the melting temperature (Tm) of allele-specific DNA fragments from individual larvae from Mexico and Australia, in which the substitution II has been exclusively found in II-S4-5 in the Australian strains and the III-S6 substitution only in the Mexican strains (Rosario-Cruz et al., 2005; Chen et al., 2009).

This shows that at least two different resistance mechanisms based on the presence of substitutions (Phe-Ile and Le-Ile) located at different positions of the sodium channel structure have developed independently between Australian and Mexican strains of *R. microplus*, which are currently not Considered to be *R. microplus*, but *R. australis* (Burguer et al., 2014).

Another modification of the target site has been identified on the acetylcholinesterase of *R. microplus,* in which a substitution of glutamine for arginine has been found on the gene for acetylcholinesterase 3. This substitution has been associated with resistance to OP compounds (Temeyer et al., 2007). In essence, resistance in the *R. microplus* tick can be metabolic or due to the modification of the target site, and both mechanisms have been widely documented, the reduced penetration of the pesticide through the cuticle has been studied but there is no evidence that demonstrate its effectiveness as a resistance mechanism.

Although resistance has confused farmers and insecticide manufacturers, it has also provided the opportunity to learn about the most relevant aspects of the biochemistry and evolution of arthropods in populations exposed to intense chemical selection pressure by xenobiotic compounds. This knowledge is undoubtedly the cornerstone to develope any insecticide resistance management program in order to reduce or delay the onset of insecticide resistance, to optimize the half-life of the insecticide as a control agent, and thus minimize the damage to human and animal populations due to the overuse of pesticides (Pimentel et al. 1990).

Catalytic Residues of Esterases

Intensive analysis of esterases has shown that the catalytic site of these enzymes encompasses three key residues involved in a charge transfer to donate protons to a peptide bond (hydrolysis reaction) (Craik et al., 1987; Rauwerdink and Kaslauskas., 2015).

These three residues are the active site of serine, basic histidine and aspartic acid and none of them is contiguous in the primary structure, but they are constituted in highly conserved regions of 8-18 amino acid residues. The relative positions of these three regions in the primary sequence are also highly conserved, with approximately 43 residues between histidine and aspartate and approximately 91 residues to serine (Rauwerdink and Kaslauskas, 2015).

The group of esterases has been considered as a family phylogenetically related to serine proteases, since these two families are similar due to the following characteristics:

1. They conserve at least three out of eight amino acids in the consensus region of the serine active site (Dayhoff et al., 1972; Drabløs and Petersen, 1997; Vath et al., 1999).
2. Diisopropyl fluorophosphate (DFP) inactivates enzymes by irreversibly reacting with the active site of serine (Krisch, 1971; Lorke et al., 2011).
3. They are inhibited by OP and carbamates (Augusteyn et al., 1969; Colovic, 2013).
4. They show broad substrate specificity (Heymann, 1980; Martinez et al., 2018).

Based on the conserved sequences characteristic of these enzymes, it has been possible to amplify sequences of esterase genes in the cattle tick *R. microplus* by the polymerase chain reaction (PCR) technique using degenerate oligonucleotides obtained from sequence alignments of enzymes belonging to the same family but from different species, which have been deposited at the GeneBank Databases.

Genes, Mutations and the Diagnosis of Acaricides Resistance

There is evidence that two different types of mutations in the same esterase gene in *Lucilia cuprina* confer specific resistance to diazinon or malathion, two types of organophosphate compounds (Campbell et al., 1998). This would explain why cases such as that of *C. quinquefasciatus* in Colombia, which was found resistant to deltamethrin and permethrin but not to lambda-cyhalothrin and cypermethrin (Bisset et al., 1998), and why a population resistant to a type of insecticide may be susceptible to another even though both of them belong to the same family. Apparently, some types of resistance are highly specific for the chemical compound, since arthropods would select mutations involved with a defense mechanism that confers them a high affinity and specificity towards an insecticide molecule (French-Constant et al., 1993). The fact that an esterase can hydrolyze one type of compound does not mean that it will hydrolyze all compounds within that family. It is necessary to understand the role that mutations play and how they contribute to the resistance process in ticks.

In the case of *R. microplus*, this information has been produced until recent years, at the molecular level regarding the basic mechanisms of resistance. Mexican strains of highly pyrethroid resistant *R. microplus* ticks possess *Kdr*-like mutations at the sodium channel gene (He et al., 1999). However, a Mexican strain also resistant to pyrethroids resulted with none mutation, but with a high prevalence of mutant alleles in an esterase gene, in addition to an increased signal in Southern hybridization, indicating a possible increase in the number of copies of the gene (Hernandez et al., 2000). On the other hand, Miller et al., (1999) reported two patterns of resistance to pyrethroids, one related to insensitivity at the target site and the other to a detoxification mechanism by metabolic enzymes, possibly involving sequestration or an increase in the rate of hydrolysis of the insecticide. Jamroz et al., (2000) working with the same strain detected an increase in the hydrolysis of permethrin by esterases. This means, that at least two different mechanisms of resistance to SP could be involved in Mexican populations of ticks.

In *R. microplus*, there are two different substitution, one located at domain III, segment 6 (III-S6) (Figure 8) (He et al., 1999), and the other is a new mutation within the gene of the sodium channel (Le-Ile) at position 190, which is located at the bridge that joins segments 4 and 5 of domain II (II-S4-5) of the sodium channel of Australian strains (Morgan et al., 2009). Both confers resistance to the family of synthetic pyrethroids,

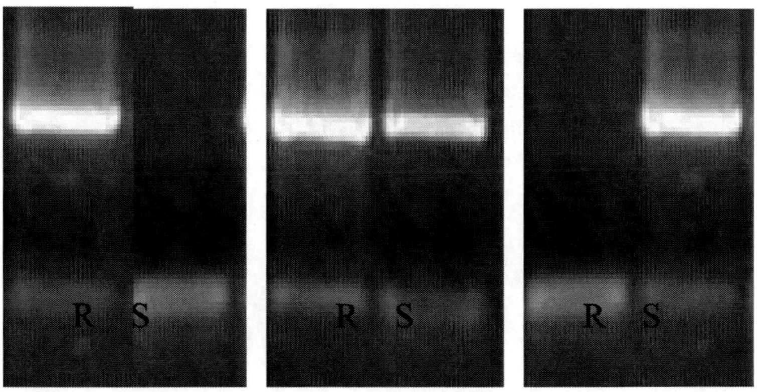

Figure 8. Allele Specific PCR for the sodium Channel mutation located at the Domine III segment 6 (Phe-Ile substitution). R = Resistant; S = Susceptible.

Tests for Ixodicides Resistance Detection in *R. microplus*

The lack of efficiency of the ixodicides applied to livestock to control infestations by *R. microplus* ticks is an indication that the genes conferring resistance to individuals of these tick populations are emerging. However, the detection of this phenomenon requires techniques and methodologies that allow us to monitor the presence of the problem in order to adopt strategies to control tick populations, prevent the spread of resistant ticks, and reduce economic losses.

There are different diagnostic methods proposed for the evaluation of resistance; the current methods are bioassays that consist of the exposure of ticks to tick products to quantify their toxicological effect, comparing the efficacy of the product with regards to a susceptible strain (Figure 9).

Bioassays, among which the larval pack test (LPT) (Stone and Haydock, 1962) (Figure 9), adult tick immersion test (AIT) Drummond et al., 1973), the larval immersion test (LIT) (Shaw, 1966) are useful tools that allow identifying and typifying the resistance problem. However, ixodicides have serious limitations such as their low sensitivity to detect resistance alleles present in the population, they require a large amount of biological material and it takes several weeks to obtain results. In addition, such tests do not provide individual information on the organisms since by the time resistance is diagnosed by these methods, the proportion of resistant alleles in the population is already highly disseminated.

In search of better alternatives to diagnose resistance, studies have been carried out to elucidate the molecular bases of resistance (Guerrero et al., 2007; Rosario et al., 2005, Aguilar et al., 2008) developed by ticks to counteract the intoxication produced. By the action of the pesticides used for its control, knowledge that has allowed the development of new diagnostic methods based on molecular strategies, fast, highly sensitive and specific in comparison with bioassays (Guerrero and Pruett, 2002).

Progresses in the development of molecular diagnostics are being studied, because they have demonstrated to be sufficiently sensitive and specific to differentiate mutant alleles using a single specimen. Which is useful to detect the presence of mutations responsible for acaricide resistance. In addition to determining low frequencies of alleles that confer resistance in field populations (Rosario-Cruz et al., 2005).

Studies on tick populations in Mexico using the polymerase chain reaction (PCR) technique to discriminate the presence of mutations in the sodium channel gene associated with resistance to the pyrethroid family (Figure 7 and figure 8), have demonstrated to have a significant association between mortality and the presence of the mutant allele in Mexican tick populations (Rosario-Cruz et al., 2009).

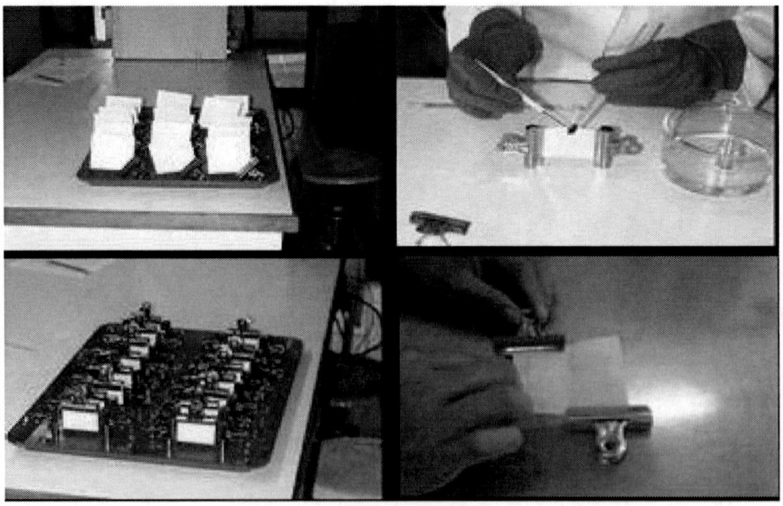

Figure 9. Larval Packet Test for acaricides resistant determination (Stone and Haydock, 1962).

CONCLUSION

Resistance is a genetic and pre-adaptive condition that implies the presence of a collection of genes within the arthropod population whose main function is the detoxification of the chemical structures used to control ticks, which explains the rapid appearance of resistance. When tick populations are exposed to toxic environments.

The plasticity of insect genomes, facilitate the development of resistance to all major classes of insecticides and allow the development of resistance to future insecticides, as long as current application techniques and used models prevail.

In strains of ticks resistant to pyrethroids, the presence of a mutation associated with the acaricide resistance, has been identified. The use of this mutation as a genetic marker detected by PCR assays, since the presence of the mutation in the sodium channel gene and resistant phenotypes to the pyrethroid family has a significant correlation. Diallelic crosses derived from RR and SS genotypes, both RS and SR show a susceptible phenotype as measured by Larval packet test discriminant dose (Figure 7 and Figure 9).

The growing problem of multiple resistance in Mexico requires research that generates in-depth knowledge about the mechanisms of resistance, which may be useful as tools to develop a rapid diagnosis and prediction of the appearance of resistance, in addition to the distribution of resistant phenotypes, in order to establish and design better strategic control programs.

It is essential to promote projects aimed at deciphering the genomes of important parasites that affect Mexican livestock, in this case the genome of *R. microplus*, in order to reduce research costs and facilitate the generation of reverse technology approaches starting from public databases instead of genetic material from cultured parasites.

REFERENCES

Aguilar-Tipacamú, G., R. J. Miller, R. Hernández-Ortiz, R. I. Rodriguez-Vivas, C. Vásquez-Peláez, Z. García-Vázquez, F. Olvera Valencia and R. Rosario-Cruz. (2008). Inheritance of pyrethroid resistance and a sodium channel gene mutation in the cattle tick *Boophilus microplus*. *Parasitology Ressearch*, 103: 633–639. doi 10.1007/s00436-008-1024-2.

Aguirre, E. J. (1987). Detección de brotes de resistencia en poblaciones de garrapatas durante una campaña de erradicación, con especial referencia a las Américas [Detection of resistance outbreaks in tick populations during an eradication campaign, with special reference to the Americas]. *Consulta de expertos sobre la erradicación de garrapatas*. México, D. F.

Alcala-Canto, Y., Figueroa-Castillo, J. A., Ibarra-Velarde, F., Vera-Montenegro, Y., Cervantes-Valencia, M. E., Salem, A. Z. M. and Cuéllar-Ordaz, J. A. 2018. Development of the first georeferenced map of *Rhipicephalus (Boophilus)* spp. in Mexico from 1970 to date and prediction of its spatial distribution. *Geospatial Health,* 13: 110-117. doi: 10. 4081/gh.2018.624.

Almazán, C., Aguilar-Tipacamu, G., Rodríguez, S., Mosqueda, J. and Pérez de León, A. (2018). Immunological control of ticks and tick-borne diseases impacting cattle health and production. *Frontiers in Bioscience.* 23:1535-1551. doi: 10.2741/4659.

Almazán, C., Torres-Torres, A., Torres-Rodríguez, L. and Soberanes-Céspedes, N. (2016). Aspectos biológicos de *Amblyomma mixtum* en el Noreste de México [Biological aspects of *Amblyomma mixtum* in Northeast Mexico]. *Quehacer Científico de Chiapas,* 11: 10-19. https://www.dgip.unach.mx/images/pdf-REVISTA-QUEHACERCIENTIFICO/2016-jul-dic/Aspectos_biologicos_de_Amblyomma _mixtum_.pdf.

Augustein, R. C., De Jersey, J., Web, E. C. and Zerner, B. (1969). On the homology of active site peptides of liver carboxilesterases. *Biochemical and Biophysics Acta,* 171: 128-137. doi: 10.1016/0005-2744(69)90112-0.

Baffi, M. A., de Souza, G. R., de Sousa, C. S., Ceron, C. R. and Bonetti, A. M. (2008). Esterase enzymes involved in pyrethroid and organophosphate resistance in a Brazilian population of *Riphicephalus* (*Boophilus*) *microplus* (Acari, Ixodidae). *Molecular and Biochemical Parasitology,* 160: 70-73. https://doi.org/10.1016/j.molbiopara.2008.03.009.

Barker, S. C. and Murrell, A. (2004). Systematics and evolution of ticks with a list of valid genus and species names. *Parasitology,* 129 Suppl: S15–S36. https://doi.org/10.1017/s0031182004005207.

Bisset, J., Rodriguez, M. M., Diaz, C., Soca, A. (1998). Resistance in a strain of *Culex quinquefasciatus* coming from Medellin, Colombia. *Revista Cubana de Medicina Tropical.* 50*:* 133-137. https://pubmed.ncbi.nlm.nih.gov/10349433/.

Burger, T. D., Shao, R., and Barker, S. C. (2014). Phylogenetic analysis of mitochondrial genome sequences indicates that the cattle tick, *Rhipicephalus (Boophilus) microplus*, contains a cryptic species. *Molecular Phylogenetics and Evolution,* 76, 241–253. https://doi.org/10.1016/j.ympev.2014.03.017.

Campbell, P. M., Newcomb, R. D., Russell, R. J. and Oakeshott J. G. (1998). Two different amino acid substitutions in the ali-esterase, E3, confer alternative types of organophosphorus insecticide resistance in the sheep blowfly, *Lucilia cuprina. Insect Biochemistry and Molecular Biology.* 28*:* 139-150. https://doi.org/10.1016/S0965-1748(97)00109-4.

Castro, M. B. and Wright, S. A. (2007). Vertebrate hosts of *Ixodes pacificus* (Acari: Ixodidae) in California. *Journal of Vector Ecology,* 32: 140-149. https://doi.org/10.3376/1081-1710(2007)32[140:VHOIPA]2.0.CO;2.

Chen, A. C., He, H., Temeyer, K. B., Jones, S., Green, P. and Barker, S. C. (2009). A survey of *Rhipicephalus microplus* populations for mutations associated with pyrethroid resistance. *Journal of Economic Entomology,* 102: 373-380. https://doi.org/10.1603/029.102.0148.

Colovic, B. M., Krstić, Z. D., Lazarević-Pašti, D. T., Bondžić, M. A. and Vasić. M. V. (2013). Acetylcholinesterase Inhibitors: Pharmacology and Toxicology. *Current Neuropharmacology,* 11(3): 315-335. doi: 10.2174/1570159X11311030006.

Craik, C. S., Roczniac, S., Largman, C. and Rutter, W.J. (1987). The catalytic role of the active site aspartic acid in serin proteases. *Science,* 237: 909-913. doi: 10.1126/science.3303334.

Dayhoff, M. O., Barker, W. C. and Hardman, J. K. (1972). Evolutionary conservatism: Enzyme estructure and function. Pp 53-57 en Dayhoff, M. O. Ed. *Atlas of protein sequence and structure.* Vol 5 Natl Biomed Res Foundation, Washington D.C. http://oasis.lib.tamuk.edu/search/?searchtype=o&searcharg=937951.

De la Fuente, J., Estrada-Pena, A., Venzal, M. J., Kocan, M. K. and Sonenshine, E. D. (2008). Overview: Ticks as vectors of pathogens that cause disease in humans and animals. *Frontiers in Bioscience,* 13: 6938-6946. doi: 10.2741/3200.

Devonshire, A. L. and L. M. Field, (1991). Gene amplification and insecticide resistance. *Annual Review of Entomology,* 36: 1-23. https://doi.org/10.1146/annurev.en.36.010191.000245.

Drabløs F. and Petersen S.B. (1997). Identification of conserved residues in family of esterase and lipase sequences. *Methods in Enzymology,* 284: 28–61. https://doi.org/10.1016/S0076-6879(97)84004-9.

Drummond, RO, Ernst, S. E., Trevino, J. L., Gladney, W. J. and Graham. O. H. (1973). *Boophilus annulatus* and *B. microplus*: laboratory tests of insecticides. *Journal of Economical Entomology,* 66 (1): 130-133. doi: 10.1093/jee/66.1.130.

Estrada-Peña, A., Guglielmone, A.A. and Mangold, A.J. (2004). The distribution and ecological 'preferences' of the tick *Amblyomma cajennense* (Acari: Ixodidae), an ectoparasite of humans and other mammals in the Americas. *Annals of Tropical Medicine and Parasitology*, 98(3): 283-292. https://doi.org/10.1179/000349804225003316.

Food and Agriculture Organization of the United Nations. (1971). Recommended methods for the detection and measurement of resistance of agricultural pests to pesticides–tentative method for larvae of cattle ticks, *Boophilus microplus* spp. FAO method No. 7. *FAO Plant Proceedings Bulletin*, 19:15–18.

Ffrench-Constant, R. H., Daborn, P. J. and Le Goff, G. (2004). The genetics and genomics of insecticide resistance. *Trends in Genetics,* 20: 163-70. https://doi.org/10.1016/j.tig.2004.01.003.

Ffrench-Constant, R. H., Steichen, J. C., Rocheleau, T. A., Aronstein, K. and Roush, R. T. (1993). A single-amino acid substitution in a gamma aminobutiric acid subtype A receptor locus is associated with cyclodiene insecticide resistance in Drosophila populations. *Proceeding of the National Academy of Science U.S.A.* 90: 1957-1961. https://doi.org/10.1073/pnas.90.5.1957.

Guerrero, F. D., Bendele, K. G., Chen, A. C., Li., A. Y., Miller, R. J., Pleasance., E. R. Varhol., M. E. Rousseau, V. M. and Nene, (2007). Serial analysis of gene expression in the southern cattle tick following acaricide treatment of larvae from organophosphate resistant and susceptible strains. *Insect Molecular Biology,* 16(1): 49-60. https://doi.org/10.1111/j.1365-2583.2007.00699.x.

Guerrero, F. D., Davey, R. B., and Miller, R. J. 2001. Use of an allele-specific Polimerase chain reaction assay to genotype pyretroid resistant strains of *Boophilus microplus. Journal of Medical Entomology*. 44-50. https://doi.org/10.1603/0022-2585-38.1.44.

Guerrero, F. D., Jamroz, R. D., Kammlah, D. and Kunz, S. E. (1997). Toxicological and molecular characterization of pyrethroid-resistant horn flies, *Haematobia irritans*: Identification of *kdr* and *super-kdr* point mutations. *Insect Biochemistry and Molecular Biology,* 27: 745-755. https://doi.org/10.1016/S0965-1748(97)00057-X.

Guerrero, F. D., Pruett, J. H. and Li, A. Y. (2002). Molecular and biochemical diagnosis of esterase-mediated pyrethroid resistance in a Mexican strain of *Boophilus microplus* (Acari: Ixodidae). *Experimental and Applied Acarology,* 28: 257-264. doi: 10.1023/a:1025319104529.

Guzmán-Cornejo, C., Robbins, R., Guglielmone, A., Montiel-Parra, G. and Perez, T. M. (2011). The *Amblyomma* (Acari: Ixodida: Ixodidae) of Mexico: Identification Keys, Distribution and Hosts. *Zootaxa,* 2998: 16-38. doi: 10.3897/zookeys.569.7221

He, H., Chen, A. C., Davey, R. D., Ivie, G. W. and George, J. E. (1999). Identification of a point mutation in the *para*-type sodium channel gene from a pyrethroid-resistant cattle tick. *Biochemistry and Biophysics Ressearch Communication,* 261: 558-561. https://doi.org/10.1006/bbrc.1999.1076.

Hemingway, J., Dunbar, S. J., Monro, A. G. and Small, G. J. (1993). Pyrethroid resistance in *German cockroaches* (Dictyoptera: Blattelidae): resistance levels and underlying mechanisms. *Journal of economic Entomology,* 86: 1631-1638. https://doi.org/10.1093/jee/86.6.1931.

Hemingway, J., Karunaratne, S. H. (1998). Mosquito carboxylesterases: a review of the molecular biology and biochemistry of a major insecticide resistance mechanism. *Medical and Veterinary Entomology,* 12: 1-12. https://doi.org/10.1046/j.1365-2915.1998.00082.x.

Hernandez, R., He, H., Chen, A.C., Waghela, S.D., Ivie, G.W., George, J.E. and Wagner, G.G. (2000). Identification of a point mutation in an esterase gene in different populations of the southern cattle tick, *Boophilus microplus*. *Insect Biochemistry and Molecular Biology,* 30: 969-977. https://doi.org/10.1016/S0965-1748(00)00069-2

Heymann, E. (1980). Carboxilesterases and amidases. Pp 291-232 en Jacoby, W. Ed. *Enzimatic Basic of detoxification*. Vol 2. Academic Press, New York. https://doi.org/10.1016/S0076-6879(81)77047-2.

Horak, I., G, Camicas, J. L. and Keirans, J. E. (2002). The Argasidae, Ixodidae and Nuttalliellidae (Acari: Ixodida): a world list of valid tick names. *Expetimental and Applied Acarology,* 28(1-4): 27-54. doi: 10.1023/a: 1025381712339.

Itoh, H., Tago, K., Hayatsu, M. and Kikuchi, Y. (2018). Detoxifying symbiosis: microbe-mediated detoxification of phytotoxins and pesticides in insects. *Natural Product Reports*, 35(5), 434–454. https://doi.org/10.1039/c7np00051k.

Jamroz, R. C., Guerrero, F. D., Pruett, J. H., Oehler, D. D. and Miller, R. J. (2000). Molecular and biochemical survey of acaricide resistance mechanisms in larvae from Mexican strains of the southern cattle tick, *Boophilus microplus*. *Journal of Insect Physiology,* 46: 685-695. https://doi.org/10.1016/S0022-1910(99)00157-2.

Jeyaprakash, A and Hoy, M. A. (2009). First divergence time estimate of spiders, scorpions, mites and ticks (subphylum: Chelicerata) inferred from mitochondrial phylogeny. *Experimental and Applied Acarology*, 47(1): 1-18. doi: 10.1007/s10493-008-9203-5.

Jongejan F. and Uilenberg, G. (2004). The global importance of ticks. *Parasitology,* 129: S3-S14. doi: 10.1017/s0031182004005967.

Jonsson, N. N., Bock, R. E. and Jorgensen, W. K. (2008). Productivity and health effects of anaplasmosis and babesiosis on *Bos indicus* cattle and their crosses, and the effects of differing intensity of tick control in Australia. *Veterinary Parasitology,* 155:1-9. https://doi.org/10.1016/j.vetpar.2008.03.022.

Keirans, J. E., Carleton, M. Clifford, H., Hoogstraal, E. and Easton, R. (1976). Discovery of *Nuttalliella namaqua* Bedford (Acarina: Ixodoidea: Nuttalliellidae) in Tanzania and Redescription of the Female Based on Scanning Electron Microcopy. *Annals of the Entomological Society of America,* 69 (5): 926–932. https://doi.org/10.1093/aesa/69.5.926.

Kessler, A. and Baldwin, I. T. (2002). Plant responses to insect herbivory: the emerging molecular analysis. *Annual Review of Plant Physiology and Plant Molecular Biology,* 53: 299–328. https://doi.org/10.1146/annurev.arplant.53.100301.135207.

Krish, K. (1971). *Carboxyl ester hydrolases*. En P. D. Boyer ed. The enzymes. Vol. 5 Academic Press New York. 43-69. https://doi.org/10.1016/S1874-6047(08)60086-4.

Lorke, E. D., Hasan, Y. M., Nurulain, M. S., Shafiullah, M., Kuča K., and Georg A. Petroianu, A.G. (2011). Pretreatment for acute exposure to diisopropylfluorophosphate: in vivo efficacy of various acetylcholinesterase inhibitors. *Journal of Applied Toxicology,* 31: 515–523. https://doi.org/10.1002/jat.1589.

Martínez, M. M., Cristina Coscolín, Gerard Santiago, Jennifer Chow, Peter J. Stogios, Rafael Bargiela, Christoph Gertler, José Navarro-Fernández, Alexander Bollinger, Stephan Thies, Celia Méndez-García, Ana Popovic, Greg Brown, Tatyana N. Chernikova, Antonio García-Moyano, Gro E. K. Bjerga, Pablo Pérez-García, Tran Hai, Mercedes V. Del Pozo, Runar Stokke, Ida H. Steen, Hong Cui, Xiaohui Xu, Boguslaw P. Nocek, María Alcaide, Marco Distaso, Victoria Mesa, Ana I. Peláez, Jesús Sánchez, Patrick C. F. Buchholz, Jürgen Pleiss, Antonio Fernández-Guerra, Frank O. Glöckner, Olga V. Golyshina, Michail M. Yakimov, Alexei Savchenko, Karl-Erich Jaeger, Alexander F. Yakunin, Wolfgang R. Streit, Peter N. Golyshin, Víctor Guallar, Manuel Ferrer, and The Inmare Consortium, (2018). Determinants and Prediction of Esterase Substrate Promiscuity Patterns. *ACS Chemical Biology.* 13 (1): 225-234. https://doi.org/10.1021/acschembio.7b00996.

Miller, R. J., Davey, R. B. and George, J. E. (1999). Characterization of pyrethroid resistance and susceptibility to coumaphos in Mexican *Boophilus microplus* (Acari: Ixodidae). *Journal of Medical Entomology,* 36: 533-538. https://doi.org/10.1093/jmedent/36.5.533.

Miller, R. J., Li, A. Y., Tijerina, M., Davey, R. B. and George, J. E. 2008. Differential response to diazinon and coumaphos in a strain of *Boophilus microplus* (Acari: Ixodidae) collected in Mexico. *Journal of Medical Entomology,* 45: 905-911. https://doi.org/10.1093/jmedent/45.5.905.

Miller, R. J., Almazán, C., Ortíz-Estrada, M., Davey, R. B., George, J. E. and De León, A. P. (2013). First report of fipronil resistance in *Rhipicephalus* (*Boophilus*) *microplus* of Mexico. *Veterinary Parasitology,* 191 (1-2): 97–101. https://doi.org/10.1016/j.vetpar.2012.08.011.

Morgan, J. A., Corley, S. W., Jackson, L. A., Lew-Tabor, A. E., Moolhuijzen, P. M. and Jonsson, N. N. (2009). Identification of a mutation in the para-sodium channel gene of the cattle tick *Rhipicephalus (Boophilus) microplus* associated with resistance to synthetic pyrethroid acaricides. *International Journal of Parasitology,* 39: 775-779. https://doi.org/10.1016/j.ijpara.2008.12.006.

Mutero, A., Pralavorio, M., Bride, J. M. and Fournier, D. (1994). Resistance-associated point mutations in insecticide-insensitive acetylcholinesterase. *Proceedings of the National Academy of Sciences of the United States of America,* 91(13): 5922–5926. https://doi.org/10.1073/pnas.91.13.5922.

Nava, S., Beati, L., Labruna, B. M., Cáceres, G. A., Mangold, J. A. and Guglielmone, A. A. (2014). Reassessment of the taxonomic status of *Amblyomma cajennense* (Fabricius, 1787) with the description of three new species, *Amblyomma tonelliae* n. sp., *Amblyomma interandinum* n. sp. and *Amblyomma patinoi* n. sp., and reinstatement of *Amblyomma mixtum* Koch, 1844, and *Amblyomma sculptum* Berlese, 1888 (Ixodida:

Ixodidae). *Ticks and Tick-Borne Diseases,* 5 (3): 252-276. https://doi.org/10.1016/j.ttbdis.2013.11.004.

Nolan, J., and H. J. Schnitzerling. (1986). Drug resistance in arthropod parasites, pp. 603-620. In W. C. Campbell and R. S. Rew [eds.], *Chemotherapy of parasitic diseases*. Plenum, New York. file:///C:/Users/INIFAP/Desktop/help%20ros/978-1-4684-1233-8_32.pdf.

Ortiz, E. M., Santamaría, V. M., Ortiz, N. A., Soberanes, C. N., Osório, M. J., Franco, B. R., Martinez, I. F., Quezada, D. R. and Fragoso, S. H. (1995). Characterization of *Boophilus microplus* resistance to ixodicides in México. In: *Seminario internacional de Parasitología Animal Acapulco*, Gro. México. Pp 58-66.

Palumbi, S. R. (2001). Humans as the world's greatest evolutionary force. *Science, 293*: 1786–1790. doi: 10.1126/science.293.5536.1786.

Peñalver, E., Arillo, A., Delclòs, X. 2017. Ticks parasitised feathered dinosaurs as revealed by Cretaceous amber assemblages. *Nature Communications,* 8: 19-24. https://doi.org/10.1038/s41467-017-01550-z.

Perez-Cogollo, L. C., Rodriguez-Vivas, R. I., Ramirez-Cruz, G. T. and Miller, R. J. (2010). First report of the cattle tick *Rhipicephalus microplus* resistant to ivermectin in Mexico. *Veterinary Parasitology,* 168(1-2): 165–169. https://doi.org/10.1016/j.vetpar.2009.10.021.

Pimentel, D., Andow, D., Dyson Hudson, R., Gallahan, D., Jacobson, S., Irish, M., Kroop, S., Moss, A., Schreiner, I., Shepard, M., Thompson, T. Y. and Vizant, B. (1990). Environmental and social costs of pesticides: a preliminary assessment. https://doi.org/10.2307/3544173. *En CRC handbook of pest management in agriculture.* (edited by Pimentel, D.) 2nd. Edn. CRC Press. Boca Raton. I. 721-742.

Pruett, J. H., Guerrero, F. D. and Hernandez, R. (2002). Isolation and identification of an esterase from a Mexican strain of *Boophilus microplus* (Acari: Ixodidae). *Journal of Economical Entomology*, 95:1001-1007. https://doi.org/10.1093/jee/95.5.1001.

Ranson, H., Claudianos, C., Ortelli, F., Abgrall, C., Hemingway, J., Sharakhova, M. V., Unger, M. F., Collins, F. H. and Feyereisen, R. (2002). Evolution of supergene families associated with insecticide resistance. *Science.* 298: 179-181. doi: 10.1126/science.1076781.

Rausher, M. D. (2001). Co-evolution and plant resistance to natural enemies. *Nature,* 411: 857–864. https://doi.org/10.1038/35081193.

Rauwerdink, A and Kaslauskas, J. R. (2015). How the Same Core Catalytic Machinery Catalyzes 17 Different Reactions: the Serine-Histidine-Aspartate Catalytic Triad of α/β-Hydrolase Fold Enzymes. *ACS Catalysis,* 5(10): 6153–6176. https://doi.org/10.1021/acscatal.5b01539.

Reck, J., Klafke, M. G., Webster, A., Dall'Agnol, B., Scheffer, R., Souza, A .U., Corassini, B. V., Vargas, R., dos Santos, S. J. and Martins, R. J. (2014). First report of fluazuron resistance in *Rhipicephalus microplus*: A field tick population resistant to six classes of acaricides. *Veterinary Parasitology,* 201(1-2): 128-136. https://doi.org/10.1016/j.vetpar.2014.01.012.

Rodríguez-Vivas, R. I., Apanaskevich, D. A., Ojeda-Chi, M. M., Trinidad-Martínez, I., Reyes-Novelo, E., Esteve-Gassent, M. D. and Pérez de León, A. A. (2016). Ticks collected from humans, domestic animals, and wildlife in Yucatan, Mexico. *Veterinary Parasitology,* 215:106-113. doi: 10.1016/j.vetpar.2015.11.010.

Rosario-Cruz, R., Almazan, C., Miller, R.J., Domínguez-Garcia, D.I., Hernandez-Ortiz, R. and de la Fuente, J. (2009). Genetic basis and impact of tick acaricide resistance. *Frontiers in Bioscience,* 14: 2657-2665. doi: 10.2741/3403.

Rosario-Cruz, R., Guerrero, F. D., Miller, R. J., Rodriguez-Vivas, R. I., Domínguez-García, D. I., Cornel, A. J., Hernandez-Ortiz, R. and George, J. E. (2005). Roles played by esterase activity and by a sodium channel mutation involved in pyrethroid resistance in populations of *Boophilus microplus* (Acari: Ixodidae) collected from Yucatan, Mexico. *Journal of Medical Entomology,* 42:1020-1025. https://doi.org/10.1093/jmedent/42.6.1020.

Servicio Nacional de Sanidad, Inocuidad y Calidad Agroalimentaria [*National Service of Agrifood Health, Safety and Quality*] (*SENASICA*). 2020. https://www.gob.mx/senasica/documentos/situacion-actual-del-control-de-la-garrapata-boophilus-spp.

Shaw, R. D. (1966). Culture of an organophosphorus-resistant strain of *Boophilus microplus* (Can.) and an assessment of its resistance spectrum. *Bulletin of Entomological Ressearch,* 56(3): 389-405. doi: https://doi.org/10.1017/S0007485300056480.

Soberanes-Céspedes, N., Santamaría, V., Fragoso, S.H., García-Vázquez. and Zeferino. (2002). Primer caso de resistencia al amitraz en la garrapata del ganado *Boophilus microplus* en México [First case of resistance to amitraz in the cattle tick Boophilus microplus in Mexico]. *Técnica Pecuaria en México,* 40(1): 81-92. [fecha de Consulta 20 de Agosto de 2020]. ISSN: 0040-1889. .https://www.redalyc.org/articulo.oa?id=613/61340106.

Stone, B. F. and Haydock, K. P. (1962). A method for measuring the acaricide susceptibility of the cattle tick *B. microplus* (Can.). *Bulletin Entomological Ressearch,* (53): 563-578. doi: https://doi.org/10.1017/S000748530004832X.

Temeyer, K. B., Pruett, J. H. and Chen, A. C. (2007). R86Q a mutation in BmAChE3 yielding a *Rhipicephalus microplus* organophosphate-insensitive acetylcholinesterase. *Journal of Medical Entomology*: 1013-1018.

Vath G. M., Earhart C. A., Monie D. D., Iandolo J. J., Schlievert P. M. and Ohlendorf D. H. (1999). The crystal structure of exfoliative toxin B: A superantigen with enzymatic activity. *Biochemistry,* 38 :10239–10246.

Vecino, J. A. C., Echeverri, J. A. B., Cárdenas, J. A. and Herrera, L. A. P. (2010). Distribución de garrapatas *Rhipicephalus (Boophilus) microplus* en bovinos y fincas del Altiplano cundiboyacense (Colombia). *Ciencia & Tecnología Agropecuaria,* 11(1): 73-84. https://doi.org/10.21930/rcta.vol11_num1_art:197.

Vulule, J. M., Beach, R. F., Atieli, F. K., McAllister, J. C., Brogdon, W. G., Roberts, J. M., Mwangi, R. W., Hawley, W. A. 1999. Elevated oxidase and esterase levels associated with permethrin tolerance in *Anopheles gambiae* from Kenyan villages using permethrin-impregnated nets. *Medical and Veterinary Entomology*. 13: 239-244. https://doi.org/10.1046/j.1365-2915.1999.00177.x.

Wheat, C. W., Vogel, H., Wittstock, U., Braby, M. F., Underwood, D., Mitchell-Olds, T. 2007. The genetic basis of a plant-insect coevolutionary key innovation. *Proceedings of the National Academy of Science of the United States of America*. 104(51): 20427-20431. https://doi.org/10.1073/pnas.0706229104.

Whittaker, R. H. y Feeny, P. P. 1971. Allelochemics: Chemical interactions between species. *Science.* 171: 757-770. Retrieved August 20, 2020. www.jstor.org/stable/1730763.

Williamson, M. S., Martinez-Torres, D., Hick, C. A., Devonshire, A. L. 1996. Identification of mutations in house-fly *para*-sodium channel gene associated with knockdown resistance (*kdr*) to pyrethroid insecticides. *Molecular and General Genetics*. 252: 51-60. https://doi.org/10.1007/BF02173204.

Wittstock, U., Gershenzon, J. 2002. Constitutive plant toxins and their role in plant defense. *Current opinions in Plant Biology*. 5: 300–307. https://doi.org/10.1016/S1369-5266(02) 00264-9.

Zhang, H. G., Ffrench-Constant, R. H., Jackson, M. B. 1994. A unique amino acid of the Drosophila GABA receptor with influence on drug sensitivity by two mechanisms. *Journal of Physiology*. 479: 65-75. doi: 10.1113/jphysiol.1994.sp020278.

In: The Entomological Guide to *Rhipicephalus*
Editors: Sachin Kumar, Raquel Cossio Bayugar et al.
ISBN: 978-1-53619-619-1

Chapter 12

PREVALENCE, RESISTANCE STATUS AND CONTROL STRATEGIES AGAINST TICKS IN WESTERN INDIA

K. P. Shyma[1,*] and Jay Prakash Gupta[2]
[1]Department of Veterinary Parasitology, College of Veterinary Science and Animal Husbandry, Sardarkrushinagar Dantiwada Agricultural University, Sardarkrushinagar, Gujarat, India
[2]Department of Animal Genetics and Breeding, College of Veterinary Science and Animal Husbandry,Sardarkrushinagar Dantiwada Agricultural University, Sardarkrushinagar, Gujarat, India

ABSTRACT

Ticks are highly specific obligate haematophagous ectoparasites of mammals, birds and reptiles, distributed globally and are of enormous medical and veterinary relevance owing to the direct damage they cause to their hosts. Moreover, they act as vectors of a large variety of human and animal pathogens. Today, most emerging infectious diseases arise from zoonotic pathogens, and many of them are transmitted by tick vectors. The cattle tick, *Rhipicephalus microplus* is one of the most damaging tick attacking bovines in the tropical and subtropical regions of the world. In this book chapter, we would like to depict the prevalence of ticks in the western India and the factors responsible for their predominance in this region. The tick control programme is focused mainly on repeated use of organophosphates, pyrethroids, formamidines, phenylpyrazoles and macrocyclic lactones. However, the application of the acaricides is poorly supervised, resulting in economic problems produced by phenomenon such as acaricide resistance. Single host life cycle with short life cycle and high fecundity, the *R. microplus* has been subjected to maximum exposure of acaricides and thus developed a significant ability to become resistant to the toxic effect of the chemicals used to manage them. Control strategies against ticks rely mainly on judicious use of acaricides. Non-chemical natural product-based tick management protocol has also shown significant success in recent years. The indigenous ethno-veterinary and remedial knowledge of Indian farmers propose a range of herbs having insecticidal and acaricidal properties. The integrated approaches

*Corresponding Author's Email: dr.shymakpvet@gmail.com.

including pasture management with limited applications of effective chemical acaricides, strategic use of phytoformulation, housing management, strengthening of dairy cattle management with alternative feeding technologies and immunization against ticks and ticks borne diseases may be the most viable option for management of acaricide failure. Integrated tick management will certainly reduce the over reliance on acaricides and lessen selection pressure on ticks thereby preserving the chemicals. The availability of different control measures against ticks, their advantages and disadvantages, and their economic impact will be discussed in this chapter with the hope that the coordinated activities will significantly improve the health status of animals and the environment by minimizing the use of non-effective acaricides which will reduce the economic burden to the farmers.

Keywords: acaricides, diagnosis, management, Pyrethroides, *Rhipicephalus microplus*

INTRODUCTION

India is mainly an agriculture-based country and has the world's highest bovine population of 303.76 million which includes 193.46 million cattle and 109.85 million buffaloes (DAHD, 2015).The Animal Husbandry sector is significantly contributing to the Indian economy by providing milk, meat, bones and hides. Livestock provides livelihood to two-third of rural community. Moreover, it also plays a major role in energy production, enhancing soil fertility and rural transportation. Although productivity per animal is very low, India stands first in the world milk production accounting for 20% of world output (BAHS, 2019). Parasitic diseases are one of the important global problems affecting the health and productivity of animals. It can be either due to endoparasites that live inside the body or ectoparasites that attack the body surfaces such as ticks, mites, flies, fleas, midges, etc. Amongst ectoparasites, ticks are very important and harmful blood sucking external parasites of mammals, birds and reptiles throughout the world (Furman and Loomis, 1984). Ticks belong to phylum, Arthropoda and make up the major collection of creatures in order Acarina (Rajput et al, 2006). The medical and economic importance of ticks had long been recognized due to their ability to transmit diseases to humans and animals.

India is a leading milk producer owing to its enormous bovine population (BAHS, 2019). Ever growing human population exerts great pressure on livestock production system. In past, efforts have been made to increase milk production by means of crossbreeding. Increased exotic inheritance led to the development of such animals which are susceptible to many diseases of tropical climate, amongst which tick-borne diseases owes a special concern (Bansal, 2005). Ticks are obligate blood-sucking arthropods found throughout most regions of the world infesting livestock and considered one of the most damaging ectoparasites to domestic and wild animals. Indigenous breeds are comparatively resistant to tick infestation but they are low producers. Ticks and tick-borne diseases (TTBDs) have created havoc in tropical country like India since they are able to transmit various pathogens and cause great damage to humans and animals all over the world (Walker, 2011). In their rough estimation, Minjauw and McLeod (2003) reported approximately 2000 crore rupees per annum losses due to tick- and tick-borne diseases (TTBDs) in India. Management of the multidimensional problems created by tick infestations is quite elaborative. Due to ideal environmental conditions in our country, almost all the economically important livestock species are suffering from ticks and tick-borne diseases (TTBDs) (Ghosh et al., 2007).

TICKS AND ITS GLOBAL IMPORTANCE

Ticks are invertebrate animals in the phylum Arthropoda which are parasitic feeders. They are highly specialized obligate haematophagous ectoparasites of mammals, birds and reptiles, distributed globally. They are of vast medical and veterinary significance due to the direct damage they cause to their hosts and as vectors of a large variety of human and animal pathogens. At present ticks are the vectors for many of the emerging infectious diseases arising out of zoonotic pathogens. Ticks are only second to mosquitoes in terms of acting as vectors for many of the human pathogens such as viruses, bacteria, rickettsia, spirochetes, etc. and the most important vector of pathogens affecting cattle worldwide (Peter et al., 2005). Numerous characteristics of ticks make them outstanding vectors of pathogenic agents, the wide host range and tendency to feed on several hosts during life cycle ensures ample opportunity to acquire and transmit pathogens. Hardiness and longevity enable them to endure in unfavorable environmental condition. High biotic potential, ensuring maintenance of a large populations and a high frequency of host-vector contact, slow feeding habits and attachment with hosts for relatively longer periods as in the case of ixodids, which allows adequate time for pathogen acquirement and transmission (Ghosh and Nagar, 2014). In humans, tick infestations typically involve few specimens and the greatest risk for people bitten by a tick lies in infection due to a tick-borne pathogen (Parola and Raoult, 2001).

There are 899 tick species those parasitize the vertebrate animals including Argasidae (185 species), Ixodidae (713 species) and Nuttalliellidae (1 species) (Barker and Murrell, 2004). Ticks are mostly belonging to two major families, the Ixodidae or hard ticks, and the Argasidae or soft ticks. The familyIxodidaecontains 14 genera (Guglielmone et al., 2010), including *Amblyomma*, *Dermacentor*, *Haemaphysalis*, *Hyalomma*, *Ixodes*, *Margaropus* and *Rhipicephalus*. The important boophilid ticks, formerly of the genus *Boophilus*, are now classified as a subgenus within the genus *Rhipicephalus* (Beati et al., 2001; Barker et al., 2004). The family Argasidae contains the important genera *Argas*, *Ornithodoros*, and *Otobius* (Taylor et al., 2016). These genera are known as soft ticks because their outer body surfaces lack hard plates. In India, seven genera of hard ticks and three genera of soft ticks have been identified. Hard ticks feed for extended periods of time on their hosts, varying from several days to weeks, depending on factors such as life stage, host type, and species of tick. The external structure i.e., cuticle, of hard ticks expands to accommodate the large volume of blood ingested (approximately, somewhere between 200 to 600 times their unfed body weight) by adult ticks (Sonenshine, 1991). Many soft ticks have an uncanny resistance to starvation, and can survive for many years without blood meal (Furman and Loomis, 1984). The outside surface, or cuticle, of soft ticks expands, but does not grow to accommodate the large volume of blood ingested, which may be anywhere from 5 to 10 times their unfed body weight (Sonenshine, 1991).

Animals can be parasitized by thousands of ticks, which either cause direct injuries or act as vector for disease transmission. Direct injuries to animals can be very serious, especially in tropical climates, and are mainly observed in infestations with ixodid ticks. In India, cattle and buffaloes are frequently heavily infested with different species of ticks, which apart from transmitting diseases such as theileriosis, babesiosis, anaplasmosis and viral diseases of livestock; also cause extensive damage to the livestock health and production. Tick-associated dermatophilosis is also a major health and management problems of livestock in many developing countries. The economically most important ixodid ticks of livestock in tropical regions belong to the genera of Hyalomma, *Boophilus*, *Rhipicephalus* and *Amblyomma* (Frans, 2000).

Ticks are involved in the transmission of a variety of disease pathogens of public health significance since time immemorial. In India, the studies of ticks received significance after the finding of Kyasanur Forest disease (KFD) transmitted by *Haemaphysalis* spp. in Shimoga district of Karnataka state in 1957 (Work et al., 1957). There was a steep rise in the number of KFD cases in Karnataka which imposed the strategic control of tick vectors (Pattnaik, 2006). Thereafter, sporadic cases of indigenous nature were reported from different parts of the country time to time. Sharma et al. (1992) reported serologically positive cases of Lyme disease from Coonoor in Nilgiri Hills of Tamil Nadu state. In 2007, an epidemic of Indian tick typhus was reported from village Deol district Kangra (Himachal Pradesh), wherein 357 cases were reported (Kumar et al., 2011). Shanmugam et al. (1976) have reported the presence of Crimean-Congo haemorrhagic fever (CCHF) virus infection in human and domestic animals from various parts of India. However, the first instance of CCHF was detected from Ahmedabad in January 2011 (Times of India, February 3, 2011) and subsequently cases/deaths were reported in 2012 and 2013. Ticks of the genus Hyalomma are well-known vectors of viruses transmissible to man namely Crimean Congo Hemorrhagic Fever (CCHF). *Hyalomma marginatum* and its sub species are the primary vector of CCHF virus where it occurs sporadically throughout Europe, Asia and Africa, ranging from Portugal eastern side to Crimea and then further deep into east including central Asia and China (Hoogstraal, 1979). Many mammalian hosts of ticks are the natural reservoir of the CCHF virus, which is perpetuated in the ticks by feeding on viremic hosts and by trans-stadial and transovarial transmission (Linthicum and Bailey, 1994). Wild mammals, ground-feeding birds and ticks etc. are acting as stable enzootic foci though, most birds are incompetent hosts for the infection (Hoogstraal, 1979). However, birds transport CCHF-infected ticks and serve as a blood meal source for tick populations (Watts et al., 1988). Hares and hedgehogs are frequently infected and are important hosts of *Hyalomma marginatum*. Cattle may also serve as reservoir of CCHF virus, although evidence is based largely on serology, with few isolates of the virus from cattle (Watts et al., 1988). Other Hyalomma ticks involved in the transmission of CCHF virus include *H. truncatum*, *H. detritum* and *H. impeltatum*. *H. anatolicum* distributed in parts of the Near East, Asia Minor, southern Europe, southern Russia and India (Goddard, 1987; Linthicum and Bailey, 1994).

Prevalence of Ticks in Western India

The distribution and prevalence of ticks are mainly attributed to the altitude, vegetation, animal activity, host availability, and other climatic factors in a particular area. Earlier studies by Geevarghese et al. (1997) and Ghosh et al. (2007) have reported that out of a total of 109 species of ticks present in India 88 belong to family Ixodidae of hard ticks. Chhabra (1992) had reported 46 tick species of family Ixodidea infesting livestock in India of which 13 were reported to commonly infest buffalo while 36 of them were reported to infest cattle.

Although a number of tick species (109) are known from India, 24 species including 18 ixodids and 6 argasids are reported from Rajasthan. First reported study about the tick fauna of Rajasthan was carried out by Joshi (1943) who observed the prevalence of *Ornithodoros savignyi* from Marwar. Later several workers (Hoogstraal and Trapido, 1963; Wattal and Srivastava, 1967; Miranpuri et al., 1975; Kaul et al., 1978; Advani and Vazirani, 1981; Miranpuri and Gill, 1983; Geeverghese and Dhanda, 1987; Frazier and Keirans, 1989 and Sanyal and De, 1996) carried out study in this regard, described as well as noted the presence of different species of ticks from the state. The list of ticks include *Amblyomma clypeolatum, A. javanense, Aponomma gervaisi, Rhipicephalus microplus, Haemaphysalis indica, H. kutchensis, Hyalomma antolicum antolicum, H. detritum, H. dromedarii, H. hussaini, H. Kumari, H. marginatum isaaci, Rhipicephalus haemaphysaloides, R. ramachandrai, R. sanguineus, R. turanicus, Ixodes vespertilionis, Argas (carios) indicus, A. reflexus hermanni, A. (Chiropteragus) wilsoni, A. vespertilionis, Argas persicus, Ornithodoros savignyi* (Sanyal and De, 2005).

The ixodid ticks in Gujarat predominated by *Hyalomma anatolicum* and *R. microplus*. The drier zones of the state showed a lower prevalence rate of ixodid ticks as compared to the zones with high annual rainfall (Patel et al., 2019). Although, *R. microplus* was the predominant ixodid tick of cattle in all agro-climatic zones of Gujarat, its prevalence was comparatively superior in areas with high annual rainfall. *Hyalomma anatolicum* was recorded more in drier zones especially in the northern districts of the state where there is an extreme of temperature (Shyma, unpublished information). *Amblyomma clypeolatum, A. integrum, A. javanense, A. leave, Boophilus annulatus, Haemaphysalis bispinosa, H. doenitzi, H. indica, H. kutchensis, Hyalomma dromedarii, H. issaci, H. hussaini, H. turanicum, Nosomma monstrosum, Rhipicephalus haemaphysaloides, R. sanguineus, R. turanicus, Argus abdussalami, A. gujaratensis, A. hoogstraali* and *Ornithodoros savignyi* are the other tick species reported from Gujarat (Ghosh et al., 2007).

Geevarghese and Dhanda (1995) studied a definite pattern in the distribution of certain species of ticks in Maharashtra. Most of the tick species recorded showed some degree of host and site preferences. The list of ticks reported from Maharashtra includes *Aponomma gervaisi, Rhipicephalus microplus, Haemaphysalis bispinosa, H. centropi, H. cuspidate, H. howletti, H. indica, H. intermedia, H. minuta, H. paraturturis, H. silvafelis, H. spinigera, Hyalomma anatolicum anatolicum, H. marginatum, H. marginatum turanicum, H. dromedarii, H. detritum, H. hussaini, H. brevipunctata, H. kumari, Ixodes ceylonensis, Nosomma monstrosum, Rhipicephalus haemaphysaloides, R. turanicus, R. sanguineus, R. ramachandrai, Dermacentor auratus* and *Amblyomma javanense* (Geevarghese and Dhanda, 1995; Maske et al., 1996; Shastri et al., 1983; Raote, 1983). *H. anatolicum* was mostly prevalent in the plateau region with a semi-arid climate and was absent, or present in negligible numbers, in the Western Ghats and the coastal regions with heavy rainfall and an equitable climate. In contrast, *H. hussaini* was totally absent in the coastal region. Its place was taken by another closely related species, *H. kumar., H. brevipunctata* was observed to have a restricted distribution, and was collected in good numbers only in Palve village in the Plateau region. *Haemaphysalis spinigera* and *Nosomma monstrosum* were seen to be distributed mostly in Western Ghats and coastal regions. Species like *R. haemaphysaloides* and *B. microplus* were common in all the regions. *B. microplus* had been reported as the predominant tick in cattle from Maharashtra (Shahardar et al., 1998).

Haemaphysalis bispinosa, *H. spinigera, H. intermedia, Amblyomma integrum, Amblyomma leave, A. varanense, B. microplus, H. a. anatolicum, H. detritum* are the tick species reported from Goa (Ghosh and Nagar, 2014, Balakrishnan et al., 2019). *Haemaphysalis spinigera* is playing an important role for the transmission of KFD in Goa (Mourya et al. 2014). *Haemaphysalis spinigera*, a major vector of KFD virus formed 0.71% of ticks on cattle. The immature of *H. spinigera* are known to parasitize certain species of wild mammal, small mammals and ground birds. Cattle have become the most important hosts for the adults of *H. spinigera* and play a prominent role in the dissemination of larval population on the forest floor.

Katuri et al. (2017) reported the presence of *Rhipicephalus (Boophilus) microplus* and *Hyalomma anatolicum* ticks in and around Jabalpur, M.P, India, out of which the *R. B. microplus* was highly abundant. They found highest tick infestation in rainy season followed by spring and winter. Age wise highest percentage of tick infestations was observed in animals <1 year of age (72.97%). Prevalence of tick infestation in relation to gender has shown higher infestation rates in male (59.97%) as compared to female animals (54.01%) irrespective of species of animals under study. Other species including *A. testudinarium, Dermacentor auratus, Haemaphysalis bispinosa, H. canestrinii, H. intermedia, H. paraturturis, H. spinigera, Hyalomma issaci, H. hussaini, Rhipicephalus haemaphysaloides, R. sanguineus, R. turanicus, Ornithodoros savignyi* and *O. megnini* were also been reported from different parts of Madhya Pradesh (Ghosh et al., 2007, Mandloi et al., 2016).

Geevarghese et al. (1997) have reported that about 10 species of hard ticks might be prevalent in Haryana infesting various livestock. Chillar et al. (2014) identified *Hyalomma anatolicum anatolicum, Hyalomma anatolicum excavatum, Rhipicephalus sanguineus, Rhipicephalus (Boophilus) microplus, Rhipicephalus (Boophilus) decoloratus* and *Dermacentor spp.* This study revealed that *Hyalomma anatolicum anatolicum* and *Rhipicephalus (Boophlilus) microplus* were the most common vector species infesting buffalo and cattle in Haryana state. Mixed infestations of *H. a. anatolicum* with *H. a. excavatum* and *R. (B.) microplus* with *R. (B.) decoloratus* were also observed in the state (Sangwan et al., 2000; Singh and Chhabra, 1999).

Tick Control Measures

Ticks are the most important ectoparasites of livestock in tropical and sub-tropical areas, and are accountable for severe economic losses. Eradication of ticks is not a realistic goal in most cases, though the eradication of *Boophilus* spp. from the USA remains an outstanding exception. Except on islands, extinction of ticks is mostly not practicable, where successful campaigns have sometimes been implemented. The different strategies to control ticks are been discussed here.

Chemical Control of Ticks and Their Mode of Action

Acaricides used to control ticks on livestock or in the environment, are applied in such a manner that the ticks are killed, but will not harm livestock or applicators, the tissues of treated animals will not contain chemical residues, and the environment will not be adversely affected (Drummond, 1983). Acaricides used for tick control include arsenicals, chlorinated hydrocarbons (e.g., dichloro diphenyl trichloroethane; DDT), organo-phosphorous compounds (e.g., Diazinon), carbamates (e.g., carbaryl), pyrethroids (e.g., permethrin, flumethrin etc.), formamidines, phenylpyrazoles and avermectins.

Arsenic

Arsenic was introduced as an acaricide around 1900. For almost 40 years it was the only product used and resistance did not appear until towards the end of that period. The liquid concentrate of arsenic trioxide (As_2O_3) was the product which was commercialised. Arsenic had the advantages of being inexpensive, stable and the concentration could be accurately determined and replenished in a simple way. Arsenical compounds cannot be used in the form of spray as the mist resulting from spraying, contaminates the surroundings and particularly the vegetation with arsenic. Arsenic is very toxic and an extremely irritating and caustic product, especially for the gastrointestinal tract. The safety margin of this chemical is very slight. There are lot many ways in which Arsenic compounds can disrupt the ATP production.

Chlorinated Hydrocarbons

The chlorinated hydrocarbons were developed in the beginning of 1940s, after the discovery (1939) of the insecticidal properties of DDT. In addition to DDT other compounds include; lindane, methoxychlor, BHC, cyclodienes (which include aldrin, dieldrin, chlordane, heptachlor, and endrin) and Chlorobenzilate. These Chlorinated hydrocarbons are fat-soluble pesticides that persist in environment for longer period of time and bio-accumulate, even though their use has been banned or restricted in the past several decades. In the general population, consumption of contaminated food is the primary source of exposure, including high levels in body fat and mother's milk. Lindane (γ-BHC) has also been banned for agricultural use in recent years, but medical uses for the topical treatment of lice and scabies continue. Mechanism of action of chlorinated hydrocarbons is not yet fully understood, the major toxic action of this group of chemicals is on the nervous system, both central and peripheral. They act by altering the electrophysiological properties of the cell membranes (particularly nerve axons) disturbing sodium and potassium ion exchange through the membrane (Kunz and Kemp, 1994). Since the procedure is reversible, organochlorine insecticides, which vary considerably in toxicity, in small dosage they have little or no obvious effect when dosages are small, even if exposure to them is of a very long duration. It should be borne in mind, however, that many of these insecticides or their degradation products are soluble in fat and can be stored in the body fat, where they remain apparently without effect if not mobilized.

Organophosphorus Compounds

Organophosphorus compounds are organic compounds containing phosphorus (Aldridge and Davison, 1952). It was primarily used in pest control as an alternative to chlorinated hydrocarbons that persist in the environment. The toxicity of organophosphorus to animals is attributed to their ability to inhibit acetylcholinesterase (AChE) which is a class of enzymes that catalyzes the hydrolysis of the neurotransmitting agent acetylcholine (ACh). The actual chemical reaction between the enzyme and the organophosphate results in inhibition of AChE and related esterases, often referred to as serine hydrolases. The phosphorylated enzyme is no longer capable of affecting the hydrolysis of ACh; this results in a buildup of the neurotransmitter at a nerve synapse or neuromuscular junction. AChE is present and has been isolated from a wide range of animals, including mammals, birds, fish, reptiles and insects (Karczmar et al., 1970). Neurotransmitter, ACh is rapidly getting hydrolysed in to the inactive compounds, choline and acetic acid by the action of AChE. Acetylcholine is one of a number of physiologically important neurotransmitting agents and is involved in the transmission of nerve impulses to effector cells at cholinergic, synaptic, and neuromuscular junctions. Infested cattle are generally dipped in the organophosphate acaricide, coumaphos (0.3% active ingredient) (Bram et al., 2002).

Carbamates

These are esters of carbamic acids and closely resemble the organophosphates (Spickett, 1998). They are a little more toxic than the organophosphates for mammals and are much more expensive. The mechanism of action is similar to that of organophosphorus compounds by inhibiting acetylcholinesterase.

Pyrethroids

Similar to the natural pyrethrins, which are produced by the flowers of pyrethrums (*Chrysanthemum cinerariaefolium* and *C. coccineum*) pyrethroid is an organic compound. Pyrethroids constitute the majority of commercial household insecticides. In the concentrations used in such products, they may also have insect repellent properties and are generally harmless to humans. Team of Rothamsted Research scientists have introduced pyrethroids in the 1960s and 1970s. Pyrethroids are axonicexcitotoxins, the toxic effects of which are mediated through preventing the closure of the voltage-gated sodium channels in the axonal membranes. Actually, sodium channel refers to membrane protein having a hydrophilic interior. This interior is shaped precisely to allow sodium ions to pass through the membrane, enter the axon, and propagate an action potential. When the toxin keeps the channels in their open state, the nerves cannot repolarize, leaving the axonal membrane permanently depolarized, thereby paralyzing the organism (Soderlund et al., 2002). Bioallethrin, tetramethrin, resmethrin, bioresmethrin are few first-generation pyrethroids, whose activities are greater than the natural pyrethrum but they are unstable in sunlight. Second generation pyrethroids are of more persistent compounds notably: permethrin, cypermethrin and deltamethrin. These compounds are significantly more resistant to degradation by light and air and that is the reason they are more suitable for use in agriculture, but often they have been significantly associated with higher mammalian toxicities too. Commercially available preparations of deltamethrin and cypermethrin were used at the dose rate of 25-50 ppm and 200-400 ppm respectively, against ticks and mites. Thereafter, over the years these pyrethroids derivatives were followed with other proprietary compounds such as fenvalerate, lambda-cyhalothrin and beta-cyfluthrin.

Formamidines

Formamidine insecticides are a comparatively new group of acaricides which are particularly useful for the control of Lepidoptera, Hemiptera, phytophagous mites and cattle ticks. The toxicity of these insecticides is very important as they are being used extensively for control of cotton insects and cattle ticks. Amitraz, chlordimeform, formetanate, formparanate, medimeform, and semiamitraz are the important members of this group. Amitraz is a broad spectrum acaricide used in veterinary medicine throughout the world. Many suggestions have been made for the underlying biochemical mechanism, including inhibition of monoamine oxidase activity, uncoupling of respiration and blockade of neuromuscular transmission (Beeman and Matsumura, 1978). Another possibility is interaction with octopamine receptors in the central nervous system (Hollingworth and Murdock, 1980) by mimicking the actions of octopamine at the locust neuromuscular junction. Commercially available preparations of Amitraz are used at the dose of 400-500ppm against ticks and mites.

Phenylpyrazoles

Fipronil is an insecticide of the phenylpyrazoles class (class of chemically-related broad-spectrum insecticides). Fipronil was discovered in 1987 and was developed initially for use in pest control in agriculture and public health. In dogs and cats, fipronil is available as a high-volume spray or a low-volume spot-on, with activity against ticks, fleas and ear mites. Mechanism of action of fipronil is complex, involving multiple interactions of both parent fipronil and its oxidation product, fipronil sulfone, on GABA-gated and glutamate-gated chloride channels in the insect nervous system. Both fipronil and fipronil sulfone inhibit GABA receptors as well as desensitizing and non-desensitizing glutamate gated chloride channels, though the activity of fipronil sulfone is much higher than fipronil for desensitizing Glutamate gated chloride channels. The net result of insect exposure to fipronil is blockade of inhibitory nerve transmission, resulting in hyperexcitability and death of susceptible parasites (Page, 2008). Mammals do not have this type of chloride channel, making them much less susceptible to its effects. Commercially available pour-on preparations of fipronil are used at the dose of 1 ml/10 kg of body weight applied on the dorsal line, between the withers and the base of the tail.

Macrocyclic Lactones

Avermectins and milbemycins (macrocyclic lactones) are products or chemical derivatives of soil microorganisms belonging to the genus *Streptomyces*. The avermectins are available commercially in the form of ivermectin, abamectin, doramectin, eprinomectin, and selamectin. Commercially available milbemycins are milbemycin oxime and moxidectin. At low dosage level, these macrocyclic lactones have broad antiparasitic spectrum. Ivermectin, eprinomectin, abamectin, doramectin, and moxidectin are variously available as oral, subcutaneously (s/c) and pour-on formulations for use in cattle. The former two formulations (oral and s/c) are given at 0.2 mg/kg, whereas the pour-on formulation is used at 0.5 mg/kg body weight. The transmission of electrical activity in invertebrate nerve and muscle cells are blocked, mostly by enhancing the effects of glutamate at the invertebrate-specific glutamate-gated chloride channel, with minor effects on gamma-aminobutyric acid receptors (Bloomquist, 1996). An influx of chloride ions into the cells is caused by this, leading to hyperpolarisation and then paralyses the invertebrate neuromuscular systems.

Modes of Application of Chemical Acaricides

Controlling the menace of tick control is of paramount importance in many countries of tropical and subtropical regions (Lodos et al., 2000). Various methods including dipping, spraying, ear tagging or pours on have been used to apply chemicals to protect livestock against ticks. The most common and popular method of acaricide application in livestock is direct application to animals (Drummond, 1983). Applications of acaricide to tick-infested cattle via dipping or sprayer can be equally effective under ideal conditions with proper handling of equipment without injuring animals and subsequent dilution of a product (George, 2000).

Dipping

Dipping is one of the methods of acaricide application in livestock and pets where, animals are dipped in a pesticide bath. If it is used in cattle, called as cattle dip. Dipping animals in vats with acaricide solution is very effective way of tick control. However, immobility of vats, high one-time cost of construction, and also the cost of the acaricides used make this system of acaricide application unfeasible for small farms. Often, ticks hidden in sheltered locations (e.g., between the toes or in the ears) are missed and survive to lay eggs and re-establish the pest population. Maintaining and managing dipping vats for optimum concentration of acaricide solution and proper dipping of cattle is challenging (Drummond, 1983).

Spray

The application of fluid acaricides to an animal by means of a spray using high-pressure sprayers has many advantages and has been successfully practiced for controlling ticks on most of the animals (Barnett, 1961). Spraying equipment is highly portable, and only small amounts of acaricides need to be mixed for a single application. However, spraying is generally less efficient in controlling ticks than immersion in a dipping vat because of problems associated with applying the acaricides thoroughly on all parts of the animal body.

Spot Treatment or Hand Dressing

There are certain places on animal body (often predilections sites for certain tick species) which are not effectively treated by spray or dips. These predilection sites *viz.* interior area of the tail, the inner side of the ear, tail switch and the groin areas in cattle with large udder, are especially liable to escape treatment. Acaricides may be applied to these sites by hand is termed as hand dressing (Barnett, 1961) or spot treatment. The application of insecticides with aerosols and in oils, smears, and dusts by hand to limited body areas is time-consuming and laborious, but in certain instances it may be more effective and economical than treating the entire animal.

Some other methods of applying acaricides are adhesive medical patches, neck bands, tail bands and pour-on, particularly for the pyrethroids with long residual activity. Proper spread of acaricide on animal's hair coat is ensured by mixing it with surfactants. Also, there are preparations in which acaricides are mixed with talc and dusted on animal's body coat or fur. The acaricides may also be incorporated in to certain kind of suitable matrices for slow release of the toxicant over a period of weeks or months to achieve long-lasting efficacy. In dogs and cats, plastic collars (flea and tick collars) are very commonly used. Similarly, acaricide-impregnated plastic ear tags are widely used for control of ear-infesting ticks on cattle and other large domestic animals. However, their effectiveness in controlling ticks in the predilection areas of these animals is low. Long-lasting and effective control by means of systemic acaricides can also be achieved. In this case, the toxicant is introduced into the host's blood to kill ticks as they feed on the treated animals. Toxicity of most of the acaricides however, is the roadblock for administering it systemically in animal body. An exception is ivermectin, which can provide excellent control of certain ticks on cattle for 2–3 months.

Phytoacaricides to Control Ticks

Ayurveda remained as backbone of medical science in ancient times, became an alternative in the modern era. The role of ethno veterinary medicine in livestock development is presently realized beyond discrepancy especially in the developing countries. This is also considered as potential tool to overcome the side-effects of chemical drugs. Previously ethnoveterinary medicine and other ethno-knowledge systems were looked upon with doubt and skepticism by the modern researchers and practitioners.

The ethno-veterinary and medical knowledge offers an array of herbs to be evaluated for their insecticidal and acaricidal properties. There is a long list of plants, herbs, shrubs showing anti-tick properties. India is one of the 12 mega biodiversity centers having 45,000 plant species however; little attention has been paid to explore the huge potentiality of the medicinal plants as acaricides. The ingredients of plants and herbs are known to possess insecticidal, growth inhibiting, anti-molting and repellent activities. The effect of a methanolic extract of *Styloxanthes scabra,* the leaves of tobacco, ethanolic extracts of *Annona squamosa* seed and *Azadirachta indica* leaves, bark and seed were evaluated for their anti-tick activities against different life stages of *H. anatolicum anatolicum* and *B. microplus,* and found to have highly efficacious against ticks (Ghosh et al., 2007). Ethanolic extract (95%) of *Ricinus communis* showed presence of quercetin, gallic acid, flavone and kaempferol which possessed synergistic acaricidal action against *Rhipicephalus (Boophilus) annulatus* ticks. (Ghosh et al., 2013). Shyma et al. (2014) studied the acaricidal activity of crude extracts of *Allium sativum* cloves and *Carica papaya* seeds and found that these extracts to have potential ovicidal properties which can be utilized in the tick control strategy. Alcoholic extracts of *Piper longum*, and *Piper nigrum* were evaluated for acaricidal activity against *Hyalomma anatolicum* (three-host ixodid tick) and found to have potential to be used as an eco-friendly acaricides (Singh et al., 2017). The ethanolic extracts of roots of *Vitex negundo* have found to have acaricidal activity against the deltamethrin resistant larvae of *H. anatolicum* (Singh et al., 2014). The bioactive sub-fractions of *Ageratum conyzoides* showed 96.2–97.5% efficacy against larvae of susceptible and 77.1–94.9% against multi-acaricide resistant larvae of *R. (B.) microplus* ticks infesting cattle and buffaloes (Kumar et al., 2015). Udayan et al. (2018) demonstrated the acaricidal activity of crude ethanolic extract of *Artemisia nilagirica* leaves and its fractions against *Rhipicephalus (Boophilus) annulatus*. It is a plant traditionally used for insect control in the southern part of India. The crude extract and hexane sub-fraction of *Artemisia nilagirica* possessed very good acaricidal activity for both adult and larval forms of *B. annulatus*. Phytol, eudesmol, 2,6-dihexadecnoate, and hexadecanoic acid ethyl ester were the major compounds identified in the hexane fraction of leaves. There are many more plants like *Acorus calamus* (Ghosh et al., 2011), *Jatropha curcas* (Julietet al., 2012), *Cassia alata* (Ravindran et al., 2011), *Leucas aspera* (Ravindran et al., 2011), *Semecarpus anacardium* and *Datura stramonium* (Ghosh et al., 2015) etc. which are reported to have acaricidal properties.

Biological Control of Ticks

An attempt to use naturally occurring species of living organisms as antagonists to reduce pest populations is referred to as biological control (Gronvold et al., 1996). Hence, augmenting populations of existing antagonists (predators, parasites, parasitoids, and pathogens) or importing exotic antagonists are included in this definition. Practically, biological control is intended to reduce the density of a tick population to an equilibrium level low enough to avoid appreciable economic or clinical effects (Gronvold et al., 1996). An efficacious agent of biological control would be estimated to be highly specific to the target organism, with no injurious effects on antagonist or benign species. Ants (Chagas et al., 2002), predatory mites (Holm and Wallace, 1989), chickens (Kohn and Norval, 1994; Dreyer et al., 1997), parasitoid wasps (Hu et al., 1998), *Bacillus thuringiensis* (Zhioua et al., 1999), entomopathogenic nematodes (Samish and Glazer, 2001), entomopathogenic fungi (Kirkland et al., 2004) and oxpeckers, *Buphagus africanus* (Mooring and Mundy, 1996) are some of the biological agents used against ticks of livestock. There are many natural predators of ticks, primarily attacking them in the free-living phase like birds, arthropods, parasitoids, infertile hybrids etc. Copping and Menn (2000) reported microbial agents and their secondary metabolites (including antibiotics), entomopathogenic nematodes, plant-derived pesticides (botanicals), arthropod pheromones, genetic modification for control of ticks.

Ticks have various natural enemies, but only a few species of entomopathogenic fungi have been evaluated as tick control agents. Some experimental results suggest that several bacteria are pathogenic to ticks, but their mode of action and their potential value as tick control agents remain unearthed. Fungi are capable of infecting a wide range of living organism including insects and ticks. Spore suspension of mold *Aspergillus terreus* at various concentrations decreases the egg conversion factor by arresting oviposition of *H. anatolicum* and may play a substantial role in tick control programs (Suliman and Mahammed, 2012). Spraying of aqueous and oil-based entomogenous fungal formulation (109 conidia/ml) depicted an 80–90% and a 40–50% mortality in adult ticks (*R. appendiculatus* and *Amblyomma variegatum*), respectively (Kaaya and Hassan, 2000). Naturally occurring soil fungus *Metarhizium anisopliae* has confirmed the effectiveness with 100% tick mortality of the most common variety of tick within two days in laboratory condition (Leemon and Jonsson, 2008). Fungal biopesticides have emerged as rational non-chemical control options for a range of pests in agriculture. Two formulations using a mixture of spores of *M. anisopliae* isolates ARI-M52 and ARI-M63 in two different dose levels provided reasonable tick control on animals which were kept below 34°C for significant time periods each day (Leemon et al., 2008). The advantages of fungal biopesticides over other biological control agents is the ease with which they can be cultured on simple artificial media like rice or other cereals with minimal specialized equipment and it can be applied using conventional acaricide fluid application systems such as hand sprays or spray races, and the fungal conidia can be even mixed with water or oil or water and oil mixtures. Till date, knowledge on pathogenesis of fungal agents against ticks of livestock, and the selection of strains for bioassays has been largely on the basis of trial and error. Substantial research is required to select appropriate strains, to develop them as biopesticides, to establish their effectiveness, and to formulate production strategies to bring them in practical use.

Some bacteria show pathogenicity to ticks. *Proteus mirabilis* is pathogenic to *Dermacentor andersoni* (Brown et al., 1970), it also attacks *Amblyomma hebraeum*, *Hyalomma marginatum*, *Rhipicephalus eversti eversti* (Hendry and Rechav, 1981) and causes Blackening disease of *Boophilus decoloratus*. Bacterium *Cedecea lapagei* (Enterobacteriaceae) is found to be pathogenic to *Rhipicephalus microplus*; this bacterium infects ticks via the genital opening and under laboratory conditions could produce cent percent mortality (Brum et al., 1991; Brum and Teixeira, 1992). Hassanain et al. (1997) found three commercial varieties of *Bacillus thuringiensis* (*B. t. kurstaki, B. t. israelensis* and *B. t. thuringiensis*) were effective when sprayed on unfed or engorged adults of *Argas persicus* or *Hyalomma dromedarii* in a Petri dish.

Entomopathogenic nematodes (EPNs) that are pathogenic to ticks can potentially control ticks, but enhanced formulations and selection of novel nematode strains are required (Samish et al., 2004). EPN penetrate engorged female *B. annulatus* ticks *via* the anus or genital pore (Samish et al., 2004). Glazer and Samish (2001) reported that Heterorhabditid nematodes could able to kill engorged *B. annulatus* females in Petri dishes within 2.5 h of exposure, whereas steinernematid nematodes required more time (4 h) to penetrate into ticks. Tick mortality caused by EPNs was discussed to be due to the rapid proliferation of the nematode symbiotic bacteria within the ticks, the nematodes do not go through their natural cycle within ticks, and the most infective juveniles die shortly after entry (Mauleon et al., 1993; Samish et al., 1995; Hill, 1998; Kocan et al. 1998; Hassanain et al. 1999).

Parasitoids used in the biological control of insect pests belong to the order Hymenoptera. The only species that has been released for biological control of ticks is *Ixodiphagus hookeri* (synonym, *Hunterellus hookeri)* which has been recorded from Asia, Africa, North America and Europe (Hu et al., 1993). *I. hookeri* uses tick odour cues to find hosts (Takasu et al. 2003). Nymphal stages of ticks were parasitized while they were engorging on vertebrates (Smith and Cole, 1943), and parasitoid egg development was found to be associated with ingestion of blood by its host tick (Hu and Hyland, 1998). Parasitoids are very rarely used for the control of ticks.

Natural predators such as ants, beetles and many bird species occasionally feed on ticks. Population of general predators of ticks can be manipulated in such a way that it will help in reducing the size of a tick population in nature. But this could also cause disturbances in natural ecosystem (Symondson et al., 2002). Chickens are the natural predators of ticks and can be used as part of an integrated tick control plan in urban cattle-management systems (Hassan et al. 1991; Hassan et al., 1992; Dreyer et al., 1997).

Tick Vaccines

Allen and Humphreys (1979) demonstrated the feasibility of controlling tick infestations through immunization of hosts with selected tick antigens. Vaccination against ticks has several advantages like cost-effectiveness, reducing environmental contamination and preventing the selection of drug-resistant ticks that result from repeated acaricide application. Moreover, development of vaccines against ticks using multiple antigens could also target a broad range of tick species thereby leading to transmission blocking of pathogens (de la Fuente and Kocan, 2006; de la Fuente et al., 2006; Willadsen, 2006; Nuttall et al., 2006).

Recombinant vaccine containing B. microplus Bm86 gut antigen were developed in the early 1990s which was a major breakthrough in tick vaccine research that induced immunological protection of vertebrate hosts against tick infestations. (Willadsen et al., 1989; Rand et al., 1989; Rodri´guez et al., 1994; Willadsen, 2004, 2006; de la Fuente and Kocan, 2003, 2006). Later, two vaccines using recombinant Bm86 were subsequently registered in Latin American countries (Gavac) and Australia (TickGARD) during 1993– 1997 (Cobon et al., 1995; Willadsen et al., 1995; Massard et al., 1995; Vanegas et al., 1995; Canales et al., 1997). These vaccines reduced the number of engorging female ticks, their weight and reproductive capacity. Thus, the greatest effect of vaccine was reduction of larval infestations in subsequent generations. Controlled field trials using Bm86-based vaccines conducted in Cuba, Mexico, Brazil, Colombia and Australia demonstrated successful control of *B. microplus* and *Boophilus annulatus* infestations (Cobon et al., 1995; Rodri´guez et al., 1995; Willadsen et al., 1995; Canales et al., 1997; de la Fuente et al., 1998, 1999; Redondo et al., 1999; Jonsson et al., 2000). A relationship between antibody responses of cattle immunized with the Bm86 vaccine and tick fertility was demonstrated in Australia (Cobon et al., 1995) and Cuba (de la Fuente et al., 1998). Subsequently, it was found that TickGARD (Hoechst Animal Health, Australia) and Gavac (Heber Biotec S. A., Havana, Cuba) confers only partial protection against phylogenetically related *Hyalomma* and *Rhipicephalus* tick genera. However, immunisation with BM86 failed to protect against the more phylogenetically distant *Amblyomma spp.* (De Vos et al., 2001). Therefore, identifying broad range vaccine candidates against tick infestations across phylogenetically distant species (Ghosh et al., 2007) is need of the hour. Jonsson et al. (2000) reported a 60% reduction in the number of acaricide treatments, besides a 56% reduction in tick numbers in a single generation, a 72% reduction in reproductive efficiency of the ticks as measured in the laboratory, and 18.6 kg higher live weight gain than unvaccinated cattle over a six months period. Similar results were described by de la Fuente et al. (2007), noting that commercial tick vaccines reduced tick infestations on cattle, as well as increasing animal production.

Recently, Subolesin was identified as a candidate tick protective antigen (Almazan et al., 2003; de la Fuente et al., 2005; de la Fuente et al., 2006). Subolesin is the ortholog (both structural and functional) of insect and vertebrate Akirin and regulates gene expression by regulating the transcription factors, thus affecting tick innate immune response, feeding, reproduction and development (Naranjo et al., 2013). Immunization trials using recombinant tick Subolesin demonstrated protective efficacy against tick infestations and reduced vectorial capacity of ticks (Canales et al., 2009; Prudencio et al., 2010) and fertility of several arthropod vector species (Harrington et al., 2009). Shakya et al. (2014) reported the efficacy of recombinant tick Subolesin vaccine as 40.65% in experimentally infected calves in India. *R. appendiculatus* 64P antigen was found to have immune cross-reactivity with *R. sanguineus*, *I. ricinus*, *A. variegatum* and *R. microplus*, which may account for the control of *R. sanguineus* and *I. ricinus* infestations in laboratory animals (Trimnell et al., 2005).

Another approach in anti-tick vaccines strategy is to target the ticks in such a way that they give protection to the host against the pathogens transmitted by them i.e., transmission-blocking effect. Anti-tick vaccines caused reduction of parasite load, which inherently lowers the possibility of transmission of haemoprotozoan parasites transmitted by them. Recombinant tick vaccines therefore could be targeted for control of both ticks and tick-borne pathogens (de la Fuente et al., 2006).

Breeding Strategies

Incorporating current breeding programmes with selection for tick resistance idea seems to be encouraging but there is either low or no correlation between tick counts and various productive and reproductive traits. Numerous efforts have been made in an attempt to isolate and validate markers for selection of resistant genotypes. Marker assisted selection could be used to pre-select young animals, shorten generation interval and increase genetic gain (Beckmann and Soller 1987; Martinez et al. 2006). These molecular markers have the capability to allow prediction of breeding value for traits that had previously been difficult to measure and hence were not counted in the selection criterion (Beckmann and Soller, 1987).

Advances in molecular genetics now offer prospects for identifying markers for parasite resistance. It is known that in many subtropical and semi-arid environments, indigenous dual-purpose breeds are highly resistant to ticks, resulting in low infestation rates that cause insignificant direct losses. But more experimental evidences are yet to be unearthed to substantiate this observation. Nevertheless, local indigenous cattle kept completely tick-free become equally susceptible. The phenomena of host resistance to ticks and enzootic stability to tick-borne diseases are well documented (Latif and Pegram, 1992). High host resistance is beneficial in any tick control program. However, its improvement has been almost entirely neglected in Asia (Frisch, 1999) including India. In India, cattle breeds of Ongole/Nellore, Sahiwal, Ponwar hill cattle, and buffaloes are recorded in the literature as resistant to tick infestation (Gaur et al., 2002). Many a times, the tick resistance status of these animals was attributed to the physical characters and behavior of these animals and no genetic basis of host resistance to ticks and tick-borne diseases have been explored.

Acosta-Rodríguez et al.(2005) reported a close association between, bovine major histocompatibility complex (MHC) (BoLA) class II alleles and tick resistance. Martinez et al. (2006) found an association between BoLA alleles and low tick number for alleles DRB 3.2 *18, *20 and *27 and suggested that BoLA–DRB 3.2 alleles could be used to help in the selection of animals, resistant to tick infestation (Shyma et al., 2015). More recently, in a linkage analysis, quantitative trait loci (QTL) were detected when a crossbred population (F2 Holstein×Gir) was scanned using 180 microsatellites covering all autosomal chromosomes (Gasparin et al.2007; Regitano et al.2008; Machado et al.2010). Different QTL were found to have significant effects on tick load in different seasons, possibly representing the environmental influence on tick counts.

Integrated Management

Integrated Pest Management (IPM) is a methodology to minimize economic and environmental harm caused by pests andminimizing the pathogen transmission by vectors. Strategic integrated acaricide application, herbal acaricides and vaccination in relation to seasonal tick populations was considered as a more cost-effective and environmentally friendly means to control tick infestations (Lodos et al., 1999, 2000). Integrated approach in combination with acaricide treatments in controlled field trials demonstrated an effective control of tick infestations with reduced use of acaricides (de la Fuente et al., 1998; de la Fuente and Kocan, 2003). Development of improved vaccines in future will augment this approach of tick control. Though the applicability of pasture management in control of ticks is meager in India, this method is a feasible choice that can be adapted in organized livestock farms where there is availability of pasture land.

Acaricide Resistance: Its Diagnosis and Status

Resistance to a given acaricide against tick can be described as a reduction in susceptibility of ticks to the acaricide when it is used at the recommended concentration and according to all of the recommendations for its use (FAO, 2004). The resistance of ticks to acaricides is observed to have an inherited component. This result in survival and reproduction of ticks even after their exposure to chemical acaricides and are not affected by the acaricide. The genes that attribute to resistance are already present at very low levels in the tick population before the introduction of a new acaricide. The rate at which a resistant allele becomes established in the population and the time it takes for the control of ticks to break down is dependent upon many factors such as the frequency of the original mutation in the population before treatment, the mode of inheritance of the resistant allele, the frequency of acaricide treatment, the concentration gradient of the acaricide and the fraction of the total tick population that is not exposed to the acaricide (refugia). There are several theories for the onset of resistance in ticks against acaricides which are discussed here.

Mechanism of Acaricide Resistance

The mechanism of acaricide resistance against ticks may be due totarget site insensitivity, increased metabolic detoxification or due to reduced penetration into the ticks. Various genetic and biochemical mechanisms for the development of resistance in ticks have been identified.

Target Site Insensitivity

Acetylcholinesterases (AChE) are the enzymes of interest to understand themolecular basis of organophosphates resistance. The insensitivity of AChE to the OP compounds in *R. (B.) microplus* was first reported by Lee and Bantham (1966) and it was considered as the principal resistance mechanism in this species. Modifications in the acetylcholinesterase structure were believed to confer resistance against OP intoxication (Nolan, 1985). According to Baxter et al. (1999), OP resistance was semi-dominant in *R. (B.) microplus* and heterozygote ticks demonstrated increased resistance to OPs. Detection of heterozygotes was considered to be necessary to check spread of resistance. Pruett (2002), while evaluating the kinetic analysis of OP-resistance in R. (B.) microplus strains, observed that the rate of inhibition of AChE by OP compounds was affected by a slower rate of enzyme phosphorylation. Villarino et al. (2002) detected resistance to coumaphos in *R. (B.) microplus* by larval packet test and reported that among two isolates showing the presence of non-inhibited acetylcholiesterase, one was resistant to OP compound. Baxter and Barker (1998) characterized AchE1 gene from Australian isolates of *R. (B.) microplus* and reported identical pattern of sequences in OP-susceptible and resistant strains (Biarra and other field strains). They opined that resistance to OPs in Australian *R. (B.) microplus* was not conferred by a mutation in alleles of the AChE gene. No nucleotide differences were recorded in the AChE2 gene from the Mexican OP resistant strain of *R. (B.) microplus* when compared to a susceptible strain from Texas (Hernandez et al., 1999). Temeyer et al. (2004) reported characterization of AChE3 gene of *R. (B.) microplus* which was comprised of 2,315 nucleotides organized as a 224-nucleotide 5'- untranslated region, an open reading frame (ORF) of 1,860 nucleotides, a 207-nucleotide 3'-untranslated region, and finally a 21-nucleotide poly (A)-tail. Considerable nucleotide differences between susceptible and resistant strains were observed in AchE3 and it was concluded that many multiple potential glycosylation, phosphorylation, and other post translational modification into altered AchE were responsible for OP resistance. Three cDNAs that encoded putative AChEs from *R. (B.) microplus* (BmAChE1, BmAChE2, and BmAChE3) have been identified and biochemical identity of only BmAChE3 cDNAs was verified with respect to expression and enzymatic properties (Temeyer et al., 2006). The inability to identify mutations associated with resistance within BmAChE1 and BmAChE2 prompted the suggestion that resistance might result from post translational modification of the acetylcholinesterase (Baxter and Barker, 1998; 2002).

In India, only a few reports are available regarding mutations in any of the AChE genes. Ghosh et al. (2015) reported four amino acid substitutions in AChE2 gene of resistant isolates of *R. microplus* collected from the state of Bihar. They detected five new SNPs (G138A, G889A, T1090A, C1234 T, and G1403A) in the AChE2 gene of organophosphate resistant reference *R. microplus* strain and in OP resistant field populations. One of the nucleotide substitutions (G138A) was asynonymous mutation, while other four substitutions were non-synonymous, replacing a valine for isoleucine in position 297 (V297I), a serine for a threonine in position 364 (S364 T), histidine for a tyrosine in 412(H412Y) and a lysine for arginine in position 468 (R468 K). They suggested that these mutations could probably be associated with organophosphates resistance but the actual contribution of these mutations in the development of insensitive AChE2 could not be ascertained. This means that some additional factors exist which contribute to insensitive acetylcholinesterase other than point mutations. Nagar et al. (2016) reported mutations in the AChE2 gene in *R. microplus* and identified the same four amino acid substitutions (viz. V297I, S364 T, H412Y and R468 K) in the resistantfield population and in reference resistant strain of *R. microplus*, as detected earlier by Ghosh et al. (2015). Meanwhile, Fular et al. (2018) identified 22 SNPs in AChE2 gene in reference multi-acaricide resistant strain (IVRI-V) of *R. microplus* and amino acid substitution was detected in 11SNPs. Jyoti et al. (2016) reported six point mutations and their corresponding amino acid changes in AChE3 gene (I48 L, I54 V, R86Q, V71A, I77 M and S79 P) of resistant *R. microplus*.

Ten mutations (SNPs) in the sodium channel gene were identified in *R. microplus* responsible for resistance against synthetic pyrethroid compound (Miller et al., 1999; He et al., 1999; Jonsson et al., 2010; Morgan et al., 2009; Stone et al., 2014; Klafke et al., 2019). Out of these ten SNPs, five SNPs (T2134A, C190A, G215 T, T170C and C2136A) were non-synonymous and their association with resistance has been proved (He et al., 1999; Morgan et al., 2009; Stone et al., 2014; Klafke et al., 2019). Although, No T2134A mutation was detected in Indian isolates (Kumar et al., 2013, 2016; Ghosh et al., 2015; Nagar et al., 2018). Similar finding was reported in pyrethroid resistant tick isolates from Australia and Brazil where there was absence of T2134A mutation (Li et al., 2007; Chen et al., 2009; Andreotti et al., 2011). Cytosine to adenine (CTC to ATC) mutation detected at position 190 (C190A) in the domain II S4-5 linker of the sodium channel gene in pyrethroids resistant isolates of *R. microplus* from several countries (Morgan et al., 2009; Andreotti et al., 2011; Lovis et al., 2012; van Wyk et al., 2016; Stone et al., 2014; Yessinoua et al., 2018). Kumar et al. (2013) initially reported the presence of C190A mutation in the sodium channel gene of the pyrethroid resistant *R. microplus* collected from four different agroclimatic zones of India. Hernandez et al. (2000) detected higher transcript of carboxylesterase in pyrethroid resistant *R. microplus* population indicative of its role in resistance development. The same tick population had a unique nucleotide substitution of adenosine (A) for guanidine (G) at position 1120 (G1120A) in CaEs encoding gene leading to a change in aminoacid at position 374 from aspartate to asparagine (D374 N) and it was postulated that the resultant mutation (D374 N) has led to pyrethroid resistance by enhancing the affinity of carboxylesterase or by augmenting its ability to hydrolyze pyrethroids (Hernandez et al., 2002).

Amitraz is a potential agonist of the octopaminergic system located in the tick synganglion. Multiple mutations in OCT/Tyr and βAOR sequences gene of *R. microplus* have been identified by several researchers (Chen et al., 2007; Baron et al., 2015; Jonssonn et al., 2018; Sungirai et al., 2018). At present, polymorphisms in octopamine-tyramine receptor and βAOR could detect the amitraz resistance to certain extent; though, the diversity of mutations suggests that no single polymorphism can be relied on (Rodriguez-Vivas et al., 2018). No data available regarding molecular mechanism of amitraz resistance in India.

Metabolic Detoxification of Enzymes

Ticks enable themselves to synthesize more quantity of detoxifying enzyme system to break down or sequester broad spectrum of acaricides due to its wide substrate specificity. Resistance mediated by detoxifying enzymes like esterases, Cytochrome P450 monoxygenases and glutathione S-transferases (GST) are apparently involved in conferring resistance against ticks (Crampton et al., 1999; Jamroz et al., 2000; Enayati et al., 2010; Temeyer et al., 2013). Miller et al. (1999) and Klafke et al. (2017) reported the enzyme, esterases play a pivotal role in development of resistance to synthetic pyrethroids and amitraz in *R. microplus*. Toxicological and biochemical studies have led to the isolation and characterization of a pyrethroid detoxifying esterase designated as CzEst9, which is responsible for moderate resistance in some of the tick strain originating from Mexico (Miller et al., 1999; Pruett, 2002). Baffi et al. (2005) compared the esterase profiles of susceptible and cypermethrin resistant strain of Brazilian *R. (B.) microplus* and also a pyrethroid susceptible reference strain (Mozzo strain) and the results demonstrated no difference in the esterase pattern between the samples from Mozzo strain and those from the cypermethrin susceptible field strain. The bands were numbered as EST-1 to EST-4. The EST-1A and EST-1B enzymes were detected only in the resistant strain. Baffi et al. (2007) developed PCR RFLP technique to detect the presence of mutations in a fragment of a putative carboxylesterase (CaEs) gene in a population of *R.* (*B.*) *microplus* with a history of resistance. A fragment of approximately 372bp was obtained in the PCR of all the strains of the tick population investigated. The restriction profile after digesting the product with EcoR1 revealed three genotypes, W, H and M observed in different frequencies. In order to establish a reliable methodology to measure monooxygenase activity levels in pyrethroids resistant Mexican ticks, Cossio- Bayugar et al. (2018) used a zymogram of monooxygenase enzymes in the presence of synthetic monooxygenase enzyme substrate. Ticks resistant to Pyrethroids showed statistically significant elevation in monooxygenase activity, when compared to acaricide susceptible reference ticks. Cossio- Bayugar et al. (2009) measured the expression level of the enzymes acetylcholinesterase and carboxylesterase by real time PCR quantification assay in acaricides resistant ticks. A quantitative assay revealed 4.35 and 2.92 times more expression of esterases in field resistant population of *R. microplus* in comparison to reference susceptible population (Kumar et al. 2013). Similar findings have been reported by other researchers also (Singh and Rath, 2014; Ghosh et al., 2015; Nandi et al., 2015; Kumar et al., 2016; Fular et al., 2018 and Chigure et al., 2018).Beside esterases and monooxygenases, over-expression of Glutathione-S-transferase in resistant population of ticks has also been reported by many researchers. A significantly higher GST activity (2–5 fold) was reported in resistant *R. microplus* collected from Punjab (Nandi et al., 2015), Bihar (Ghosh et al., 2015 and Uttar Pradesh (Chigure et al., 2018) and from states of India.

Reduced penetration into ticks is yet another mechanism of resistance in ticks. Penetration resistance in ticks is possible when there is alteration in the ability of an acaricide to infiltrate into ticks that are treated with chemical acaricide. This mechanism has been identified in some of the arthropods (Noppun et al., 1989) however, only a few reports are available on occurrence of this mechanism in *R. (B.) microplus* (Schnitzerling et al., 1983).

Diagnosis of Resistance

According to FAO (2004) the appropriate laboratory test for detection of acaricide resistance should be sensitive enough to identify resistance early in its emergence, simple and inexpensive, provide a speedy and reliable result, and suitable for standardization among laboratories in many countries. An ideal test should cover the full range of chemical groups that are in use, including the most recently developed active ingredients. The most widely used i*n vitro* tests are bioassays applied to larvae and engorged female ticks.

Bioassays

Bioassay, being easy and cheap, is a main tool for categorization of resistant tick populations. This test gives indication of phenotype data on the level of resistance in field ticks without the use of expensive reagents and equipment (Kumar, 2019).Currently, four bioassays are available to test for acaricide resistance each with its own advantages and disadvantages.

Adult Immersion Test (AIT) is the most widely used *in vitro* test for the diagnosis of resistance. Although this test provides results within 7days, requires a large number of engorged females which may be a drawback sometimes (Drummond et al. 1973). The Larval Packet Test (LPT) (Stone and Haydock 1962) and LIT (Sabatini et al. 2001; Shaw 1966) involves the use of larvae and require 5–6weeks to complete. LIT is not so widely used for the diagnosis of resistance and has not been promoted by FAO. Bioassay allow measuring resistance to compounds whose resistance mechanisms are unknown, but it does not give the clear picture of the mechanism involved in acaricide resistance. In these tests, tick larvae are exposed to chemically impregnated filter papers and their subsequent mortality is quantified after 24 hours. Adequate training is essential in order to achieve a high degree of confidence in these techniques. More recently, protocols have been added for macrocyclic lactones (MLs). There are some problems with the application of the standard LPT and AIT protocols for susceptibility to amitraz, requiring specific modifications for this acaricide. Nevertheless, when a bioassay is undertaken along with more sensitive biochemical and molecular assay, more information and solid evidence come out about the acaricide resistance status and its possible mechanism (Enayati and Haghi, 2007).

Biochemical Assays

The most common mechanism of the development of resistance is the metabolic detoxification of chemicals by enzymes which may confer resistance to multiple classes of acaricides. Ticks use their detoxifying enzyme system to break down or sequester foreign compounds including acaricides. As discussed earlier, an elevated activity of esterases, monooxygenases and GST are linked with resistance. Biochemical estimation of enzymes is often considered as one of the indicators in addition to bioassays for resistance monitoring. Different enzyme assays are available for quantitative estimation of detoxifying enzymes such as acetylcholinesterase (AChE) assay, esterase assay, monoxygenase assay, Glutathione S-Transferase assay. In acetylcholinesterase (AChE) assay, inhibition of AChE activity in the presence of propoxur is measured (Hemingway et al. 1998; Baxter et al. 1999). The esterase assay measures the esterase activity directly using a substrate (napthyl acetate) and product formed is measured spectroscopically at 570 nm (Hemingway et al. 1998; Jamroz et al., 2000). Monoxygenase assay measures the haem content of tick with presumption that majority of haem in arthropod is associated with cyt P450 (Brogdon et al., 1997). Glutathione S-Transferase Assay: The glutathione - S-transferase assay utilizes 1-Chloro-2,4-dinitrobenzene (CDNB) and upon conjugation of the thiol group of glutathione to the CDNB substrate, there is a surge in the absorbance at 340 nm (Habig et al., 1974).

Molecular Assays

Molecular tools are available presently to detect resistance SNPs and reveal the underlying molecular mechanisms of acaricide resistance in *R. microplus* which supports the tick eradication programme. The advantages of assay are that the resistance is diagnosed in early stage and quickly. Several molecular techniques such as target gene amplification, allele-specific PCR, sequencing techniques with DNA, transcriptomic analysis with RNA, probe-based quantitative PCR, pyroseqencing have been established for detecting acaricide resistance in ticks which has greatly improved the resistance management (Guerrero et al. 2001; Hernandez et al. 2002; Chen et al. 2007; Rodriguez-Vivas et al. 2011; Lovis et al. 2012; Sungirai et al. 2018). These molecular based assays are sensitive and rapid than that of bioassays for determining the status of acaricide resistance within a population to enable rational selection of appropriate treatments for parasite control (Jonsson et al., 2010; Kumar, 2019).

Status of Acaricide Resistance in Western India

Tick control strategies are principally based on extensive chemical treatments which are flooded with the problems like acaricide resistance and environmental issues. Shyma et al. (2015) evaluated the acaricide resistance status of *R. microplus* collected from organized and unorganized farms of North Gujarat state to deltamethrin, flumethrin, and fipronil. Through AIT, a low-grade resistance (level I, RF > 5) was detected against both deltamethrin and fipronil. Sharma et al. (2018) studied the comparative resistance status of *H. anatolicum* and *R. microplus* ticks of North Gujarat against synthetic pyrethroids (deltamethrin and cypermethrin) by Larval Packet Test revealed that low level of resistance in *H. anatolicum* tick isolates from Vasda, Shihori, and Palanpur against deltamethrin (level I; RF = 2.39–3.79) and that for cypermethrin Vasda, Vav, Soneth, Tick isolates from Palanpur were reported to be resistant (level I; RF = 1.5–4.6). The *R. (B.) microplus* isolates from Vav revealed highest resistance level IV (RF = 43.8) and level I resistance (RF = 1.77–3.52) from Mahadevpura, Vasda, Rooppura, Bhadath against deltamethrin. Resistance level for cypermethrin in Ranpur isolate was I (RF = 1.62); however, in Mahadevpura and Vasda tick isolate it was up to level

II (RF = 17.26 and 8.16, respectively). *R. (B.) microplus* tick isolates from Mahadevpura as well as Vasda were resistant against both deltamethrin and cypermethrin. Sharma et al. (2012) reported significantly decreased efficacy of deltamethrin against tick population of Rajasthan. Acaricide resistance status of *R. microplus* and *Hyalomma anatolicum* ticks against deltamethrin and diazinon was ascertained in Hisar (Haryana) and its adjoining district Churu (Rajasthan) using adult immersion test (AIT) and larval packet test (LPT) by Gaur et al. (2016). They further reported that *R. (B.) microplus* ticks from Tohana (Hisar) were found resistant to both the acaricides while the ticks of Agroha (Hisar) were found to be susceptible using AIT. The *H. anatolicum* tick isolates from Churu were found to be resistant using AIT for both the acaricides. Kumar et al. (2014) reported AIT based Amitraz resistance ranging from level I to III with a maximum RF of 27.3 in Sikar isolates from Rajasthan. Later amitraz resistance was also detected from Gujarat state (Singh et al., 2015). Sagar et al. (2020) reported resistance in *R. (B.) microplus* tick population from different areas of Madhya Pradesh against synthetic pyrethroid, deltamethrin and cypermethrin, organophosphate compound, coumaphos and ivermectin at different levels of resistance. Bharkhad (2019) demonstrated resistance in *R. microplus* ticks collected from different areas of Maharashtra like Marathwada, Konkan, Vidarbha, North Maharashtra and Western Maharashtra and detected high levels of resistance against deltamethrin and amitraz in these areas.

Status of Acaricide Resistance in Other Parts of India

Interest and importance of acaricide resistance have been boosted with the article published by Chaudhari and Naithani (1964) reporting resistant *R. microplus* population-infesting cattle against benzene hexachloride. Later, resistance to carbaryl, a broad-spectrum carbamate pesticide, was reported by Basu and Haldar (1997). In subsequent years there were numerous reports of resistance from different part of the country, against different acaricide compounds (Khan and Srivastava, 1990; Jani et al., 1991; Ghosh and Nanda, 1997; Sinha et al., 2010, Vatsya and Yadav, 2011; Sharma et al., 2012; Kumar et al., 2013; Khajuria et al., 2014; Singh and Rath, 2014; Ahanger et al., 2015; Gaur et al., 2016; Lenka et al., 2016; Gaur et al., 2017; Dutta et al., 2017; Godara et al., 2019).

Ghosh et al. (2015) studied the acaricide resistance status in *R. microplus* ticks of Bihar, an eastern Indian state. It was observed that most of the tick isolates were resistant to deltamethrin, diazinon and cypermethrin. Pradeep et al. (2012) detected resistance to deltamethrin and cypermethrin in *R. microplus* population from Bangalore, southern India using tea bag method, filter paper impregnation and immersion method. A low level of deltamethrin resistance was described in the tick population of *R. annulatus* collected from different districts of Kerala using LPT (Jyothimol et al., 2014; Ravindran et al., 2014). The first report of resistance in *R. microplus* population against malathion was documented from the Punjab state (Rath et al., 2006) and later many researchers reported the similar findings (Kumar et al., 2011, Jyoti et al., 2015). Resistance to diazinon in *R. microplus* ticks collected from Bihar, Haryana and Rajasthan has been documented using both larval and adult stages (Ghosh et al., 2015; Gaur et al., 2016). The development of resistance against synthetic pyyrethroid and organophosphorus acaricides has compelled the livestock owners to rely on formamidine class of acaricides, Sooner, due to indiscriminate use of this chemical, reports of resistance of Amitraz also started surging in different parts of India (Surbhi et al., 2018; Jyoti et al., 2019; Dutta et al., 2017). Jyoti et al. (2016) reported resistance in *R. microplus* of various districts of Punjab state, India against fenvalerate. The resistance factors were in range of 1.56–54.34 Two field isolates (Jalandhar and Ludhiana) showed level I resistance; three (Patiala, Fatehgarh Sahib and Amritsar) showed level II and Kapurthala isolate showed level IV resistance against fenvalerate. Macrocyclic lactones (MLs) were introduced into the market to compact the resistance produced by SP, OP, formamidine compounds. Though the drug was giving an impressive result in tick control, the selection pressure due to overuse of this drugs in past few years has elicited the development of resistance to ivermectin. In India, first report of resistance in *R. microplus* against ivermectin was documented from Punjab state (Singh et al., 2015). Further, Nandi et al. (2018) reported different level of resistance in *R. microplus* collected from different states of India.

Management of Acaricide Resistance

The tick control measures should have the prospective to decrease the tick populations over successive generations and eventually reducing the need for regular application of chemicals (George, 1990; Honer and Gomes, 1990). Periodic and suitable tests of monitoring of resistance are need of the hour to analyze comprehensively, the status of resistance against acaricide throughout the country. These activities may significantly improve the health status of animals, reduces the environment pollution and able to conserve these chemicals for future.

Single application of the acaricides will not always eliminate all the ticks on an animal. The surviving ticks eventually contribute to the development of resistance. The control measures depend on the tick load on and off the host as well the favorable climatic condition. Monitoring the resistance status help to identify the acaricides to which the parasites are resistant. Rotation and use of combination of acaricides may be helpful in preserving the existing compounds. Unnecessary acaricide treatments should be discouraged to reduce the development of resistance (Ghosh et al., 2007).

Natural plant-based product for tick management has revealed a bunch of scope. The indigenous ethno-veterinary and medicinal knowledge of Indian farmers has established a range of herbs for their insecticidal and acaricidal properties. Many plants and herbs with acaricidal, growth-inhibiting, antimoulting and repellent activities have been scientifically recognized. The most feasible option for management of acaricide failure is to encourage integrated tick management. The integrated approaches include pasture management combined with restricted applications of effective chemical acaricides, strategic use of phytoacaricides, housing management, nutritional enrichment of the animal, strengthening of dairy cattle management with alternative feeding technologies such as silage and hay to reduce contact between ticks and cattle, immunization against ticks and ticks borne diseases (de la Fuente et al., 2000; Stutzer et al., 2013). Integrated tick management will significantly reduce the greater dependence on acaricides and lessen selection pressure on ticks thereby preserving the efficacy of chemicals and delaying the onset of acaricide resistance (Jongejan and Uilenberg, 2004).

Conclusion

Tick-borne diseases are growing rapidly due to the establishment of the tick vector posing serious threat to the animal and human population. The occurrence of KFD (Kyasanur forest disease) in Karnataka and Crimean-Congo haemorrhagic fever (CCHF) in Gujarat, India emphasizes the importance of ticks and its control strategies. The indiscriminate and persistent use of acaricides with inadequate concentrations has possibly contributed to the development of resistance. The tick population has immense potential for rapidly developing resistance due to their biological and behavioral characteristics. The availability and afordability of DNA-based molecular assays for diagnosis of resistance could assist the management of resistance in a very effective way. Researches are now focused on exploring the phytochemistry of acaricidal plants, which would help identify other species with similar active ingredients. The present state of tick control with alternative methods, does not allow complete dismissal of chemical acaricide use. New inventions are awaited in these fields, with a hope that someday the use of chemical acaricides would plunge below a certain point, thus eliminating the copious undesirable effects.

References

(FAO) Food and Agriculture Organization, 2004. *Resistance management and integrated parasite control in ruminants–guidelines, module 1–ticks: acaricide resistance: diagnosis, management and prevention.*

Acosta-Rodriguez, R., Alonso-Morales, R., Balladares, S., Flores-Aguilar, H., Garcia-Vazquez, Z. and Gorodezky, C. (2005). Analysis of BoLA class II microsatellites in cattle infested with *Boophilus microplus* ticks: class II is probably associated with susceptibility. *Veterinary Parasitology*, 127(3-4): 313-321.

Advani, R. and Vazirani, T. G. (1981). Studies on ectoparasites of bats of Rajasthan and Gujarat (India). Zoological Survey of India. *Controller of Publications, ZSI, Calcutta*, 22: 155.

Ahanger, R. R., Godara, R., Katoch, R., Yadav, A., Bhutyal, A. D. S., Katoch, M., Singh, N. K. and Bader, M. A. (2015). Deltamethrin resistance in field populations of *Rhipicephalus (Boophilus) microplus* (Acari: Ixodidae) in Jammu and Kashmir, India. *Experimental and Applied Acarology*, 67(3): 467-475.

Aldridge, W.N. and Davison, A. N. (1952). The inhibition of erythrocyte cholinesterase by tri-esters of phosphoric acid. 1. Diethyl p-nitrophenyl phosphate (E 600) and analogues. *Biochemical Journal*, 51(1): 62-70.

Allen, J. R. and Humphreys, S. J. (1979). Immunisation of guinea pigs and cattle against ticks. *Nature*, 280 (5722): 491-493.

Almazán, C., Kocan, K. M., Bergman, D. K., Garcia-Garcia, J.C., Blouin, E. F. and de la Fuente, J. (2003). Identification of protective antigens for the control of Ixodes scapularis infestations using cDNA expression library immunization. *Vaccine*, 21(13-14): 1492-1501.

Andreotti, R., Guerrero, F.D., Soares, M. A., Barros, J.C., Miller, R.J. and Léon, A.P. D. (2011). Acaricide resistance of *Rhipicephalus (Boophilus) microplus* in state of Mato Grosso do Sul, Brazil. *Revista Brasileira de Parasitologia Veterinária*, 20(2): 127-133.

Baffi, M. A., de Souza, G. R. L., Vieira, C. U., de Sousa, C. S., Gourlart, L. R. and Bonetti, A. M. (2007). Identification of point mutations in a putative carboxylesterase and their association with acaricide resistance in *Rhipicephalus (Boophilus) microplus* (Acari: Ixodidae). *Veterinary parasitology*, 148(3-4): 301-309.

Baffi, M. A., Pereira, C. D., Souza, G. R. L. D., Bonetti, A. M., Ceron, C. R. and Gourlart, L. R. (2005). Esterase profile in a pyrethroid-resistant Brazilian strain of the cattle tick *Rhipicephalus microplus* (Acari, Ixodidae). *Genetics and molecular biology*, 28(4): 749-753.

BAHS. (2019). *Basic Animal Husbandry & Fisheries Statistics*. Ministry of Agriculture, Department of Animal Husbandry, Dairying & Fisheries, Krishi Bhawan, New Delhi.

Balakrishnan, N., Srilatha, K. P. and Thomas, T. G. (2019). Prevalence of Ixodid Ticks in Some States of the Country and its Public Health Importance. *Journal of Communicable Diseases,* 51(1): 18-21.

Bansal G. C. (2005). Bovine theileriosis in India: an overview. *Proceedings of National Academy of Science India* 75(B) - special issue: 134-143.

Barker, S. C. and Murrell, A. (2004). Systematics and evolution of ticks with a list of valid genus and species names. *Parasitology*, 129(S1): S15-S36.

Barnett, S. F. (1961). *The Control of Ticks on Livestock*. Food and Agriculture Organization of the United Nations, Rome, Italy

Baron, S., van der Merwe, N. A., Madder, M. and Maritz-Olivier, C. (2015). SNP analysis infers that recombination is involved in the evolution of amitraz resistance in *Rhipicephalus microplus*. *PloS one*, 10(7): e0131341.

Baxter, G.D. and Barker, S. C. (1998). Acetylcholinesterase cDNA of the cattle tick, *Rhipicephalus microplus*: characterisation and role in organophosphate resistance. *Insect biochemistry and molecular biology*, 28(8): 581-589.

Baxter, G.D. and Barker, S. C. (2002). Analysis of the sequence and expression of a second putative acetylcholinesterase cDNA from organophosphate-susceptible and organophosphate-resistant cattle ticks. *Insect biochemistry and molecular biology*, 32(7): 815-820.

Baxter, G. D., Green, P., Stuttgen, M. and Barker, S. C. (1999). Detecting resistance to organophosphates and carbamates in the cattle tick *Rhipicephalus microplus*, with a propoxur-based biochemical test. *Experimental & applied acarology*, 23(11): 907-914.

Beati, L., Keirans, J. E. (2001). Analysis of the systematic relationships among ticks of the genera *Rhipicephalus* and *Boophilus* (Acari: Ixodidae) based on mitochondrial 12S ribosomal DNA gene sequences and morphological characters. *Journal of Parasitology*, 13: 32–48.

Beckmann, J. S. and Soller, M. (1987). Molecular markers in the genetic improvement of farm animals. *Biotechnology*, 5(6): 573-576.

Beeman, R. W. and Matsumura, F. (1978). Formamidine pesticides-actions in insects and acarines. In *Pesticide and Venom Neurotoxicity* (pp. 179-188). Springer, Boston, MA.

Bloomquist, J. R. (1996). Ion channels as targets for insecticides. *Annual review of entomology*, 41(1): 163-190.

Bram, R. A., George, J. E., Reichard, R. E. and Tabachnick, W. J. (2002). Threat of foreign arthropod-borne pathogens to livestock in the United States. *Journal of medical entomology*, 39(3): 405-416.

Brogdon, W. G., Mcallisteri, J. C., Vulule, J. (1997). Heme peroxidase activity measured in single mosquitoes identifies individuals expressing an elevated oxidase for insecticide resistance. *Journal of the American Mosquito Control,* 13: 233–237.

Brown, R. S., Reichelderfer, C. F. and Anderson, W. R. (1970). An endemic disease among laboratory populations of *Dermacentor andersoni* (=*D. venustus*) (Acarina: Ixodidae). *Journal of Invertebrate Pathology*, 16: 142–143.

Brum, J. G. W. and Teixeira, M. O. (1992). Acaricidal activity of *Cedea lapagei* on engorged females of *Rhipicephalus microplus* exposed to the environment. *Arquivos Brasileiro de Medicina Veterinaria i Zoologia*, 44: 543–544.

Brum, J. G. W., Faccini, J. L. H. and Do Amaral, M. M. (1991). Infection in engorged females of *Rhipicephalus microplus* (Acari: Ixodidae). II. Histopathology and in vitro trials. *Arquivos Brasileiro Medicina Vetrinaria Zootecnologia*, 43: 35–37.

Canales, M., Enriquez, A., Ramos, E., Cabrera, D., Dandie, H., Soto, A., Falcón, V., Rodríguez, M. and de la Fuente, J. (1997). Large-scale production in Pichia pastoris of the recombinant vaccine Gavac™ against cattle tick. *Vaccine*, 15(4): 414-422.

Canales, M., Naranjo, V., Almazán, C., Molina, R., Tsuruta, S. A., Szabó, M. P., Manzano-Roman, R., de la Lastra, J. M. P., Kocan, K. M., Jiménez, M. I. and Lucientes, J. (2009). Conservation and immunogenicity of the mosquito ortholog of the tick-protective antigen, subolesin. *Parasitology Research*, 105(1): 97-111.

Chagas, A. C. D. S., Passos, W. M., Prates, H. T., Leite, R. C., Furlong, J. and Fortes, I. C. P. (2002). Efeito acaricida de óleos essenciais e concentrados emulsionáveis de Eucalyptus spp em *Rhipicephalus microplus* [Acaricidal effect of essential oils and emulsifiable concentrates of Eucalyptus spp on *Rhipicephalus microplus*]. *Brazilian Journal of Veterinary Research and Animal Science*, 39(5): 247-253.

Chen, A.C., He, H. and Davey, R. B. (2007). Mutations in a putative octopamine receptor gene in amitraz-resistant cattle ticks. *Veterinary parasitology*, 148(3-4): 379-383.

Chen, A. C., He, H., Temeyer, K. B., Jones, S., Green, P. and Barker, S. C. (2009). A survey of *Rhipicephalus microplus* populations for mutations associated with pyrethroid resistance. *Journal of economic entomology*, 102(1): 373-380.

Chhabra, M. B. (1992). Tick infestations of livestock in India and their control. *International Journal of Tropical Insect Science*, 13(4): 649-655.

Chhillar, S., Chhilar, J. S. and Kaur, H. (2014). Investigations on some hard ticks (Acari: Ixodidae) infesting domestic buffalo and cattle from Haryana, India. *Journal of Entomology and Zoology Studies,* 2(4): 99–104.

Chigure, G. M., Sharma, A. K., Kumar, S., Fular, A., Sagar, S. V., Nagar, G., Upadhaya, D., Saravanan, B. C., Kumar, R. and Ghosh, S. (2018). Role of metabolic enzymes in conferring resistance to synthetic pyrethroids, organophosphates, and phenylpyrazole compounds in *Rhipicephalus microplus*. *International Journal of Acarology*, 44(1): 28-34.

Cobon, G., Hungerford, J., Woodrow, M., Smith, D. and Willadsen, P. (1995). *Vaccination against Rhipicephalus microplus: The Australian field experience*.

Coddard, J. (1987). *Ticks of medical importance occurring in the Western Hemisphere*. Final report for period June 1986- June 1987. U.S. Air Force School of Aerospace Medicine, Brooks Air Force Base, San Antonio, Tex.

Copping, L. G. and Menn, J. J. (2000). Biopesticides: a review of their action, applications and efficacy. *Pest Management Science: Formerly Pesticide Science*, 56(8): 651-676.

Cossío-Bayúgar, R., Martínez-Ibañez, F., Aguilar-Díaz, H. and Miranda-Miranda, E. (2018). Pyrethroid acaricide resistance is proportional to P-450 cytochrome oxidase expression in the cattle tick *Rhipicephalus microplus*. *BioMed research international*, 2018.

Cossío-Bayúgar, R., Miranda-Miranda, E., Ortiz-Nájera, A., Neri-Orantes, S. and Olvera, F. (2008.) Cytochrome P-450 monoxygenase gene expression supports multifactorial origin for acaricide resistance in *Rhipicephalus microplus*. *Journal of Parasitology Research*, 3: 59-66.

Crampton, A. L., Baxter, G. D. and Barker, S. C. (1999). Identification and characterisation of a cytochrome P450 gene and processed pseudogene from an arachnid: the cattle tick, *Rhipicephalus microplus*. *Insect biochemistry and molecular biology*, 29(4): 377-384.

DAFD, MOA, (2019). *20th Livestock Census All India Report*, Department of Animal Husbandry, Dairying and Fisheries, Ministry of Agriculture, Government of India.

De la Fuente, J. and Kocan, K. M. (2006). Strategies for development of vaccines for control of ixodid tick species. *Parasite immunology*, 28(7): 275-283.

De la Fuente, J., Almazán, C., Blouin, E. F., Naranjo, V. and Kocan, K. M. (2006). Reduction of tick infections with *Anaplasma marginale* and *A. phagocytophilum* by targeting the tick protective antigen subolesin. *Parasitology research*, 100(1): 85-91.

De la Fuente, J., Almazán, C., Blouin, E. F., Naranjo, V. and Kocan, K. M. (2005). RNA interference screening in ticks for identification of protective antigens. *Parasitology research*, 96(3): 137-141.

De la Fuente, J., Kocan, K. M. and Blouin, E. F. (2007). Tick vaccines and the transmission of tick-borne pathogens. *Veterinary research communications*, 31(1): 85-90.

De la Fuente, J., Rodriguez, M. and García-Garí, J. C. (2000). Immunological control of ticks through vaccination with *Rhipicephalus microplus* gut antigens. *Annals of the New York Academy of Sciences*, 916(1): 617-621.

De La Fuente, J., Rodrıguez, M., Montero, C., Redondo, M., Garcıa-Garcıa, J. C., Méndez, L., Serrano, E., Valdés, M., Enrıquez, A., Canales, M. and Ramos, E. (1999). Vaccination against ticks (*Boophilus* spp.): the experience with the Bm86-based vaccine Gavac™. *Genetic Analysis: Biomolecular Engineering*, 15(3-5): 143-148.

De La Fuente, J., Rodríguez, M., Redondo, M., Montero, C., García-García, J., Méndez, L., Serrano, E., Valdés, M., Enriquez, A., Canales, M. and Ramos, E. (1998). Field studies and cost-effectiveness analysis of vaccination with Gavac™ against the cattle tick *Rhipicephalus microplus.Vaccine*, 16(4): 366-373.

De Vos, S., Zeinstra, L., Taoufik, A., Willadsen, P. and Jongejan, F. (2001). Evidence for the utility of the Bm86 antigen from *Rhipicephalus microplus* in vaccination against other tick species. *Experimental & applied acarology*, 25(3): 245-261.

Dreyer, K. A. R. I. N., Fourie, L. J. and Kok, D. J. (1997). Predation of livestock ticks by chickens as a tick-control method in a resource-poor urban environment. *Onderstepoort Journal of Veterinary Research*, 64: 273-276.

Drummond, R. O. (1983). Tick-borne livestock diseases and their vectors. Chemical control of Ticks. *World Animal Review*, (FAO). 36: 28–33.

Drummond, R. E. A., Ernst, S. E., Trevino, J. L., Gladney, W. J. and Graham, O. H. (1973). *Boophilus annulatus* and *B. microplus*: laboratory tests of insecticides. *Journal of Economic Entomology*, 66(1): 130-133.

Dutta, S., Godara, R., Katoch, R., Yadav, A., Katoch, M. and Singh, N. K. (2017). Detection of amitraz and malathion resistance in field populations of *Rhipicephalus (Boophilus) microplus* (Acari: Ixodidae) in Jammu region of India. *Experimental and Applied Acarology*, 71(3): 291-301.

Enayati, A. A., Asgarian, F., Amouei, A., Sharif, M., Mortazavi, H., Boujhmehrani, H. and Hemingway, J. (2010). Pyrethroid insecticide resistance in *Rhipicephalus bursa* (Acari, Ixodidae). *Pesticide biochemistry and physiology*, 97(3):.243-248.

Enayati, A. A. and Motevali, H. F. (2007). Biochemistry of pyrethroid resistance in German cockroach (Dictyoptera, Blatellidae) from hospitals of Sari, Iran. *Iranian Biomedical Journal,* 11: 251–258.

Frans, J. (2000). *Final Report, Integrated Control of Ticks and Tick-Born Diseases* (*ICTTD*).

Frazier, J. G. and Keirans, J. E. (1989). Ticks (Acari: Ixodidae) collected on chelonians (Reptilia) from India and Burma. *Journal Bombay Natural History Society*, 87: 287-289.

Frisch, J. E. (1999). Towards a permanent solution for controlling cattle ticks. *International journal for parasitology*, 29(1): 57-71.

Fular, A., Sharma, A. K., Kumar, S., Nagar, G., Chigure, G., Ray, D. D. and Ghosh, S. (2018). Establishment of a multi-acaricide resistant reference tick strain (IVRI-V) of *Rhipicephalus microplus*. *Ticks and tick-borne diseases*, 9(5): 1184-1191.

Furman, D. P. and Loomis, E. C. (1984). *The ticks of California (Acari: Ixodida)* (Vol. 25). University of California Press.

Gasparin, G., Miyata, M., Coutinho, L. L., Martinez, M. L., Teodoro, R. L., Furlong, J., Machado, M. A., Silva, M. V. G. B., Sonstegard, T. S. and Regitano, L. C. A. (2007). Mapping of quantitative trait loci controlling tick [*Riphicephalus (Boophilus) microplus*] resistance on bovine chromosomes 5, 7 and 14. *Animal Genetics*, 38(5): 453-459.

Gaur, G. K., Kaushik, S. N. and Garg, R. C. (2002). Ongole cattle status in India. *Animal Genetic Resources/Resources génétiquesanimales/Recursosgenéticosanimales*, 32: 27-34.

Gaur, R. S., Sangwan, A. K., Sangwan, N. and Kumar, S. (2016). Acaricide resistance in *Rhipicephalus (Boophilus) microplus* and *Hyalomma anatolicum* collected from Haryana and Rajasthan states of India. *Experimental and Applied Acarology*, 69(4): 487-500.

Gaur, R. S., Sangwan, A. K., Sangwan, N., Ghosh, M. and Kumar, S. (2017). Comparative study of esterases in deltamethrin and diazinon resistant *Rhipicephalus microplus* and *Hyalomma anatolicum* ticks collected from the Trans-Gangetic plains of India. *Experimental and Applied Acarology*, 73(1): 115-127.

Geevarghese, G. and Dhanda, V. (1987). *Check-list of world Hyalomma ticks. The Indian Hyalomma Ticks,* Indian Council of Agricultural Research, New Delhi, pp.12-15.

Geevarghese, G. and Dhanda, V. (1995). Ixodid ticks of Maharashtra state, India. *Acarologia*, 36(4): 309-313.

Geevarghese, G., Fernandes, S. and Kulkarni, S. M. (1997). A checklist of Indian ticks (Acari: Ixodoidea). *Indian Journal of Animal Sciences*, 67(7): 566-574.

George, J. E. (2000). Present and future technologies for tick control. *Annals of the New York Academy of Sciences*, 916: 583–588.

George, J. E. (1990). Summing-up of strategies for the control of ticks in regions of the world other than Africa. *Parassitologia (Roma)*, 32(1): 203-209.

Ghosh, S. and Nagar, G. (2014). Problem of ticks and tick-borne diseases in India with special emphasis on progress in tick control research: a review. *Journal of Vector Borne Diseases*, 51(4):259-270.

Ghosh, S., Bansal, G. C., Gupta, S. C., Ray, D., Khan, M. Q., Irshad, H., Shahiduzzaman, M. D., Seitzer, U. and Ahmed, J. S. (2007). Status of tick distribution in Bangladesh, India and Pakistan. *Parasitology Research*, 101(2): 207-216.

Ghosh, S., Kumar, R., Nagar, G., Kumar, S., Sharma, A. K., Srivastava, A., Kumar, S., Kumar, K. A. and Saravanan, B. C. (2015). Survey of acaricides resistance status of *Rhipiciphalus (Boophilus) microplus* collected from selected places of Bihar, an eastern state of India. *Ticks and tick-borne diseases*, 6(5): 668-675.

Ghosh, S., Sharma, A. K., Kumar, S., Tiwari, S. S., Rastogi, S., Srivastava, S., Singh, M., Kumar, R., Paul, S., Ray, D. D. and Rawat, A. K. S. (2011). *In vitro* and *in vivo* efficacy of *Acorus calamus* extract against *Rhipicephalus (Boophilus) microplus*. *Parasitology research*, 108(2): 361-370.

Ghosh, S., Tiwari, S. S., Srivastava, S., Sharma, A. K., Nagar, G., Kumar, K. A., Kumar, R. and Rawat, A. K. S. (2015). *In vitro* acaricidal properties of *Semecarpus anacardium* fruit and *Datura stramonium* leaf extracts against acaricide susceptible (IVRI-I line) and resistant (IVRI-V line) *Rhipicephalus (Boophilus) microplus*. *Research in Veterinary Science*, 101: 69-74.

Ghosh, S., Tiwari, S. S., Srivastava, S., Sharma, A. K., Kumar, S., Ray, D. D. and Rawat, A. K. S. (2013). Acaricidal properties of *Ricinus communis* leaf extracts against organophosphate and pyrethroids resistant *Rhipicephalus (Boophilus) microplus*. *Veterinary Parasitology*, 192(1-3): 259-267.

Godara, R., Katoch, R., Rafiqi, S. I., Yadav, A., Nazim, K., Sharma, R., Singh, N. K. and Katoch, M. (2019). Synthetic pyrethroid resistance in *Rhipicephalus (Boophilus) microplus* ticks from north-western Himalayas, India. *Tropical Animal Health and Production*, 51(5): 1203-1208.

Grønvold, J., Henriksen, S. A., Larsen, M., Nansen, P. and Wolstrup, J. (1996). Biological control aspects of biological control—with special reference to arthropods, protozoans and helminths of domesticated animals. *Veterinary Parasitology*, 64(1-2): 47-64.

Guerrero, F. D., Davey, R. B. and Miller, R. J. (2001). Use of an allele-specific polymerase chain reaction assay to genotype pyrethroid resistant strains of *Rhipicephalus microplus* (Acari: Ixodidae). *Journal of Medical Entomology*, 38(1):44-50.

Guglielmone, A. A., Robbins, R. G., Apanaskevich, D. A., Petney, T. N., Estrada-Pena, A., Horak, I. G., Shao, R. and Barker, S. C. (2010). *The Argasidae, Ixodidae and Nuttalliellidae* (*Acari: Ixodida*) *of the world: a list of valid species names*.

Habig, W. H., Pabst, M. J. and Jakoby, W. B. (1974). Glutathione S-transferases the first enzymatic step in mercapturic acid formation. *Journal of Biological Chemistry*, 249(22): 7130-7139.

Harrington, D., Canales, M., de la Fuente, J., de Luna, C., Robinson, K., Guy, J. and Sparagano, O. (2009). Immunisation with recombinant proteins subolesin and Bm86 for the control of *Dermanyssus gallinae* in poultry. *Vaccine*, 27(30): 4056-4063.

Hassan, S. M., Dipeolu, O. O. and Munyinyi, D. M. (1992). Influence of exposure period and management methods on the effectiveness of chickens as predators of ticks infesting cattle. *Veterinary Parasitology*, 43: 301–309.

Hassan, S. M., Dipeolu, O. O., Amoo, A. O. and Odhiambo, T. R. (1991). Predation on livestock ticks by chickens. *Veterinary Parasitology*, 38: 199–204.

Hassanain, M. A., Derbala, A. A., Abdel-Barry, N. A., El-Sherif, M. A. and El-Sadawy, H. A. E. (1999). Biological control of ticks (Argasidae) by entomopathogenic nematodes. *Egyptian Journal of Biological Pest Control*, 7: 41–46.

Hassanain, M. A., El Garhy, M. F., Abdel-Ghaffar, F. A., El-Sharaby, A. and Megeed, K. N. A. (1997). Biological control studies of soft and hard ticks in Egypt, I. The effect of Bacillus thuringiensis varieties on soft and hard ticks (Ixodidae). *Parasitology Research*, 83: 209–213.

He, H., Chen, A. C., Davey, R. B., Ivie, G. W. and George, J. E. (1999). Identification of a point mutation in the para-type sodium channel gene from a pyrethroid-resistant cattle tick. *Biochemical and biophysical research communications*, 261(3): 558-561.

Hemingway, J., Hawkes, N., Prapanthadara, L. A., Jayawardenal, K. I. and Ranson, H. (1998). The role of gene splicing, gene amplification and regulation in mosquito insecticide resistance. *Philosophical Transactions of the Royal Society of London. Series B: Biological Sciences*, 353(1376): 1695-1699.

Hendry, D. A. and Rechav, Y. (1981). Acaricidal bacteria infecting laboratory colonies of the tick *Boophilus decoloratus* (Acarina: Ixodidae). *Journal of Invertebrate Pathology*, 38(1): 149-151.

Hernandez, R., Guerrero, F. D., George, J. E. and Wagner, G. G. (2002). Allele frequency and gene expression of a putative carboxylesterase-encoding gene in a pyrethroid resistant strain of the tick *Rhipicephalus microplus*. *Insect biochemistry and molecular biology*, 32(9): 1009-1016.

Hernandez, R., He, H., Chen, A. C., Ivie, G. W., George, J. E. and Wagner, G. G. (1999). Cloning and sequencing of a putative acetylcholinesterase cDNA from *Rhipicephalus microplus* (Acari: Ixodidae). *Journal of medical entomology*, 36(6): 764-770.

Hernandez, R., He, H., Chen, A. C., Waghela, S. D., Ivie, G. W., George, J. E. and Wagner, G. G. (2000). Identification of a point mutation in an esterase gene in different populations of the southern cattle tick, *Rhipicephalus microplus*. *Insect Biochemistry and Molecular Biology*, 30(10): 969-977.

Hill, D. E. (1998). Entomopathogenic nematodes as control agents of developmental stages of the black-legged tick, Ixodes scapularis. *Journal of Parasitology*, 84: 1124–1127.

Hollingworth, R. M. and Murdock, L. L. (1980). Formamidine pesticides: octopamine-like actions in a firefly. *Science*, 208(4439): 74-76.

Holm, E. and Wallace, M. M. H. (1989). Distribution of some anystid mites (Acari: Anystidae) in Australia and Indonesia and their role as possible predators of the cattle tick, Boophilis microplus (Acari: Ixodidae). *Experimental & Applied Acarology*, 6(1): 77-83.

Honer, M. R. and Gomes, A. (1990). The integrated management of horn fly, botflies and ticks in beef cattle. *Circular Técnica-Centro Nacional de Pesquisa de Gado de Corte*, (22).

Hoogstraal, H. and Trapido, H. (1963). *Haemaphysalis silvafelis* sp. n., a parasite of the jungle cat in southern India (Ixodoidea, Ixodidae). *The Journal of Parasitology*, 49(2): 346-349.

Hoogstraal, H. (1979). The epidemiology of tick-borne Crimean-Congo hemorrhagic fever in Asia, Europe, and Africa. *Journal of medical entomology*, 15(4): 307-417.

Hu, R. and Hyland, K. E. (1997). Prevalence and seasonal activity of the wasp parasitoid, *Ixodiphagus hookeri* (Hymenoptera: Encyrtidae) in its tick host, *Ixodes scapularis* (Acari: Ixodidae). *Systematic and Applied Acarology*, 2: 95–100.

Hu, R., Hyland, K. E. and Mather, T. N. (1993). Occurrence and distribution in Rhode Island of *Hunterellus hookeri* (Hymenoptera: Encyrtidae), a wasp parasitoid of Ixodes dammini. *Journal of Medical Entomology*, 30: 277–280.

Hu, R., Hyland, K. E. and Oliver, J. H. (1998). A review on the use of Ixodiphagus wasps (Hymenoptera: Encyrtidae) as natural enemies for the control of ticks (Acari: Ixodidae). *Systematic and Applied Acarology*, 3(1): 19-28.

Jamroz, R. C., Guerrero, F. D., Pruett, J. H., Oehler, D. D. and Miller, R. J. (2000). Molecular and biochemical survey of acaricide resistance mechanisms in larvae from Mexican strains of the southern cattle tick, *Rhipicephalus microplus*. *Journal of Insect Physiology*, 46(5): 685-695.

Jongejan, F. and Uilenberg, G. (2004). The global importance of ticks. *Parasitology-Cambridge*, 129: S3.

Jonsson, N. N., Cutullè, C., Corley, S. W. and Seddon, J. M. (2010). Identification of a mutation in the para-sodium channel gene of the cattle tick *Rhipicephalus microplus* associated with resistance to flumethrin but not to cypermethrin. *International Journal of Parasitology*, 40(14): 1659-1664.

Jonsson, N. N., Klafke, G., Corley, S. W., Tidwell, J., Berry, C. M. and Koh-Tan, H. H. (2018). Molecular biology of amitraz resistance in cattle ticks of the genus *Rhipicephalus*. *Frontiers in Bioscience*, 23(2): 796-810.

Jonsson, N. N., Mayer, D. G. and Green, P. E. (2000). Possible risk factors on Queensland dairy farms for acaricide resistance in cattle tick (*Rhipicephalus microplus*). *Veterinary Parasitology*, 88(1-2): 79-92.

Joshi, B. B. (1943). The tampan ticks of Marwar. *Indian Fmg.*,4(3): 141-142.

Juliet, S., Ravindran, R., Ramankutty, S. A., Gopalan, A. K. K., Nair, S. N., Kavillimakkil, A. K., Bandyopadhyay, A., Rawat, A. K. S. and Ghosh, S. (2012). *Jatropha curcas* (Linn) leaf extract–a possible alternative for population control of *Rhipicephalus (Boophilus) annulatus*. *Asian Pacific Journal of Tropical Disease*, 2(3): 225-229.

Jyothimol, G., Ravindran, R., Juliet, S., Ajithkumar, K. G., Suresh, N. N., Vimalkumar, M. B., Lenka, D. R., Varghese, S. and Ghosh, S. (2014). Low level deltamethrin resistance in ticks from cattle of Kerala, a south Indian state. *Veterinary Parasitology*, 204(3-4): 433-438.

Jyoti, N. K. S., Prerna, M., Singh, H. and Rath, S. S. (2015). Detection of malathion resistance in *Hyalomma anatolicum anatolicum* from Bathinda district, Punjab. *Toxicology International*, 22(1): 125.

Jyoti, Singh, N. K., Singh, H. and Rath, S. S. (2019). Modified larval packet test based detection of amitraz resistance in *Hyalomma anatolicum* Koch (Acari: Ixodidae) from Punjab districts of India. *International Journal of Acarology*, 45(6-7): 391-394.

Jyoti, Singh, N. K., Singh, H., Rath, S. S. (2016). Multiple mutation in the acetylcholinesterase 3 gene associated with organophosphate resistance in *Rhipicephalus (Boophilus) microplus* tick from Punjab, India. *Veterinary Parasitology*, 216: 108–117.

Kaaya, G. P. and Hassan, S. (2000). Entomogenous fungi as promising biopesticides for tick control. *Experimental and Applied Acarology*, 24(12): 913-926.

Karczmar, A. G., Usdin, E. and Wills, J. H. (1970). *Anticholinesterase agents* (Vol. 1). Pergamon.

Katuri, R. N., Das, G., Singh, A. K., Chalhotra, S. K. and Nath, S. (2017). Comparative efficacy of deltamethrin and chlorpyriphos in bovine ticks in and around Jabalpur. *Journal of Parasitic Diseases*, 41(3): 713-715.

Kaul, H. N., Mishra, A. C., Dhanda, V., Kulkarni, S. M. and Guttikar, S. N. (1978). Ectoparasite arthropods of birds and mammals from Rajasthan State, India. *Indian Journal of Parasitology*, 2(1): 19-25.

Khajuria, V., Godara, R., Yadav, A. and Katoch, R. (2014). Deltamethrin resistance in *Rhipicephalus microplus* of Jammu region. *Indian Veterinary Journal*, 91: 21-24.

Kirkland, B. H., Cho, E. M. and Keyhani, N. O. (2004). Differential susceptibility of *Amblyomma maculatum* and *Amblyomma americanum* (Acari: Ixodidea) to the entomopathogenic fungi *Beauveria bassiana* and *Metarhizium anisopliae*. *Biological Control*, 31(3): 414-421.

Klafke, G. M., Miller, R. J., Tidwell, J. P., Thomas, D. B., Sanchez, D., Arroyo, T. P. F. and de León, A. A. P. (2019). High-resolution melt (HRM) analysis for detection of SNPs associated with pyrethroid resistance in the southern cattle fever tick, *Rhipicephalus (Boophilus) microplus* (Acari: Ixodidae). *International Journal for Parasitology: Drugs and Drug Resistance*, 9: 100-111.

Klafke, G., Webster, A., Agnol, B. D., Pradel, E., Silva, J., de La Canal, L. H., Becker, M., Osório, M. F., Mansson, M., Barreto, R. and Scheffer, R. (2017). Multiple resistance to acaricides in field populations of *Rhipicephalus microplus* from Rio Grande do Sul state, Southern Brazil. *Ticks and tick-borne diseases*, 8(1): 73-80.

Kocan, K. M., Pidherney, M. S., Blouin, E. F., Claypool, P. L., Samish, M. and Glazer, I. (1998). Interaction of entomopathogenic nematodes (Steinernematidae) with selected species of ixodid ticks (Acari: Ixodidae). *Journal of Medical Entomology*, 35: 514–520.

Kocan, K. M., De la Fuente, J., Guglielmone, A. A. and Meléndez, R. D. (2003). Antigens and alternatives for control of *Anaplasma marginale* infection in cattle. *Clinical Microbiology Reviews*, 16(4): 698-712.

Kohn, W. K. and Norval, R. A. I. (1994). Predation of ticks by chickens in Zimbabwe. *Zimbabwe Veterinary Journal*, 25(2): 33–37.

Kumar, K., Jain, S. K. and Kumar, A. (2011). Outbreak Indian tick typhus amongst residents of Deol village, district, Kangra, Himachal Pradesh (India). *International Journal of Medicine and Public Health*, 1(3): 67-71.

Kumar, R. (2019). Molecular markers and their application in the monitoring of acaricide resistance in *Rhipicephalus microplus*. *Experimental and Applied Acarology*, 78: 149-172.

Kumar, R., Nagar, G., Sharma, A. K., Kumar, S., Ray, D. D., Chaudhuri, P. and Ghosh, S. (2013). Survey of pyrethroids resistance in Indian isolates of *Rhipicephalus (Boophilus) microplus*: identification of C190A mutation in the domain II of the para-sodium channel gene. *Acta Tropica*, 125(2): 237-245.

Kumar, S., Paul, S., Sharma, A. K., Kumar, R., Tewari, S. S., Chaudhuri, P., Ray, D. D., Rawat, A. K. S. and Ghosh, S. (2011). Diazinon resistant status in *Rhipicephalus (Boophilus) microplus* collected from different agro-climatic regions of India. *Veterinary Parasitology*, 181(2-4): 274-281.

Kumar, S., Sharma, A. K., Nagar, G., Rawat, S. S., Tiwari, S. S., Kumar, R., Dhakad, M. L., Sharma, R. K., Saxana, R. K., Mehraniya, R. S. and Singh, R. S. (2016). Characterization of acaricide resistance in tick isolates collected from Rajasthan, India. *Indian Journal of Animal Sciences*, 86(1): 14-23.

Kunz, S. E. and Kemp, D. H. (1994). Insecticides and acaricides: resistance and environmental impact. *Revue Scientifique et Technique-Office International des Epizooties*, 13(4): 1249-1286.

Latif, A. A. and Pegram, R. G. (1992). Naturally acquired host resistance in tick control in Africa. *International Journal of Tropical Insect Science*, 13(4): 505-513.

Lee, R. M., Bantham, P. (1966). The activity and organophosphate inhibition of cholinesterase from susceptible and resistant ticks (Acari). *Entomologia Experimentalis et Applicata,* 9: 13–24.

Leemon, D. M. and Jonsson, N. N. (2008). Laboratory studies on Australian isolates of *Metarhizium anisopliae* as a biopesticide for the cattle tick *Rhipicephalus microplus*. *Journal of Invertebrate Pathology*, 97(1): 40-49.

Leemon, D. M., Turner, L. B. and Jonsson, N. N. (2008). Pen studies on the control of cattle tick (*Rhipicephalus (Boophilus) microplus*) with *Metarhizium anisopliae* (Sorokin). *Veterinary Parasitology*, 156(3-4): 248-260.

Lenka, D. R., Ravindran, R., Jyothimol, G., Udaykumar, M., Reddy, P. M. T., Sathish, N., Palla, I., Chandramohan, B., Ajithkumar, K. G., Nair, S. N. and Chandrasekhar, L. (2016). Deltamethrin resistance in south Indian isolates of *Rhipicephalus (Boophilus) microplus*. *Veterinary Parasitology: Regional Studies and Reports*, 5: 37-41.

Li, A. Y., Chen, A. C., Miller, R. J., Davey, R. B. and George, J. E. (2007). Acaricide resistance and synergism between permethrin and amitraz against susceptible and resistant strains of *Rhipicephalus microplus* (Acari: Ixodidae). *Pest Management Science: formerly Pesticide Science*, 63(9): 882-889.

Linthicum, K. J. and Bailey, C. L. (1994). Ecology of Crimean-Congo hemorrhagic fever. *Ecological Dynamics of Tick-borne Zoonoses*, 392: 437.

Lodos, J., Boue, O. and de la Fuente, J. (2000). A model to simulate the effect of vaccination against *Boophilus* ticks on cattle. *Veterinary Parasitology*, 87(4): 315-326.

Lodos, J., Ochagavia, M. E., Rodriguez, M. and De la Fuente, J. (1999). A simulation study of the effects of acaricides and vaccination on Boophilus cattle–tick populations. *Preventive Veterinary Medicine*, 38(1): 47-63.

Lovis, L., Guerrero, F. D., Miller, R. J., Bodine, D. M., Betschart, B. and Sager, H. (2012). Distribution patterns of three sodium channel mutations associated with pyrethroid resistance in *Rhipicephalus (Boophilus) microplus* populations from North and South America, South Africa and Australia. *International Journal for Parasitology: Drugs and Drug Resistance*, 2: 216-224.

Machado, M. A., Azevedo, A. L. S., Teodoro, R. L., Pires, M. A., Peixoto, M. G. C., de Freitas, C., Prata, M. C. A., Furlong, J., da Silva, M. V. G., Guimarães, S. E. and Regitano, L. C. (2010). Genome wide scan for quantitative trait loci affecting tick resistance in cattle (*Bos taurus*× *Bos indicus*). *BMC genomics*, 11(1): 280.

Mandloi, U. K., Jayraw, A. K., Haque, M., Jamra, N. (2016). Prevalence of ixodid ticks in cattle population of Indore, Madhya Pradesh. *The Indian Journal of Veterinary Sciences & Biotechnology*. 12(2): 62-65.

Martinez, M. L., Machado, M. A., Nascimento, C. S., Silva, M. V. G. B., Teodoro, R. L., Furlong, J., Prata, M. C. A., Campos, A. L., Guimarães, M. F. M., Azevedo, A. L. S. and Pires, M. F. A. (2006). Association of BoLA-DRB3. 2 alleles with tick (*Rhipicephalus microplus*) resistance in cattle. *Genetics and Molecular Research* 5(3): 513-524.

Maske, D. K., Bhilegaonkar, N. G. and Jangde, C.R. (1996). Treatment of tick infestation in cattle with pestoban. *Indian Journal of Indigenous Medicines*, 17(2): 81–83.

Massard, C. L., da Fonseca, A. H., Bittencourt, V. R. E. P., de Oliveira, J. B. and da Silva, K. M. (1995). Evaluation of recombinant rBm86-" Gavac" vaccine against the tick *Rhipicephalus microplus* in Brazil. *Revista Brasileira de Medicina Veterinaria (Brazil).*

Mauleon, H., Barre, N. and Panoma, S. (1993). Pathogenicity of 17 isolates of entomophagous nematodes (Steinernematidae and Heterorhabditidae) for the ticks *Amblyomma variegatum* (Fabricius), *Rhipicephalus microplus* (Canestrini) and *Boophilus annulatus* (Say). *Experimental and Applied Acarology*, 17: 831–838.

Miller, R. J., Davey, R. B. and George, J. E. (1999). Characterization of pyrethroid resistance and susceptibility to coumaphos in Mexican *Rhipicephalus microplus* (Acari: Ixodidae). *Journal of medical entomology*, 36(5): 533-538.

Minjauw, B. and Mc Leod, A. (2003). Research report, DFID animal health programme. *Centre for Tropical Veterinary Medicine,* University of Edinburgh, UK.

Miranpuri, G. S. and Gill, H. S. (1983). *Ticks of India.* Centre for Tropical Veterinary Medicine.

Miranpuri, G. S., Bindra, O. S. and Prasad, V. (1975). Tick fauna of north-western India (Acarina: Metastigmata). *International Journal of Acarology*, 1(1), pp.31-54.

Mooring, M. S. and Mundy, P. J. (1996). Interactions between impala and oxpeckers at Matobo National Park, Zimbabwe. *African Journal of Ecology*, 34(1): 54-65.

Morgan, J. A., Corley, S. W., Jackson, L. A., Lew-Tabor, A. E., Moolhuijzen, P. M. and Jonsson, N. N. (2009). Identification of a mutation in the para-sodium channel gene of the cattle tick *Rhipicephalus (Boophilus) microplus* associated with resistance to synthetic pyrethroid acaricides. *International Journal of Parasitology*, 39(7): 775-779.

Mourya, D. T., Yadav, P. D. and Patil, D. Y. (2014). Highly infectious tick-borne viral diseases: Kyasanur forest disease and Crimean–Congo haemorrhagic fever in India. *WHO South-East Asia Journal of Public Health*, 3(1): 8-21.

Nagar, G., Sharma, A. K., Chigure, G., Manjunathachar, H. V., Saravanan, B. C., Rai, A. and Ghosh, S. (2016). Identification of mutations in acetylcholinesterase 2 gene of acaricide resistant isolates of *Rhipicephalus (Boophilus) microplus. International Journal of Science Environment and Technology*, 5: 3440-3447.

Nagar, G., Sharma, A. K., Kumar, S., Saravanan, B. C., Kumar, R., Gupta, S., Kumar, S. and Ghosh, S. (2018). Molecular mechanism of synthetic pyrethroid and organophosphate resistance in field isolates of *Rhipicephalus microplus* tick collected from a northern state of India. *Experimental and Applied Acarology*, 75(3): 319-331.

Nandi, A., Sagar, S. V., Chigure, G., Fular, A., Sharma, A. K., Nagar, G., Kumar, S., Saravanan, B. C. and Ghosh, S. (2018). Determination and validation of discriminating concentration of ivermectin against *Rhipicephalus microplus*. *Veterinary parasitology*, 250: 30-34.

Nandi, A., Singh, H. and Singh, N. K. (2015). Esterase and glutathione S-transferase levels associated with synthetic pyrethroid resistance in *Hyalomma anatolicum* and

Rhipicephalus microplus ticks from Punjab, India. *Experimental and Applied Acarology*, 66(1): 141-157.

Naranjo, V., Ayllon, N., de la Lastra, J. M. P., Galindo, R. C., Kocan, K. M., Blouin, E. F., Mitra, R., Alberdi, P., Villar, M. and de la Fuente, J. (2013). Reciprocal regulation of NF-kB (Relish) and Subolesin in the tick vector, Ixodes scapularis. *PLoS One*, 8(6): p.e65915.

Nolan, J. (1985). Mechanisms of resistance to chemicals in arthropod parasites of veterinary importance. *Veterinary Parasitology*, 18(2): 155-166.

Noppun, V., Saito, T. and Miyata, T. (1989). Cuticular penetration of S-fenvalerate in fenvalerate-resistant and susceptible strains of the diamondback moth, Plutella xylostella (L.). *Pesticide Biochemistry and Physiology*, 33(1): 83-87.

Nuttall, P. A., Trimnell, A.R., Kazimirova, M. and Labuda, M. (2006). Exposed and concealed antigens as vaccine targets for controlling ticks and tick-borne diseases. *Parasite immunology*, 28(4): 155-163.

Page, W. S. (2008). *Antiparasitic drugs, Small Animal Clinical Pharmacology* (Second Edition): 198-260.

Parola, P. and Raoult, D. (2001). Ticks and tickborne bacterial diseases in humans: an emerging infectious threat. *Clinical infectious diseases*, 32(6): 897-928.

Patel, D.C., Solanki, J. B. and Kumar, N. (2019). Risk factors associated prevalence of hard tick in large ruminants of coastal areas of South Gujarat, India. *Indian Journal of Animal Research*, 53(11): 1514-1517.

Pattnaik, P. (2006). Kyasanur forest disease: an epidemiological view in India. *Reviews in Medical Virology*, 16(3):151-165.

Peter, R. J., Van den Bossche, P., Penzhorn, B. L. and Sharp, B. (2005). Tick, fly, and mosquito control—lessons from the past, solutions for the future. *Veterinary Parasitology*, 132(3-4): 205-215.

Pradeep, B. S., Renukaprasad, C. and Souza, P. E. (2012). Evaluation of the commonly used acaricides against different stages of the cattle tick *Rhipicephalus microplus* by using different in vitro tests. *Indian Journal of Animal Research*, 46(3): 248-252.

Prudencio, C. R., Marra, A. O., Cardoso, R. and Goulart, L. R. (2010). Recombinant peptides as new immunogens for the control of the bovine tick, *Rhipicephalus (Boophilus) microplus*. *Veterinary Parasitology*, 172(1-2): 122-131.

Pruett, J. H. (2002). Comparative inhibition kinetics for acetylcholinesterases extracted from organophosphate resistant and susceptible strains of *Rhipicephalus microplus* (Acari: Ixodidae). *Journal of Economic Entomology*, 95(6): 1239-1244.

Rajput, Z. I., Hu, S. H., Chen, W. J., Arijo, A. G. and Xiao, C. W. (2006). Importance of ticks and their chemical and immunological control in livestock. *Journal of Zhejiang University Science B*, 7(11): 912-921.

Rand, K. N., Moore, T., Sriskantha, A., Spring, K., Tellam, R., Willadsen, P. and Cobon, G. S. (1989). Cloning and expression of a protective antigen from the cattle tick *Rhipicephalus microplus. Proceedings of the National Academy of Sciences*, 86(24): 9657-9661.

Raote, Y. V. (1983). *Incidence of parasitic infestation on organised cattle farms in Bombay Region*. M.V.Sc. thesis submitted to Konkan Krishi Vidyapeeth, Dapoli, Ratnagiri.

Rath, S. S., Kumar, S. and Joia, B. S. (2006). Resistance to diazinon and malathion in *Rhipicephalus microplus* (Acari: Ixodidae) populations from Punjab, India. *Journal of Insect Science-Ludhiana*, 19(I): 74.

Ravindran, C., Varatharajan, G. R. and Karthikeyan, A. (2011). Role of alkaline-tolerant fungal cellulases in release of total antioxidants from agro-wastes under solid state fermentation. *Bioresources*, 6(3): 3142-3154.

Ravindran, R., Ramankutty, S. A., Juliet, S., Palayullaparambil, A. K. T., Gopi, J., Gopalan, A. K. K., Nair, S. N. and Ghosh, S. (2014). Comparison of *in vitro* acaricidal effects of commercial preparations of cypermethrin and fenvalerate against *Rhipicephalus (Boophilus) annulatus*. *SpringerPlus*, 3(1): 90.

Redondo, M., Fragoso, H., Montero, C., Lona, J., Medellín, J. A., Fría, R., Hernández, V., Franco, R., Machado, H., Rodríguez, M. and De La Fuente, J. (1999). Integrated control of acaricide-resistant *Rhipicephalus microplus* populations on grazing cattle in Mexico using vaccination with Gavac™ and amidine treatments. *Experimental & Applied Acarology*, 23(10): 841-849.

Regitano, L. C. A., Ibelli, A. M. G., Gasparin, G., Miyata, M., Azevedo, A. L. S., Coutinho, L. L., Teodoro, R. L., Machado, M. A., Silva, M. V. G. B., Nakata, L. C. and Zaros, L. G. (2008). On the search for markers of tick resistance in bovines. In *Animal Genomics for Animal Health* (Vol. 132, pp. 225-230). Karger Publishers.

Rodríguez, M., Rubiera, R., Penichet, M., Montesinos, R., Cremata, J., Falcón, V., Sánchez, G., Bringas, R., Cordovés, C., Valdés, M. and Lleonart, R. (1994). High level expression of the B. microplus Bm86 antigen in the yeast Pichia pastoris forming highly immunogenic particles for cattle. *Journal of Biotechnology*, 33(2): 135-146.

Rodriguez-Vivas, R. I., Jonsson, N. N. and Bhushan, C. (2018). Strategies for the control of *Rhipicephalus microplus* ticks in a world of conventional acaricide and macrocyclic lactone resistance. *Parasitology Research*, 117(1): 3-29.

Rodriguez-Vivas, R. I., Trees, A. J., Rosado-Aguilar, J. A., Villegas-Perez, S. L. and Hodgkinson, J. E. (2011). Evolution of acaricide resistance: phenotypic and genotypic changes in field populations of *Rhipicephalus (Boophilus) microplus* in response to pyrethroid selection pressure. *International Journal for Parasitology*, 41(8): 895-903.

Sabatini, G. A., Kemp, D. H., Hughes, S., Nari, A. and Hansen, J. (2001). Tests to determine LC50 and discriminating doses for macrocyclic lactones against the cattle tick, *Rhipicephalus microplus*. *Veterinary Parasitology*, 95(1): 53-62.

Samish, M. and Glazer, I. (2001). Entomopathogenic nematodes for the biocontrol of ticks. *TRENDS in Parasitology*, 17(8): 368-371.

Samish, M., Alekseev, E. A. and Glazer, I. (1995). The development of entomopathogenic nematodes in the tick *B. annulatus*. In the *Second International Conference on Tick Borne Pathogens at the Host Vector Interface: A Global Perspective* (ed. Coons, L. & Rothschild, M.), pp. 2–4. Krugel National Park, South Africa: Proceedings and Abstracts. S.I.: The conference, 1998.

Samish, M., Ginsberg, H. and Glazer, I. (2004). Biological control of ticks. *Parasitology*, 129(S1): S389:S403.

Sangwan, A. K., Sangwan, N. and Goel, M. C. (2000). Progressive displacement of Hyalomma ticks by *Rhipicephalus microplus* in Haryana. *Journal of Parasitic Diseases*, 24(1): 95-96.

Sanyal, A. and De, K. (2005). Status of Ticks (ACARI: METASTIGMATA) of Rajasthan. *Records of Zoological Survey of India*, 104 (Part 3-4): 129-136.

Sanyal, A. K. and De, S. K. (1996). Ticks (Acrina: Metastigmata) fauna in the Thar desert. *Ghosh, AK, Baqri, QH*.

Schnitzerling, H. J., Nolan, J. and Hughes, S. (1983). Toxicology and metabolism of some synthetic pyrethroids in larvae of susceptible and resistant strains of the cattle tick *Rhipicephalus microplus* (Can.). *Pesticide Science*, 14(1): 64-72.

Shahardar R. A., Niphadkar, S. M., Narsapur, V. S., Gatne M. L. (1998). Ixodid ticks of cattle and buffaloes in coastal districts of Maharashtra state. *Indian Veterinary Journal*, 75: 503–506.

Shakya, M., Kumar, B., Nagar, G., de la Fuente, J. and Ghosh, S. (2014). Subolesin: A candidate vaccine antigen for the control of cattle tick infestations in Indian situation. *Vaccine*, 32(28): 3488-3494.

Shanmugam, J., Smirnova, S.E. and Chumakov, M. P. (1976). Presence of antibody to arboviruses of the Crimean Haemorrhagic Fever-Congo (CHF-Congo) group in human beings and domestic animals in India. *Indian Journal of Medical Research*, 64(10):1403-1413.

Sharma, N., Singh, V., Shyma, K. P., Solanki, V. and Gupta, J. P. (2018). Comparative resistance status of *Hyalomma anatolicum* and *Rhipicephalus (Boophilus) microplus* ticks against Synthetic Pyrethroids (deltamethrin and cypermethrin) from Banaskantha, Gujarat, India. *International Journal of Acarology*, 44(6): 268-275.

Sharma, R.S., Verghese, T., Gupta, R. S. and Chattopadhya, D. (1992). Epidemiology of Lyme disease in the Nilgiri hills—1992—a premilinary report. In *Proceedings of the 3rd Symposium of Vectors and Vector Borne Diseases*. Haffkine's Institute.

Shastri U. V., Deshpande M. S. and Deshpande P. D. (1983) Prevalence of hard ticks of cattle and buffaloes in Marathwada region (Maharashtra state) with notes on their habits. *Cheiron*, 12: 318–322.

Shaw, R. D. (1966). Culture of an organophosphorus-resistant strain of *Rhipicephalus microplus* (Can.) and an assessment of its resistance spectrum. *Bulletin of Entomological Research*, 56(3): 389-405.

Shyma, K. P., Gupta, J. P. and Singh, V. (2015). Breeding strategies for tick resistance in tropical cattle: a sustainable approach for tick control. *Journal of Parasitic Diseases*, 39(1): 1-6.

Shyma, K. P., Gupta, J. P., Ghosh, S., Patel, K. K. and Singh, V. (2014). Acaricidal effect of herbal extracts against cattle tick *Rhipicephalus (Boophilus) microplus* using in vitro studies. *Parasitology research*, 113(5): 1919-1926.

Singh, N. K. and Rath, S. S. (2014). Esterase mediated resistance against synthetic pyrethroids in field populations of *Rhipicephalus (Boophilus) microplus* (Acari: Ixodidae) in Punjab districts of India. *Veterinary Parasitology*, 204(3-4): 330-338.

Singh, N. K., Haque, M., Singh, H., Rath, S. S. and Ghosh, S. (2014). A comparative study on cypermethrin resistance in *Rhipicephalus (Boophilus) microplus* and *Hyalomma anatolicum* from Punjab (India). *Ticks and Tick-borne Diseases*, 5(2): 90-94.

Singh, N. K., Saini, S. P. S., Singh, H., Sharma, S. K. and Rath, S. S. (2017). *In vitro* assessment of the acaricidal activity of *Piper longum, Piper nigrum, and Zingiber officinale* extracts against *Hyalomma anatolicum* ticks. *Experimental and Applied Acarology*, 71(3): 303-317.

Singh, N. K., Singh, H., Prerna, M. and Rath, S. S. (2015). First report of ivermectin resistance in field populations of *Rhipicephalus (Boophilus) microplus* (Acari: Ixodidae) in Punjab districts of India. *Veterinary Parasitology*, 214(1-2): 192-194.

Singh, S. and Chhabra, M. B. (1999). A note on ticks of camels in Haryana (India). *Journal of Camel Practice and Research*, 6(1): 77-78.

Sinha, S. R. P., Sinha, S., Kumar, P., Jayachandran, C. and Mandal, K. G. (2010). Therapeutic efficacy of deltamethrin and piperonyl butoxide (Butox plus) against cattle tick *Rhipicephalus microplus* and brown dog tick *Rhipicephalus sanguineus*. *Journal of Veterinary Parasitology*, 24(1): 75-77.

Smith, C. N. and Cole, M. M. (1943). Studies of parasites of the American dog tick. *Journal of Economic Entomology*, 36: 469–472.

Soderlund, D. M., Clark, J. M., Sheets, L. P., Mullin, L. S., Piccirillo, V. J., Sargent, D., Stevens, J. T. and Weiner, M. L. (2002). Mechanisms of pyrethroid neurotoxicity: implications for cumulative risk assessment. *Toxicology*, 171(1): 3-59.

Sonenshine, D. E. (1991). Tick life cycles. *Biology of Ticks*, 1: 51-66.

Spickett, A. M. (1998). Acaricides and resistance. *Veterinary Ectoparasitology and Protozoology*, 1: 1-13.

Stone, B. F. and Haydock, K. P. (1962). A method for measuring the acaricide-susceptibility of the cattle tick *Rhipicephalus microplus* (Can.). *Bulletin of entomological research*, 53(3): 563-578.

Stone, N. E., Olafson, P. U., Davey, R. B., Buckmeier, G., Bodine, D., Sidak-Loftis, L. C., Giles, J. R., Duhaime, R., Miller, R. J., Mosqueda, J. and Scoles, G. A. (2014). Multiple mutations in the para-sodium channel gene are associated with pyrethroid resistance in *Rhipicephalus microplus* from the United States and Mexico. *Parasites & vectors*, 7(1): 456.

Stutzer, C., van Zyl, W. A., Olivier, N. A., Richards, S. and Maritz-Olivier, C. (2013). Gene expression profiling of adult female tissues in feeding *Rhipicephalus microplus* cattle ticks. *International journal for parasitology*, 43(7): 541-554.

Suliman, E.A. and Mohammed, Y. O. (2012). The activity of *Aspergillus terreus* as entomopathogenic fungi on different stages of *Hyalomma anatolicum anatolicum* under experimental conditions. *Journal of Entomology*, 9(6): 343-351.

Sungirai, M., Baron, S., Moyo, D. Z., De Clercq, P., Maritz-Olivier, C. and Madder, M. (2018). Genotyping acaricide resistance profiles of *Rhipicephalus microplus* tick populations from communal land areas of Zimbabwe. *Ticks and Tick-borne Diseases*, 9(1): 2-9.

Surbhi, G. S. and Gupta, S. K. (2018). Acaricide resistance mechanisms and monitoring tools available for *Rhipicephalus (Boophilus) microplus*. *Pharma Innovation Journal*, 7(7): 398-405.

Symondson, W. O. C., Sunderland, K. D. and Greenstone, M. H. (2002). Can generalist predators be effective biocontrol agents? *Annual Review of Entomology,* 47: 561–594.

Takasu, K., Takano, S. I., Sasaki, M., Yagi, S. and Nakamura, S. (2003). Host recognition by the tick parasitoid *Ixodiphagus hookeri* (Hymenoptera: Encyrtidae). *Environmental Entomology*, 32: 614–617.

Taylor, S., Durden, L. A., Foley, E. H. and Reeves, W. K. (2016). *The bat tick Carios azteci (Acari: Argasidae) from Belize, with an endosymbiotic Coxiellaceae* (No. AFRL-SA-WP-JA-2015-0066). USAFSAM/FHT Wright-Patterson AFB United States.

Temeyer, K. B., Olafson, P. U., Brake, D. K., Tuckow, A. P., Li, A. Y. and de León, A. A. P. (2013). Acetylcholinesterase of *Rhipicephalus (Boophilus) microplus* and *Phlebotomus papatasi*: gene identification, expression, and biochemical properties of recombinant proteins. *Pesticide Biochemistry and Physiology*, 106(3): 118-123.

Temeyer, K. B., Pound, J. M., Miller, J. A., Chen, A. C., Pruett, J. H., Guerrero, F. D., Davey, R. B., Untalan, P. M., Lohmeyer, K. H., Li, A. Y. and Miller, R. J. (2004). Organophosphate resistance in Mexican strains of *Rhipicephalus microplus*: a major threat to the US cattle industry. *SAAS Bull. Biochem. Biotechnol*, 17: 43-51.

Temeyer, K. B., Pruett, J. H., Untalan, P. M. and Chen, A. C. (2006). Baculovirus expression of Bm AChE3, a cDNA encoding an acetylcholinesterase of *Rhipicephalus microplus* (Acari: Ixodidae). *Journal of Medical Entomology*, 43(4): 707-712.

Trimnell, A. R., Davies, G. M., Lissina, O., Hails, R. S. and Nuttall, P. A. (2005). A cross-reactive tick cement antigen is a candidate broad-spectrum tick vaccine. *Vaccine*, 23(34): 4329-4341.

Udayan, A., Kathiresan, S. and Arumugam, M. (2018). Kinetin and Gibberellic acid (GA3) act synergistically to produce high value polyunsaturated fatty acids in Nannochloropsisoceanica CASA CC201. *Algal Research*, 32: 182-192.

van Wyk, R. D., Baron, S. and Maritz-Olivier, C. (2016). An integrative approach to understanding pyrethroid resistance in *Rhipicephalus microplus* and R. decoloratus ticks. *Ticks and Tick-borne Diseases*, 7(4): 586-594.

Vanegas, A., Keller, C., Krüger, A., Manchang, T. K., Hagen, R. M., Frickmann, H., Veit, A., Achukwi, M. D., Krücken, J. and Poppert, S. (2018). Molecular detection of spotted fever group rickettsiae in ticks from Cameroon. *Ticks and Tick-borne Diseases*, 9(4): 1049-1056.

Vatsya, S. and Yadav, C. L. (2011). Evaluation of acaricide resistance mechanisms in field populations of *Rhipicephalus (Boophilus) microplus* collected from India. *International Journal of Acarology*, 37(5): 405-410.

Villarino, M. A., Waghela, S. D. and Wagner, G. G. (2003). Biochemical detection of esterases in the adult female integument of organophosphate-resistant *Rhipicephalus microplus* (Acari: Ixodidae). *Journal of Medical Entomology*, 40(1): 52-57.

Walker, A. R. (2011). Eradication and control of livestock ticks: biological, economic and social perspectives. *Parasitology*, 138(8): 945-959.

Wattal, L. and Srivastava, S. P. (1967). Ectoparasitic forms of small mammals and domestic animals in the neighourhood of Alwar city, Rajasthan. *Bulletine of Indian Society of Malaria and Communicable diseases*, 4: 191-192.

Watts, D. M., Ksiazek, T. G., Linthicum, K. J. and Hoogstraal, H. (1988). Crimean-Congo hemorrhagic fever. *The Arboviruses: Epidemiology and Ecology*, 2: 177-260.

Willadsen, P. (2006). Tick control: thoughts on a research agenda. *Veterinary parasitology*, 138(1-2): 161-168.

Willadsen, P. (2004). Anti-tick vaccines. *Parasitology*, 129(S1): S367.

Willadsen, P., Riding, G. A., McKenna, R. V., Kemp, D. H., Tellam, R. L., Nielsen, J. N., Lahnstein, J., Cobon, G. S. and Gough, J. M. (1989). Immunologic control of a parasitic arthropod. Identification of a protective antigen from *Rhipicephalus microplus*. *The Journal of Immunology*, 143(4): 1346-1351.

Work, T. H. and Trapido, H. (1957). Kyasanur Forest disease: a new infection of man and monkeys in tropical India by a virus of the Russian spring-summer complex. In *Proceedings IXth Pacific Science Congress,* Bangkok, 17: 80-84.

Yessinou, R. E., Akpo, Y., Sidick, A., Adoligbe, C., Karim, I. Y. A., Akogbeto, M. and Farougou, S. (2018). Evidence of multiple mechanisms of alphacypermethrin and deltamethrin resistance in ticks *Rhipicephalus microplus* in Benin, West Africa. *Ticks and Tick-borne Diseases*, 9(3): 665-671.

Zhioua, E., Heyer, K., Browning, M., Ginsberg, H. S. and LeBrun, R. A. (1999). Pathogenicity of *Bacillus thuringiensis* variety kurstaki to *Ixodes scapularis* (Acari: Ixodidae). *Journal of Medical Entomology*, 36(6): 900-902.

In: The Entomological Guide to *Rhipicephalus*
Editors: Sachin Kumar, Raquel Cossio Bayugar et al.
ISBN: 978-1-53619-619-1

Chapter 13

ETHNOVETERINARY PLANTS AND PRACTICES TO CONTROL TICKS WITH SPECIAL REFERENCE TO *RHIPICEPHALUS* IN INDIA

K. P. Shyma[1] and Veer Singh[2,*]
[1]Department of Veterinary Parasitology, College of Veterinary Science & Animal Husbandry, Sardarkrushinagar Dantiwada Agricultural University, Sardarkrushinagar, Gujarat, India
[2]Bihar Veterinary College Campus, Patna, India

ABSTRACT

Ticks and tick-borne diseases (TTBDs) of cattle pose a serious threat to the growth of thedairy industry and cause a significant reduction in profit by severe loss in lactation. The problem of TTBDs has been ranked high in terms of their adverse impact on the livelihood of poor farming communities in developing countries including India. The tick, *Rhipicephalus (Boophilus) microplus* is the most widely distributed ticks and serves as the vector of bacterial and protozoan diseases causing heavy economic losses directly and indirectly in animal husbandry. At present, tick control in India is based on large scale repeated use of synthetic acaricides belonging to three major classes, namely organophosphates, carbamates, and pyrethroids. Besides various indirect effects like environmental pollution due tothe large-scale use of synthetic acaricides, there is an emerging problem of thedevelopment of resistance in the ticks to these acaricides. Emphasis should be made on a reliable compound against the ticks which are effective against all the stages of the ticks and does not leave any residues in the environment. Compounds of plant origin constitute an important segment of such effort and have been used in therecent past against all the stages of economically important tick species with encouraging results. This chapter will focus on the different herbs and their active principles which are used to control ticks in India. Several medicinal plants have been screened for acaricidal activity against the female tick, *Rhipicephalus microplus* throughout the world. There are different herbs like *Allium sativum, Carica papaya, Datura stramonium, Azadirachta indica, Calotropis procera, Annona squamosa*,

* Corresponding Author's Email: veersinghgau@gmail.com.

Cymbopogon winterianus and many more are found to have anti-tick activity against *R. microplus.* The established plants extracts with acaricidal activity will be eco-friendly and will be beneficial for controlling chemical-resistant tick infestation in animals.

Keywords: environment, herbs, phyto-acaricide, resistance, *Rhipicephalus microplus*

Introduction

Ayurvedic medicines particularly, ethno-veterinary practices are now anintegral part of veterinary medicine and its use is not limited only for human diseases. Ethnoveterinary medicines are not restricted to thetreatment of animal diseases alone, but involve a myriad of disciplines and all aspects of people's knowledge and practices in animal healthcare, productivity and performances including diagnostic and etiological understanding, therapeutic, prophylactic, and management skills (Lans et al., 2007). Animals are anessential component of Indian culture and many of the rural culture they are given very high regard considered as sacred (Raikwar et al., 2015). Attention in medicinal botanicals owing to their efficiency, lack of their counterpart, low cost, and integration with human culture (Tabuti et al., 2003) are been given from time immemorial in India. Use of folk medicines and medicines based on traditional knowledge has been recently getting the impetus in tropical countries, as many such chemical drugs created havoc and the tropical flora were suitable for getting ethno-botanicals (Fokunang et al., 2008). The unsystematic use of veterinary antimicrobials to treat livestock has possibly created resistance in animal pathogens and therefore alternatives to treat veterinary infections are in need today.

Genesis of Ethnoveterinary Medicines

The recognition of ethnoveterinary medicines in scientific research is not older than 30 years, though these traditional practices are thousands of years old (Lans et al., 2007). Knowledge of behavioral, anatomical and physiological aspects of animals like cattle and sheep, which were used for divination and appeasement of gods in ancient cultures was crucial from thereligious point of view resulted in theacquisition of rudimentary skills of healing art. The traditional knowledge in the form of selective breeds and breeding, animal feeds and feeding, animal behavior, ritualism, herbalism, spiritualism and ethno-epidemiology of livestock diseases originated in different ancient civilizations to raise superior livestock and keep them healthy for better production and performance. Generally, the traditional knowledge was transmitted verbally; however codified system also existed in the form of written scripts such as books, papyrus, stele, clay tablets etc.

The approaches of treatment and control of diseases changed with the changing concepts in different civilizations (Thrusfield et al., 2005). The development of ethnoveterinary knowledge has been guided by the changing beliefs about the causes of diseases over the period ofantiquity. The ancient Indian healers used minerals, animal products and earth, besides magico-religious practices to cure diseases. Therefore, the uses of medicinal herbs and dietary supplements or dietary changes remained important inveterinary ayurvedic practices.

CURRENT STATUS OF VETERINARY AYURVEDA

Ayurveda remained as the backbone of medical science in ancient times, became an alternative in the modern era. The role of ethnoveterinary medicine in livestock development is presently realized beyond dispute especially in developing countries. This is also considered as apotential tool to overcome the side-effects of modern drugs and organic farming. Previously ethnoveterinary medicine and other ethno-knowledge systems were looked upon with disbelief and skepticism by modern researchers and practitioners. Improper scientific documentation and validation, concealment, lack of dissemination of traditional knowledge, non-availability of raw materials, declining herbal resource slow therapeutic response in acute conditions and lack of interest in theyounger generation have slowed down the popularity of ethnoveterinary practice.

TICK AND TICK BORNE DISEASES (TTBDS) AND ITS SIGNIFICANCE IN LIVESTOCK PRODUCTION

Ticks and tick-borne diseases (TTBD) of cattle pose serious threats to the progress of thedairy industry (FAO, 1996) and cause a substantial reduction in profit by severe loss in lactation (McLeod and Kristjanson, 1999). Moreover, the direct effect of tick infestation has atremendous effect on the availability of good quality hides for the leather industry which is suffering from ahuge shortfall of 3,000 million pieces of hides and skin per year (Leatherware 2002). Ticks which are highly specialized obligate haematophagousectoparasites are extensively distributed around the world (Magnarelli, 2009) but they have a tendency to flourish more in countries with warm, humid climates, for the reason that they require a certain amount of moisture in the air to undergometamorphosis and low temperatures hinder their development from eggs to larvae (Nuttal, 1905). Even though, India accounts for a significant stake of theworld's livestock resources, livestock production is greatly affected by ticks and tick-borne diseases (TTBDs). The problem of TTBD has been ranked high in terms of their adverse impact on the livelihood of poor farming communities in developing countries including India. Ticks of domestic animalscause considerable harm to livestock by transmission of many species of pathogens, as well as causing anaemia, damaging wool and hides.Some of the species which exert their most devastating effect are prevalent in tropical countries. *Rhipicephalus (Boophilus) microplus* is the most widely distributed ticks and serves as the major vector of economically important pathogens like *Babesia bigemina* and *Anaplasma marginale* (Ghosh et al. 2007a,b). The other species of tick-borne pathogens infecting Indian farm animals include *Theileria annulata*, *T. buffeli*, *T. lestocardi* (hirci), *Babesia motasi*. *Anaplasma marginale*, *Ehrlichia bovis* and *E. phagocytophila.* The incidence of *T. annulata* and *B. bigemina* is reported from different parts of the country however the prevalence of *A. marginale*, *E. bovis* and *E. phagocytophila* is limited to certain zones. Despite, cattle and buffaloes co-exist in nature, the incidence of tick borne diseases in buffaloes is comparatively lesser than in cattle. Devendra (1995) estimated a potential loss due to TTBDs is over US$ 800 million per annum while a recent estimate of 498.7 million US$ per annum has been calculated as the cost of TTBDs in cattle in India (Minjauw and Mc

Leod 2003). The epidemiology of tick infestations, and their vectorial potential varies with the varying environmental conditions due to global warming.

Ticks are next to mosquitoes in the transmission of a variety of disease pathogens of public health and veterinary importance since time immemorial.The studies on the prevalence of ticks, their vector potentiality received researcher's curiosity and public interest only after the discovery of Kyasanur Forest disease (KFD) transmitted by *Haemaphysalis* spp. in Shimoga district of Karnataka state. The Animal Husbandry Department undertook stringently measures to control ticks, after the report of incidence of Crimean-Congo haemorrhagic fever (CCHF) virus infection in human and domestic animals from Gujarat in 2011 (Mourya et al., 2012) and some other parts of the country (Yadav et al., 2016). The control of tick infestation is considered a major factor for sustainable livestock production worldwide and especially in tropical country like India.

CURRENT PRACTICES FOR TICK CONTROL

Tick control in India is based on large-scale repeated use of synthetic acaricides mostly of three major classes, namely synthetic pyrethroids, organophosphates, formamidines, phenylpyrazoles, carbamates and macrocyclic lactones. Large-scale use of synthetic acaricides causes serious environmental pollution, besides there is an emerging problem of development of resistance in the ticks to these acaricides. The development of acaricidal resistance in ticks arrests the progress of tick control programs in a country (Coroson et al., 2001). In the Indian subcontinent, the incidence of acaricide resistance has been widely reported in one host ticks of the genus *Rhipicephalus* (*Boophilus)* (Singh et al., 2012; Kumar et al., 2013; Shyma et al., 2013) and in multihost tick *Hyalomma anatolicum* (Shyma et al., 2012). This occurs when larger fraction of the total population of such species remains under chemical challenge. The loss of acaricidal effectiveness as a sequel to the development of resistance entails increased application frequencies and dosage of the acaricides, and finally, adoption of more expensive replacement compound as new acaricides becomes increasingly more cost intensive and difficult to discover (NRC, 1986). Emphasis should be made on a reliable compound against the ticks which are effective against all the stages of the ticks and do not leave any residues in the environment as per therequirement of theWorld Trade Organization.

ETHNOVETERINARY MEDICINES IN MANAGEMENT OF TICKS

Currently, special emphasis has been given to the development of plant based ectoparasiticidals. Compounds of plant origin comprise a chief segment of alternate control strategies and have been used inthe recent past with promising results against all the stages (adult, nymph, and larva) of economically important tick species (Mehlhorn et al., 2005; Mehlhorn, 2008; Kamaraj et al., 2010; Ghosh et al., 2011; Semmler et al., 2011; Singh et al., 2013). The active ingredients of plants (Phytochemicals) probably act by inhibiting egg development, defying growth regulating hormones, interfering in reproduction and mating, and checking the chitin formation in the insects (Katoch et al., 2007; Chagas et al., 2012).

Plant-derived antiparasitic drugs like Quinine, the oldest antimalarial drug which is obtained from the South American plant, *Cinchona officinalis* L., and artemisinin from *Artemisia annua*L. (Roth and Acton, 1987) has already made major contributions to human and animal health. The dried flower heads of *Chrysanthemum cinerariafolium* (Trev.) Vis and *Chrysanthemum coccineum* derived from pyrethrum has been used for centuries for the treatment of the problems of insects and lice in the Middle East (Casida, 1980). Pyrethrum provided the backbone for the synthesis of more potent synthetic pyrethroids which is the most commonly used, nonsynthetic insecticide allowed in certified organic agriculture (Pottorff, 2010) owing to its lower mammalian toxicity and unusually fast biodegradability.

More meticulous evaluation of plants for their acaricidal activities was started by Khaidarov (1971), who evaluated 84 plant species out of which, 21 had *in vitro* acaricidal activity against larvae and adults of *Rhipicephalus bursa*, *Hyalomma anatolicum* and *Hyalomma marginatum.* Inthe recent past, both repellent as well as acaricidal properties of many of the plant products, their crude extracts and essential oils have been tested for their action against all the stages (adult, nymph, larva and egg) of many tick species with encouraging results (Chungsamarnyart et al., 1988, 1990, 1991; Kaaya, 2000; Copping and Menn, 2000; Flamini, 2003; Mehlhorn et al., 2005; Kamaraj et al., 2010; Zorloni et al., 2010; Nerio et al., 2010; Ghosh et al., 2011; Zoubiri and Baaliouamer, 2011; Maia and Moore, 2011; Borges et al., 2011; Monteiro et al., 2012; Singh et al., 2014, Andreotti et al., 2014; Shyma et al., 2014).

The extracts of *Annona squamosa* and *Azadirachta indica* seeds showed a high level of mortality (70.8%) in adult *R. (B.) microplus* after 24 h of treatment and also assessed the significant reduction in oviposition of the survived ticks (Magadum et al., 2009, Srivastava et al., 2008). The extract prepared from therhizome of *Acorus calamus* proved highly efficacious and 100% final mortality within 14 day post treatment when used against laboratory-reared *R. (B.) microplus* adults using the adult immersion method (Ghosh et al., 2011). Ghosh et al. (2011) further reported the LC_{85} value of the extract as 11.26. *In vivo* experiments confirmed the efficacy of the extract up to 42%, and suggested for a repeat application seven days post treatment. On animals, even after the application of five times of the concentration used for *in vivo* studies i.e., with 50% of the concentration, the extract was found to be safe with no adverse reaction. Similar studies were also carried by Chungsamarnyart et al. (1988) using different formulation of *A. calamus* extracts and reported a wide variation of 8-99% efficacy against tropical cattle ticks. *Leucasaspera*, which grows as a weed on waste lands and roadsides all over India, has been reported to have potential to kill the adult ticks as well as produces significant inhibition of fecundity at the highest concentration (100mg/ml) (Ravindran et al., 2011). Using adult immersion test, the acaricidal activity of ethanolic extracts of leaves of *Cassia alata* was studied by Ravindran et al. (2012) against *R. (B.) annulatus* adults and found 45.8% mortality and 10.9% inhibition of fecundity at the highest concentration tested (100 mg/ml).

Ghosh et al. (2013) demonstrated the castor bean (*Ricinus communis*) leaf extracts were toxic to organophosphate and pyrethroids resistant *R. (B.) microplus* adults. The study reported mortality of ticks in the range of 55.0 ± 9.6–95.0 ± 5.0 when used in the range of concentrations, i.e., 6–10%. The highest percentage of ticks died when treated at 10% concentration and the survived ticks laid fewer eggs. Preliminary phytochemical test showed presence of alkaloids, flavonoids, steroids and saponins in the extract.

Shyma et al. (2014) reported significant acaricidal activities of *Datura stramonium* leaf extracts against *R. (B.) microplus.* Similar findings were reported by Ghosh et al. (2015). In their venture to explore the plants for its acaricidal activity, reported 95% ethanolic extracts of *Datura metel* fruits and *Argemone mexicana* whole plant were found effective showing more than 50% mortality of cattle tick, *Rhipicephalus (Boophilus) microplus.* Phytochemical studies indicated the presence of alkaloids and glucosides in *D. metel* fruits and alkaloids, terpenoids, flavonoids and phenolics in *A. mexicana* whole plant extracts.

Kumar et al. (2015) evaluated the acaricidal potential of *Ageratum conyzoides* plants against acaricides-resistant *R. (B.) microplus* ticks infesting cattle and buffaloes. The ethanolic extract (leaves and stems) of this plant found efficacious against 76.7–90% acaricides-resistant field ticks and adversely affected oviposition showing 7.04–31.3% reduction in egg laying capacity. *In vitro* efficacy of up to 52.5 and 76.7% could also be achieved with this extract against reference resistant IVRI-4 and 5 lines. The profiling of hexane extract, two bioactive sub-fractions and essential oils revealed the presence of 6,7-dimethoxy2,2-dimethyl-2H-1-benzopyran (precocene II) as a major phyto-compound. 77.1–94.9% against multi-acaricide resistant larvae of IVRI-5 line and 96.2–97.5% efficacy against larvae of IVRI-1 of *R. (B.) microplus* could be recorded with the bioactive sub-fractions. Ghosh et al. (2015) evaluated the acaricidal property of fruits of *Semecarpus anacardium* and leaves of *Datura stramonium* and observed 50% and 20% mortality, respectively within 72 h of treatment by adult immersion test. The extract of *S. anacardium* was also found efficacious (73.3% ± 3.3%) against the multi-acaricide-resistant IVRI-V line of *R. (B.) microplus*. It significantly affected the reproductive physiology of treated ticks by inhibiting the oviposition and was found safe. The phytochemical profiling revealed the presence of pyrocatechol as a marker compound in this extract. Larval immersion test on 14–21 days old unfed larvae was conducted with an alcoholic and various aqueous extracts of *Piper longum*, *Piper nigrum* and *Zingiber officinale* as well as their combinations to evaluate the acaricidal activity against the three-host ixodid tick, *Hyalomma anatolicum* (Singh et al., 2017). The highest acaricidal activity was noted with alcoholic extract of *P. longum* seeds having the minimum LC_{50} and LC_{95} (95% CL) values of 0.071% (0.07–0.072) and 0.135% (0.13–0.14), respectively, followed by alcoholic combinations. There was no acaricidal activity in the extracts prepared from the rhizome of *Z. officinale*. The results indicated that the ethanolic extracts of *P. longum* and *P. nigrum* and combinations thereof, can be used effectively under integrated tick control programme.

Jyoti et al. (2019) evaluated the *in vitro* acaricidal activity of essential oils (EOs) of cloves (*Syzygium aromaticum*), bark of cinnamon (*Cinnamomum zeylanicum*) and leaves of lemon grass (*Cymbopogon citratus*) against unfed larvae of *R.* (*B.*) *microplus* by larval packet test. The cinnamon EO and cinnamon EO+lemon grass EO combination registered highest acaricidal activity against *R.* (*B.*) *microplus* larvae. The combination of all three EOs (1:1:1) showed high synergism (CI value of 0.376) and favorable dose reduction against *R.* (*B.*) *microplus* larvae. Udayan et al. (2020) described the acaricidal property of crude ethanolic extract of *Artemisia nilagirica* leaves and its fractions against *Rhipicephalus (Boophilus) annulatus*. Phytochemical characterization of ethanolic extract of *A. nilagirica* have showed the presence of terpenoids, flavonoids, steroids, saponins, fixed oils and fats, tannins, and glycosides. Hexane fraction of the plant is much more effective than the crude fraction, since it possesses maximum acaricidal effect even at lower concentration of 2.5%. At 10% level, it was more effective (33.37 ± 1.71%) than deltamethrin at its standard dose.

Currently, nanobiotechnology represents a novel approach in livestock science, with special reference to the treatment of parasitic infections and arthropod vectors (Irache et al., 2011; Rai and Ingle, 2012; Benelli and Lukehart, 2017). The silver, gold, zinc oxide, selenium oxide, and chitosan nanoparticles have high efficacy of inhibition against vectors of parasitic infections (Elmi et al., 2013; Sujitha et al., 2015). Zinc oxide nanoparticles synthesized with the leaf extract of *Lobelia leschenaultiana* (Ll-ZnO NPs) were effective against *R. (B.) microplus*. *R. (B.) microplus* exposed to 40, 40, 8 mg/L of *L. leschenaultiana* leaf extract, zinc acetate and Ll-ZnO NPs showed 93.33, 70, 100% mortality with LC50 of 5, 19.2, and 1.7 mg/L, respectively (Banumathi et al., 2016). Marimuthu et al. (2011) evaluated the antiparasitic activity efficacy of silver nanoparticles synthesized using the aqueous leaf extract *Mimosa pudica* against the larvae of *R. (B.) microplus*. The highest percentage mortality, i.e., 89%, was reported at 20 mg/L concentration and the LC50 was 8.98 mg/L. The antiparasitic activity of aqueous leaf extracts of *Manilkara zapota* and *M. zapota*-synthesized silver nanoparticles on *R. (B.) microplus* have also been reported.

As an important constituent of integrated tick management system, continuous attempts are been made to identify herbs having anti-tick activity as they are easily degradable and safer than synthetic acaricides. Nevertheless, chemical analyses of only a few of the identified plants were attempted to identify the principal components in these botanicals which are active against ticks. Researches have to be focused to develop a promising herbal formulations for the sustained control of tick populations.

LIMITATIONS AND STRENGTH OF ETHNOVETERINARY MEDICINES

In the modern era, many professionals question the usefulness, applicability and adaptability of ethnoveterinary medicines. These medicines are cheap, easily available, selective, biodegradable and cause low mammalian toxicity (George et al, 2010; Ghosh et al., 2012) and are effective even against resistant tick strains as there is usually a mixture of different active agents with different mechanisms of action (Chagas et al., 2003; Olivo et al., 2009).Some practices are harmful and also not readily available. Certainly, many of these opinions are true to some extent. Ethno-veterinary medicines have both limitations and strengths. These medicines relatively slow to act and may not be as piquant as the medicines having chemical origin, making them less suitable for controlling and treating rapidly spreading contagious diseases (e.g., foot-and-mouth disease, rinderpest, haemorrhagic septicaemia, anthrax, black quarter, rabies etc.) and emerging infectious diseases. Some of the remedies of ethnoveterinary medicine are difficult to prepare and administer. Some herbal preparations are safe for some animal species may be toxic to others like garlic is recommended in human to reduce the blood cholesterol but cause anaemia in dogs. The plant materials if improperly collected and processed may be injurious to health. Lack of documentation, validation and failure to disseminate and promote, poorly researched and understood, adversely affect development and utilization of ethnoveterinary medicine by the end users. There is rapid decline in experienced traditional healers and pastoralist communities and depleting plant resources limits the trend of ethnoveterinary medicines.

A lot of chemical differences have been reported between plants from different varieties, area to area, season, type of extraction etc. The onset of action of herbal acaricides is slow or these compounds having variable efficacy (Kiss et al., 2012). Another difficulty is in marketing due to lack of access to the latest technological, market information, trained personnel and equipment and constraints with clinical trials due to regulatory issues (Khare et al., 2019). Standardization and quality control of herbal drugs are much complex because it is based on the evaluation of identity, purity and quality of the active herbal ingredients (Patwekar et al., 2016). The major challenges in the development of these drugs are quality assurance of herbal drugs including safety, efficacy, toxicity and their interactions (Shulammithi et al., 2016). The correct identity of the cured herbal material is of prime importance in establishing the quality control of herbal drugs. To assess the identity and purity based on criteria such as type of preparation, sensory properties, physical constants, adulteration, contaminants, moisture, ash content and solvent residues are to be checked (Dhiman et al., 2019). Some of herbal drugs interact with other compounds leading to the harmful effects on the host body (Khare et al., 2019). Since, all herbal medicines are complex mixtures of one or more active ingredient, multitude of active ingredients will increase the possibilities of interaction between herbal medicines and conventional drugs. Example-piperonylbutoxide (PBO) also act as synergistic which potentiate the toxicity of several natural pyrethrum and synthetic pyrethroids compounds (Fazolin et al., 2016).

The prime limitations are like indiscriminate pre and poor post harvesting practices of herbal drugs. There are a lot of chemical variabilities reported between essential oils from different varieties, area to area, season, type of extraction etc. For the identification of individual bioactive ingredients: The technical grading parameters should be fixed by different manufacturers for neem, terpenes (thymol), eugenol etc. (George et al., 2009). These compound having short residual activity. Therefore, it requires a dose of the same crude drug even after a certain period to maintain the minimal inhibitory concentration (MIC) of those particular drugs (Isman, 2006; George et al., 2010).

However, most research in the development of natural products for pest control have usually ended in the laboratory However, veterinary traditional healthcare and management practices are increasingly accepted as a viable solution to compact animal health problems. Farmers are more comfortable to receive healthcare from known and trusted people who speaks their native language and better understanding of the local conditions. Most traditional practices are easy to adopt compared to that of modern technologies which may not be compatible with the existing situation of the farmers. Ethnoveterinary medicine research and development have practical applications to ensure cost-effective ways to control economically important health problems. It may be an effective resource for thecommunity development and to protect the right of ethnovets and owners of traditional knowledge at community level. Ethnoveterinary medicine supports emerging agribusiness opportunities like organic animal husbandry and herbal farming. It acts as a bridge to connect natural resources and their human management for the future.

CONCLUSION

Control of ticks and mites that adversely affect milk and meat production is a daunting challenge. Chemical acaricides such as organophosphates, synthetic pyrethroids, amitraz, and ivermectin are commonly used for tick control. The involvement of high cost, environmental hazards, and development of acaricidal resistance are some of the drawbacks of these chemical acaricides. However, the inherent problems of chemical control such as the development of resistance and thepossibility of acaricide contamination in the environment necessitate the need for strategic control of tick vectors. Such future strategies may encompass thedevelopment of the vaccine, development of a newer generation of chemical acaricides, herbal acaricides and transgenic animals. Success in vaccine development and production of transgenic animals was limited. Plant-based formulations are one of the promising approaches for the control of ticks. One advantage of using herbal acaricide is that there is either no development or very slow rate (if it is there) of development of resistance as there is a mixture of different active ingredients with different mechanisms of action. However, in order to harness these, herbal products should work to their full target potential. Taking in to account of the modes of action and metabolism, there is need to further reconnoiter these compounds While using herbal insecticide/acaricide, environmental safety is of primary importance. So that the drug does not causing any undesirable persistence in the environment, should be of negligible resistance and residue in the final product and free from any type of toxicity. With the help of proper standard technique, drugs should be assessed properly before coming into the market. There should be need of collaborative research by comparing the efficacy of different plants material in the same insect's species repeatedly. So, by this way, we have to know about the content of any plant material in respect of any challenges and constraints. Currently, there is a need of molecular farming for the production of biopharmaceuticals and edible vaccines in plants which is a future approach to combat these ectoparasitic infestations.

REFERENCES

Andreotti, R. and Barros, J. (2015). Diflubenzuron Effectiveness in Cattle Tick (*Rhipicephalus Boophilus microplus*) Control in Field Conditions. *Embrapa Gado de Corte-Artigoemperiódicoindexado (ALICE)*, 6(6): 1000373.

Banumathi, B., Malaikozhundan, B. and Vaseeharan, B. (2016). *In vitro* acaricidal activity of ethnoveterinary plants and green synthesis of zinc oxide nanoparticles against *Rhipicephalus (Boophilus) microplus*. *Veterinary Parasitology*, 216: 93-100.

Benelli, G. and Lukehart, C. M. (2017). Applications of green-synthesized nanoparticles in pharmacology, parasitology and entomology. *Journal of Cluster Science*, 28(1): 1-2.

Borges, L. M. F., Sousa, L. A. D. D. and Barbosa, C. D. S. (2011). Perspectives for the use of plant extracts to control the cattle tick *Rhipicephalus (Boophilus) microplus*. *Revista Brasileira de Parasitologia Veterinária*, 20(2): 89-96.

Casida, J. E. (1980). Pyrethrum flowers and pyrethroid insecticides. *Environmental Health Perspectives*, 34: 189-202.

Chagas, A. C. S., Domingues, L. F., Fantatto, R. R., Giglioti, R., Oliveira, M. C., Oliveira, D. H., Mano, R. A. and Jacob, R. G. (2014). *In vitro* and *in vivo* acaricide action of juvenoid analogs produced from the chemical modification of *Cymbopogon* spp. and *Corymbia citriodora* essential oil on the cattle tick *Rhipicephalus (Boophilus) microplus*. *Veterinary Parasitology*, 205(1-2): 277-284.

Chungsamarnyart, N., Jiwajinda, S. and Jansawan, W. (1990). Effects of plant crude-extracts on the cattle tick (*Boophilus microplus*) insecticidal action I. *Agriculture and Natural Resources*, 24(5): 28-31.

Chungsamarnyart, N., Jiwajinda, S., Jansawan, W., Kaewsuwan, U. and Buranasilpin, P. (1988). Effective Plant Crude-Extracts on the Tick (*Boophilus microplus*) I. Larvicidal Action. *Agriculture and Natural Resources*, 22(5): 37-41.

Chungsamarnyart, N., Jiwajinda, S., Ratanagreetakul, C. and Jansawan, W. (1991). Practical extraction of sugar apple seeds against tropical cattle ticks. *Agriculture and Natural Resources*, 25(5): 101-105.

Copping, L. G. and Menn, J. J. (2000). Biopesticides: a review of their action, applications and efficacy. *Pest Management Science: Formerly Pesticide Science*, 56(8): 651-676.

Corson, M. S., Teel, P. D. and Grant, W. E. (2001). Influence of acaricide resistance on cattle-fever tick (*Boophilus* spp.) infestations in semi-arid thorn shrublands: a simulation approach. *Experimental and Applied Acarology*, 25(2): 171-184.

deMonteiro, C. M., Maturano, R., Daemon, E., Catunda-Junior, F. E. A., Calmon, F., de Souza Senra, T., Faza, A. and de Carvalho, M. G. (2012). Acaricidal activity of eugenol on *Rhipicephalus microplus (Acari: Ixodidae)* and *Dermacentornitens* (Acari: Ixodidae) larvae. *Parasitology Research*, 111(3): 1295-1300.

de Souza Chagas, A. C., de Barros, L. D., Cotinguiba, F., Furlan, M., Giglioti, R., de Sena Oliveira, M. C. and Bizzo, H. R. (2012). *In vitro* efficacy of plant extracts and synthesized substances on *Rhipicephalus (Boophilus) microplus* (Acari: Ixodidae). *Parasitology Research*, 110(1): 295-303.

Devendra, C. (1995). Animal production systems in South East and East Asia: potential challenges for research. In: *Proc. Global agenda for livestock research*, ILRI, January 18–20, pp 41–48.

Dhiman, A. and Bhalla, D. (2019). Development and evaluation of lycopene loaded chitosan nanoparticles. *Current Nanomedicine (Formerly: Recent Patents on Nanomedicine)*, 9(1): 61-75.

Elmi, T., Gholami, S., Fakhar, M. and Azizi, F. (2013). A review on the use of nanoparticles in the treatment. *Journal of Mazandaran University of Medical Sciences*, 23(102): 126-133.

FAO. (1996). *Food requirement and population growth*. WFS96/TECH/10. Provisional version of technical paper for the world food summit, 60 pp.

Fazolin, M., Estrela, J. L. V., Medeiros, A. F. M., Silva, I. M. D., Gomes, L. P. and Silva, M. S. D. F. (2016). Synergistic potential of dillapiole-rich essential oil with synthetic pyrethroid insecticides against fall armyworm. *Ciência Rural*, 46(3): 382-388.

Flamini, G. (2006). Acaricides of natural origin. Part 2. Review of the literature (2002–2006). *Natural Product Communications*, 1(12): p.1934578X0600101216.

Fokunang, C. N., Ndikum, V., Tabi, O. Y., Jiofack, R. B., Ngameni, B., Guedje, N. M., Tembe-Fokunang, E. A., Tomkins, P., Barkwan, S., Kechia, F. and Asongalem, E. (2011). Traditional medicine: past, present and future research and development prospects and integration in the National Health System of Cameroon. *African Journal of Traditional, Complementary and Alternative Medicines*, 8(3): 284-295.

George, D. R., Olatunji, G., Guy, J. H., Sparagano, O. A. E. (2010). Effect of plant essential oils as acaricides against the poultry red mite, *Dermanyssus gallinae*, with special focus on exposure time. *Veterinary Parasitology*, 169: 222-225.

George, D. R., Sparagano, O. A. E., Port, G., Okello, E., Shiel, R. S. and Guy, J. H. (2009). Repellence of plant essential oils to *Dermanyssus gallinae* and toxicity to the non-target invertebrate Tenebriomolitor. *Veterinary Parasitology*, 162(1-2): 129-134.

Ghosh, S., Azhahianambi, P. and Yadav, M. P. (2007). Upcoming and future strategies of tick control: a review. *Journal of Vector Borne Diseases*, 44(2): 79-89.

Ghosh, S., Bansal, G. C., Gupta, S. C., Ray, D., Khan, M. Q., Irshad, H., Shahiduzzaman, M. D., Seitzer, U. and Ahmed, J. S. (2007). Status of tick distribution in Bangladesh, India and Pakistan. *Parasitology Research*, 101(2): 207-216.

Ghosh, S., Sharma, A. K., Kumar, S., Tiwari, S. S., Rastogi, S., Srivastava, S., Singh, M., Kumar, R., Paul, S., Ray, D. D. and Rawat, A. K. S. (2011). In vitro and in vivo efficacy of *Acorus calamus* extract against *Rhipicephalus (Boophilus) microplus*. *Parasitology research*, 108(2): 361-370.

Ghosh, S., Tiwari, S. S., Srivastava, S., Sharma, A. K., Kumar, S., Ray, D. D. and Rawat, A. K. S. (2013). Acaricidal properties of *Ricinus communis* leaf extracts against organophosphate and pyrethroids resistant *Rhipicephalus (Boophilus) microplus*. *Veterinary Parasitology*, 192(1-3): 259-267.

Ghosh, S., Tiwari, S. S., Srivastava, S., Sharma, A. K., Nagar, G., Kumar, K. A., Kumar, R. and Rawat, A. K. S. (2015). *In vitro*acaricidal properties of *Semecarpusanacardium* fruit and *Daturastramonium* leaf extracts against acaricide susceptible (IVRI-I line) and resistant (IVRI-V line) *Rhipicephalus (Boophilus) microplus*. *Research in Veterinary Science*, 101: 69-74.

Irache, J. M., Esparza, I., Gamazo, C., Agüeros, M. and Espuelas, S. (2011). Nanomedicine: novel approaches in human and veterinary therapeutics. *Veterinary Parasitology*, 180(1-2): 47-71.

Isman, M. B. (2006). Botanical insecticides, deterrents, and repellents in modern agriculture and an increasingly regulated world. *Annual Review Entomology*, 51: 45-66.

Jyoti, Singh, N. K., Singh, H., Mehta, N., Rath, S. S. (2019). *In vitro* assessment of synergistic combinations of essential oils against *Rhipicephalus (Boophilus) microplus Canestrini* (Acari: Ixodidae). *Experimental Parasitolology*, 201: 42-48.

Kaaya, G. P. and Hassan, S. (2000). Entomogenous fungi as promising biopesticides for tick control. *Experimental and Applied Acarology*, 24(12): 913-926.

Kamaraj, C., Rahuman, A. A., Bagavan, A., Elango, G., Rajakumar, G., Zahir, A. A., Marimuthu, S., Santhoshkumar, T. and Jayaseelan, C. (2010). Evaluation of medicinal plant extracts against blood-sucking parasites. *Parasitology Research*, 106(6): 1403-1412.

Katoch, R., Katoch, M., Yadav, A. and Srivastava, A. K. (2007). Formulation of herbal ectoparasiticidals. *Incompendium of 18th National Congress of Veterinary Parasitology*, pp. 7-9.

Khaidarov, K. M. (1971). Sensitivity of ixodid ticks to some species of plants. In *Conference proceedings Problems of Veterinary Sanitation,* 40: 341-343.

Khare, R. K., Das, G., Kumar, S., Bendigeri, S., Sachan, S., Saiyam, R., Banerjee, D. K. and Khare, D. S. (2019). Herbal insecticides and Acaricides: Challenges and constraints. *IJCS*, 7(4): 118-125.

Kiss, T., Cadar, D. and Spînu, M. (2012). Tick prevention at a crossroad: new and renewed solutions. *Veterinary Parasitology*, 187(3-4): 357-366.

Kumar, K. A., Tayade, A. B., Kumar, R., Gupta, S., Sharma, A. K., Nagar, G., Tewari, S. S., Kumar, B., Rawat, A. K. S., Srivastava, S. and Kumar, S. (2016). Chemo-profiling and bioassay of phytoextracts from *Ageratum conyzoides* for acaricidal properties against *Rhipicephalus (Boophilus) microplus* (Acari: Ixodidae) infesting cattle and buffaloes in India. *Ticks and Ttick-borne Diseases*, 7(2): 342-349.

Kumar, R., Nagar, G., Sharma, A. K., Kumar, S., Ray, D. D., Chaudhuri, P. and Ghosh, S. (2013). Survey of pyrethroids resistance in Indian isolates of *Rhipicephalus (Boophilus) microplus*: identification of C190A mutation in the domain II of the para-sodium channel gene. *Actatropica*, 125(2): 237-245.

Lans, C., Turner, N., Khan, T., Brauer, G. and Boepple, W. (2007). Ethnoveterinary medicines used for ruminants in British Columbia, Canada. *Journal of Ethnobiology and Ethnomedicine*, 3(1): 11.

Leatherware (2002). *Leather goods sector poised to take on new challenges*. 16(10): 10–12.

Magadum, S., Mondal, D. B. and Ghosh, S. (2009). Comparative efficacy of *Annona squamosa* and *Azadirachta indica* extracts against *Boophilus microplus* Izatnagar isolate. *Parasitology Research*, 105(4): 1085-1091.

Magnarelli, L. A. (2009). Global importance of ticks and associated infectious disease agents. *Clinical Microbiology Newsletter*, 31(5): 33-37.

Maia, M. F. and Moore, S. J. (2011). Plant-based insect repellents: a review of their efficacy, development and testing. *Malaria Journal*, 10(S1): S11.

Marimuthu, S., Rahuman, A.A., Rajakumar, G., Santhoshkumar, T., Kirthi, A. V., Jayaseelan, C., Bagavan, A., Zahir, A. A., Elango, G. and Kamaraj, C. (2011). Evaluation of green synthesized silver nanoparticles against parasites. *Parasitology Research*, 108(6): 1541-1549.

McLeod, R. and Kristjanson, P. (1999). *Economic impact of ticks and tick-borne diseases to livestock in Africa, Asia and Australia*. Report to the International Livestock Research Institute, Nairobi, Kenya.

Mehlhorn, H. (ed.) (2008). *Encyclopedia of Parasitology*, 3rd edn. Springer-Heidelberg, New York.

Mehlhorn, H., Schmahl, G. and Schmidt, J. (2005). Extract of the seeds of the plant *Vitexagnus castus* proven to be highly efficacious as a repellent against ticks, fleas, mosquitoes and biting flies. *Parasitology Research*, 95(5): 363-365.

Minjauw, B. and Mc Leod, A. (2003). Research report, DFID animal health programme. *Centre for Tropical Veterinary Medicine,* University of Edinburgh, UK.

Mourya, D. T., Yadav, P. D., Shete, A. M., Gurav, Y. K., Raut, C. G., Jadi, R. S., Pawar, S. D., Nichol, S. T. and Mishra, A. C. (2012). Detection, isolation and confirmation of Crimean-Congo hemorrhagic fever virus in human, ticks and animals in Ahmadabad, India, 2010–2011. *PLoS Neglected Tropical Diseases*, 6(5): p.e1653.

National Research C. (1986). *Pesticide resistance: strategies and tactics for management.* National Academic Press, Washington, DC, p 471.

Nerio, L. S., Olivero-Verbel, J. and Stashenko, E. (2010). Repellent activity of essential oils: a review. *Bioresource Ttechnology*, 101(1): 372-378.

Nuttall, G.H. (1905). Ticks and tick-transmitted diseases. *Transactions. Epidemiological Society of London*, 24:12.

Olivo, C. J., Heimerdinger, A., Ziech, M. F., Agnolin, C. A., Meinerz, G. R., Both, F. and Charão, P. S. (2009). Rope tobacco aqueous extract on the control of cattle ticks. *Ciência Rural*, 39(4): 1131-1135.

Patwekar, S. L. (2016). Nanobiocomposite: a new approach to drug delivery system. *Asian Journal of Pharmaceutics (AJP): Free full text articles from Asian J. Pharm*, *10*(04): S646-656.

Pottorff, L. P. (2010). Some Pesticides permitted in organic gardening. *Colorado State University/Denver County Extesion Master Gardener*, 888.

Rai, M. and Ingle, A. (2012). Role of nanotechnology in agriculture with special reference to management of insect pests. *Applied Microbiology and Biotechnology*, 94(2): 287-293.

Raikwar, A. and Maurya, P. (2015). Ethnoveterinary medicine: in present perspective. *International Journal of Agricultural Sciences and Veterinary Med*icine, 3(1): 44-49.

Ravindran, C., Varatharajan, G. R. and Karthikeyan, A. (2011). Role of alkaline-tolerant fungal cellulases in release of total antioxidants from agro-wastes under solid state fermentation. *Bio Resources*, 6(3): 3142-3154.

Ravindran, R., Juliet, S., Sunil, A. R., Kumar, K. G. A., Nair, S. N., Amithamol, K. K., Bandyopadhyay, A., Rawat, A. K. S. and Ghosh, S. (2012). Acaricidal activity of *Cassia alata* against *Rhipicephalus (Boophilus) annulatus*. *Experimental and Applied Acarology*, 56(1): 69-74.

Roth, R. J. and Acton, N. (1987). Isolation of arteannuic acid from *Artemisia annua. PlantaMedica*, 53(05): 501-502.

Semmler, M., Abdel-Ghaffar, F., Al-Rasheid, K. A. and Mehlhorn, H. (2011). Comparison of the tick repellent efficacy of chemical and biological products originating from Europe and the USA. *Parasitology research*, 108(4): 899-904.

Shulammithi, R., Sharanya, M., Tejaswini, R. and Kiranmai, M. (2016). Standardization and quality evaluation of herbal drugs. *Asian Journal of Pharmacuitics*, 11(5): 89-100.

Shyma, K. P., Gupta, J. P., Ghosh, S., Patel, K. K. and Singh, V. (2014). Acaricidal effect of herbal extracts against cattle tick *Rhipicephalus (Boophilus) microplus* using *in vitro* studies. *Parasitology Research*, 113(5): 1919-1926.

Shyma, K. P., Kumar, S., Sangwan, A. K., Sharma, A. K., Nagar, G., Ray, D. D. and Ghosh, S. (2013). Acaricide resistance status of *Rhipicephalus (Boophilus) microplus* and *Hyalomma anatolicum* collected from Haryana. *Indian Journal of Animal Sciences*, 83(6): 591-594.

Shyma, K. P., Kumar, S., Sharma, A. K., Ray, D. D. and Ghosh, S. (2012). Acaricide resistance status in Indian isolates of *Hyalomma anatolicum*. *Experimental and Applied Acarology*, 58(4): 471-481.

Singh, N. K., Gelot, I. S., Singh, V. and Rath, S. S. (2015). Detection of amitraz resistance in *Rhipicephalus (Boophilus) microplus* from North Gujarat, India. *Journal of Parasitic Diseases*, 39(1): 49-52.

Singh, N. K., Haque, M., Singh, H., Rath, S. S. and Ghosh, S. (2014). A comparative study on cypermethrin resistance in *Rhipicephalus (Boophilus) microplus* and *Hyalomma anatolicum* from Punjab (India). *Ticks and Tick-borne Diseases*, 5(2): 90-94.

Singh, N. K., Saini, S. P. S., Singh, H., Sharma, S. K. and Rath, S. S. (2017). *In vitro* assessment of the acaricidal activity of *Piper longum, Piper nigrum, and Zingiber officinale* extracts against *Hyalomma anatolicum* ticks. *Experimental and Applied Acarology*, 71(3): 303-317.

Srivastava, R., Ghosh, S., Mandal, D. B., Azhahianambi, P., Singhal, P. S., Pandey, N. N. and Swarup, D. (2008). Efficacy of *Azadirachta indica* extracts against *Boophilus microplus*. *Parasitology Research*, 104(1): 149-153.

Sujitha, V., Murugan, K., Paulpandi, M., Panneerselvam, C., Suresh, U., Roni, M., Nicoletti, M., Higuchi, A., Madhiyazhagan, P., Subramaniam, J. and Dinesh, D. (2015). Green-synthesized silver nanoparticles as a novel control tool against dengue virus (DEN-2) and its primary vector *Aedesaegypti*. *Parasitology Research*, 114(9): 3315-3325.

Tabuti, J. R., Lye, K. A. and Dhillion, S. S. (2003). Traditional herbal drugs of Bulamogi, Uganda: plants, use and administration. *Journal of Ethnopharmacology*, 88(1): 19-44.

Thrusfield, M. (2005). *Veterinary epidemiology*. 3rd ed. Blackwell Science. Ltd. London, UK. pp: 228-246.

Udayan, D., Nair, S. N., Juliet, S., Ravindran, R., Athalathil, S., Adarshkrishna, T. P., Ajithkumar, K. G., Sreelekha, K. P., Chandrashekar, L. and Ghosh, S. (2020). Acaricidal Activity of *Artemisia nilagirica* Leaves Against *Rhipicephalus (Boophilus) annulatus* Ticks. *Planta Medica*. doi: 10.1055/a-1161-0220.

Yadav, P. D., Patil, D. Y., Shete, A. M., Kokate, P., Goyal, P., Jadhav, S., Sinha, S., Zawar, D., Sharma, S. K., Kapil, A. and Sharma, D. K. (2016). Nosocomial infection of CCHF among health care workers in Rajasthan, India. *BMC Infectious Diseases*, 16(1): p.624.

Zorloni, A., Penzhorn, B. L. and Eloff, J. N. (2010). Extracts of *Calpurnia aurea* leaves from southern Ethiopia attract and immobilise or kill ticks. *Veterinary Parasitology*, 168(1-2): 160-164.

Zoubiri, S. and Baaliouamer, A. (2011). Chemical composition and insecticidal properties of some aromatic herbs essential oils from Algeria. *Food Chemistry*, 129(1): 179-182.

In: The Entomological Guide to *Rhipicephalus*
Editors: Sachin Kumar, Raquel Cossio Bayugar et al.
ISBN: 978-1-53619-619-1

Chapter 14

ALLOMONES IN RESISTANT HOSTS: FEATURES AND USE IN THE CONTROL OF *RHIPICEPHALUS* TICKS

Lígia Miranda Ferreira Borges*[*] *and Luiz Gustavo Gardinassi
Universidade Federal de Goiás, Instituto de Patologia Tropical e Saúde Pública,
Goiânia, Goiás, Brazil

ABSTRACT

It has been well known that some hosts are resistant to the cattle tick *Rhipicephalus microplus*, which occurs due to immunological, physical and mechanical attributes of the skin of resistant animals. On the other hand, the development of resistance in dogs to the tick *Rhipicephalus sanguineus* sensu lato (s.l.) was refuted until some years ago. However, evidences accumulated over recent years indicate that some breeds, such as beagles and probably miniature pinschers, develop resistance against *R. sanguineus*s.l. Ticks that feed on resistant hosts have impairment in their biological parameters, such as reduced numbers of ticks feeding to engorgement, reduced egg production and egg viability. Despite the fact that hosts resistant to ticks have been known since the beginning of the last century, it was recently understood that ticks can distinguish resistant from susceptible hosts using their sense organs. In this chapter we will briefly describe immunological features and address the behavioural, electrophysiological and chemical features that are involved in how *R. sanguineus*s.l. and *R. microplus* perceive resistant dogs and bovines, respectively. The use of repellent chemical compounds (allomones) identified on resistant animals to control *R. sanguineus*s.l. and *R. microplus* on susceptible animals will be discussed.

Keywords: Allomones, dogs, bovines, *Rhipicephalus microplus*, *Rhipicephalus sanguineus* sensu lato

[*]Corresponding Author's Email: ligia@ufg.br.

LIST OF ABBREVIATIONS

GC–MS: coupled high-resolution gas chromatography–mass spectrometry
GPCR: G-protein coupled receptor s.l.:sensulato
VOC: volatile organic compounds

INTRODUCTION

Rhipicephalus microplus and *Rhipicephalus sanguineus* sensu lato (s. l.) are two of the most important species of ticks in the world, parasitizing mainly bovines and dogs, respectively. *R. sanguineus*s.l. can also parasitize humans (Dantas-Torres et al., 2006; Loulyet al., 2006). Both species can cause direct injury to the animals because of blood spoliation. *R. microplus* is a vector of *Babesia bigemina*, *Babesia bovis* and *Anaplasma marginale* (de Castro, 1997), while *R. sanguineus*s.l. is a vector of several pathogens for dogs such as *Babesia vogeli,* and *Ehrlichia canis* (Otranto et al., 2012); and also transmits *Rickettsia conorii* and *Rickettsia rickettsia* to humans (Raoult and Parola, 2007; Parolaet al., 2009). Due to the prevalent acaricide resistance of both species (Abbas et al., 2014; Becker et al., 2019), finding new alternatives to control them is mandatory. Semiochemicals are organic compounds that carry chemical messages. Ticks, like other animals, communicate through chemical and physical stimuli. The study of semiochemicals and their biological roles can be a useful starting point for the development of alternative and specific ways of controlling ticks to reduce the use of acaricides and damage to the environment.

Semiochemicals can play a role within individuals of a species, as pheromones or between different species, such as allelochemicals, which include kairomones, allomones and synomones (Vilela, 2001). Pheromones are chemical substances released by individuals of a certain species that promote an excitation or stimulus, influencing individuals of the same species. As obligatory hematophagous, ticks need to use chemical clues produced by their hosts, the kairomones, to start the feeding process. When a host produces a compound that repels ticks and is therefore beneficial to the emitter, this compound is classified as an allomone. Ectoparasiterepellency in vertebrates due to the production of allomones can be categorized as avoidance of individuals that are closely but not taxonomically linked or those that fall within the same species (Pickett et al., 2010). Synomones are compounds that benefit both emitters and receptors, and to our knowledge, they have not been described in ticks.

The perception of semiochemicals occurs through chemosensilla present in Haller's organ, fourth palpal article and the digits of chelicerae. Haller's organ is located on the tarsus of the first pair of legs and is composed of an anterior depression that contains approximately 6-7 olfactory sensilla innervated by 2-9 bipolar sensory neurons and a posterior capsule with 4-13 sensilla, each innervated by 4-5 bipolar sensory neurons. The number of sensilla and receptors varies between different tick species and even between males and females (Waladde and Rice, 1982; Sonenshine, 1993; Sonenshine and Roe, 2014; Joseket al., 2018a). Unlike Haller's organ, the palpal sensory organ present in the fourth palpal article has a similar structure in all tick species (Leonovich, 2013). There is a grouping of 10 to 12 contact sensilla (gustatory) that receive double innervation (Foelix and Chu-Wang, 1972). The palpal organ can participate in the discrimination of the tick attachment sites, although their sensilla do not enter the feeding injury they can still validate or not the probe of the chelicera for penetration in the host (Waladde, 1977). In the internal digits of the chelicerae, there are one to two porosensilla (Waladde and Rice, 1978; Soareset al., 2013), and in *R. microplus*, there is the presence of only one neuron associated with one porosensillum and eleven neurons in the other porosensillum (Waladde and Rice, 1978). They are responsible for the perception of phagostimulants such as blood, plasma, ATP, reduced glutathione, salts and glucose, which stimulate the tick to perform the blood meal (Waladde and Rice, 1978; Soareset al., 2013) and deterrent compounds (Ferreira et al., 2015).

The walls of the olfactory sensilla contain numerous pores or grooves through which odour molecules pass. On the other hand, gustatory sensilla have a terminal pore through which the stimulus that is perceived by contact enters. Carr et al. (2017) suggested that olfaction in ticks is accomplished by a G-protein coupled receptor (GPCR) signal cascade and that they lack the odorant binding proteins described in insects. However, Renthal et al. (2017) noted the expression of some Microplusin-like proteins in tick forelegs and proposed that they might be odour binding proteins involved in tick olfaction. Josek et al. (2018b) followed Renthal et al.'s (2017) proposition and suggested that ionotropic receptors and perhaps gustatory receptors play a role in olfaction processing in ticks.

After the perception of the semiochemicals by the sensory organs, the information needs to be transmitted to the synganglion where sensory information will be integrated through the rest of the neuronal network. In *A. americanum* Borges et al. (2016) observed that the neuronal arborisations from Haller's organ were confined to the olfactory lobes and the first pedal ganglion. Sixteen to 22 olfactory glomeruli per olfactory lobe were counted in the females; however, the number of glomeruli was not counted in males because they were densely packed. The palpal ganglion in the synganglion received input from the sensory neurons associated with sensilla at the distal end of the palpal organ through the palpal nerve. Gustatory sensory neurons associated with palpalsensilla projected into a commissure with several bulges, which are confined in the palpal ganglion. Similar results related to Haller organ projection were observed for *A. sculptum* (Menezes, 2017). The number of glomeruli ranged from 26-30 in males and 24-29 in females. A positive correlation regarding the glomeruli volumes (μm^3) was observed both in individuals of the same sex and in intersexual comparisons.

Differences in host resistance of cattle have been well known for many years in *R. microplus* (Jonssonet al., 2014) and more recently in dogs to *R. sanguineus*s.l. (Inokumaet al., 1997; Jittapalaponget al., 2000; Loulyet al., 2009). Studies on host resistance in early investigations focused on the physical and mechanical attributes of the skin of indicine and taurine animals. Subsequent investigations focused on the acquired immune response to infestation, and more recent work has highlighted again the potential role of physical attributes of the skin, including the capacity for wound healing (Jonssonet al., 2014). Recently, we observed that breeds of dogs that are resistant to *R. sanguineus*s.l. produce allomones that repel the tick. In this chapter, we will introduce some aspects of immunological resistance of bovines and dogs to *R. microplus* and *R. sanguineus*s.l., respectively. We will also explore behavioural, electrophysiological and chemical features that are involved in how both species of ticks perceive resistant hosts. The possibility of the use of repellent chemical compounds identified to control *R. sanguineus*s.l. and *R. microplus* on susceptible animals will be discussed.

IMMUNOLOGICAL RESISTANCE TO *RHIPICEPHALUS* TICKS

The three-way interactions between ticks, hosts and pathogens induce the activation of innate and adaptive immunological pathways, which operate to maintain local (skin) and systemic host homeostasis. Immunological resistance to ticks promotes reduced blood-meal volume, reduced engorgement, decreased production and viability of ova, suppression of moulting and increased mortality of feeding ticks (Trager, 1939; Wikel, 1996). These outcomes result from complex arrays of events triggered by tick feeding. Hard ticks feed for days to weeks and thus maintain long-term interactions with components of host defense, which they must evade to complete their life cycle.

The success of infestations with ticks of the *R. microplus* species complex depends on genotypes of bovine hosts (Porto Netoet al., 2011). Resistant cattle *Bos taurus indicus* (indicine) co-evolved with *R. microplus*, whereas susceptible *Bos taurus taurus* (taurine) hosts may harbor debilitating infestations (Wamburaet al., 1998). Members of the *R. sanguineus*s.l. species complex preferentially feed on dogs. Interestingly, guinea pigs develop immunological resistance to consecutive infestations with *R. sanguineus*s.l. (Szabóet al., 1995), but only dogs of the beagle breed seem to be resistant to infestations by *R. sanguineus*s.l. (Inokumaet al., 1997; Loulyet al., 2010). In this section, we will briefly discuss biochemical and cellular mechanisms involved in immunological resistance to *R. microplus* and *R. sanguineu s* s.l.

INNATE IMMUNE RESISTANCE TO *R. MICROPLUS* AND *R. SANGUINEUS*S.L.

Ticks attach to their host´s skin and create a blood feeding pool to continue their life cycle. The skin promotes a physical barrier to ectoparasites, and some aspects have been associated with resistance to tick infestations, including thickness, skin pigmentation and density of the fur coat (Tabor et al., 2017). As larval stages attach to the host´s skin, they start saliva secretion, which mediates important functions, including extracellular matrix digestion and evasion of coagulation and complement cascades (Chmelaret al., 2012). Some of these immunological components have been associated with resistance to *R. microplus*, including increased levels of histamine and complement in resistant cattle (Wamburaet al., 1998; Zhao et al., 2013). Furthermore, acute-phase proteins are produced in response to tissue damage, and the levels of haptoglobulin, alpha-1 acid glycoprotein and serum amyloid A also differed between resistant and susceptible hosts infested by *R. microplus* (Carvalhoet al., 2008).

Innate immune cells, including macrophages and dendritic cells (DCs) that reside in the skin, induce an inflammatory response to the tissue damage caused by tick feeding (Kotálet al., 2015). Salivary gland extract of *R. microplus* reduces the activation of bovine macrophages stimulated with lipopolysaccharide, suppressing the production of inflammatory cytokines (Brake and Pérez de León, 2012). Furthermore, *R. sanguineus*s.l. saliva has been shown to modulate the differentiation, migration, and maturation of DCs (Cavassaniet al., 2005; Oliveira et al., 2011). These studies suggest critical roles of skin resident innate immune cells in the resistance to *Rhipicephalus* ticks, whereas saliva is important to inhibit/delay their function during tick feeding.

Activation of resident cells is important for recruitment of other leukocytes, such as granulocytes, which execute effector functions during tick infestations. Granulocytes are characterized by cytoplasmic granules such as cytotoxic molecules, enzymes, antimicrobial peptides, vasodilators, and histamine, among others. These cells include neutrophils, basophils, eosinophils and mast cells (Robbertseet al., 2017).

Resistant indicine bovines develop more intense cutaneous delayed type hypersensitivity reactions to *R. microplus* larvae extracts, while susceptible hosts are associated with more intense immediate hypersensitivity reactions (Marufuet al., 2013). These findings are interesting because of the nature of the hypersensitivity reaction, in which the immediate reaction is mediated by antibodies and granulocytes such as basophils and mast cells, while the delayed reaction is mediated by T lymphocytes previously exposed to the same antigens, indicating the development of robust cellular-mediated immunity in resistant hosts. Therefore, resistant and susceptible bovine hosts preferentially engage in different responses to counteract tick infestation, which are tightly associated with the success of tick feeding.

A recent study reported that the total cell count of cellular infiltrates did not differ between resistant and susceptible bovines but increased significantly as *R. microplus* ticks progressed from the larval to nymph stage (Franzinet al., 2017). This study also showed that skin from *R. microplus*-infested resistant bovines exhibited reduced numbers of eosinophils at the larval stage but increased numbers of eosinophils and basophils at the nymph stage (Franzinet al., 2017). Corroborating these data, the numbers of eosinophils and basophils were also increased in the skin of resistant cattle during infestation with adult *R. microplus* (Carvalhoet al., 2010), while basophil numbers negatively correlated with tick counts (Marufuet al., 2014). In previous studies, resistant bovine hosts were associated with higher mast cell degranulation (Schlegeret al., 1976). This suggested that histamine from mast cells stimulated self-grooming, which is associated with resistance to *R. microplus* (Bennett, 1969). However, histamine was shown to induce larval detachment, which correlated with higher eosinophil infiltration into the bite site (Kemp and Bourne, 1980). Both resistant and susceptible cattle present neutrophils at bite sites; however, their role during infestation to *R. microplus* remains controversial (Jonssonet al., 2014).

Consecutive infestations of mongrel dogs to *R. sanguineus*s.l. resulted in the accumulation of neutrophils in the skin, whereas the same exposure of guinea pigs resulted in the accumulation of mononuclear cells, eosinophils and basophils (Szabó and Bechara, 1999). Guinea pigs are not the preferred host of *R. sanguineus*s.l., but they have been used as a proxy of resistance to this tick (Ferreira et al., 2003). Blood leucocytes did not change in mongrel dogs during primary and secondary infestations with *R. sanguineus*s.l., while guinea pigs exhibited basophilia that increased further after the secondary infestation (Szabóet al., 2003). The findings that the beagle dog breed is resistant to infestations by *R. sanguineus*s.l. suggest the development of protective innate immune responses. Therefore, it remains to be elucidated whether Beagles and other dog breeds present with differential innate immune cell activation and distinct dynamics of granulocyte influx into sites of larvae attachment.

ADAPTIVE IMMUNE RESISTANCE TO *R. MICROPLUS* AND *R. SANGUINEUS*S.L.

Adaptive immune responses are coordinated by T and B lymphocytes and play crucial roles during the development of resistance to ticks. Consecutive infestations by *R. microplus* were associated with increased cellular accumulation of MHC class II-expressing cells, $CD25^+$ cells and T cells at the bite site in the skin in all bovine breeds, whereas resistant indicine cattle presented increased numbers of gamma-delta T cells throughout infestations (Constantinoiuet al., 2010). Another study demonstrated that resistant indicine cattle had increased percentages of $CD4^+$, $CD25^+$ and gamma-delta T lymphocytes and increased gene expression of the gene coding for interleukin-2 (Piper et al., 2009). Comparison between resistant and susceptible hosts infested with *R. microplus* revealed higher counts of $CD3^+$ and gamma-delta T lymphocytes at sites of nymph attachment (Franzinet al., 2017). Furthermore, a recent report demonstrated that three cattle breeds with varying degrees of resistance to *R. microplus* differ significantly in several biochemical and immunological parameters after infestation, including lymphocyte counts in the peripheral blood (Marimaet al., 2020).

Upon consecutive infestations with *R. sanguineus*s.l., only leukocytes isolated from guinea pigs (resistant hosts), but not mongrel dogs (susceptible hosts), exhibited high proliferation when stimulated with saliva or tick extract (Ferreira et al., 2003). Interestingly, saliva from *R. sanguineus*s.l. contains immunomodulatory metabolites, such as adenosine, which suppress the production of inflammatory cytokines by T lymphocytes (Anatrielloet al., 2017). This suggests that susceptible hosts are capable of mounting protective responses, but they are suppressed by tick saliva. In this context, the expression of salivary components from *R. microplus* depends not only on the tick developmental stage but also on whether they feed on resistant or susceptible hosts (Garcia et al., 2020). The differential expression of salivary antigens of *R. microplus* fed in hosts with varying degrees of resistance formed the basis underlying an experimental vaccine to this tick (Maruyama et al., 2017). Therefore, it is reasonable to speculate that salivary components of *R. sanguineus*s.l. are also modulated when fed on resistant hosts, such as guinea pigs and beagle dogs, in which the expression of different salivary components could favour the development of effective immune responses.

Activated B cells can differentiate into plasma cells and produce antigen-specific antibodies to counteract salivary components of parasitism and mediate cytotoxic activity. Diverse studies have investigated the humoral immune responses associated with resistance to *Rhipicephalus*. Interestingly, susceptible hosts present with differential regulation of B lymphocytes in lymph nodes accompanied by a reduction of T cells upon *R. microplus* infestation (Robbertseet al., 2018). Upon heavy infestation, indicine-resistant cattle presented significantly higher levels of specific IgG1 and IgG2 antibodies to tick saliva compared to susceptible taurinehosts (Kashinoet al., 2005). Other studies found that tick-specific IgG or IgG1 correlated with tick loads in susceptible hosts (Cruz et al., 2008; Piper et al., 2009). A comparative study showed that levels of total IgG1 and IgG2 were higher in tick-susceptible hosts before infestation, while total IgG1 increased throughout the infestations of both breeds (Garcia et al., 2017). Moreover, the study reported that resistant hosts exhibited higher levels of saliva-specific antibodies before infestation, but this pattern changed, with susceptible hosts developing higher levels of saliva-specific IgG1 and IgG2 antibodies (Garcia et al., 2017).

Mongrel dogs and guinea pigs presented with low levels of IgG against salivary gland antigens from *R. sanguineus*s.l. across three successive infestations (Szabóet al., 2003). In addition, the dogs exhibited increased levels of total IgE throughout the infestations by *R. sanguineus*s.l. (Szabóet al., 2003). Despite these results, future studies are required to define the role of antibody responses in resistance to infestations by *R. sanguineus*s.l.

*R. SANGUINEUS*S.L RELATIONSHIP WITH RESISTANT AND SUSCEPTIBLE HOSTS

One of the first evidence that *R. sanguineus*s.l. can differentiate resistant from susceptible hosts before attachment was obtained by Louly et al. (2010). To ascertain whether the brown dog tick infested resistant (beagle) and susceptible (English cocker spaniel) dogs differently, five animals of each breed were maintained in a kennel whose walls were infested, with larvae, nymphs, males and females. Five times more ticks were found on cocker spaniels than on beagles. The lower number of ticks found on beagles could be because they were grooming themselves. It is known that self-grooming is an effective way used by resistant hosts to decrease parasitism by ticks (Jonsson et al., 2014). To establish that the difference in parasitism was not just related to grooming but also to the perception of volatile or contact compounds on the skin of the dogs, substances were collected and tested for arrestment and attractiveness of ticks. When allowed to choose between substances rubbed from dogs and a control, more ticks were arrested by extracts from the cockers than from beagles. To test for attraction/repellence, odours from both breeds were collected, and the tests were carried out in a Y-olfactometer. Ticks were not attracted to the odour of either breed; however, the odour of the beagle was repellent. These results indicate that *R. sanguineus*s.l. can use substances from dogs to differentiate susceptible English cocker spaniels from resistant beagles.

Many studies have been performed to determine the presence of natural repellents in certain hosts, such as in the waterbuck *Kobus defassa* in relation to tsetse fly *Glossina morsitans* (Gikonyoet al., 2000; Saini et al., 2017); cattle in relation to horn fly *Haematobia irritans* (Jensen et al., 2004; Birkettet al., 2004); salmon in relation to the salmon louse *Lepeophtheirus salmonis* (Bailey et al., 2006, 20; Mordue (Luntz) and Birkett, 2009) and human beings in relation to highland midge *Culicoides impunctatus* (Diptera: Ceratopogonidae) (Logan et al., 2009) and yellow fever mosquito *Aedes aegypti* (Diptera: Culicidae) (Logan et al., 2008). To confirm the hypothesis that resistant dogs produce repellent compounds, Borges et al. (2015a) accomplished comparative chemical analysis on beagle and English cocker spaniel extracts using GC–MS. The beagle odour extracts contained almost three times as many chemical compounds as cocker spaniel samples. Several nonhost compounds were identified in higher amounts in beagle extracts, such as 2-hexanone, benzaldehyde, nonane, decane and undecane. Behavioural tests performed on Petri-dish and olfactometer assays showed that VOCs 2-hexanone, benzaldehyde, and undecane were weak repellents for *R. sanguineus*s.l. Nonane and decane did not show any significant repellency at any concentration or time evaluated. When 2-hexanone and benzaldehyde were combined, an increase in the repellency rate was observed, with activity comparable or better than N,N-diethyl-3-methylbenzamide (DEET). Undecane did not improve this repellence. This study identified nonhost semiochemicals that mediate avoidance of the beagle dog breed by *R. sanguineus*s.l.

Silva (2016), while ranking dogs based on their ticks and transmitted hemoparasites, found that English cocker spaniel had the highest prevalence (34.25%), while miniature pinscher was the least parasitized (4.72%). Although ticks were not identified in this study, *R. sanguineus*s.l. has the highest prevalence on urban dogs in Brazil, more than any other tick species (Szabóet al., 2001, 2010; Silva et al., 2017). Knowing that resistant dogs produce allomones against *R. sanguineus*s.l., Zeringóta et al. (2021) compared the semiochemicals produced by the putative tick-resistant breed of dog, miniature pinscher, with known tick-resistant (beagle) and tick-susceptible (English cocker spaniel) breeds. The mean amounts of the VOCs benzaldehyde, 2-hexanone and 1,2,4-trimethylbenzene were similar for beagles and miniature pinschers. Furthermore, two compounds, 6-methyl-5-hepten-2-one and 1,2,4-trimethylbenzene, were found in higher amounts in volatile samples collected from miniature pinscher dogs. In Petri-dish assays with adult *R. sanguineus*, 6-methyl-5-hepten-2-one was repellent for all observation periods evaluated for the two highest concentrations. These results support the hypothesis that miniature pinschers are a tick-resistant dog breed, although artificial infestations are needed to confirm this supposition. It can be inferred that the differential production of semiochemicals by miniature pinschers, beagles and English cocker spaniels is a consequence of genetic differences between the breeds, since the production of semiochemicals is thought to be under genetic control (Pickett et al., 2010).

To use semiochemicals to control arthropods in addition to knowing the identity of the compound(s) mediating certain behaviour, it is necessary to know its natural release rate. Ectoparasite location in animal hosts is affected by variation in these odour components and their ratios (Gikonyoet al., 2002, 2003). As mentioned above, the repellence by the tick *R. sanguineus*s.l. dog breed, the beagle, is mediated by the VOCs 2-hexanone and benzaldehyde released on their odours. The odour of three beagles was collected over one week and identified using GC-MS, and both compounds were found in all dogs on all days. The amount of benzaldehyde was always higher than that of 2-hexanone, and the actual benzaldehyde:2-hexanone ratio, based on quantitative readings, ranged from 1.1:1 to 3.7:1, with an average of 2.2:1 (de Oliveira Filhoet al., 2016).

A slow-release formulation releasing a ratio of 2-hexanone and benzaldehyde broadly matched the ratio of natural release from beagles was tested by Oliveira Filho et al. (2017) to reduce the burden of *R. sanguineus* s. l. on English cocker spaniel dogs. Twelve dogs were randomly assigned to two groups, treated and control, with six dogs each. After five environmental infestations, the mean tick infestation in the treated group was significantly lower (approximately 50% for immatures and 20% for adults) than in the control group. This study highlighted for the first time the potential use of a novel allomone (repellent)-based formulation for the reduction of tick infestation in susceptible dogs. Recently, Oliveira Filho et al. (2018) tested the hypothesis that an increase in repellent dose and release rate could increase repellent efficacy and persistence; however, they could not find any significant increase in repellent efficacy. During the first three weeks of the study, the ratio of benzaldehyde:2-hexanone release from the formulations was roughly similar to that found naturally in beagles (de Oliveira Filho et al., 2016). However, from the 4th week, the ratio of release was markedly different, which coincided with a reduction in tick repellency. They observed a loss of repellent activity after the third week of testing, which coincided with a marked change in the relative release rates for the two compounds.

Recently, Barreto et al. (unpublished data) tested the efficacy of the same slow-release formulation used by Oliveira Filho et al. (2017) on 15 dogs naturally infested by *R. sanguineus*s.l. When the total number of ticks accumulated for 30 days of evaluation was considered, the dogs from the control group had significantly higher loads of ticks than the dogs from the treatment group; conversely, the counts of larvae, nymphs and adults were similar between the control and the treatment groups on each tick collection day within the 30 days of evaluation,.The efficacy of the treatment with repellent collars targeting the adult ticks varied from 18.4% to 41.8% among intervals, with a mean efficacy of 30.8%, whereas the mean efficacy of the treatment was extremely low in reducing larva or nymph infestation.

Although it has been shown that slow-release neck collars releaseallomones, 2-hexanone and benzaldehyde can control natural or artificial infestations of dogs by *R. sanguineus*s.l., it is necessary to improve the efficacy of this type of control. One possibility would be to include in this formulation the nonhostsemiochemical 6-methyl-5-hepten-2-one identified in the odour of miniature pinschers (Zeringótaet al., 2021). Another possibility would be the adoption of the push-pull strategy. This strategy consists of two distinct phases: the push phase, which aims to repel or deter parasites from resources such as crops and animals through the use of stimuli that act as repellents for these parasites or mask the presence of the host. While simultaneously in the pull phase, highly attractive stimuli that may indicate the presence of the host, or pheromones, which indicate conspecifics, are used to attract and divert parasites from the resources to be protected (Agelopouloset al., 1999; Cook et al., 2007). In this case, ticks repelled from the dogs would be attracted by arrestment pheromones to traps on the walls (Gowrishankaret al., 2019). On the traps they could be killed by an acaricide or any other biological agent effective against this tick.

R. microplus Relationship with Resistant and Susceptible Hosts

Despite the fact that host resistance to *R. microplus* has been known since the beginning of last century, little is known about how ticks relate to resistant hosts. Waladde et al. (1979) suggested that ticks could perceive different compounds present in the blood of resistant bovines because when they were fed on an artificial membrane, they fed for a longer period of time in the blood of sensitive cattle than in resistant cattle. To address this topic, Ferreira et al. (2015) conducted taste electrophysiology experiments with cheliceral pit sensilla preparations of *R. microplus* females. Solutions of five known ixodid tick phagostimulants were tested at different concentrations and serum samples from susceptible and resistant bovines. A dose-dependent response of gustatory neurons associated with the chelicerae sensillum was observed to all tested phagostimulants, such as NaCl, glucose, GSH, and ATP. In addition, chemoreceptors in the chelicera of *R. microplus* were more responsive to serum of *R. microplus*-resistant bovines than to serum samples from susceptible cattle. These results allowed us to conclude that through chelicera pit sensilla, *R. microplus* detected phagostimulants present in the sera of both susceptible and resistant bovines, and they also detected deterrents in the sera of resistant bovines. Bovines that are resistant to ticks contain several components contributing to such phenotypes, such as acute phase proteins, antibodies, and other elements associated with the inflammatory process elicited by tick blood feeding (Wamburaet al., 1998; Kashinoet al., 2005; Carvalhoet al., 2008, 2010; Piper et al., 2009).

Thus, these substances could be perceived by taste sensilla in the chelicerae to limit hematophagy, as suggested by Waladde et al. (1979).

R. microplus uses anambush strategy to find the host. The general behaviour in such a strategy is to climb vegetation, spread the first pair of legs, and wait for the host. This behaviour is stimulated by carbon dioxide, vibration, shadowing and odour stimuli and promotes an immediate reaction by the tick to attach the host with its claws (Apanaskevich and Oliver, 2014). Using an olfactometer bioassay, where the larva could express this questing strategy, Borges et al. (2015b) aimed to determine whether *R. microplus* exhibits different behaviours depending on the bovine breed. Larvae were individually exposed to the odours of Holstein-Friesian cattle (susceptible), Nelore cattle (resistant), hexane (negative control) and 2-nitrophenol (positive control). The highest questing responses were observed to 2-nitrophenol and Holstein odours. The lowest responses were observed to the solvent and to Nelore odours. It is possible to observe that *R. microplus* expressed different questing behaviours depending on the odour of the breed perceived. However, based on these results, it was not possibleto distinguish whether the tick was been repelled or failing to perceive a key attractant in Nelore cattle odour. Franzin et al. (2017) also found that *R. microplus* larvae prefer to arrest to susceptible bovine sebum over sebum from resistant animals.

Despite the evidence of the presence of repellents in cattle against *R. microplus*, the chemical compounds responsible for this behaviour have not yet been identified. Therefore, it is impossible to attempt to control *R. microplus* with the slow-release formulation used to control *R. sanguineus*s.l. (de Oliveira Filhoet al., 2017; Oliveira Filhoet al., 2018). It is also necessary to note that *R. microplus* uses the ambushing strategy to find the host (Apanaskevich and Oliver, 2014), which can hinder the adoption of this type of control. Tabor et al. (2017) suggest that volatile compounds from susceptible animals may be influenced by their microbiome and, in turn, by their diet, which could be a future option for controlling this tick. Since *R. microplus* shows a decrease in the questing posture when exposed to the odour of resistant cattle, which in turn results in fewer ticks ready to attach the host with its claws, it is reasonable to assume that this strategy would be successful.

References

Abbas, R. Z., Zaman, M. A., Colwell, D. D., Gilleard, J. and Iqbal, Z. (2014). Acaricide resistance in cattle ticks and approaches to its management: The state of play. *Veterinary Parasitology,* 203: 6–20.

Agelopoulos, N., Birkett, M. A., Hick, A. J., Hooper, A. M., Pickett, J. A., Pow, E. M., Smart, L. E., Smiley, D. W. M., Wadhams, L. J. and Woodcock, C. M. (1999). Exploiting semiochemicals in insect control. *Pesticide Science,* 55: 225–235.

Anatriello, E., Oliveira, C. J. F., Oliveira, N. B., Fisch, A., Milanezi, C. M., da Silva, J. S., de Miranda-Santos, I. K. F. and Ferreira, B. R. (2017). Interaction between saliva's adenosine and tick parasitism: effects on feeding and reproduction. *Parasites & Vectors*, 10: 326.

Apanaskevich, D. and Oliver, J. (2014). Life cycles and natural history of ticks. In *Biology of Ticks*, pp. 59–73. Oxford University Press.

Bailey, R. J. E., Birkett, M. A., Ingvarsdottir, A., Mordue, A. J., Mordue, W., O'Shea, B. A., Pickett, J. A. and Wadhams, L. J. (2006). The role of semiochemicals in host location and non-host avoidance by salmon louse (Lepeophtheirussalmonis) copepodids. *Canadian Journal of Fisheries and Aquatic Sciences,* 63: 448–456.

Barreto, L.P., Ribeiro-Silva, C.S., Machado, L.D.C., Lima, V.H., Pinto, S. M. N., Pereira-Júnior, R. A., Carreira, D.N., Mascarin, G.M., Fernandes, E. K. K, Borges, L.M.F. (2020). Efficacy of collars with allomones on dogs to control *Rhipicephalus sanguineus sensu lato* infestations under field conditions. *International Journal of Acarology* (submitted).

Becker, S., Webster, A., Doyle, R. L., Martins, J. R., Reck, J. and Klafke, G. M. (2019). Resistance to deltamethrin, fipronil and ivermectin in the brown dog tick, *Rhipicephalus sanguineus sensu stricto*, Latreille (Acari: Ixodidae). *Ticks and Tick-borne Diseases*, 10: 1046–1050.

Bennett, G. F. (1969). *Boophilus microplus* (Acarina: Ixodidae): Experimental infestations on cattle restrained from grooming. *Experimental Parasitology*, 26: 323–328.

Birkett, M. A., Agelopoulos, N., Jensen, K.-M. V., Jespersen, J. B., Pickett, J. A., Prijs, H. J., Thomas, G., Trapman, J. J., Wadhams, L. J. and Woodcock, C. M. (2004). The role of volatile semiochemicals in mediating host location and selection by nuisance and disease-transmitting cattle flies. *Medical and Veterinary Entomology*, 18: 313–322.

Borges, L. M. F., de Oliveira Filho, J. G., Ferreira, L. L., Louly, C. C. B., Pickett, J. A. and Birkett, M. A. (2015a). Identification of non-host semiochemicals for the brown dog tick, *Rhipicephalus sanguineus* sensu lato (Acari: Ixodidae), from tick-resistant beagles, Canis lupus familiaris. *Ticks and Tick-Borne Diseases,* 6: 676–682.

Borges, L. M. F., Duarte, S. C., Louly, C. C. B., Borges, L. M. F., Duarte, S. C. and Louly, C. C. B. (2015b). Cattle tick differentiates between the odors of Holstein Friesian and Nelore cattle. *Ciência Rural*, 45: 2023–2025.

Borges, L. M. F., Li, A. Y., Olafson, P. U., Renthal, R., Bauchan, G. R., Lohmeyer, K. H. and León, A. A. P. de (2016). Neuronal projections from the Haller's organ and palp sensilla to the synganglion of *Amblyommaamericanum. Revista Brasileira de Parasitologia Veterinária,* 25: 217–224.

Brake, D. K. and Pérez de León, A. A. (2012). Immunoregulation of bovine macrophages by factors in the salivary glands of *Rhipicephalus microplus*. *Parasites & Vectors,* 5: 38.

Carr, A. L., Mitchell III, R. D., Dhammi, A., Bissinger, B. W., Sonenshine, D. E. and Roe, R. M. (2017). Tick Haller's organ, a new paradigm for arthropod olfaction: how ticks differ from insects. *International Journal of Molecular Sciences,* 18.

Carvalho, W. A., Bechara, G. H., Moré, D. D., Ferreira, B. R., da Silva, J. S. and de Miranda Santos, I. K. F. (2008). *Rhipicephalus (Boophilus) microplus*: Distinct acute phase proteins vary during infestations according to the genetic composition of the bovine hosts, *Bos taurus* and *Bos indicus*. *Experimental Parasitology,* 118: 587–591.

Carvalho, W. A., Franzin, A. M., Abatepaulo, A. R. R., de Oliveira, C. J. F., Moré, D. D., da Silva, J. S., Ferreira, B. R. and de Miranda Santos, I. K. F. (2010). Modulation of cutaneous inflammation induced by ticks in contrasting phenotypes of infestation in bovines. *Veterinary Parasitology*, 167: 260–273.

Cavassani, K. A., Aliberti, J. C., Dias, A. R. V., Silva, J. S. and Ferreira, B. R. (2005). Tick saliva inhibits differentiation, maturation and function of murine bone-marrow-derived dendritic cells. *Immunology*, 114: 235–245.

Chmelar, J., Calvo, E., Pedra, J. H. F., Francischetti, I. M. B. and Kotsyfakis, M. (2012). Tick salivary secretion as a source of antihemostatics. *Journal of Proteomics*, 75: 3842–3854.

Constantinoiu, C. C., Jackson, L. A., Jorgensen, W. K., Lew-Tabor, A. E., Piper, E. K., Mayer, D. G., Venus, B. and Jonsson, N. N. (2010). Local immune response against larvae of *Rhipicephalus* (*Boophilus*) *microplus* in *Bos taurusindicus* and *Bos taurustaurus* cattle. *International Journal for Parasitology,* 40: 865–875.

Cook, S. M., Khan, Z. R. and Pickett, J. A. (2007). The Use of Push-Pull Strategies in Integrated Pest Management. *Annual Review of Entomology*, 52: 375–400.

Cruz, A. P. R., Silva, S. S., Mattos, R. T., Da Silva Vaz, I., Masuda, A. and Ferreira, C. a. S. (2008). Comparative IgG recognition of tick extracts by sera of experimentally infested bovines. *Veterinary Parasitology*, 158: 152–158.

Dantas-Torres, F., Figueredo, L. A. and Brandão-Filho, S. P. (2006). *Rhipicephalus sanguineus* (Acari: Ixodidae), o carrapatovermelho do cão, parasitandohumanos no Brasil [*Rhipicephalus sanguineus* (Acari: Ixodidae), the dog's red tick, parasitizing humans in Brazil]. *Revista da Sociedade Brasileira de Medicina Tropical*, 39: 64–67.

de Castro, J. J. (1997). Sustainable tick and tickborne disease control in livestock improvement in developing countries. *Veterinary Parasitology,* 71: 77–97.

de Oliveira Filho, J. G., Sarria, A. L. F., Ferreira, L. L., Caulfield, J. C., Powers, S. J., Pickett, J. A., de León, A. A. P., Birkett, M. A. and Borges, L. M. F. (2016). Quantification of brown dog tick repellents, 2-hexanone and benzaldehyde, and release from tick-resistant beagles, Canis lupus familiaris. *Journal of Chromatography B*, 1022: 64–69.

de Oliveira Filho, J. G., Ferreira, L. L., Sarria, A. L. F., Pickett, J. A., Birkett, M. A., Mascarin, G. M., de León, A. A. P. and Borges, L. M. F. (2017). Brown dog tick, *Rhipicephalus sanguineus* sensu lato, infestation of susceptible dog hosts is reduced by slow release of semiochemicals from a less susceptible host. *Ticks and Tick-Borne Diseases,* 8: 139–145.

Ferreira, B. R., Szabó, M. J. P., Cavassani, K. A., Bechara, G. H. and Silva, J. S. (2003). Antigens from *Rhipicephalus sanguineus* ticks elicit potent cell-mediated immune responses in resistant but not in susceptible animals. *Veterinary Parasitology*, 115: 35–48.

Ferreira, L. L., Soares, S. F., de Oliveira Filho, J. G., Oliveira, T. T., Pérez de León, A. A. and Borges, L. M. F. (2015). Role of *Rhipicephalus microplus* cheliceral receptors in gustation and host differentiation. *Ticks and Tick-borne Diseases,* 6: 228–233.

Foelix, R. F. and Chu-Wang, I. W. (1972). Fine structural analysis of palpal receptors in the tick *Amblyomma americanum* (L.). *Zeitschriftfür Zellforschung und Mikroskopische Anatomie,* 129: 548–560.

Franzin, A. M., Maruyama, S. R., Garcia, G. R., Oliveira, R. P., Ribeiro, J. M. C., Bishop, R., Maia, A. A. M., Moré, D. D., Ferreira, B. R. and Santos, I. K. F. de M. (2017). Immune and biochemical responses in skin differ between bovine hosts genetically susceptible and resistant to the cattle tick *Rhipicephalus microplus*. *Parasites & Vectors, 10.*

Garcia, G. R., Maruyama, S. R., Nelson, K. T., Ribeiro, J. M. C., Gardinassi, L. G., Maia, A. A. M., Ferreira, B. R., Kooyman, F. N. J. and de Miranda Santos, I. K. F. (2017). Immune recognition of salivary proteins from the cattle tick *Rhipicephalus microplus* differs according to the genotype of the bovine host. *Parasites & Vectors*, 10: 144.

Garcia, G. R., Chaves Ribeiro, J. M., Maruyama, S. R., Gardinassi, L. G., Nelson, K., Ferreira, B. R., Andrade, T. G. and de Miranda Santos, I. K. F. (2020). A transcriptome and proteome of the tick *Rhipicephalus microplus* shaped by the genetic composition of its hosts and developmental stage. *Scientific Reports,* 10: 12857.

Gikonyo, N. K., Hassanali, A., Njagi, P. G. N. and Saini, R. K. (2000). Behaviour of *Glossinamorsitansmorsitans* Westwood (Diptera: Glossinidae) on waterbuck Kobus defassaRuppel and feeding membranes smeared with waterbuck sebum indicates the presence of allomones. *Acta Tropica*, 77: 295–303.

Gikonyo, N. K., Hassanali, A., Njagi, P. G. N., Gitu, P. M. and Midiwo, J. O. (2002). Odor composition of preferred (buffalo and ox) and nonpreferred (waterbuck) hosts of some Savanna tsetse flies. *Journal of Chemical Ecology*, 28: 969–981.

Gikonyo, N. K., Hassanali, A., Njagi, P. G. N. and Saini, R. K. (2003). Responses of *Glossinamorsitansmorsitans* to blends of electroantennographically Active compounds in the odors of Its preferred (Buffalo and Ox) and nonpreferred (waterbuck) Hosts. *Journal of Chemical Ecology*, 29: 2331–2345.

Gowrishankar, S., Latha, B. R., Sreekumar, C. and Leela, V. (2019). Innovative way to dispense pheromones for off-host control of *Rhipicephalus sanguineus sensu lato* ticks. *Veterinary Parasitology,* 275: 108936.

Inokuma, H., Tamura, K. and Onishi, T. (1997). Dogs develop resistance to *Rhipicephalus sanguineus*. *Veterinary Parasitology,* 68: 295–297.

Jensen, K. M. V., Jespersen, J. B., Birkett, M. A., Pickett, J. A., Thomas, G., Wadhams, L. J. and Woodcock, C. M. (2004). Variation in the load of the horn fly, *Haematobiairritans*, in cattle herds is determined by the presence or absence of individual heifers. *Medical and Veterinary Entomology*, 18: 275–280.

Jittapalapong, S., Stich, R. W., Gordon, J. C., Wittum, T. E. and Barriga, O. O. (2000). Performance of female *Rhipicephalus sanguineus* (Acari: Ixodidae) fed on dogs exposed to multiple infestations or immunization with tick salivary gland or midgut tissues. *Journal of Medical Entomology,* 37: 601–611.

Jonsson, N. N., Piper, E. K. and Constantinoiu, C. C. (2014). Host resistance in cattle to infestation with the cattle tick *Rhipicephalus microplus*. *Parasite Immunology,* 36: 553–559.

Josek, T., Allan, B. F. and Alleyne, M. (2018a). Morphometric analysis of chemoreception organ in male and female Ticks (Acari: Ixodidae). *Journal of Medical Entomology*, 55: 547–552.

Josek, T., Walden, K. K. O., Allan, B. F., Alleyne, M. and Robertson, H. M. (2018b). A foreleg transcriptome for Ixodesscapularis ticks: Candidates for chemoreceptors and binding proteins that might be expressed in the sensory Haller's organ. *Ticks and Tick-Borne Diseases*, 9: 1317–1327.

Kashino, S. S., Resende, J., Sacco, A. M. S., Rocha, C., Proença, L., Carvalho, W. A., Firmino, A. A., Queiroz, R., Benavides, M., Gershwin, L. J. and De Miranda Santos, I. K. F. (2005). *Boophilus microplus*: the pattern of bovine immunoglobulin isotype responses to high and low tick infestations. *Experimental Parasitology*, 110: 12–21.

Kemp, D. H. and Bourne, A. (1980). *Boophilus microplus*: the effect of histamine on the attachment of cattle-tick larvae-studies in vivo and in vitro. *Parasitology*, 80: 487–496.

Kotál, J., Langhansová, H., Lieskovská, J., Andersen, J. F., Francischetti, I. M. B., Chavakis, T., Kopecký, J., Pedra, J. H. F., Kotsyfakis, M. and Chmelař, J. (2015). Modulation of host immunity by tick saliva. *Journal of proteomics*, 128: 58–68.

Leonovich, S. A. (2013). The main evolutionary trends in sensory organs and questing behavior of Parasitiform ticks and mites. *Entomological Review*, 93: 1190–1195.

Logan, J. G., Birkett, M. A., Clark, S. J., Powers, S., Seal, N. J., Wadhams, L. J., MordueLuntz, A. J. and Pickett, J. A. (2008). Identification of human-derived volatile chemicals that interfere with attraction of *Aedesaegypti* mosquitoes. *Journal of Chemical Ecology*, 34: 308–322.

Logan, J. G., Seal, N. J., Cook, J. I., Stanczyk, N. M., Birkett, M. A., Clark, S. J., Gezan, S. A., Wadhams, L. J., Pickett, J. A. and Mordue, A. J. (2009). Identification of human-derived volatile chemicals that interfere with attraction of the Scottish biting midge and their potential use as repellents. *Journal of Medical Entomology*, 46: 208–219.

Louly, C. C. B., Fonseca, I. N., Oliveira, V. F. de and Borges, L. M. F. (2006). Ocorrência de *Rhipicephalus sanguineus* emtrabalhadores de clínicasveterinárias e canis, no município de Goiânia, GO [Occurrence of *Rhipicephalus sanguineus* in veterinary clinic and kennel workers, in the city of Goiânia, GO]. *Ciência Animal Brasileira,* 7: 103–106.

Louly, C. C. B., Soares, S. F., Silveira, D. N., Neto, O. J. S., Silva, A. C. and Borges, L. M. F. (2009). Differences in the susceptibility of two breeds of dogs, English cocker spaniel and beagle, to *Rhipicephalus sanguineus* (Acari: Ixodidae). *International Journal of Acarology,* 35: 25–32.

Louly, C. C. B., Soares, S. F., da Nóbrega Silveira, D., Guimarães, M. S. and Borges, L. M. F. (2010). Differences in the behavior of *Rhipicephalus sanguineus* tested against resistant and susceptible dogs. *Experimental & Applied Acarology,* 51: 353–362.

Marima, J. K., Nel, C. L., Marufu, M. C., Jonsson, N. N., Dube, B. and Dzama, K. (2020). A genetic and immunological comparison of tick-resistance in beef cattle following artificial infestation with *Rhipicephalus* ticks. *Experimental & Applied Acarology*, 80: 569–590.

Marufu, M. C., Chimonyo, M., Mans, B. J. and Dzama, K. (2013). Cutaneous hypersensitivity responses to *Rhipicephalus* tick larval antigens in pre-sensitized cattle. *Ticks and Tick-borne Diseases,* 4: 311–316.

Marufu, M. C., Dzama, K. and Chimonyo, M. (2014). Cellular responses to *Rhipicephalus microplus* infestations in pre-sensitised cattle with differing phenotypes of infestation. *Experimental and Applied Acarology,* 62: 241–252.

Maruyama, S. R., Garcia, G. R., Teixeira, F. R., Brandão, L. G., Anderson, J. M., Ribeiro, J. M. C., Valenzuela, J. G., Horackova, J., Veríssimo, C. J., Katiki, L. M., Banin, T. M., Zangirolamo, A. F., Gardinassi, L. G., Ferreira, B. R. and de Miranda-Santos, I. K. F. (2017). Mining a differential sialotranscriptome of *Rhipicephalus microplus* guides antigen discovery to formulate a vaccine that reduces tick infestations. *Parasites & Vectors*, 10: 206.

Menezes, K. M. F. (2017). *Reconstrução tridimensional do lobo olfativo do carrapato Amblyommasculptum* (*Acari: Ixodidae*) [*Three-dimensional reconstruction of the olfactory lobe of the tick Amblyommasculptum* (*Acari: Ixodidae*)] - Master´s Dissertation at Federal University of Goiás.

Mordue (Luntz), A. J. and Birkett, M. A. (2009). A review of host finding behaviour in the parasitic sea louse, *Lepeophtheirussalmonis* (Caligidae: Copepoda). *Journal of Fish Diseases*, *32*: 3–13.

Oliveira, C. J. F., Sá-Nunes, A., Francischetti, I. M. B., Carregaro, V., Anatriello, E., Silva, J. S., de Miranda Santos, I. K. F., Ribeiro, J. M. C. and Ferreira, B. R. (2011). Deconstructing tick saliva. *The Journal of Biological Chemistry,* 286: 10960–10969.

Oliveira Filho, J. G. de, Ferreira, L. L., Silva, F. de O., Menezes, K. M. F., Muniz, E. R., Paula, L. G. F. de, Sarria, A. L. F., Pickett, J. A., Birkett, M. A., Mascarin, G. M., Borges, L. M. F., Oliveira Filho, J. G. de, Ferreira, L. L., Silva, F. de O., Menezes, K. M. F., Muniz, E. R., Paula, L. G. F. de, Sarria, A. L. F., Pickett, J. A., Birkett, M. A., Mascarin, G. M. and Borges, L. M. F. (2018). Persistence and efficacy of a new formulation based on dog allomonal repellents against *Rhipicephalus sanguineus sensu lato* tick. *Revista Brasileira de Parasitologia Veterinária,* 27: 313–318.

Otranto, D., Dantas-Torres, F., Tarallo, V. D., Ramos, R. A. do N., Stanneck, D., Baneth, G. and de Caprariis, D. (2012). Apparent tick paralysis by *Rhipicephalus sanguineus* (Acari: Ixodidae) in dogs. *Veterinary Parasitology,* 188: 325–329.

Parola, P., Socolovschi, C. and Raoult, D. (2009). Deciphering the relationships between *Rickettsia conoriiconorii* and *Rhipicephalus sanguineus* in the ecology and epidemiology of Mediterranean spotted fever. *Annals of the New York Academy of Sciences,* 1166: 49–54.

Pickett, J. A., Birkett, M. A., Dewhirst, S. Y., Logan, J. G., Omolo, M. O., Torto, B., Pelletier, J., Syed, Z. and Leal, W. S. (2010). Chemical ecology of animal and human pathogen vectors in a changing global climate. *Journal of Chemical Ecology*, 36: 113–121.

Piper, E. K., Jonsson, N. N., Gondro, C., Lew-Tabor, A. E., Moolhuijzen, P., Vance, M. E. and Jackson, L. A. (2009). Immunological profiles of *Bos taurus* and *Bos indicus* cattle infested with the cattle tick, *Rhipicephalus* (*Boophilus*) microplus. *Clinical and Vaccine Immunology,* 16: 1074–1086.

Porto Neto, L. R., Jonsson, N. N., D'Occhio, M. J. and Barendse, W. (2011). Molecular genetic approaches for identifying the basis of variation in resistance to tick infestation in cattle. *Veterinary Parasitology*, 180: 165–172.

Raoult, D. and Parola, P. (2007). *Rickettsial Diseases*. CRC Press.

Renthal, R., Manghnani, L., Bernal, S., Qu, Y., Griffith, W. P., Lohmeyer, K., Guerrero, F. D., Borges, L. M. F. and Pérez de León, A. (2017). The chemosensory appendage proteome of *Amblyomma americanum* (Acari: Ixodidae) reveals putative odorant-binding and other chemoreception-related proteins. *Insect Science,* 24: 730–742.

Robbertse, L., Richards, S. A. and Maritz-Olivier, C. (2017). Bovine Immune Factors Underlying Tick Resistance: Integration and Future Directions. *Frontiers in Cellular and Infection Microbiology*, 7.

Robbertse, L., Richards, S. A., Clift, S. J., Barnard, A.-C., Leisewitz, A., Crafford, J. E. and Maritz-Olivier, C. (2018). Comparison of the differential regulation of T and B-lymphocyte subsets in the skin and lymph nodes amongst three cattle breeds as potential mediators of immune-resistance to *Rhipicephalus microplus. Ticks and Tick-Borne Diseases,* 9: 976–987.

Saini, R. K., Orindi, B. O., Mbahin, N., Andoke, J. A., Muasa, P. N., Mbuvi, D. M., Muya, C. M., Pickett, J. A. and Borgemeister, C. W. (2017). Protecting cows in small holder farms in East Africa from tsetse flies by mimicking the odor profile of a non-host bovid. *PLOS Neglected Tropical Diseases*, 11, e0005977.

Schleger, A. V., Lincoln, D. T., McKenna, R. V., Kemp, D. H. and Roberts, J. A. (1976). *Boophilus microplus*: cellular responses to larval attachment and their relationship to host resistance. *Australian Journal of Biological Sciences,* 29: 499–512.

Silva, S. R. (2016). *Levantamento epidemiológico de carrapatosem cães assistidosemdua sunidadesprivad as de serviço de saúde animal na Zona Oeste da cidade do Rio de Janeiro* [*Epidemiological survey of ticks in dogs assisted in two private animal health service units in the West Zone of the city of Rio de Janeiro*].

Silva, B. R. da, Garcia, M. V., Rodrigues, V. da S., Andreotti, R., Dittrich, R. L., Silva, B. R. da, Garcia, M. V., Rodrigues, V. da S., Andreotti, R. and Dittrich, R. L. (2017). Ixodidae fauna of domestic dogs in Parana, southern Brazil. *Revista Brasileira de Parasitologia Veterinária,* 26: 375–377.

Soares, S. F., Louly, C. C. B., Marion-Poll, F., Ribeiro, M. F. B. and Borges, L. M. F. (2013). Study on cheliceralsensilla of the brown dog tick *Rhipicephalus sanguineus* (Latreille, 1806) (Acari: Ixodidae) involved in taste perception of phagostimulants. *Acta Tropica,* 126: 75–83.

Sonenshine, D. E. (1993). *Biology of Ticks*. Oxford University Press.

Sonenshine, D. E. and Roe, R. M. (2014). *Biology of Ticks*, Second Edition. Oxford University Press.

Szabó, M. P. J. and Bechara, G. H. (1999). Sequential Histopathology at the *Rhipicephalus Sanguineus* tick feeding site on dogs and guinea pigs. *Experimental & Applied Acarology*, 23: 915–928.

Szabó, M. P. J., Mukai, L. S., Rosa, P. C. S. and Bechara, G. H. (1995). Differences in the acquired resistance of dogs, hamsters, and guinea pigs to repeated infestations with adult ticks *Rhipicephalus sanguineus* (Acari: Ixodidae). *Brazilian Journal of Veterinary Research and Animal Science*, 32: 43–50.

Szabó, M. P., Marquez Cunha, T., Pinter, A. and Vicentini, F. (2001). Ticks (Acari: Ixodidae) associated with domestic dogs in Franca Region, São Paulo, Brazil. *Experimental & Applied Acarology*, 25: 909–916.

Szabó, M. P. J., Aoki, V. L., Sanches, F. P. S., Aquino, L. P. T. C. T., Garcia, M. V., Machado, R. Z. and Bechara, G. H. (2003). Antibody and blood leukocyte response in *Rhipicephalus sanguineus* (Latreille, 1806) tick-infested dogs and guinea pigs. *Veterinary Parasitology,* 115: 49–59.

Szabó, M. P. J., de Souza, L. G. A., Olegário, M. M. M., Ferreira, F. A. and de Albuquerque PajuabaNeto, A. (2010). Ticks (Acari: Ixodidae) on dogs from Uberlândia, Minas Gerais, Brazil. *Transboundary and Emerging Diseases*, 57: 72–74.

Tabor, A. E., Ali, A., Rehman, G., Rocha Garcia, G., Zangirolamo, A. F., Malardo, T. and Jonsson, N. N. (2017). Cattle tick *Rhipicephalus microplus*-host interface: A review of resistant and susceptible host responses. *Frontiers in Cellular and Infection Microbiology,* 7.

Trager, W. (1939). Acquired immunity to ticks. *The Journal of Parasitology,* 25: 57–81.

Vilela, E. F. (2001). *Feromônios de insetos: biologia, química e emprego no manejo de pragas* [*Insect pheromones: biology, chemistry and use in pest management*], 2nd Edn. Holos.

Waladde, S. M. (1977). The Sensory Nervous System of the adult aattle Tick *Boophilus microplus* (Canestrini) Ixodidae Part Ii. Scanning Electron Microscopy. *Australian Journal of Entomology*, 16: 73–79.

Waladde, S. M. and Rice, M. J. (1978). The sensory nervous system of the sdult cattle tick *Boophilus microplus* (Canestrini) Ixodidae. Part III. Ultra-structure and electrophysiology of cheliceral receptors. *Australian Journal of Entomology*, 16: 441–453.

Waladde, S. M. and Rice, M. J. (1982). CHAPTER 3 - The Sensory Basis of Tick Feeding Behaviour. In *Physiology of Ticks* (ed. Obenchain, F. D. and Galun, R.), pp. 71–118. Pergamon.

Waladde, S. M., Kemp, D. H. and Rice, M. J. (1979). Feeding electrograms and fluid uptake measurements of cattle tick *Boophilus microplus* attached on artificial membranes. *International Journal for Parasitology,* 9: 89–95.

Wambura, P. N., Gwakisa, P. S., Silayo, R. S. and Rugaimukamu, E. A. (1998). Breed-associated resistance to tick infestation in *Bos indicus* and their crosses with *Bos taurus*. *Veterinary Parasitology*, 77: 63–70.

Wikel, S. K. (1996). Host immunity to ticks. *Annual Review of Entomology,* 41:1–22.

Zeringóta, V., Pereira-Junior, R. A., Sarria, A. L. F., Henrique, A. C. C., Birkett, M. A. and Borges, L. M. F. (2021). Identification of a non-host semiochemical from miniature pinscher, *Canis lupus familiaris*, that repels *Rhipicephalus sanguineus* sensu lato (Acari: Ixodidae). *Ticks and Tick-Borne Diseases,* 12: 101582 .

Zhao, G., Yu, M., Cui, Q. W., Zhou, X., Zhang, J. C., Li, H. X., Qu, K. X., Wang, G. L. and Huang, B. Z. (2013). Association of bovine Toll-like receptor 4 with tick infestation rates and blood histamine concentration. *Genetics and molecular research: GMR*, 12: 2783–2793.

In: The Entomological Guide to *Rhipicephalus* ISBN: 978-1-53619-619-1
Editors: Sachin Kumar, Raquel Cossio Bayugar et al.

Chapter 15

MEDICAL AND VETERINARY IMPORTANCE OF *RHIPICEPHALUS* TICKS

Consuelo Almazán[1,*] and Shawgi Mohamed Hassan[2]
[1]Facultad de Ciencias Naturales, Universidad Autónoma de Querétaro, Juriquilla, Querétaro, México, Angers, France
[2]Department of Parasitology, Faculty of Veterinary Medicine, University of Khartoum, Sudan

ABSTRACT

Ticks are obligate hematophagous ectoparasites of humans, domestic, and wild animals, and have been recognized as one of the most important vectors of diseases in terms of diversity of pathogen transmission. The organisms known to be transmitted by ticks include viruses, bacteria, fungi, and protozoans. The Ixodidae family (hard ticks) comprises fourteen genera, with *Rhipicephalus* representing the most widespread genus of ticks affecting animals. In this chapter we review the medical and veterinary importance of the most important *Rhipicephalus* ticks such as the ones included in the complex *R. (Boophilus) microplus; R. appendiculatus,* and *R. sanguineus.* The cattle tick *R. microplus*, and the brown ear tick *R. appendiculatus* are the vectors of babesiosis, anaplasmosis, and theileriosis; the most devastating tick-borne diseases that constrain the development of livestock in tropical and subtropical regions of the world. The brown dog tick, *R. sanguineus* is the most widely distributed species of ticks worldwide and is involved in the transmission of several diseases to humans and dogs.

Keywords: *Rhipicephalus*, tick-borne diseases, livestock, humans, dogs

ABBREVIATIONS

CD	Corridor disease
CTP	Canine tropical pancytopenia

* Corresponding Author's Email: c_almazan_g@hotmail.com.

CCT	Canine cyclic thrombocytopenia
ECF	East Coast fever
FAO	Food and Agriculture Organization of the United Nations
NSD	Nairobi Sheep Disease
LSD	Lumpy skin disease
MSF	Mediterranean spotted fever
RMSF	Rocky Mountain spotted fever
TVD	Thogoto virus disease

INTRODUCTION

Ticks are obligate hematophagous ectoparasites of humans, domestic, and wild animals (Jongejan and Uilenberg, 1994; Parola and Raoult, 2001) and are involved in the transmission of a widely diverse array of pathogens, including viruses, bacteria, fungi, and protozoans (Shonenshine and Roe, 2014). In humans, ticks rank as the second most important vectors after mosquitoes, and in animals they are themost important vectors of diseases (de la Fuente et al., 2008). They are divided into three families: The Ixodidae (hard ticks) with 692 species, the Argasidae (soft ticks) with 186 species, and the Nuttallielidae with 1 species (Shonenshine and Roe, 2014). The Ixodidae family comprises most of the ticks of human and veterinary importance grouped into 14 genera (Guglielmone et al., 2014). The genus *Rhipicephalus* comprises 84 species (Guglielmone et al., 2014; Nicholson et al., 2019), and some ticks from this genus are highly adapted to different environments, thus they are widely distributed. *Rhipicephalus* ticks are important ectoparasites of cattle, dogs, and other domestic and wild animals (Durden and Beati, 2014).

The cattle tick *Rhipicephalus microplus* is the most devastating pest of livestock. According to the Food and Agricultural Organization (FAO, 2004), 80% of the world cattle population is exposed to tick infestations. Approximately 1 billion cattle areat risk of being parasitized by *R. microplus* (Betancourt and Giraldo-Rios, 2019). By the end of the twentieth century, economic losses from ticks and tick-borne diseases globally were calculated to be US $13.9-18.7 billion per year (de Castro, 1997). The losses attributable to ticks are caused either directly (tick worry, blood consumption, damage to hides and udders and the injection of toxins), or indirectly through mortality or debility caused by tick-borne diseases (FAO, 2004). The economic impact of *R. microplus* to livestock has also been evaluated in several countries. For instance, in Brazil, the country with the largest commercial cattle herd, yearly economic losses due to *R. microplus* were estimated in US $ 3.24 billion (Grisi et al., 2014). In Australia, one of the biggest cattle exporters in the world, the impact of ticks, tick-borne diseases, and treatments Cost Aus $ 175 million per year (Lew-Tabor and Rodriguez-Valle 2016). In India, with a cattle inventory of 199 million and buffalo of 105 million, the economic loss was estimated at US $ 498.7 million per year (Gosh and Nagar, 2014). In South Africa, the Cost due to tick borne diseases was US $ 21 million per year (McLeod and Kristjanson, 1999). In Mexico, *R. microplus* tick infestations limits cattle exportations and produces economical losses estimated in US $573.6 million per year (Rodríguez-Vivas et al., 2017). In Colombia, approximately US $168 million per year were recently attributed to *R. microplus* (Betancourt and Giraldo-Rios, 2019).

THE *RHIPICEPHALUS MICROPLUS* COMPLEX

R. microplus ticks (Figure 1) are native of Asia from where they evolved as ectoparasites of zebu cattle (*Bos indicus*) and were dispersed to the African, American continent and Oceania by commercial cattle transportation (Walker et al., 2003). *R. microplus* ticks have been established as common cattle ectoparasites in the regions where livestock is developed because of its ready adaptation to tropical and subtropical regions with climatological conditions suitable for its development, and because of the availability of susceptible hosts.

Following the reasignment of the genus *Boophilus* into *Rhipicepahlus* by Barker and Murrel (2003), some studies demonstrated that the crossbreeding between *R. microplus* strains from Africa and Australia produced infertile offspring (Labruna et al., 2009). Morphological and phylogenetical analysis of 12S and 16S ribosomal sequences allowed investigators to conclude that *R. microplus* from Africa and Australia were different species, and cattle ticks from Australia were denominated *R. Australis* (Estrada-Peña et al., 2012). Currently, *R. microplus* is considered a complex that includes five taxa: *R. australis* (former *R. microplus* ticks from Australia, Indonesia, and New Caledonia), *R. annulatus,* and *R. microplus* clades A, B, C. Clade A comprises ticks from southeast Asia, America, and Africa; B includes *R. microplus* ticks from China, and C comprises ticks from India and Malasyia (Low et al., 2015). Due to the large amount of existing literature on the genus *Boophilus* and to avoid confusions with the new species belonging to *Rhipicephalus*, the former genus *Boophilus* remains as a subgenus of *Rhipicephalus.* The cattle ticks group include *Rhipicephalus (Boophilus) annulatus, R.(B) australis, R.(B) decoloratus, R. (B)geigy, R.(B) microplus* and its clades.

Although few differences may exist in the life cycle of tick species included in the *R. microplus* complex, all of them are one-host ticks that preferentially feed on cattle and other ungulates (Guglielmone et al; 2014). The developmental stages of larvae, nymphs and adults including feeding, mating and female engorgement occurs on the same host. Fully engorged females detach to undergo oviposition in the soil. Each female lays approximately 4400 eggs in one batch and then dies (Strickland, 1976). In laboratory-reared ticks performed by our group, oviposition starts after 2 to 5 days of incubation, and lasts approximately 2 weeks. After a period of 14 to 21 days of incubation, hatching occurs and larvaemature for approximately 14 to 21 days, at which point they become ready to infest a new host.

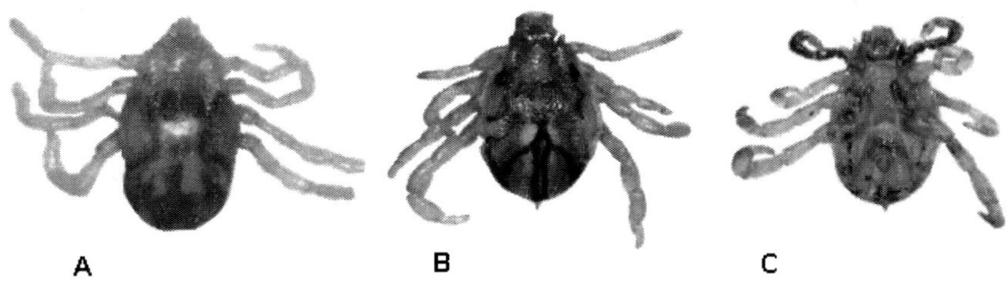

Figure 1. *Rhipicephalus microplus* adults. A, recently molted female; B, male, dorsal view; C, male, ventral view.

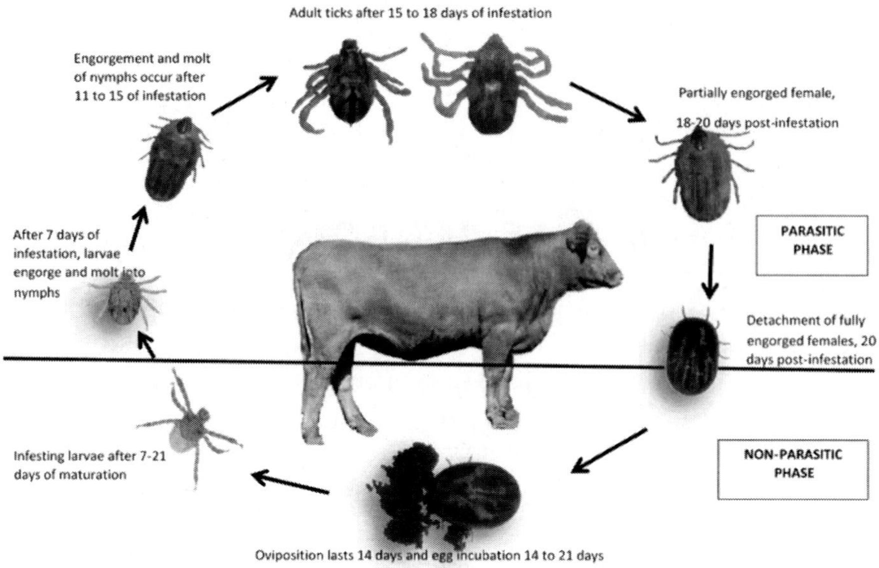

Figure 2. Life cycle of *Rhipicephalus microplus* under laboratory conditions. The life cycle composed of: A) Parasitic phase, which lasts 20 to 23 days, starts with larvae attachment and ends with the detachment of engorged females; and B) Non-parasitic phase, which lasts 44-61 days, includes the preoviposition and oviposition periods, hatching, and maduration of larvae. The complete life cycle lasts 65 to 84 days.

In experimental infestations in northern Mexico, it was observed that 20-day-old larvae are in optimal conditions to infest cattle. The parasitic cycle normally lasts 20 to 23 days, the non-parasitic cycle 44 to 61 days, and (under laboratory conditions) the whole life cycle lasts from 65 to 84 days (Figure 2). The survival of questing larvae is dependent on environmental conditions. It has been documented that *R. microplus* larvae can survive 240 days without feeding and under optimal conditions, *R. microplus* may produce 4 generations per year (Strickland, 1976).

Geographical Distribution

The cattle tick *R. microplus* represents a threat to cattle health and production in tropical and subtropical regions worldwide. In Asia, it is a common ectoparasite of cattle and water buffaloes in southern China, Malaysia, Cambodia (Low et al., 2015), and northern India and Pakistan (Ghosh and Nagar, 2014; Roy et al., 2018).

In the American continent, *R. microplus* is considered an emergent threat of cattle in the USA, where after an intensive eradication program, the country was declared free of *R. microplus* and *R. annulatus* in 1943, except for a permanent quarantine zone in south Texas on the border with Mexico along the Rio Grande where surveillance is maintained to buffer cattle tick incursions (George, 1990). Currently, *R. microplus* is geographically located from southern USA to Argentina in the American continent including the Caribbean Islands (George, 1990; Guglielmone et al., 2014) and without doubt is the most serious problem that limits genetic improvement of livestock in countries like Colombia, Mexico, Argentina, and Brazil, where livestock represents one of the most important economic activities.

During the last decade it was known that the geographical distribution of *R. microplus* in Africa was delimited to the eastern and southern Africa and Madagascar (Walker et al., 2003). However, currently *R. microplus* is present in the West African region including the countries of Ivory Coast, Benin, Burkina Faso, Mali, Togo, and Nigeria (De Clercq et al., 2015), where it arrived in cattle imported from Brazil, and have found optimal conditions for survival. In fact, this tick is replacing other local tick species and is becoming a serious problem of control due to resistance to acaricides in the imported tick strains (Madder et al., 2011). Recently, it was reported that *R. microplus* has invaded the entire territory of Ivory Coast and currently it is the most prevalent tick species (Boka et al., 2017). The presence of *R. microplus* in Cameroon (Silatsa et al., 2019) is an indication of the rapid dispersion and adaptation of this species.

R. annulatus is geographically distributed in areas with Mediterranean and savanna climates and is mainly a tick of west and North Africa, but has been found also in southeast Sudan, Central African Republic, and Democratic Republic of Congo (Walker et al., 2003). In Europe, it has been recorded from Portugal, Spain, Italy, Greece, and Turkey (Estrada-Peña et al., 2004). In Asia it is found in southern Russia, The Middle East, and India (Estrada-Peña and Benzal, 2006). In the American continent it is stablished in northeastern Mexico and southwest USA (George, 1990; Estrada-Peña and Benzal, 2006; Almazán et al., 2018).

R. australis, the former *R. microplus* is distributed in Australia, in a broad band following the general outline of the coast from northeastern New South Wales to northeastern western Australia; New Caledonia, Borneo, Philippines, Sumatra, Java, New Guinea, Cambodia, and Tahiti (Estrada-Peña et al., 2012; Guglielmone et al., 2014; Barker and Walker, 2014).

R. decoloratus, the African blue tick, is endemic in dry regions of Africa, savanna and temperate climates, specifically in Sub Saharan and southern Africa where it is the most common, widespread and frequent tick of cattle (Walker et al., 2003). This tick has remained in Africa because exportation of cattle and horses out of the continent has not been significant (George, 1990).

R. geigy is confined in West Africa, sharing habitat with *R. decoloratus.* Both ticks are morphologically similar. There is no information regarding the role of this tick on transmission of any disease (Walker et al., 2003).

Diversity of Hosts Affected by *R. microplus*

R. microplus preferentially parasitizes cattle, but other domestic animals like horses, mules, goats and sheep (George, 1990), and wild ungulates (George, 1990; Walker et al., 2003; Guglielmone et al., 2014) may act as occasional hosts and *R. microplus* can persist for years using deer as the sole host (George, 1990). *R. microplus* also feeds on the pampa's deer *Ozoctoceros bezoarticus,* which is more susceptible to tick infestations than cattle in the region of Pantanal, Brazil (Cançado et al., 2009). In northern Mexico and southwest USA, infestations of *R. microplus* on wild animals such as the white-tailed deer, *Odocoileus virginianus* (Lohmeyer et al., 2011), and the red deer *Cervus elaphus* in southern Mexico have been documented (Rodríguez-Vivas et al., 2013). Currently, the white-tailed deer plays an important role in regions where this species cohabits with cattle promoting the maintenance of the life cycle of *R. microplus* (Figure 3) by feeding on hosts that are not treated against ticks (Rev by Almazán et al., 2018). In addition, because of the uncontrolled movement of white-tailed deer, these animals contribute to tick dispersion and reinfestation of cattle from tick-free areas, which threatened the tick eradication program along the Mexican-US border (Pound et al., 2010; Lohmeyer et al., 2011). Recently another cervid, the mazama deer (*Mazama temama*) was found to be infested with *R. microplus* in Yucatan, Mexico (Ojeda-Chi, et al., 2019).

Other cervids, such as the rusa deer (*Cervus timorensis*), have been shown to be susceptible hosts that maintain the tick population of *R. microplus*. Soon after its introduction to New Caledonia, the rusa deer became a host for *R. microplus,* where it contributed to pasture infestation, which complicates the tick control program on that island (De Meeûs et al., 2010).

The nilgai antelope, *Boselaphus tragocamelus,* originally from India was imported to Texas, USA for hunting purposes. By 1992, the free-ranging nilgai antelope population was estimated at more than 37,000 head and currently is the most abundant exotic ungulate in Texas, where it contributes to dispersion and maintenance of *R. microplus* (Lohmeyer et al., 2018). The increase in population of this and other wild species is of big concern for the cattle tick eradication program in the USA (Lohmeyer et al., 2018).

The water buffalo *Bubalus bubalis* is increasingly exploited in several Latin American countries due its resistance to high temperatures and ectoparasites. However, infestations by *R. microplus* have been found in Argentina (Benitez et al., 2012) and Brazil (Correa et al., 2012), and evidence that *R. microplus* can successfully complete its life cycle when fed on this host has been demonstrated (Benitez et al., 2012). Previously, *R. australis* (formerly *R. microplus)* feeding on water buffaloes and other wild animals like marsupials, feral pigs, and dogs in Australia, had been reported (Roberts, 1970).

Regarding infestations by *R. microplus* to other domestic animals than cattle, George (1990), mentions that horses, mules, goats, and sheep can act as hosts for this tick. Tick surveys in several animals have allowed the identification of R. microplus parasitizing goats (Dantas-Torres, 2009), horses (Dantas-Torres, 2009; Volpato et al., 2013), sheep (Rodríguez-Vivas et al., 2016; Coronel-Benedett et al., 2018), and dogs (Ojeda-Chi et al., 2019). Figure 4 shows a high infestation of *R. microplus* in a lamb.

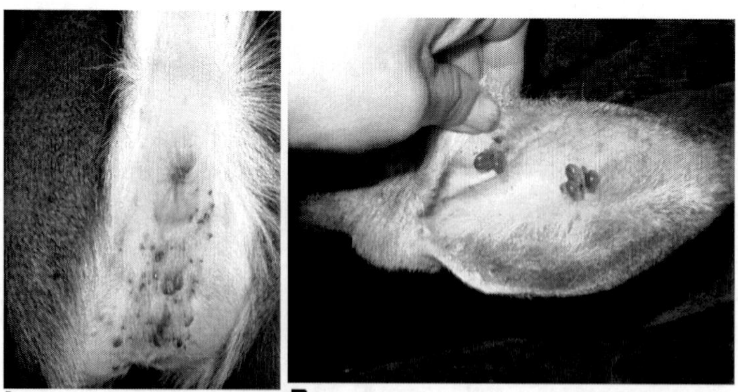

A B

Figure 3. Infestation of *Rhipicephalus microplus* in a white-tailed deer from northern Mexico. Semi-engorged and engorged females attached to the perinneum (A) and inner ear surface (B) are observed.

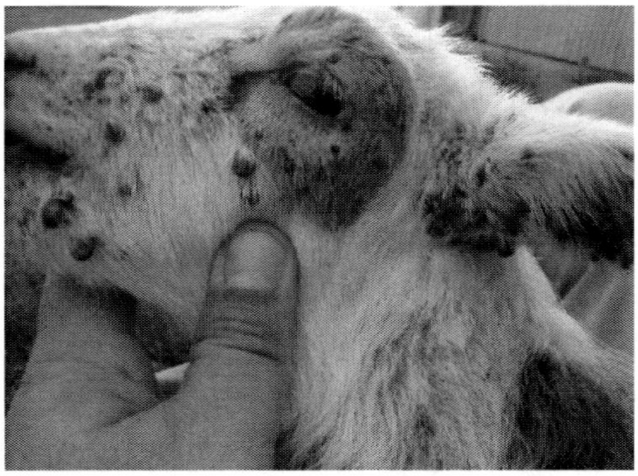

Figure 4. Heavy infestation of *Rhipicephalus microplus* in a lamb in northern Mexico. Image courtesy of Dr Martin Ortiz-Estrada.

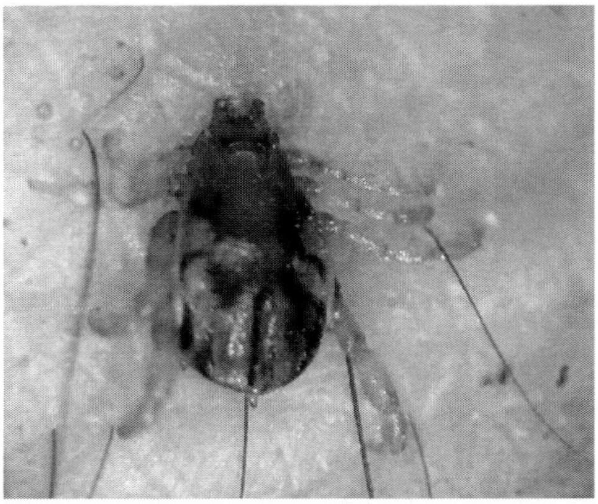

Figure 5. *Rhipicephalus microplus* male attached to the hand of a person after cattle manipulation.

As reviewed by Estrada-Peña and Jongejan (1999), sporadic cases of *R. microplus* parasitizing humans in Cuba, Argentina, and China have been reported. Recently, the infestation by *R. microplus* larvae in a person in India was described (Kaur et al., 2019). However, infestations in humans are incidental and occur mainly in personal related with animal manipulation and husbandry. Figure 5 shows a *R. microplus* male found attached to the skin of a veterinarian after 3 hours that he had been inspecting cattle.

With respect to *R. decoloratus,* the African blue tick, it is a parasite of cattle, but it also feeds on horses, donkeys, sheep, goats, and wild ungulates (Walker et al., 2003). In a recent study by Moyo et al. (2018), *R. decoloratus* was the most prevalent tick parasitizing the African buffalo *Syncerus caffer.* Other wild animals like the impala (*Aepyceros melampus*) and the elephant (*Loxodonta africana*) have been recorded as hosts for this tick (Moyo et al., 2018).

Direct Effects of *R. microplus* on Cattle

The effects of parasitism of *R. microplus* in cattle are divided into direct and indirect. The direct effects rely on the physical damage of skin and the anemia caused by the tick blood consumption, which in ixodid ticks can be two or three times the weight of an engorged female (Jongejan and Uilenberg, 1994). Tick feeding causes significant damage to the skin and this damage is related to the introduction of the tick's mouth parts and saliva over several days and the modulation of wound healing which is essential for maintaining a feeding site. This results in a significant decrease in quality of hides from infested animals (Guglielmone and Nava, 2013; Wikel 2014). *R. microplus* normally attaches to the soft and thin skin in areas with abundant blood such as the dewlap, brisket, neck, axillae, groins, abdomen, and genitals (Strickland, 1976). However, during the highest seasons, in absence of treatment or due to fail in tick control, general infestations around the body can be observed (figure 6). High infestations of *R. microplus* cause damage to hides by the formation of scar tissue, or granuloma at the feeding sites (Barker and Walker, 2014). A study performed in Australia determined that on average, each engorging female tick is responsible for the loss of 1.37 ± 0.25 g bodyweight in *B. taurus* cattle, while in crossed cattle (*B. taurus* × *B. indicus)* the loss of weight decreases to 1.18 ± 0.21 g per engorged tick (Jonsson, 2006). In addition, it has been observed that heifers with infestations by *R. microplus* need more time to reach the appropriate weight required for reproduction (Guglielmone and Nava, 2013).

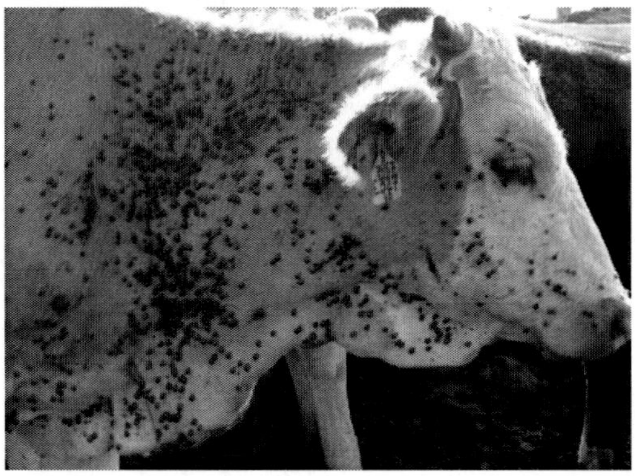

Figure 6. Heavy infestation of *Rhipicephalus microplus* in cattle. Image courtesy of Dr. Martin Ortiz-Estrada.

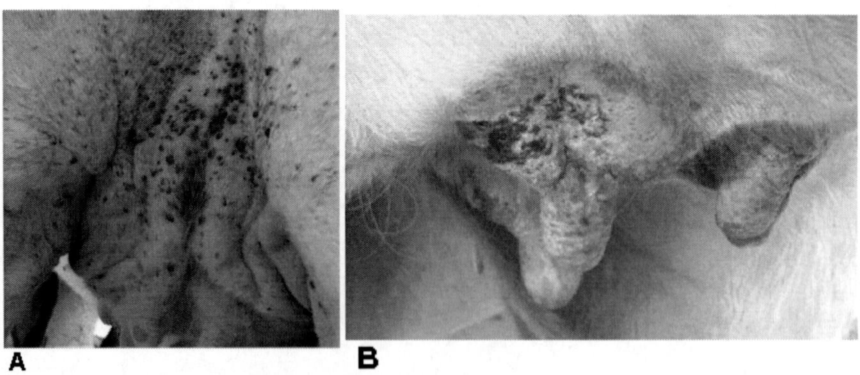

Figure 7. *Rhipicephalus microplus* infesting the udder and teats of a cow. A) Many adult ticks attached to the udder are observed; B) Lesions of the udder and teats due to tick infestation.

With reference to the effects on milk reduction due to parasitism with *R. microplus*, it was estimated that each engorging female tick is responsible for the loss of 8.9 ml of daily milk in Holstein-Friesian cows in Australia (Jonsson et al., 1998). A further estimation indicated that an average of 71.67 ticks per cow (3/4 Holstein x Zebu) causes a reduction of 90.24 L of milk per year in Brazil (Rodrigues and Leite, 2013). In semi-intensive or extensive livestock systems, the tick infestations cause lesions to the udders and teats, causing mastitis, secondary infections, attraction of insects to the bloody wound sites, with decrease or interruption of milk production affecting the lactation and development of heifers (Figure 7).

Indirect Effects of *R. microplus* on Cattle

Table 1. *Rhipicephalus* ticks as vectors of pathogens to humans and animals

Tick species	Pathogen	Disease	Affected species
R. microplus complex	*Babesia bovis* and *B. bigemina*	Bovine babesiosis	Cattle
	Anaplasma marginale[1]	Bovine anaplasmosis	Cattle
	Borrelia theleri	Bovinie borreliosis	Cattle, sheep, goats, horses
R. appendiculatus	*Theileria parva*	Theileriosis	Cattle, buffaloes
	Anaplasma bovis	Bovine ehrlichiosis	Cattle
	Capripoxvirus	Lumpy skin disease	Cattle, buffaloes
	Nairovirus	Nairobi sheep disease	Sheep, humans?
	Thogotovirus	Thogoto virus disease	sheep, cattle, camels, humans
R. e. eversti[2]	*Babesia caballi* and *Theileria equi*	Equine piroplasmosis	Horses, donkeys, mules
	Borrelia theleri	Bovinie borreliosis	Cattle, sheep, goats, horses
R. bursa	*Babesia ovis*	Ovine babesiosis	Sheep, goats
	Anaplasma ovis	Ovine anaplasmosis	Sheep, goats
*R. sanguineus*s.l.	*Rickettsia rickettsii*[3]	Rocky Mountain spotted fever	Humans, dogs
	R. connorii[4]	Mediterranean spotted fever	Humans, dogs
	R. conorii var. israelensis	Israeli spotted fever	Humans, dogs
	R. conorii var. indica	Indian tick typhus	Humans, dogs
	Rickettsia typhi	Murine typhus	Humans, dogs?
	Coxiella burnetii[5]	Q fever	Humans, cattle, sheep, goats
	Ehrlichia canis	Canine tropical pancytopenia	Dogs, humans
	Anaplasma platys	Canine cyclic thrombocytopenia	Dogs, humans
	B. vogeli and *B. gibsoni*	Canine babesiosis	Dogs
	Hepatozoon canis	Hepatozoonosis	Dogs, cats

[1] Besides *R. microplus*, many other ticks and blood sucking arthropods transmit *A. marginale* mechanically.
[2] *R. e. eversti* also causes "Spring lamb paralysis" to sheep.
[3] *Dermacentor* spp.and *Amblyomma*s pp. also transmit *R. rickettsi.*
[4] *Dermacentor marginatus* and *D. reticulatus* also transmit *R. connorii.*
[5] Several other ticks may transmit *C. burnetti,* although the main way of transmission is by aerosols.

The indirect effects associated with *R. microplus* ticks is the transmission of pathogens such as *Babesia bigemina* and *B. bovis,* the causative agents of cattle babesiosis, and *A. marginale*, the causative agent of cattle anaplasmosis. Together, babesiosis and anaplasmosis are the most important diseases affecting livestock worldwide. Regarding *R. decoloratus, this tick* is involved in the transmission of *B. bigemina, and A. marginale* to cattle in Africa, and *Borrelia theileri*, the causative agent of spirochaetosis to cattle, sheep, goats, and horses (Baron et al., 2018), (Table 1).

Bovine babesiosis

Bovine babesiosis, also known as piroplasmosis, Texas fever, red water disease, and cattle tick fever, is an infectious disease caused by intraerythrocytic protozoans of the genus *Babesia*, specifically *Babesia bovis* (Babes, 1888), and *B. bigemina* (Smith and Kilbourne, 1893) that are transmitted transovarially by *R. microplus* and *R. annulatus* in tropical regions of the world, and *R. decoloratus* in dried areas of Southern Africa, where this tick is responsible for the transmission of *B. bigemina* (De Vos et al., 2004; Nicholson et al., 2019). Although the tick bite is the natural way of transmission of babesiosis, cattle are also infected with blood transferred from infected animals by contaminated fomites. Bovine babesiosis is characterized by fever, hemolytic anemia, tachycardia, lethargy, anorexia, and nervous signs, but several forms of the disease that go from subclinical to hyperacute can be presented. The clinical signs vary according to pathogenicity of the *Babesia* species and strains involved. The protozoan *B. bovis* is more pathogenic, causing nervous signs and higher mortality than *B. bigemina*, which can develop and spread rapidly, and cattle may present with severe hemolytic anemia, jaundice and death (Bock et al., 2004). The disease is widely distributed and is common in tropical and subtropical regions, between the parallels 32° south and 40° north of Equator, where the tick vectors are distributed (Vega et al., 1999).

While *B. bovis* is transmitted by larval ticks upon attachment, both nymphs and adults are capable of transmitting *B. bigemina*, which leads to higher prevalence of *B. bigemina* in areas where both pathogens are present (De Vos et al., 2004).

The parasitemia in cattle occurs by a complicated process which takes place after tick inoculation of *Babesia* sporozoites into the blood stream, which penetrate erythrocytes and develop into trophozoites. A reproduction by binary fission takes place and trophozoites divide to form two "pear shapes" called merozoites, which are released from erythrocytes and then infect new erythrocytes. Intraerythrocytic forms not released become gametocytes, the infective forms for tick vectors (Reviewed by Pérez de León et al., 2014). Merozoites continue invading erythrocytes until the host dies or the parasite is eliminated and the destruction of erythrocytes in this phase causes anemia and hemoglobinuria (Reviewed by Almazán et al., 2018). During this process, alterations of mechanical and adhesive properties of infected red blood cells appear to be critical for the parasite. The immune-mediated response is involved in the pathogenesis and contributes to disease progress in naïve cattle (Gohil et al., 2010). The severity of babesiosis depends on the natural resistance and immune status or acquired resistance (Nicholson et al., 2019).

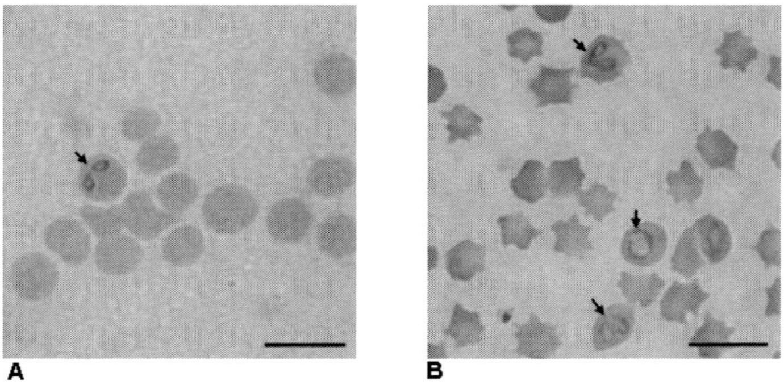

Figure 8. Bovine erythrocytes of cattle showing inclussions corresponding to merozoites of *Babesia.* The classic small and large pear shapes of A) *B. bovis* and B) *B. bigemina* are shown. Magnification = 1000, bar = 10 μm.

The incubation period of babesiosis varies from 6 to 15 days in *B. bovis* and *B. bigemina* respectively (De Vos et al., 2004). During the acute phase, the first clinical sign is fever, and it is during this period when *Babesia* organisms are detected in blood. The parasitemia in blood smears during the acute phase may be as low as 1% but can be as high as 36% (Reviewed by Pérez de León et al., 2014). As the number of organisms increase in blood, the temperature increases as well, and 40 to 41.5°C rectal temperature can be presented during 2 to 3 days. In this phase, other signs such as anorexia, hemoglobinemia, hemoglobinuria, jaundice, constipation, and dehydration may be seen, as well as abortion associated with high fever in pregnant cattle (Bock et al., 2004). As the disease progresses, the animals become lethargic, with muscle wasting, tremors, recumbency, terminal coma and death (Bock et al., 2004; Nicholson et al., 2019). When the infections are due to *B. bovis,* nervous signs are presented due to adhesion of infected erythrocytes to the brain capillaries.

The gross lesions associated with babesiosis include enlarged soft and pulpy spleen, swollen liver, distented of gall bladder, congested dark-colored kidneys and generalized anemia and jaundice (Figure 9). Other lesions may include petechial hemorrhages and pulmonary edema, and the grey matter surface of the brain can appear pink (Bock et al., 2004).

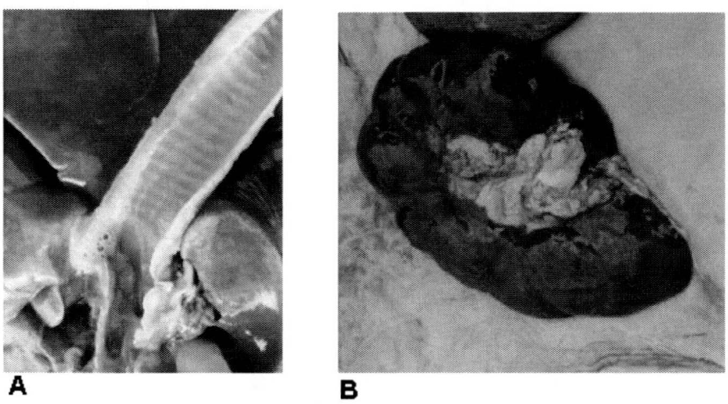

Figure 9. Necropsy lesions from a cow that died of babesiosis A. Icteric tracheal mucous membranes. B congested, dark-colored kidney. Images courtesy of Dr Ned Ivan de la Cruz-Hernandez, University of Tamaulipas, Mexico.

Young animals are generally more resistant than naïve adults because they are protected by acquired antibodies in the colostrum of immune cows, thus when young animals from a local herd are exposed to infected ticks no clinical signs are presented. Furthermore, as consequence of the prolonged exposure to the ticks and the tick-borne diseases, animals become tolerant, with high prevalence of disease without clinical sigs; this condition is called enzootic stability (Figueroa et al., 1998, Vega et al., 1999). However, when naïve animals from 9 to 24-months old are exposed to ticks infested areas for the first time, an extraordinarily strong reaction is developed resulting in very high mortality rates (Reviewed by Pérez de León et al., 2014; Almazán et al., 2018). After recovery from *B. bovis* infection, animals become protected and can resist further infections, while animals that recover from *B. bigemina* remain infective for ticks for 4 to 7 weeks and carriers for several months (Bock et al., 2004).

The most widely used compounds for the treatment of cattle babesiosis include imidocarb dipropionate and diminazene aceturate (Vial and Gorenflot, 2006). However, although chemoprophylaxis is useful, it is expensive and unpractical to employ it as definitive or unique control strategy. Therefore, prevention by controlling the tick vectors is recommended and includes control of cattle movement to avoid the introduction of infected animals into babesiosis free areas, the use of resistant animals, tick control, and immunization (Reviewed by Almazán et al., 2018). The most common method of immunization against cattle babesiosis in several regions of the world is using attenuated, live organisms, which confer strong immunity (Figueroa et al., 1998). However, since infected red blood cells are used, the risk of contamination with other pathogens, makes this procedure impractical and unsafe (Florin-Christensen et al., 2014). In addition, protective cross immunity between *B. bovis* and *B. bigemina* does not occur (Vega et al., 1999; Bock et al., 2004), thus using live vaccines becomes complicated and impractical in areas where infections with both *Babesia* species co-exist. Despite efforts to the develop recombinant vaccines against cattle babesiosis (Florin-Christensen et al., 2014, Mosqueda et al., 2019), none have been effective and immunoprophylaxis against babesiosis currently depends on the use of live, attenuated pathogens produced in splenectomized cattle or from *in vitro* cultures (Figueroa et al., 1998; Vega et al., 1999).

Bovine Anaplasmosis

Anaplasmosis is an infectious disease caused by *A. marginale*, a gram-negative bacterium that belongs to the order Rickettsiales that measures 0.3 μm diameter and is spherical in shape (Reviewed by Gaff et al., 2014). In blood smears, *A. marginale* organisms are observed as membrane-bound inclusions (Figure 10) or initial bodies of 4 to 8 rickettsiae in mature erythrocytes (Kocan et al., 2010). *A. marginale* is transmitted biologically by ticks when the organism multiplies in the tick and then is transmitted during feeding; and mechanically by ticks and other hematophagous arthropods, and fomites such as needles and surgical tools contaminated with infected blood (Kocan et al., 2010; Aubry and Geale, 2011; Gaff et al., 2014). Anaplasmosis represents a serious problem in livestock from tropical and subtropical regions of the world and has been estimated to be responsible for up to 25% of the total losses among beef cattle in genetic improvement programs (Rodríguez et al., 2009). Anaplasmosis is characterized clinically by recumbency, fever, jaundice, anemia, severe loss of weight, weakness, reduction of growth and milk production, abortion in the last trimester of pregnancy and death of animals over 2 years old (Rodríguez et al., 2009; Aubry and Geale, 2011).

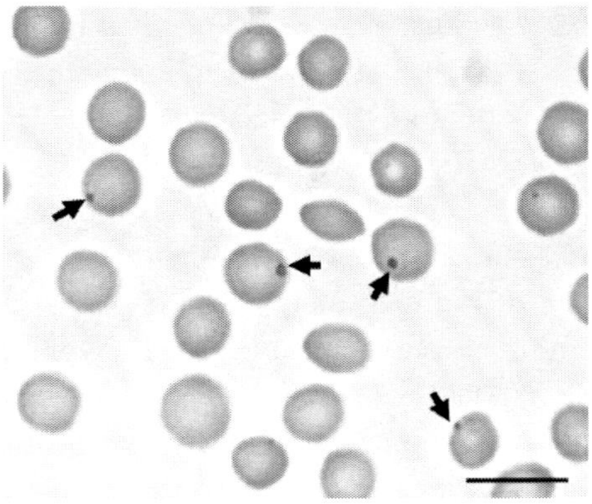

Figure 10. *Anaplasma marginale* inclusion bodies in bovine erythrocytes. Magnification 1000x, Bar = 10 μm. Image courtesy of Dr Sergio D. Rodríguez.

The incubation period of *A. marginale* ranges from 7 to 60 days, although it varies depending on the infective dose. After the organism enters blood stream, the number of infected erythrocytes increases exponentially to reach its maximum level in a period of 6 to 8 days (Reviewed by Gaff et al., 2014). Experimentally, the clinical signs are observed from 15 to 30 days after inoculation of *A. marginale* (Rodríguez-Camarillo, 2007). During this phase, the rickettsemia may reach 70% and mild to severe anemia and jaundice are produced (Gaff et al., 2014). The destruction of erythrocytes is produced as a mechanism of response to infected erythrocytes which are removed by the spleen, resulting in extravascular hemolytic anemia and splenomegaly (Rodriguez-Camarillo, 2007; Gaff et al., 2014). Anemia, icterus, splenomegaly, and hepatomegaly are the most important findings at the necropsy in cattle with anaplasmosis. Mortalities of 10 to 14% have been reported in cattle introduced to *A. marginale* enzootic regions (Rodríguez-Camarillo, 2007).

Animals that survive the initial phase develop persistent infection characterized by intermitent rickettsemia and become reservoirs for ticks and cattle, a particularly important factor in the epidemiology of the disease. Persistent infections have also been observed in animals treated with tetracyclines, the antibiotic of choice, which decreases the reckettsemia but does not eliminate the infection completely and animals remain as asymptomatic carriers (Kocan et al., 2010). In carrier animals the levels of *A. marginale* can beas low as 10^2 to10^7 organisms/ml and that infection cannot be detected microscopically. When no treatment is applied, *A. marginale* produces destruction of 50% of erythrocytes, and cattle lose up to 20% of their body weight, requiring several months to recover (Rodríguez-Camarillo, 2007). Transplacental transmission also occurs, and some infected calves suffer a hyperacute form of anaplasmosis resulting in death, while others do not show signs and remain as carriers.

Because of the different ways of transmission and the presence of asymptomatic carriers, it is difficult to eliminate *A. marginale* from a herd or a determined region. Therefore, prevention is essential, and is based on vector control, good sanitary practices (avoiding the use of contaminated fomites such as re-use of needles), and, in some cases, the use of tetracyclines as a food additive (Kocan et al., 2010). Immunization of cattle with recombinant vaccines would seem to be an excellent method of prevention. However, despite intensive research, a recombinant vaccine has not been developed, which has been attributed to the high genetic diversity of *A. marginale* among geographical isolates (de la Fuente et al., 2007), including strains from the same region or herd (Almazán et al., 2008; Castañeda-Ortiz et al., 2015). In some countries live vaccines using *A. centrale*, a related non-pathogenic organism, produced in splenectomized cattle that confers protection against *A. marginale* has been used for decades (Wilson et al., 1980), but this vaccine cannot be used in countries where *A. centrale* is exotic (Kocan et al., 2010). In addition, live vaccines have the risk of transporting other pathogens such as bovine leukemia virus and hemolytic diseases in calves born to immunized cows (Rodríguez-Camarillo, 2007). The inactivated or killed vaccines using local strains have the disadvantage of the cost of purification of *A. marginale* from erythrocytes and the lack of cross protection among isolates from geographically distant areas (Reviewed by Quiroz-Castañeda et al., 2016). It is expected that the sequencing of the genome of *A. marginale* and comparison with genomes from different regions may contribute to the discovery of antigens with specific and protective response to develop novel vaccines against cattle anaplasmosis (Martinez-Ocampo et al., 2020).

Bovine borreliosis

Bovine borreliosis is caused by *Borrelia theileri,* a spirochete of 20 to 30 μm long, found in the peripheral blood. The disease is characterized by a mild and variable illness with low mortality in large domestic mammals such as cattle, goats, sheep, horses, and antelopes. Several species of ticks from the genus *Rhipicephalus* have been involved in the transmission. In Africa, *R. evertsi evertsi, R. decoloratus,* and *R. annulatus* are the main vectors (Bishop, 1994); while R. australis and R. microplus are the vectors in Australia and South America respectively (Morel et al., 2019).

Ticks become infected with *B. theileri* transovarially and transstadially. In cattle, the prepatent period takes two to three weeks, and spirochaetes are usually found in peripheral blood. Fever is presented at this time, followed by anorexia, apathy, depression, and anemia in splenectomized calves. Reports from South Africa have shown an association with sporadic abortions in cattle. Lesions include slight lymph node enlargement, pale mucous membranes, and petechiae (Bishop, 1994).

RHIPICEPHALUS APPENDICULATUS

The brown ear tick *R. appendiculatus*, is the main vector of East Coast Fever (ECF). The animal hosts of *R. appendiculatus* are cattle, goats, dogs, sheep, buffaloes, elands, waterbucks, nyalas, greater kudus and antelopes. This tick has become adapted to the presence of domestic cattle and can be maintained by all stages feeding on cattle. Immature ticks can feed on small antelopes and scrub hares. The adults prefer to feed on the ear pinna. On cattle, the immature stages attach mainly on the neck and dewlap, the cheeks, eyelids, muzzle, and ears (Hassan, 1997). *R. appendiculatus* is a three-host tick and the whole life cycle can be completed in three months under the most favourable conditions which allows more than one generation per year in Kenya. In southern Africa, *R. appendiculatus* has a strictly seasonal, single generation per year. In hot regions close to the equator with rainfall more evenly spread during the year, several overlapping generations can be completed annually, and no clear pattern of seasonal abundance may be evident.

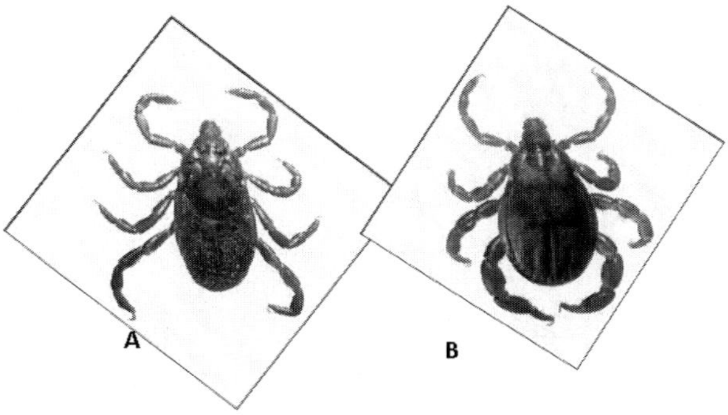

Figure 11. *Rhipicephalus appendiculatus* adults A, female; B, female. Images courtesy of Dr Ladislav Šimo, National Institute of Agricultural Research (INRAE), France.

Other tick species like *R. appendiculatus* are *R. punctatus*, *R. pravus*, *R. zambeziensis,* and *R. duttoni* (Walker et al., 2003). *R. duttoni* is a common tick species in Angola, where its distribution is sympatric with that of *R. zambeziensis.* Gomes and Wouters (1991) reported that *R. duttoni,* which is similar than *R. appendiculatus* occurs separately in western Angola and into the neighbouring countries and suggested *R. duttoni* as a distinct species of the *R. appendiculatus* group. These ticks are difficult to distinguish morphologically and could be assigned to *R. duttoni* or *R. zambeziensis* based on their isoenzyme patterns. No *R. appendiculatus* isoenzyme pattern was found in samples from southwestern Angola (Lawrence et al., 2004).

Geographical Distribution

The brown ear tick, *R. appendiculatus,* is found in the following African countries: Central African Republic, Sudan, Democratic Republic of Congo, Uganda, Kenya, Rwanda, Burundi, Tanzania, Zambia, Malawi, Mozambique, Zimbabwe, Botswana, Swaziland, and South Africa (Norval et al., 1992). It is distributed from the south of Sudan, through to the southeastern coast of South Africa, and there is controversy over how far to the west it is distributed. In South Africa, it is present in Gauteng and Mpumalanga Provinces, the coastal regions of Kwa Zulu-Natal and Eastern Cape Province down to Grahamstown. Walker et al., (2003) mention that *R. zambeziensis* is found in savanna and steppe climatic regions replacing *R. appendiculatus* in the hot, dry river valley systems of south eastern Africa. The distribution of *R. zambeziensis* and *R. appendiculatus* overlap where gradual transitions between wet and dry areas exist. This occurs in parts of the eastern and southern provinces of Zambia bordering the Zambezi Valley, eastern Botswana and in the northwest and Mpumalanga Province, South Africa. In these situations, some interspecific hybridization occurs. *R. zambeziensis* is absent from semi-desert and desert areas. Amzati et al., (2018) investigated the genetic distribution of *R. appendiculatus* across agro-ecological zones in the African

Great Lakes region, observing significant genetic variation segregating two lineages and low structure among populations with high degree of migration, concluding that two sympatric *R. appendiculatus* lineages occur in the Great Lakes region. These two lineages are sympatric in central and eastern Africa and allopatric in southern Africa. In Sudan, several reports of *R. appendiculatus* exist (Hoogstraal, 1956; Morzaria et al., 1981; Julla et al., 1989). A survey performed by Marcellino et al., (2011) documented *R. appendiculatus*, *R. decoloratus*, *R. evertsi evertsi*, *R. praetextatus* and *R. sanguineus* group in central Equatoria and confirmed that *R. appendiculatus* was abundant in the Juba area.

Direct Effects of *R. appendiculatus* on Cattle

Heavy losses in weight of cattle were attributed to *R. appendiculatus* tick infestation by Norval et al. (1988), who observed that the mean damage caused by a standard female of *R. appendiculatus* in *Bos taurus* cattle was about 4.4 g. In Africa, about 11,000 *R. appendiculatus* females per host per year complete engorgement (Kaiser et al., 1982).

Indirect Effects of *R. appendiculatus*

The most important pathogen transmitted by *R. appendiculatus* is indeed, *Theileria parva*, the causative agent of theileriosis in cattle (Drummond, 1976). However, this tick is also involved in the transmission of other pathogens such as *T. taurotragi* causing benign bovine theileriosis, *Anaplasma bovis* (formerly *Ehrlichia bovis*) causing bovine ehrlichiosis, and *Rickettsia conorii* causing tick typhus in humans. *R. appendiculatus* is also involved in the transmission of several viruses that cause disease in domestic animals such as Lumpy Skin Disease (LSD), Nairobi sheep disease (NSD), and Thogoto virus disease (TVD), (Strauss and Strauss, 2008; Nutall, 2014; Tuppurainen et al., 2011). LSD and NSD cause significant losses to livestock, and NSD, and TVD are both considered zoonosis (Strauss and Strauss, 2008; Nutall, 2014) (Table 1).

Theileriosis

Theileriosis is a highly pathogenic disease caused by *Theileria parva*, an intra-cellular protozoon, that infects leukocytes and erythrocytes of wild and domestic ruminants (Bishop et al., 2004). *T. parva* causes a lympho-proliferative disease known as East Coast fever (ECF), corridor disease (CD), and "January disease" in east and central Africa, where *R. appendiculatus* is prevalent and acts as the main vector. However, ECF is also transmitted by *R. zambeziensis* in Zimbabwe and *R. duttoni* in Angola (Gomes and Wouters, 1991). Uilenberg (1981) and Ochanda et al. (1996) reported that *T. parva* is transstadially transmitted by *R. appendiculatus* and *R. zambieziensis.* After infection with *Theileria* spp. via tick bites, sporozoites infect leukocytes, where multi nucleate schizonts are developed, then schizonts differentiate into merozoites, which infect erythrocytes. Also, intra-cellular stages of *Theileria* survive free in the cytoplasm. The intraerythrocytic forms are taken up by ticks during the blood meal, where they develop in gut epithelial cells and salivary glands.

Theileria undergoes an obligate sexual cycle, involving fusion of gametes in the tick gut, to produce a zygote (Bishop et al., 2004). The existence of sexual recombination in *T. parva* has been confirmed in the laboratory and is presumed to contribute to the extensive polymorphism observed in field isolates. The risk of infection with *T. parva* is related to the duration of tick attachment. Konnai et al. (2007) investigated the relationship between the attachment duration of *R. appendiculatus* and the transmission of *T. parva.* Infected adult ticks were allowed to attach to naive mice for variable lengths of time. Attached ticks and host skin biopsies from the tick attachment site were collected daily starting 24 hours after tick attachment and used for detection of *T. parva* by seminested PCR. *T. parva*-infected ticks started to transmit the parasites from 72 hours post-tick attachment, and the transmission of *T. parva* from ticks to mouse increased with duration of tick attachment.

The incubation period of *T. parva* is 8 to 12 days. The nodular skin lesions as result of the swelling of superficial lymph nodes is the first clinical sign in cattle, and fever may reach 42°C. Other signs are anorexia, diarrhea, emaciation, congestion of mucous membranes, petechiae and ecchymosis in conjunctive and oral mucosa, corneal opacity, photophobia, nasal discharge, recumbency, coma, and death. When cerebral theileriois occurs, the clinical manifestations are known as "turning sickness" with fatal results (Reviewed by Pérez de Leon et al., 2014).

In Angola, the most important disease caused by *Theileria* spp. is due to *T. parva* infection. Acute, frequently fatal ECF is transmitted among cattle by *R. appendiculatus* ticks, whereas CD causing almost 100% morbidity is transmitted from its asymptomatic carrier, the African buffalo *Syncerus caffer* to cattle also by *R. zambeziensis* and *R. duttoni* (Lawrence et al., 2004). Mortality due to ECF has been estimated in 30-50% among calves and up to 90% for adults without previous exposure to the disease (Drummond, 1976). Twenty-four million out of the 64 million cattle in central and east Africa are at risk of theileriosis due to *T. parva* (Mukhebi et al., 1992). These authors estimated the annual losses due to *T. parva* in central and east Africa at US $ 168 million which include an estimated mortality of 1.1 million cattle. On Rusinga Island in western Kenya, among the 33% of calf mortality was attributed to ECF (Latif et al., 1995).

In Zambia, theileriosis is presented in the form of a severe lympho-proliferative disease (Morel et al., 1981; Uilenberg et al., 1982; 1981). Amongst the five *Theileria* species and subspecies that are known to exist in Zambia, the most economically important are *T. parva parva* and *T. parva lawrencei* (Geysen, 1997). Molecular DNA studies have, however, shown that *T. parva* subspecies are indistinguishable from each other at the molecular level, although the disease they cause are quite distinct (Geysen, 1997); Jongejan et al., 1986; Musisi et al., 1984). Taurine (*Bos taurus*) cattle, their crosses, and improved zebu (*Bos indicus*) cattle originating in non-endemic areas are the most severely affected. CD is widely spread in the southern, central, and Lusaka provinces and was recorded in Copperbelt province of Zambia (Musisi et al., 1984).

Control of ECF in Kenya has relied predominantly on tick control using chemical acaricides and chemotherapy, while immunization against ECF is increasing. In the production systems (except marginal areas), immunization against ECF play an essential role in the overall control of the disease (Gachohi et al., 2014). Because of immunization against ECF in large endemic areas, endemic stability and control of the disease are expected.

Lumpy Skin Disease

Lumpy skin disease (LSD) is a disease of cattle and water buffaloes caused by a virus from the Poxviridae family, genus *Capripoxvirus* (Elhaigh et al., 2017). It is transmitted by blood feeding insects and several tick species including *R. appendiculatus* and *R. decoloratus*. LSD is endemic in many African countries (Tuppurainen et al., 2011) and has been detected in the Middle East region (Elhaigh et al., 2017). LSD is characterized by high fever, formation of circumscribed skin lesions and ulcerative lesions on the mucous membranes of the mouth, respiratory and digestive tracts. It is an economically important disease due to the permanent damage to hides, the reduction in productivity and trade restrictions imposed on affected areas (Tuppurainen et al., 2013).

The role of several tick species in the transmission LSD virus was investigated by Tuppurainen (2011). The study showed evidence of transstadial and transovarial transmission of LSD virus by *R. decoloratus* ticks and mechanical or intrastadial transmission by *R. appendiculatus* and *A. hebraeum*. In another study, laboratory reared *R. appendiculatus* males feed on experimentally infected LSD viraemic 'donor' cattle, transmitted the disease to recipient cattle, which presented mild clinical signs. In addition, *R. appendiculatus* males were found to transmit LSD virus through feeding on skin lacking visible lesions, demonstrating that viraemic animals without lesions may be a source of infection (Tuppurainen et al., 2013).

Nairobi Sheep Disease

Nairobi sheep disease (NSD) is a zoonosis caused by a virus from the family *Bunyaviridae*, genus *Nairovirus*, *that was* first identified in 1917 (Strauss and Strauss, 2008). *R. appendiculatus* transmits NSD, which is probably the most pathogenic virus known for sheep and goats causing acute gastroenteritis. The virus is transmitted either transstadially and transovarially. In susceptible populations, mortality rates of over 90% regularly occur. The disease is known in east Africa, Somalia and Rwanda, and may occur in southeast Ethiopia, while countries like Zambia, Zimbabwe and Botswana appear to be free (Davies, 1997). Clinical signs include high fever and diarrhea, depression, inappetence, and mucopurulent discharge (Nutall, 2014). Humans can be infected by *Nairovirus,* presenting a mild illness (Strauss and Strauss, 2008).

Thogoto Virus Disease

Thogoto virus disease (TVD) is caused by a virus of the family Orthomyxoviridae; genus *Thogotovirus* (Booth et al., 1989) that affects domestic animals, such as sheep, cattle, and camels, causing neurologic disease and abortion. It is also a zoonotic disease and is transmitted by *R. appendiculatus* and *Amblyomma variegatum* (Booth et al., 1989). The infection in humans is characterized by bilateral optic neuritis and meningitis (Nutall, 2014). Studies conducted by Booth et al., (1989) on TVD from cell culture and in *R. appendiculatus* ticks, found that morphogenesis of TVD virus in cell culture was similar to influenza virus, with polymorphic virus particles budding at the plasma membrane. In this study, the TVD viral infection caused no obvious pathology, and virions or budding profiles were not observed in electron micrographs; although replication, transstadial persistence and transmission to a susceptible host occurred. The virus was detected in the salivary glands of transstadially infected ticks 7 days after feeding. Jones et al. (1990) studied the vector efficiency of *R. appendiculatus* and *A. variegatum* in transmission of TVD, finding that when nymphs of these two tick species were employed as donors, there was no significant difference in the number of infected recipient nymphs. In contrast, 77% of recipient ticks which co-fed with *R. appendiculatus* donor adults acquired Thogoto virus compared to 44.7% of recipient ticks which co-fed with *A. variegatum* donors. Thus, adults of *R. appendiculatus* are more efficient than *A. variegatum* in mediating non-viraemic transmission to their host ranges.

THE *RHIPICEPHALUS SANGUINEUS* COMPLEX

Morphological, biological, and phylogenetic studies on R. sanguineus have generated evidence to support that what was considered a species is indeed a group of several morphologically related species including *R. sanguineus, R. turanicus and other tick species and lineages (*Dantas-Torres, 2010). *Despite this controversy, the R. sanguineus* sensu stricto (s.s.) is as a well-defined taxon in comparison with other species in the *R. sanguineus* group (Nava et al., 2018). *R. sanguineus* sensu lato (s.l.) includes two lineages called temperate and tropical in the American Continent (Moraes-Filho et al., 2015; Nava et al., 2018). Furthermore, other lineages have also been documented in Europe (Dantas-Torres et al., 2018).

R. turanicus, which is included in the *R. sanguineus* complex has been identified as *R. sanguineus* in some cases. However, R. turanicus is a common livestock tick from Mediterranean Europe and Northern Africa, and it is adapted to savanna, steppe, and desert climates. In Africa, it is distributed in Morocco, Algeria, and Tunisia. In Europe it is distributed in Portugal, Spain, France, Italy, Greece, and Turkey (Estrada-Peña et al., 2004). In Asia, it occurs in India, Pakistan, Russia, and China (Walker et al., 2003). This tick is a three-host tick, and parasitizes a wide number of hosts, including humans, dogs, sheep, and cattle, sometimes horses, and wild animals (Estrada-Peña et al., 2004). However, the implication of R. turanicus in the transmission of diseases is unknown.

R. *sanguineus*s.l. (Figure 12), the brown dog tick, also known as the kennel tick or the pan-tropical dog tick is the most widespread tick in the world (Walker et al., 2003, Dantas-Torres 2010). Because *R. sanguineus* survives in sheltered kennels, houses, and peri-domestic areas when dogs are kept indoors (Nicholson et al., 2019). The brown dog tick is also found in cool temperate climatic regions, and due to its association with dogs, it is practically found in all areas of the world where humans and dogs inhabit. However, the highest populations are found in regions with warm and moist climates (Walker et al., 2003).

Although the domestic dog is the main host for *R. sanguineus* in both urban and rural areas, *R. sanguineus* may feed on other hosts such as cats, rodents, and birds (Walker et al., 2003; Dantas-Torres, 2010). Immature phases feed on rodents, hedgehogs, and other small mammals, while adults feed on large animals, dogs, wild carnivorous, and ungulates (Estrada-Peña and Jongejan 1999; Marcellino et al., 2011). Humans are likely to be infested with immature stages of this tick (Estrada-Peña et al., 2004). The risk factors associated to *R. sanguineus* parasitism include dog ownership, presence of infested dogs indoors and high level of environmental infestation (Dantas-Torres, 2010).

R. sanguineus is a three-host tick that is found normally on the ears, neck, and shoulders of dogs (Figure 13) (Walker et al., 2003). It can also be found in inter-digital spaces, back, inguinal region, and axilla (Dantas-Torres, 2010). After infestation, larvae become engorged in 3 to 8 days and molt within 2 to 6 weeks. The nymphs engorge in 4 to 10 days and molt in 2 to 26 weeks. The females may engorge in 7 days, but they can remain on the dog for 3 weeks if unmated and the males can remain for several months for repeated mattings. The engorged female detaches and lays approximately 3200 eggs, which hatch within 3 to 10 weeks. The molting into engorged larvae and nymphs, as well as oviposition of engorged females take place in hidden places like cracks and crevices in the walls (Figure 14), between rocks, and sometimes, almost inside the ground (Dantas-Torres, 2010).

Under normal conditions, the life cycle of *R. sanguineus* can be completed in 10 weeks (Walker et al., 2003). In one year, 4 generations may be produced and seasonal activity peaks in summer. In tropical and subtropical areas, *R. sanguineus* occurs throughout the year (Dantas-Torres et al; 2010), especially if they are parasitizing dogs maintained in shelters or indoors (Nicholson et al., 2019). In some temperate areas of North America, they can overwinter in the environment and even infest dogs in winter (Reviewed by Dantas-Torres et al., 2010).

A B

Figure 12. *Rhipicephalus sanguineus* adults. A, engorged female dorsal view; B, male, dorsal view. Images courtesy of Biol. Jorge Osorio-Miranda, Cenasica, Sader, Mexico.

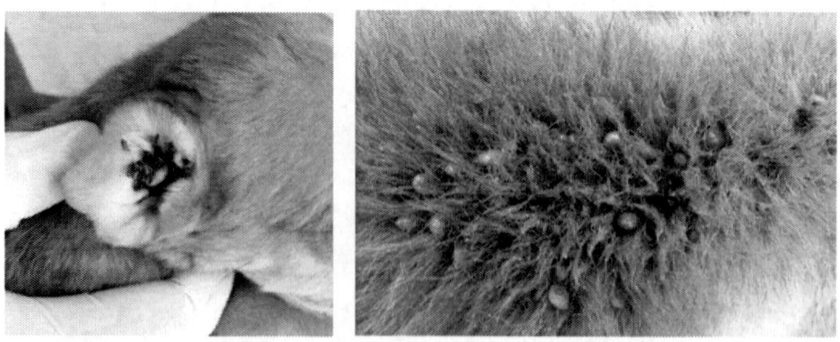

A B

Figure 13. Infestation of *Rhipicephalus sanguineus* in a young dog. Adult and engorged ticks attached to the internal surface of the ear and the back are observed. Images courtesy of Dr V. Homero González-álvarez, University of Guerrero, Mexico.

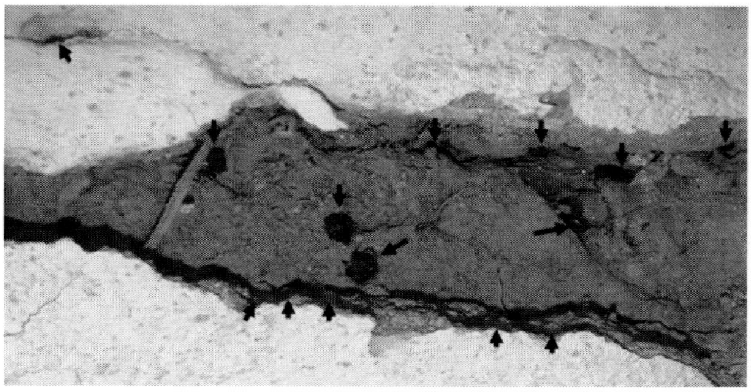

Figure 14. *Rhipicephalus sanguineus* in a crevise of a wall house. Several adult ticks are indicated. Image courtesy of Dr V. Homero González-álvarez, University of Guerrero, Mexico.

Importance of *R. sanguineus* as Vector of Pathogens

The *R. sanguineus* group includes the most widely distributed of all ticks (Estrada-Peña and Jongejan, 1999). Goddard (1989) reported 15 actively biting cases in humans due to *R. sanguineus*. In another report about ticks on humans, *R. sanguineus* accounted for 7% of the ticks collected (Campbell and Bowles, 1994). Felz *et al.,* (1996) found that *R. sanguineus* represented 0.7% of ticks collected from humans in Georgia and South Carolina with a high incidence of immature *R. sanguineus.*

The importance of *R. sanguineus* as vector of diseases in both humans and animals is increasing (Dantas- Torres, 2010). This tick is involved in the transmission of several diseases to humans and dogs such as rickettsiosis and Q fever (Walker et al., 2003, Dantas-Torres, 2010); ehrlichiosis and anaplasmosis (Nicholson et al., 2019). Rickettsiosis is caused by *Rickettsia rickettsii,* also known as Rocky Mountain Spotted Fever (RMSF), which is considered the most severe tick-borne rickettsiosis in humans on the American continent and causes fatal disease in dogs. Another ricketsial disease is the Mediterranean Spotted Fever or "boutonneuse fever" caused by *R. connori* (Nicholson et al., 2019). *R. sanguineus* and several other tick species are involved in the transmission of Q fever caused by *Coxiella burnetii.* In addition, this tick transmits Enrlichiosis and Anaplasmosis, caused by *Ehrlichia canis* and *Anaplasma platys* in dogs. All previously mentioned diseases are of zoonotic concern (Dantas- Torres, 2010). *R. sanguineus* is also the vector of protozoan diseases such as canine babesiosis caused by several species of *Babesia* and hepatozoonosis caused by *Hepatozoon canis* to dogs, cats and other carnivorous (Walker et al., 2003; Baneth et al., 2007; Solano and Baneth, 2011) (Table 1).

Rickettsiosis

Rocky Mountain Spotted Fever

Rickettsiosis is caused by *Rickettsia rickettsii,* a gram-negative, obligate intracellular, aerobic bacterium of 0.7 to 2.0 μm by 0.3 to 0.5 μm in size (Macaluso and Paddock, 2014). Several groups of Rickettsia and at least 20 species that produce disease in humans are recognized. In North America, *Dermacentor andersoni and D. variabilis* transmit

R. rickettsii, the causative agent of Rocky Mountain Spotted Fever (RMSF), while several *Amblyomma* species may be involved in the transmission in South America (Szabó et al., 2013). *R. sanguineus* is the vector RMSF in the southwestern United States, northwest Mexico (Tinoco-Gracia et al., 2018), and some parts of central and south America (Nicholson et al., 2019), where infected ticks exist in peridomestic environments and stray and free-ranging dogs play an important role in the epidemiology of this disease (Macaluso and Paddock, 2014, Tinoco-Gracia et al., 2018).

The infection of ticks occurs when *Rickettsia* organisms are acquired with the blood meal and colonize the midgut and the salivary glands. The bacteria are then transmitted to the vertebrate host via saliva during feeding (Reviewed in Galleti et al., 2016). In natural vertebrate hosts, infections may result in a rickettsemia that allows noninfected ticks to become infected, perpetuating the natural cycle (Parola et al., 2005).

The incubation period of RMSF following tick bite is 7 days (Parola et al., 2005). During the infection in humans rickettsia disseminate to endothelial cells which become dysfunctional and lose integrity. Microvascular thrombosis and increased permeability of microvasculature resulting in edema are common histologic findings in rickettsial infection (Adem, 2019). The infection of endothelial cells by *R. rickettsii* causes an intense vasculitis, which can lead to the failure of important organs, including the brain, lungs, and kidneys (Macaluso and Paddock, 2014). Death normally occurs before day 7 of infection due to organ failure (Adem, 2019).

In humans, clinical signs include high fever, headache, nausea, vomiting, anorexia, and generalized myalgia, followed by a characteristic small rash with pink, blanching macules that evolve to maculo-papules on the wrists, ankles, forearms, and in some cases, is generalized (Reviewed by Macaluso and Paddock, 2014). Lesions such as hemorrhage, cutaneous necrosis, and gangrene can be present in rare cases (Ulloa-García et al., 2018). Petechial lesions can also be found in palms (Figure 15) and soles of the feet, although not in all cases (Ulloa-García et al., 2018).

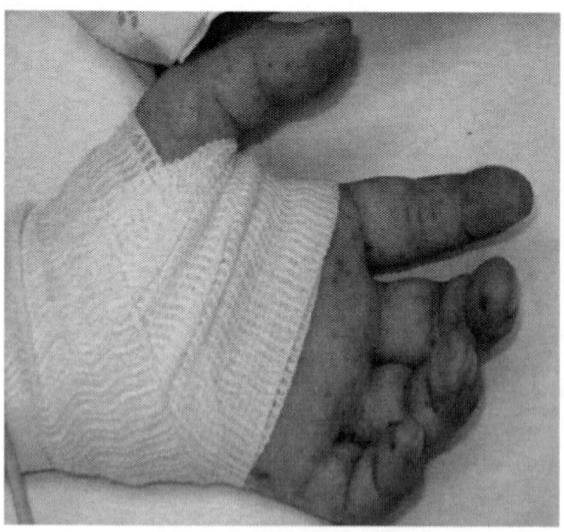

Figure 15. Maculo-papular rash caused by Rickettsia rickettsi on the palm of the hand of an infected child. Image courtesy of Dr Carmen Soria Rodríguez, General Hospital of Mexicali, Baja California; Mexico.

Rickettsiosis in dogs is normally presented with signs of general malaise, including fever, depression, subconjunctival and oral petechia, nasal secretion, limping due to polyarthritis, and abdominal pain. In some cases, cutaneous exanthema occurs, but the rash may not be detected due to the overlying haircoat (Tinoco-Gracia, personal communication). In the late stage of the disease, necrosis of extremities, ears, feet, scrotum, and lips may be found as well as neurological signs, myalgia, and multiorgan failure with a poor prognosis (Herrin, 2020). In experimentally infected dogs, fever was the first detectable sign, appearing at 3 days post-infection (dpi); petechia in ocular and oral mucosa were seen at 6 to 11 dpi; neurological signs occurred at 6 to 8 dpi; and maculopapular rash and petechia in oral mucous appeared at 11 and 12 dpi. The hematological analysis revealed leukocytosis and granulocytosis at 7 to 10 dpi (Levin et al., 2014). Reoccurrence of fever, testicular edema and the return of oral petechial lesions and skin rash may occur several days after treatment with antibiotics (Levin et al., 2014).

Although rickettsiosis can be treated with tetracyclines, (mainly doxycycline) (Adem, 2019), 23% of mortality due to RMSP in the USA has been reported (Macaluso and Paddock, 2014). In Mexico, the health authorities have reported 35% of mortality in cases of rickettsiosis in the northwest region (Ulloa-García et al., 2018), where an average of 180 to 250 cases per year are recorded. When cases are not treated promptly, mortality may reach 80-90% in children, teenagers, and immune compromised patients (Reviewed by Ulloa-García et al., 2018). Although other species of *Rickettsia* have been identified, in Mexico, *R. rickettsii* is the most prevalent, with *R. sanguineus* identified as the most common vector. In Brazil, the fatality rates of RMSF are also high, reaching 30 to 40% (Szabó et al., 2013).

Mediterranean Spotted Fever

Mediterranean spotted fever (MSF) is caused by *R. conorii, a* gram-negative, obligate intracellular bacterium that is transmitted transovarially and transtadially by *R. sanguineus* (Macaluso and Paddock, 2014). Socolovschi1 and colleagues (2009) demonstrated the vertical transmission of *R. conorii* in naturally infected *R. sanguineus* ticks over several generations. Furthermore, the transovarian transmission rate was 100%, and the proportion of infected progenie was up to 99% for the fourth generations of infected ticks. Also, *R. conorii conorii* was detected in the ovaries of infected ticks, supporting the mechanism of transmission in infected *R. sanguineus*.

MSF disease *was reported for the first time in Tunisia in 1910 as* "boutonneuse fever" because of a papular rather than macular rash, and currently MSF is endemic in the Mediterranean regions of northern Africa and southern Europe. Sporadic cases are also reported in northern and central Europe, including Belgium, Switzerland, and France (Reviewed by Parola et al., 2005).

In MSF, an inoculation eschar or black spot is formed at the tick-bite site within 4 to 9 days in 60 to 75% of the cases (Macaluso and Paddock, 2014). Typical cases present with high fever and flu-like symptoms. After 1 to 7 days, a generalized maculopapular rash develops, and often involves the palms and soles of the feet but spares the face (Reviewed by Parola et al., 2005). Other signs include conjuntivitis, vomiting, diarrhea, pneumonitis, and myocarditis. In addition, thrombocytopenia, hyponatremia, and elevated hepatic transaminase levels have been detected in more than 30% of patients (Macaluso and Paddock, 2014). Even though patients normally recover within 10 days, severe forms showing neurological signs may occur in 5 to 6% of the cases, and mortality in 2.5% of cases has been estimated (Parola et al., 2005).

Israeli Spotted Fever

A more pathogenic subspecies of *R. conorii* (var. *israelensis)* has been detected in Israel, Italy, and Portugal where a kind of spotted fever without an eschar at the inoculation site in more than 90% of cases is presented. Instead of the eschar, a small pinkish papule is observed, with mortalities ranking from 20 to 29% of infected patients. This illness is called Israeli spotted fever and is also transmitted by *R. sanguineus (Rev*iewed *by* Macaluso and Paddock, 2014; Parola et al., 2005).

Indian tick typhus

Another subspecies of *R. conorii* with similar manifestations as seen with RMSF occurs in India. This disease is called Indian tick typhus, and *R. sanguineus,* and other ticks such as *Haemaphysalis* and *Hyalomma* are involved in the transmission (Ghosh and Nagar, 2014).

Murine typhus

Murine typhus is a human disease caused by *Rickettsia typhi.* This disease was studied by Dzul-Rosado et al., (2017) finding that *R. typhi* is carried out by rat fleas in urban settlements from Yucatan, Mexico. Blood samples and *R. sanguineus* ticks collected from domestic dogs were molecularly analyzed finding that *Rickettsia typhi* was presente in 100% of the analyzed samples. This evidence suggests for the first time that *R. sanguineus* ticks may be playing a role in the transmission of murine typhus.

Q Fever

Q fever or query fever is a highly contagious disease caused by *Coxiella burnetii*, an obligate intracellular Gram-negative bacterium, that develops in the phagolysosomes of monocytes and macrophages (Dehhaghi et al., 2019). Sporulation of this bacterium gives it the capacity to survive for long periods in hostile environments such as tick feces (Parola and Raoult, 2001; Dehhaghi et al., 2019). The pathogen is transmitted transstadially from larvae to nymphs and from nymphs to adults in several tick species including *R. sanguineus* (Dehhaghi et al., 2019). Although several ticks can carry C. burnetti and acquire the pathogen from an infected animal during blood meals, and transmit it to an uninfected one, it is known that aerosols are the most significant way of transmission (Reviewed by Alvarez-Alonso et al., 2020). More research regarding the role of ticks in the transmission of this pathogen is required.

Domestic animals serve as reservoirs of Q fever. Although animals are generally asymptomatic, *C. burnetii* infection causes abortions in small ruminants, and in goats the abortions may be as high as 70-90%. *C. burnetii* can be found in high numbers in birth products, vaginal fluids, feces, milk, and urine for several weeks after abortion or normal parturition, which are an important source of infection to other animals and people (Reviewed by Alvarez-Alonso et al., 2020).

In humans, the incubation period is 20 days, and the most common signs are persistent fever, chills, sweats, diarrhea, sore throat, sensitivity to light, muscle pain, headache, and fatigue and cough (Nicholson et al., 2019). In acute cases fever, pneumonia or hepatitis may be present, while in chronic cases endocarditis in immunocompromised hosts and pregnant women may occur (Dehhaghi et al., 2019). According to the OIE, Q fever is the second most reported laboratory infection worldwide (Sanchez-Vizcaíno et al., 2019).

Diseases Caused by Anaplasmatacea in Dogs

Canine Tropical Pancytopenia

Canine tropical pancytopenia is caused by *Ehrlichia canis,* a small, not-motile, gram-negative intracellular bacterium that develops in cellular endosomes of monocytes and macrophages of dogs and in colonies or morulae in vertebrates acting as tick hosts (Gaff et al., 2014). Because dogs infected with *E. canis* show a reduction of erythrocytes, leucocytes and platelets, the produced disease is known as canine tropical pancytopenia (CTP) (Nicholson et al., 2019). In addition to domestic dogs, *E. canis* affects other canids such as wolves and jackals (Gaff et al., 2014), and has been detected using molecular techniques in humans with clinical signs in Venezuela (Perez et al., 2006), and in blood donors in Costa Rica (Bouza-Mora et al., 2017). Therefore *E. canis* is considered a zoonosis in Central America.

CTP is transmitted by *R. sanguineus* transstadially, when the pathogen is acquired during the phase of larvae or nymphs and transmitted to dogs by nymphs or adults (Stich et al., 2008). Clinically, dogs show fever, lethargy, apathy, anorexia, pale mucous membranes, splenomegaly, hemorrhages, and uveitis (Nakaghi et al., 2008). In the acute phase of infection leukopenia, thrombocytopenia and death may occur. In the chronic phase, anemia and hemorrhagic nasal bleeding may be present. Dogs may remain subclinical persistent carriers for prolonged periods (Nicholson et al., 2019).

Other canine ehrlichiosis are caused by *E. ewingii* and *E. chanfeensis* and are transmitted by the lone star tick *Amblyomma americanum* in the USA (Gaff et al., 2014).

Canine Cyclic Thrombocytopenia

Canine cyclic thrombocytopenia is caused by *Anaplasma platys*, a gram-negative intracellular bacterium, that infects platelets and causes canine cyclic thrombocytopenia (CCT) in dogs (Gaff et al., 2014), but has also been identified in cats (Lima et al., 2010), and foxes (Cardoso et al., 2015). In addition, molecular identification of *A. platys* has also been reported in humans (Maggui et al., 2013), which suggests that *A. platys* is a zoonotic pathogen.

The transmission of *A. platys* by ticks remains unclear, although it is likely that *R. sanguineus* is the vector since several studies have shown the identification of *A. platys* in this tick (Latrofa et al., 2014, Almazán et al., 2016, Lara et al., 2020), and evidence of the transstadial transmission by *R. sanguineus* has been reported (Aktas and Ozubek, 2018). Vertical transmission may also occur in dogs (Latrofa et al., 2016).

The infection of *A. platys* are mostly subclinical or asymptomatic, and clinical signs include lymphadenomegaly, pale mucous membranes, lethargy, anorexia, weight loss, fever, epistaxis, and thrombocytopenia with hemorrhage (Cardoso et al., 2015). A more pathogenic strain has been proposed due to more severe disease attributed to *A. platys* infections in dogs in Europe (Maggui et al., 2013). Thrombocytopenia is common in dogs infected with both

A. platys and *E. canis* and is thought to be a consequence of platelet phagocytosis by macrophages, either because of injury to platelets by the organism or immune-mediated destruction (Lara et al., 2020). Both *A. platys* and *E. canis* frequently occur in the same geographical areas, and even co-infections are common, which is attributed to the fact that the same tick vector transmits the two pathogens.

Diseases Caused by Protozoans in Dogs

Canine Babesiosis

Canine babesiosis, also called canine piroplasmosis is caused by protozoan parasites that belong to the phylum Apicomplexa, order Piroplasmida and family *Babesiidae* that affect erythrocytes and are transmitted by ticks to dogs and other canids (Solano and Baneth, 2011; Baneth, 2018). Several species of *Babesia* affecting dogs are recognized and they are divided in large and small species.

Large *Babesia* Species

B. canis, is distributed mostly in central and northern Europe and transmitted by the tick *Dermacentor reticulatus* (Solano et al., 2016). Among the *Babesia species, B. canis* is the less virulent (Baneth, 2018).

B. vogeli is distributed mostly in tropical and subtropical regions including the Mediterranean basin, the Middle East, Asia, Australia, and the American continent. It is transmitted by *R. sanguineus*s.l. (Solano and Baneth, 2011).

Small *Babesia* Species

B. gibsoni is distributed in Southeast Asia and Japan, Australia, Europe, and north and South America (Baneth, 2018). It is transmitted by *Haemaphysalis longicornis* and possibly by *Haemaphysalis bispinosa* and *R. sanguineus*s.l. (Baneth, 2018). In addition, direct transmission of this species by dog bites in bull terriers was reported in Australia (Jefferies et al., 2007). In dogs the infection is often subclinical (Boozer and Macintire, 2003).

B. rossi is a very pathogenic species affecting dogs in South Africa (Penzhorn et al., 2017), and other sub-Saharan African countries including Sudan and Nigeria, with jackals acting as subclinical carriers. Haemapysalis ticks act as vectors (Baneth, 2018).

B. conradae was identified in California and *B. vulpes* in northern Spain. They infect dogs and wild red-foxes respectively (Solano et al., 2016). The tick vectors for both *B. conradae,* and *B. vulpes* species have not been determined (Baneth, 2018).

Clinical signs of *canine babesiosis* vary from subclinical infections to multi-organ failure, with a risk of death (Solano and Baneth, 2011; Solano et al., 2016). The signs depend on the species of *Babesia* causing infection, the age of dogs, the immune status, the presence of concurrent infections or other diseases. Some clinical signs include apathy, weakness, anorexia, pale mucous membranes, and a poor general condition. All *Babesia* species can cause fever, enlarged lymph nodes and spleen, anemia, thrombocytopenia, and jaundice. The anemia is caused by a combination of intravascular and extravascular hemolysis resulting from parasite-injury and rupture of red blood cells, the increased of osmotic fragility of cells, and the activity of secondary immune-mediated processes (Reviewed by Solano-Gallego et al., 2016; Baneth, 2018). Hemostatic abnormalities including coagulation and electrolyte imbalances may be present as well.

Canine hepatozoonosis

Canine hepatozoonosis is caused by protozoan parasites that belong to the phylum Apicomplexa, family Hepatozoidae that infect leukocytes and parenchymal tissues. Two species affect dogs, *Hepatozoon americanun,* transmitted by *Amblyomma maculatum*, the Gulf Coast tick, which is distributed in southeastern USA (Ewing and Panciera, 2003), and *H. canis* transmitted by *R. sanguineus,* a worldwide distributed tick (Baneth et al., 2007). Other *Hepatozoon* species such as *H. felis* and *H. silvestris* have been identified in cats from several countries (Reviewed by Basso et al., 2019).

R. sanguineus nymphs and adults containing *H. canis* sporozoites within oocysts can infect dogs and other canids such as foxes that may act as reservoirs (Cardozo et al., 2014). The infection in dogs occurs by ingestion of the infected tick vectors, which carry the sporulated oocysts (Ewing and Panciera, 2003). After ingestion, the released sporozoits are spread via the bloodstream and lymph to the spleen, bone marrow, lung, liver, and kidney, where merogony occurs and merozoits are released to infect leucocytes (preferentially neutrophils) to form gamonts (Reviewed by Otranto et al., 2011). Gamonts are the forms observed in stained blood smears from infected dogs and are the infective phase for ticks (Figure 16). Transmission in ticks may occur transstadially from larvae to nymphs and adults (Cardozo et al., 2014).

Vertical transmission has been suggested for both *H. canis* and *H. americanum* (Ewing and Panciera, 2003). Experimentally, gamonts were detected in blood from dog after 28 of infestation by nymphs. During the parasitemia, skeletal pain and recumbency were seen in infected dogs (Baneth et al., 2007). Other clinical signs such as fever, lethargy, anemia, lymphadenomegaly and emaciation in dogs with high parasitemia can be observed. However, dogs with low-level parasitemia may be asymptomatic (Baneth et al., 2007). In cats, hepatozoonosis is associated with infection of skeletal muscle and myocardium (Basso et al., 2019).

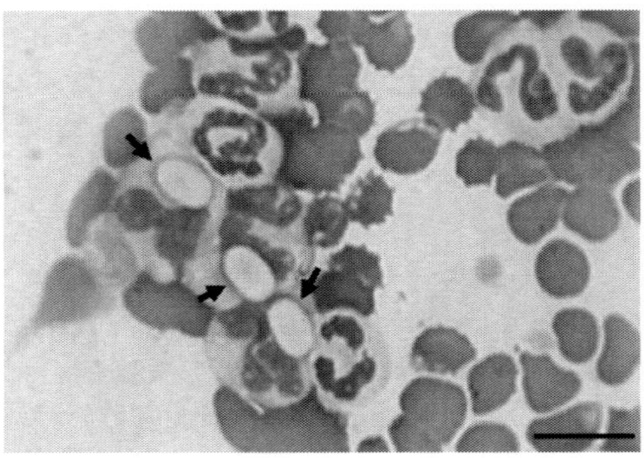

Figure 16. *Hepatozoon canis* gamonts in a blood smear from an infected dog. The arrows indicate three intracytoplasmatic ellipsoid-shaped inclusions in neutrophils. Bar = 10 μm. Image courtesy of Dr Carlos A. Barron, University of Tamaulipas, Mexico.

Other *Rhipicephalus* Ticks of Veterinary Importance

Rhipicephalus eversti eversti

R. e. evertsi is known as red-legged tick and it is the largest and most distinctive *Rhipicephalus* species. It is geographically distributed in Africa and confined to the Afrotropical zoogeographical region in sub-Saharan Africa (Walker et al., 2003). It is a two-host tick. Larvae and nymphs engorge on the first host, and adults on the second host, and parasitizes horses, donkeys, cattle, and sheep. *R. e. evertsi* transmits the protozoans *Babesia caballi* and *Theileria equi* (formerly *Babesia equi*) to horses, both causing forms of equine piroplasmosis. This tick is also involved in the transmission of *B. theleri* that causes bovine borreliosis to cattle (Table 1). It is also the cause of tick paralysis known as "spring lamb paralysis" due to toxins produced by adult ticks that feed for at least 5 days on lambs (Estrada-Peña and Mans, 2014). The number of female ticks per kg of weight of host is important in the cause of paralysis, and the effect is reversed by tick removal (Walker et al., 2003).

Importance of *R. e. eversti* as Vector of Diseases

Equine piroplasmosis

Equine piroplasmosis is a disease caused by the protozoan parasites *Theileria equi* and *Babesia caballi* that affect horses, mules, donkeys, and zebras (Wilson 2011). It has also recently been reported in camels (Sanchez-Vizcaíno et al., 2019). Merozoites of *T. equi* are smaller than *B. caballi* measuring 2-3 µm long, and may be pyriform, round, or ovoid. In some erythorocytes, four *T. equi* merozoites forming a tetrad known as "Maltese cross" may be observed (Sanchez-Vizcaíno et al., 2019).

Besides *R. e. evertsi*, other ticks such as *Dermacentor* and *Hyalomma* may act as vectors of piroplasmosis. Equine piroplasmosis is presented as peracute, acute and chronic forms, but the acute form is more common, however, the disease caused by *T. equi* is more pathogenic. During the acute phase of the disease, animals present with haemolytic anaemia. Other signs include fever, anorexia, pale to icteric mucous membranes, tachycardia, and jaundice. Abortions and neonatal death may also occur (Wilson, 2011).

Rhipicephalus bursa

R. bursa is distributed in the Mediterranean region, and northern Africa where it is confined to the coastal areas of Morocco and Libya. This tick is found in the basins of the Mediterranean, Black and Caspian Seas (Estrada-Peña et al., 2004). This tick uses goats, sheep, horses, and cattle as hosts (Walker et al., 2003). However, it may also parasitize wild ungulates, small mammals, and humans. *R. bursa* is a two-host tick, and the immature stages can feed on rodents and rabbits (Walker et al., 2003). The life cycle of this tick is completed in one year (Estrada-Peña et al., 2004).

Importance of R. bursa as Vector of Diseases

The importance of *R. bursa* in veterinary medicine concerns to the transmission of anaplasmosis caused by *Anaplasma ovis* and babesiosis caused by *Babesia ovis and B. motasi* to sheep and goats (Walker et al., 2003; Jiménez et al., 2019; Ajith et al., 2017), (See Table 1).

Ovine and caprine anaplasmosis

Ovine and caprine anaplasmosis are caused by *Anaplasma ovis*, an obligate intraerythrocytic tick transmitted bacterium that causes anemia, weight loss and weakness to sheep, goats, and wild ruminants. Sheep experimentally infected with *A. ovis* manifested a severe normocytic and normochromic anaemia, ruminal atony, paleness of mucous membranes, weakness, and fever. During the chronic condition, a loss of body weight, cachexia, and severe anemia were the most relevant clinical signs. Transplacental transmission was not demonstrated (Jiménez et al., 2019).

Ovine and caprine babesiosis

Ovine and caprine babesiosis are caused by intra-erythrocytic protozoan parasites belonging to the phylum Apicomplexa and the order Piroplasmida, *Babesia ovis* and *B. motasi* in sheep and goats. *B. ovis* is a small (1-2.5 μm) rounded organism observed in the periphery of the erythrocyte, and *B. motasi* is a large (2.5-4 × 2 μm), piriform in shape organism, found in pairs at acute angle inside erythrocytes. Both pathogens are transmitted by *R. bursa* ticks and the disease is clinically characterized by anaemia, jaundice and haemoglobinuria (Reviewed by Ajith et al., 2017).

Clinical signs in a naturally infected goat included muco-purulent nasal discharge, dyspnea, cough, icteric sclera and oral mucosa, bloated abdomen, diarrhea, hematochezia, and brown coloured urine. Laboratory studies resulted in anemia, neutrophilia, thrombocytopenia, hypoproteinemia, hyperbilirubinemia and haemoglobinuria (Ajith et al., 2017).

Acknowledgments

We thank Dr Martin Ortiz-Estrada (Consultant on Tick control, Mexico); Dr Ladislav Šimo (National Institute of Agricultural Research, France); Dr Sergio D. Rodríguez (INIFAP, Mexico); Dr Ned Ivan de la Cruz-Hernández and Dr Carlos Alberto Barrón Vargas (University of Tamaulipas, Mexico); Dr Vicente Homero González-Álvarez (University of Guerrero, Mexico); Dr Carmen Soria Rodríguez, General Hospital of Mexicali, Baja California; Mexico; Dr Luis Tinoco-Gracia (University of Baja California, Mexico); and Biol. Jorge Osorio-Miranda (SENASICA, Mexico), for providing images and fruitful comments.

Dr Jeffrey L. Blair is acknowledged by reading and editing this chapter.

Thanks to Isabelle Almazán Blair, for helping organize the references.

References

Adem, P. V. (2019). Emerging and re-emerging rickettsial infections. In: *Seminars in Diagnostic Pathology*, (Vol. 36, No. 3, pp. 146-151), WB Saunders.

Ajith, Y., Rajeswari, T. U., Siji, S. R., & Dilip, C. (2017). Case report on babesiosis associated pre-hepatic jaundice in a Malabari goat. *Veterinary and Animal Science*, 3: 1-3.

Aktas, M., and Ozubek, S. (2018). Molecular evidence for trans-stadial transmission of *Anaplasma platys* by *Rhipicephalus sanguineus* sensu lato under field conditions. *Medical and Veterinary Entomology*, 32(1): 78-83.

Almazán, C., Gonzalez-Alvarez, V. H., de Mera, I. G. F., Cabezas-Cruz, A., Rodríguez-Martínez, R. and de la Fuente, J. (2016). Molecular identification and characterization of *Anaplasma platys* and *Ehrlichia canis* in dogs in Mexico. *Ticks and Tick-borne Diseases,* 7(2): 276-283.

Almazán, C., Medrano, C., Ortiz, M. and de la Fuente, J. (2008). Genetic diversity of *Anaplasma marginale* strains from an outbreak of bovine anaplasmosis in an endemic area. *Veterinary Parasitology,* 158(1-2): 103-109.

Almazán, C., Tipacamu, G. A., Rodríguez, S., Mosqueda, J. and Pérez de León, A. (2018). Immunological control of ticks and tick-borne diseases that impact cattle health and production. *Frontiers in Bioscience,* 23*:* 1535-1551.

Alvarez-Alonso, R., Zendoia, I. I., Barandika, J. F., Jado, I., Hurtado, A., López, C. M. and Garcia-Perez, A. L. (2020). Monitoring *Coxiella burnetii* infection in naturally infected dairy sheep flocks throughout four lambing seasons and investigation of viable bacteria. *Frontiers in Veterinary Science,* 7*:* 352.

Amzati, G. S., Pelle, R., Muhigwa, J. B. B., Kanduma, E. G., Djikeng, A., Madder, M. and Marcotty, T. (2018). Mitochondrial phylogeography and population structure of the cattle tick *Rhipicephalus appendiculatus* in the African Great Lakes region. *Parasites & Vectors,* 11(1): 329.

Aubry, P. and Geale, D. W. (2011). A review of bovine anaplasmosis. *Transboundary and Emerging Diseases,* 58(1): 1-30.

Babes, V. (1888). Sur l'hemoglobinurie bacterienne du boeuf [On bacterial hemoglobinuria in beef]. *Comptes Rendus Hebdomadaires des Seances de l'Academie des Sciences*, 107*:* 692-694.

Baneth, G. (2018). Antiprotozoal treatment of canine babesiosis. *Veterinary Parasitology,* 254*:* 58-63.

Baneth, G., Samish, M. and Shkap, V. (2007). Life cycle of *Hepatozoon canis* (Apicomplexa: Adeleorina: Hepatozoidae) in the tick *Rhipicephalus sanguineus* and domestic dog (*Canis familiaris*). *Journal of Parasitology*, 93(2): 283-299.

Barker, S. C. and Murrell, A. (2003). Phylogeny, evolution and historical zoogeography of ticks: a review of recent progress. In *Ticks and Tick-borne pathogens* (pp. 55-68). Springer, Dordrecht.

Barker, S. C. and Walker, A. R. (2014). Ticks of Australia. The species that infest domestic animals and humans. *Zootaxa,* 3816(1): 1-144.

Baron, S., Barrero, R. A., Black, M., Bellgard, M. I., van Dalen, E. M., Fourie, J. and Maritz-Olivier, C. (2018). Differentially expressed genes in response to amitraz treatment suggests a proposed model of resistance to amitraz in *R. decoloratus* ticks. *International Journal for Parasitology: Drugs and Drug Resistance*, 8(3): 361-371.

Basso, W., Görner, D., Globokar, M., Keidel, A. and Pantchev, N. (2019). First autochthonous case of clinical *Hepatozoon felis* infection in a domestic cat in Central Europe. *Parasitology International*, 72: 101945.

Benitez, D., Cetrá, B. and Florin-Christensen, M. (2012). *Rhipicephalus (Boophilus) microplus* ticks can complete their life cycle on the water buffalo (*Bubalus bubalis*). *Journal of Buffalo Science*, 1(2).

Betancourt-Hurtado, O. J. & Giraldo-Ríos, C. (2018). Economic and health impact of the ticks in production animals. In *Ticks and Tick-Borne Pathogens. Intech Open*.

Bishop, R., Musoke, A., Morzaria, S., Gardner, M., and Nene, V. (2004). *Theileria*: intracellular protozoan parasites of wild and domestic ruminants transmitted by ixodid ticks. *Parasitology*, 129(S1): S271.

Bishop, G. C. (1994). *Borrelia theileri* infection. *Infectious Diseases of Livestock*. JAW Coetzer, GR Thomson, and RC Tustin, eds. Oxford University Press, Oxford, 992-993.

Bock, R., Jackson, L., De Vos, A. and Jorgensen, W. (2004). Babesiosis of cattle. *Parasitology*, 129 (S1): S247.

Boka, O. M., Achi, L., Adakal, H., Azokou, A., Yao, P., Yapi, Y. G. and Kaboret, Y. Y. (2017). Review of cattle ticks (Acari, Ixodida) in Ivory Coast and geographic distribution of *Rhipicephalus (Boophilus) microplus*, an emerging tick in West Africa. *Experimental and Applied Acarology*, 71(4): 355-369.

Booth, T. F., Davies, C. R., Jones, L. D., Staunton, D. and Nuttall, P. A. (1989). Anatomical basis of Thogoto virus infection in BHK cell culture and in the ixodid tick vector, *Rhipicephalus appendiculatus*. *Journal of General Virology*, 70(5): 1093-1104.

Boozer, A. L. and Macintire, D. K. (2003). Canine babesiosis. The Veterinary Clinics of North America. *Small Animal Practice*, 33(4): 885-904.

Boozer, A. L. and Macintire, D. K. (2003). Canine babesiosis. The Veterinary Clinics of North America. *Small Animal Practice*, 33(4): 885-904.

Bouza-Mora, L., Dolz, G., Solórzano-Morales, A., Romero-Zuñiga, J. J., Salazar-Sánchez, L., Labruna, M. B. & Aguiar, D. M. (2017). Novel genotype of *Ehrlichia canis* detected in samples of human blood bank donors in Coasta Rica. *Ticks and tick-borne Diseases*, 8(1): 36-40.

Campbell, B. S., and Bowles, D. E. (1994). Human tick bite records in a United States Air Force population, 1989–1992: implications for tick-borne disease risk. *Journal of Wilderness Medicine*, 5(4): 405-412.

Cançado, P. H. D., Zucco, C. A., Piranda, E. M., Faccini, J. L. H. and Mourão, G. M. (2009). *Rhipicephalus (Boophilus) microplus* (Acari: Ixodidae) as a parasite of pampas deer (*Ozoctoceros bezoarticus*) and cattle in Brazil's Central Pantanal. *Revista Brasileira de Parasitologia Veterinaria*, 18(1), 42-46.

Cardoso, L., Cortes, H. C., Eyal, O., Reis, A., Lopes, A. P., Vila-Viçosa, M. J., and Baneth, G. (2014). Molecular and histopathological detection of *Hepatozoon canis* in red foxes (*Vulpes vulpes*) from Portugal. *Parasites & Vectors*, 7(1): 1-6.

Cardoso, L., Gilad, M., Cortes, H. C., Nachum-Biala, Y., Lopes, A. P., Vila-Viçosa, M. J., and Baneth, G. (2015). First report of *Anaplasma platys* infection in red foxes (*Vulpes vulpes*) and molecular detection of *Ehrlichia canis* and *Leishmania infantum* in foxes from Portugal. *Parasites & Vectors*, 8(1): 144.

Castañeda-Ortiz, E. J., Ueti, M. W., Camacho-Nuez, M., Mosqueda, J. J., Mousel, M. R., Johnson, W. C., and Palmer, G. H. (2015). Correction: Association of *Anaplasma marginale* Strain Superinfection with Infection Prevalence within Tropical Regions. *PloS one*, 10(5): e0129415.

Coronel-Benedett, K. C., Ojeda-Robertos, N. F., González-Garduño, R., Ibañez, F. M., and Rodríguez-Vivas, R. I. (2018). Prevalence, intensity and population dynamics of hard ticks (Acari: Ixodidae) on sheep in the humid tropics of Mexico. *Experimental and Applied Acarology*, 74(1): 99-105.

Corrêa, F. D. N., Cunha, N. C. D., Rangel, C. P., and Fonseca, A. H. D. (2012). Ticks on buffaloes (*Bubalus bubalis*) in the state of Rio de Janeiro, Brazil. *Revista Brasileira de Parasitologia Veterinaria,* 21(3): 313-314.

Couto, J., Villar, M., Mateos-Hernández, L., Ferrolho, J., Sanches, G. S., Sofia Santos, A., and de la Fuente, J. (2020). Quantitative Proteomics Identifies Metabolic Pathways Affected by *Babesia Infection* and Blood Feeding in the Sialoproteome of the Vector *Rhipicephalus bursa*. *Vaccines*, 8(1): 91.

Dantas-Torres, F. (2009). Ticks on domestic animals in Pernambuco, Northeastern Brazil. *Revista Brasileira de Parasitologia Veterinaria*, 18(3): 22-28.

Dantas-Torres, F. (2010). Biology and ecology of the brown dog tick, *Rhipicephalus sanguineus*. *Parasites & Vectors*, 3(1): 1-11.

Dantas-Torres, F., Latrofa, M. S., Ramos, R. A. N., Lia, R. P., Capelli, G., Parisi, A., and Otranto, D. (2018). Biological compatibility between two temperate lineages of brown dog ticks, *Rhipicephalus sanguineus (sensu lato)*. *Parasites & Vectors*, 11(1): 398.

Dantas-Torres, F., Latrofa, M. S., Ramos, R. A. N., Lia, R. P., Capelli, G., Parisi, A. and Otranto, D. (2018). Biological compatibility between two temperate lineages of brown dog ticks, *Rhipicephalus sanguineus (sensu lato)*. *Parasites & Vectors*, 11(1): 398.

Davies, F. G. (1988). Nairobi sheep disease. The arboviruses: *Epidemiology and Ecology*, 3: 191-203.

de Castro, J. J. (1997). Sustainable tick and tickborne disease control in livestock improvement in developing countries. *Veterinary Parasitology*, 71(2-3): 77-97.

De Clercq, E. M., Leta, S., Estrada-Peña, A., Madder, M., Adehan, S. and Vanwambeke, S. O. (2015). Species distribution modelling for *Rhipicephalus microplus* (Acari: Ixodidae) in Benin, West Africa: Comparing datasets and modelling algorithms. *Preventive Veterinary Medicine*, 118(1): 8-21.

de la Fuente, J., Estrada-Peña, A., Venzal, J. M., Kocan, K. M. and Sonenshine, D. E. (2008). Overview: ticks as vectors of pathogens that cause disease in humans and animals. *Frontiers in Bioscience*, 13(13): 6938-6946.

de La Fuente, J., Ruybal, P., Mtshali, M. S., Naranjo, V., Shuqing, L., Mangold, A. J. and Torina, A. (2007). Analysis of world strains of *Anaplasma marginale* using major surface protein 1a repeat sequences. *Veterinary Microbiology*, 119(2-4): 382-390.

De Meeûs, T., Koffi, B. B., Barré, N., de Garine-Wichatitsky, M. and Chevillon, C. (2010). Swift sympatric adaptation of a species of cattle tick to a new deer host in New Caledonia. *Infection, Genetics and Evolution*, 10(7): 976-983.

De Vos, A. J., De Waal, D. T. and Jackson, L. A. (2004). Bovine babesiosis. *Infectious Diseases of Livestock*, Vol 1, (Ed. 2): 406-424.

Dehhaghi, M., Kazemi Shariat Panahi, H., Holmes, E. C., Hudson, B. J., Schloeffel, R. and Guillemin, G. J. (2019). Human tick-borne diseases in Australia. *Frontiers in Cellular and Infection Microbiology*, 9*:* 3.

Durden, L. and Beati, L. A. (2013). Modern tick systematics. *Biology of Ticks*, 17-58.

Dzul-Rosado, K., Lugo-Caballero, C., Tello-Martin, R., López-Avila, K. and Zavala-Castro, J. (2017). Direct evidence of *Rickettsia typhi* infection in *Rhipicephalus sanguineus* ticks and their canine hosts. *Open Veterinary Journal*, 7(2): 165-169.

Elhaig, M. M., Selim, A. & Mahmoud, M. (2017). Lumpy skin disease in cattle: Frequency of occurrence in a dairy farm and a preliminary assessment of its possible impact on Egyptian buffaloes. *Onderstepoort Journal of Veterinary Research*, 84(1): 1-6.

Elhaig, M. M., Selim, A. and Mahmoud, M. (2017). Lumpy skin disease in cattle: Frequency of occurrence in a dairy farm and a preliminary assessment of its possible impact on Egyptian buffaloes. *Onderstepoort Journal of Veterinary Research*, 84(1): 1-6.

Estrada-Peña, A. and Mans, B. J. (2013). *Tick-induced paralysis and toxicoses* (pp. 313-332). Sonenshine DE, Roe RM. USA: Oxford University Press.

Estrada-Peña, A. and Jongejan, F. (1999). Ticks feeding on humans: a review of records on human biting Ixodoidea with special reference to pathogen transmission. *Experimental and Applied Acarology*, 23(9): 685-715.

Estrada-Peña, A. and Venzal, J. M. (2006). High-resolution predictive mapping for *Boophilus annulatus* and *B. microplus* (Acari: Ixodidae) in Mexico and Southern Texas. *Veterinary Parasitology*, 142(3-4): 350-358.

Estrada-Peña, A., Bouattour, A., Camicas, J. L. and Walker, A. R. (2004). *Ticks of domestic animals in the Mediterranean region.* University of Zaragoza, Spain, 131.

Estrada-Peña, A., Venzal, J. M., Nava, S., Mangold, A., Guglielmone, A. A., Labruna, M. B. and De La Fuente, J. (2012). Reinstatement of *Rhipicephalus (Boophilus) australis* (Acari: Ixodidae) with redescription of the adult and larval stages. *Journal of Medical Entomology*, 49(4): 794-802.

Ewing, S. A. and Panciera, R. J. (2003). American canine hepatozoonosis. *Clinical Microbiology Reviews*, 16(4): 688-697.

Drummond, R. O. (1976). Tick-borne livestock diseases and their vectors. 4. Chemical control of ticks. *World Animal Review (Italia)* no. 19: 28-33.

FAO, Food and Agriculture Organization. (2004). *Resistance management and integrated parasite control in ruminants–guidelines, module 1–ticks: acaricide resistance: diagnosis, management and prevention.*

Felz, M. W., Durden, L. A. and Oliver Jr, J. H. (1996). Ticks parasitizing humans in Georgia and South Carolina. *The Journal of Parasitology*, 505-508.

Figueroa, J. V., Cantó, G. J., Alvarez, J. A., Lona, R., Ramos, J. A. and Vega, C. A. (1998). Capacidad protectora en bovinos de una cepa de *Babesia bigemina* derivada de cultivo *in vitro* [Protective capacity in cattle of a *Babesia bigemina* strain derived from *in vitro* culture]. *Revista Mexicana de Ciencias Pecuarias*, 36(2): 95-08.

Florin-Christensen, M., Suarez, C. E., Rodriguez, A. E., Flores, D. A., and Schnittger, L. (2014). Vaccines against bovine babesiosis: where we are now and possible roads ahead. *Parasitology*, 141(12): 1563.

Gachohi, J., Skilton, R., Hansen, F., Ngumi, P., and Kitala, P. (2012). Epidemiology of East Coast fever (*Theileria parva* infection) in Kenya: past, present and the future. *Parasites & Vectors*, 5(1): 194.

Gaff, H. D., Kocan, K. M., and Sonenshine, D. E. (2014). Tick-borne rickettsioses II (Anaplasmataceae). *Biology of ticks*, 2: 251-77.

Galletti, M. F. B., Fujita, A., Rosa, R. D., Martins, L. A., Soares, H. S., Labruna, M. B. and Fogaça, A. C. (2016). Virulence genes of *Rickettsia rickettsii* are differentially modulated by either temperature upshift or blood-feeding in tick midgut and salivary glands. *Parasites & Vectors*, 9(1): 331.

George, J. E. (1990). Wildlife as a constraint to the eradication of *Boophilus* spp. (Acari: Ixodidae). *Journal of Agricultural Entomology*, 7(2): 119-125.

Geysen D. (1997). Contribution on the characterization of *Theileria* isolates and stocks in Zambia, *International Consortium on Ticks and Tick-borne Diseases* (*ICTTD*) *Newsletter* 4 11-12.

Ghosh, S. and Nagar, G. (2014). Problem of ticks and tick-borne diseases in India with special emphasis on progress in tick control research: a review. *Journal of Vector Borne Diseases*, 51(4): 259.

Goddard, J. (1989). Focus of human parasitism by the brown dog tick, *Rhipicephalus sanguineus* (Acari: Ixodidae). *Journal of Medical Entomology*, 26(6): 628-631.

Gohil, S., Kats, L. M., Sturm, A. and Cooke, B. M. (2010). Recent insights into alteration of red blood cells by *Babesia bovis*: moovin'forward. *Trends in Parasitology*, 26(12): 591-599.

Gomes, A. F. and Wouters, G. (1991). Species identification of *Rhipicephalus duttoni* in relation to the *Rhipicephalus appendiculatus* group (Acarina: Ixodidae). *Journal of Medical Entomology*, 28(1): 16-18.

Grisi, L., Leite, R. C., Martins, J. R. D. S., Barros, A. T. M. D., Andreotti, R., Cançado, P. H. D. and Villela, H. S. (2014). Reassessment of the potential economic impact of cattle parasites in Brazil. *Revista Brasileira de Parasitologia Veterinaria*, 23(2): 150-156.

Guglielmone, A. A. and Nava, S. (2013). *Epidemiología y control de las garrapatas de los bovinos en la Argentina. Enfermedades parasitarias con importancia clínica y productiva en rumiantes: fundamentos epidemiológicos para su diagnostico y control* [*Cattle tick epidemiology and control in Argentina. Parasitic diseases with clinical and productive importance in ruminants: epidemiological foundations for their diagnosis and control*], 441-456. ed. A. Nari and C. Fiel, Editorial Hemisferio Sur, Buenos Aires.

Guglielmone, A. A., Robbins, R. G., Apanaskevich, D. A., Petney, T. N., Estrada-Peña, A. and Horak, I. G. (2014). *The hard ticks of the world*. New York: Springuer.

Guglielmone, A. A., R. G. Robbins, D. A. Apanaskevich, T. N. Petney, A. Estrada Peña, and I. G. Horak. (2014). The genus *Rhipicephalus*. Pages 527–603 in: *The hard ticks of the world*. Springer, Dordrecht, The Netherlands.

Hassan SM (1997). *Ecological studies on Rhipicephalus appendiculatus and Amblyomma variegatum (Acari: Ixodidae): Drop-off rhythms, development, survival and seasonal population dynamics*. PhD thesis. Department of Parasitology, Faculty of Veterinary Medicine, Kenyatta University, Nairobi, Kenya.

Herrin, B. Tick-borne Rickettsial Infections of Dogs. *Today's Veterinary Practice*. Retrieved from https://todaysveterinarypractice.com/tick-borne-rickettsial-infections-of-dogs/

Hogstral, H. (1956). Ticks of the Sudan (with special reference to equatorial province and with preliminary review of the genera *Boophilus, Margaropus* and *Hyalomma*, 5-912.

Hurtado, O. J. B., and Giraldo-Ríos, C. (2018). Economic and health impact of the ticks in production animals. In: *Ticks and Tick-Borne Pathogens*. IntechOpen.

Jefferies, R., Ryan, U. M., Jardine, J., Broughton, D. K., Robertson, I. D. and Irwin, P. J. (2007). Blood, bull terriers and babesiosis: further evidence for direct transmission of *Babesia gibsoni* in dogs. *Australian Veterinary Journal*, 85(11): 459-463.

Jiménez, C., Benito, A., Arnal, J. L., Ortín, A., Gómez, M., López, A., & Lacasta, D. (2019). *Anaplasma ovis* in sheep: Experimental infection, vertical transmission and colostral immunity. *Small Ruminant Research*, 178, 7-14.

Jones, L. D., Davies, C. R., Williams, T., Cory, J. and Nuttall, P. A. (1990). Non-viraemic transmission of Thogoto virus: vector efficiency of *Rhipicephalus appendiculatus* and *Amblyomma variegatum*. *Transactions of the Royal Society of Tropical Medicine and Hygiene*, 84(6): 846-848.

Jongejan, F. and Uilenberg, G. (1994). Ticks and control methods. *Revue scientifique et technique* (International Office of Epizootics), 13(4): 1201-1226.

Jongejan, F., Musisi, F. L., Moorhouse, P. D. S., Snacken, M. & Uilenberg, G. (1986). *Theileria* taurotragi in Zambia. *Veterinary Quarterly*, 8(3): 261-263.

Jonsson, N. N., Mayer, D. G., Matschoss, A. L., Green, P. E. and Ansell, J. (1998). Production effects of cattle tick (*Boophilus microplus*) infestation of high yielding dairy cows. *Veterinary Parasitology*, 78(1): 65-77.

Julla, I. I.; Tingwa, N. O. and Kwajok, V. L. (1989). Theileriosis in Eastern, Central and Southern Africa: *Proceedings of a Workshop on East Coast Fever Immunization*, Held in Lilongwe, Malawi 20-22 September 1988. Dolan, T. T. (Ed.). ILRI (aka ILCA and ILRAD).

Kaiser, M. N., Sutherst, R. W. and Bourne, A. S. (1982). Relationship between ticks and zebu cattle in southern Uganda. *Tropical Animal Health and Production*, 14(2): 63-74.

Kaur, N., Prasher, P., Kumar, K. and Dhingra, S. (2019). *Rhipicephalus (Boophilus) microplus* (Arachnida: Ixodidae) larvae infestation of human eyelids. *A Rare Case. Acarologia*, 59 (1): 21-25.

Kocan, K. M., de la Fuente, J., Blouin, E. F., Coetzee, J. F. & Ewing, S. A. (2010). The natural history of *Anaplasma marginale*. *Veterinary Parasitology*, 167(2-4): 95-107.

Konnai, S., Yamada, S., Imamura, S., Simuunza, M., Chembensof, M., Chota, A. and Onuma, M. (2007). Attachment duration required for *Rhipicephalus appendiculatus* to transmit *Theileria parva* to the host. *Vector-Borne and Zoonotic Diseases*, 7(2): 241-248.

Labruna, M. B., Naranjo, V., Mangold, A. J., Thompson, C., Estrada-Peña, A., Guglielmone, A. A. and de La Fuente, J. (2009). Allopatric speciation in ticks: genetic and reproductive divergence between geographic strains of *Rhipicephalus (Boophilus) microplus. BMC Evolutionary Biology*, 9(1): 46.

Lara, B., Conan, A., Thrall, M. A., Ketzis, J. K., Branford, G. C. and Rajeev, S. (2020). Serologic and Molecular Diagnosis of *Anaplasma platys* and *Ehrlichia canis* Infection in Dogs in an Endemic Region. *Pathogens*, 9(6): 488.

Latif, A. A., Rowlands, G. J., Punyua, D. K., Hassan, S. M. and Capstick, P. B. (1995). An epidemiological study of tick-borne diseases and their effects on productivity of zebu cattle under traditional management on Rusinga Island, western Kenya. *Preventive Veterinary Medicine*, 22(3): 169-181.

Latrofa, M. S., Dantas-Torres, F., de Caprariis, D., Cantacessi, C., Capelli, G., Lia, R. P. and Otranto, D. (2016). Vertical transmission of *Anaplasma platys* and *Leishmania infantum* in dogs during the first half of gestation. *Parasites & Vectors*, 9(1): 269.

Latrofa, M. S., Dantas-Torres, F., Giannelli, A. and Otranto, D. (2014). Molecular detection of tick-borne pathogens in *Rhipicephalus sanguineus* group ticks. *Ticks and Tick-Borne Diseases, 5*(6): 943-946.

Lawrence, J. A., Perry, B. D. and Williamson, S. (2004). *Corridor disease*. Oxford University Press.

Levin, M. L., Killmaster, L. F., Zemtsova, G. E., Ritter, J. M., and Langham, G. (2014). Clinical presentation, convalescence, and relapse of Rocky Mountain spotted fever in dogs experimentally infected via tick bite. *PLoS One*, 9(12): e115105.

Lew-Tabor, A. E. and Valle, M. R. (2016). A review of reverse vaccinology approaches for the development of vaccines against ticks and tick-borne diseases. *Ticks and Tick-Borne Diseases*, *7*(4): 573-585.

Lima, M. L. F., Soares, P. T., Ramos, C. A. N., Araújo, F. R., Ramos, R. A. N., Souza, I. I. F. and Alves, L. C. A. (2010). Molecular detection of *Anaplasma platys* in a naturally-infected cat in Brazil. *Brazilian Journal of Microbiology*, 41(2): 381-385.

Lohmeyer, K. H., May, M. A., Thomas, D. B. and Pérez de León, A. A. (2018). Implication of nilgai antelope (Artiodactyla: Bovidae) in reinfestations of *Rhipicephalus (Boophilus) microplus* (Acari: Ixodidae) in South Texas: a review and update. *Journal of Medical Entomology*, 55(3): 515-522.

Lohmeyer, K. H., Pound, J. M., May, M. A., Kammlah, D. M. and Davey, R. B. (2011). Distribution of *Rhipicephalus (Boophilus) microplus* and *Rhipicephalus (Boophilus) annulatus* (Acari: Ixodidae) infestations detected in the United States along the Texas/Mexico border. *Journal of Medical Entomology*, 48(4): 770-774.

Low, V. L., Tay, S. T., Kho, K. L., Koh, F. X., Tan, T. K., Lim, Y. A. L., Ong, B. L., Panchadcharam, C., Norma-Rashid, Y. and Sofian-Azirun, M. (2015). Molecular characterisation of the tick *Rhipicephalus microplus* in Malaysia: new insights into the cryptic diversity and distinct genetic assemblages throughout the world. *Parasites & Vectors*, 8(1): 1-10.

Macaluso, K., Paddock, C. D., (2014). Tick-borne spotted fever group rickettsioses and Rickettsia species. *Biology of Ticks*, 2: 211-250.

Madder, M., Thys, E., Achi, L., Touré, A. and De Deken, R. (2011). *Rhipicephalus (Boophilus) microplus*: a most successful invasive tick species in West-Africa. *Experimental and Applied Acarology*, 53(2): 139-145.

Maggi, R. G., Mascarelli, P. E., Havenga, L. N., Naidoo, V. & Breitschwerdt, E. B. (2013). Co-infection with *Anaplasma platys, Bartonella henselae* and *Candidatus Mycoplasma haematoparvum* in a veterinarian. *Parasites & Vectors*, 6(1): 103.

Marcellino, W. L., Julla, I. I., Salih, D. A. and El Hussein, A. R. (2011). Ticks infesting cattle in the Central Equatoria region of South Sudan. *Onderstepoort Journal of Veterinary Research*, 78(1): 1-5.

Martínez-Ocampo, F., Quiroz-Castañeda, R. E., Amaro-Estrada, I., Dantán-González, E., de la Torre, J. F. P. and Rodríguez-Camarillo, S. (2020). Whole-Genome Sequencing of Mexican Strains of *Anaplasma marginale*: An Approach to the Causal Agent of Bovine Anaplasmosis. *International Journal of Genomics*, 2020.

McLeod, R., and Kristjanson, P. (1999). *Final report of joint esys/ILRI/ACIAR TickCoast project-Economic impact of ticks and tick-borne diseases to livestock in Africa, Asia and Australia*. International Livestock Research Institute, Nairobi.

Moraes-Filho, J., Marcili, A., Nieri-Bastos, F. A., Richtzenhain, L. J. and Labruna, M. B. (2011). Genetic analysis of ticks belonging to the *Rhipicephalus sanguineus* group in Latin America. *Acta tropica*, 117(1): 51-55.

Morel, N., De Salvo, M. N., Cicuttin, G., Rossner, V., Thompson, C. S., Mangold, A. J. and Nava, S. (2019). The presence of *Borrelia theileri* in Argentina. *Veterinary Parasitology: Regional Studies and Reports*, 17*:* 100314.

Morel, P. C. and Uilenberg, G. (1981). The nomenclature of some *Theileria* species (Sporozoa, Babesioidea) of domestic ruminants. *Revue d'élevage et de médecine vétérinaire des pays tropicaux*, 34(2): 139-143.

Morzaria, S. P., Tatchell, R. J., Minor, R., Pedersen, V., Julla, I., Rahim, A. and Van Aarle, P. A. M. (1981). Preliminary studies on the epidemiology of theileriosis in Eastern Equatoria Province of the Sudan. In *Advances in the control of theileriosis* (pp. 83-85). Springer, Dordrecht.

Mosqueda, J., Hidalgo-Ruiz, M., Calvo-Olvera, D. A., Hernandez-Silva, D. J., Ueti, M. W., Mercado-Uriostegui, M. A. and Igarashi, I. (2019). RON2, a novel gene in *Babesia bigemina*, contains conserved, immunodominant B-cell epitopes that induce antibodies that block merozoite invasion. *Parasitology*, 146(13): 1646-1654.

Moyo, D. Z., Chakuya, J. and Sungirai, M. (2018). Ixodid ticks of African buffalo (*Syncerus caffer*), impala *(Aepyceros melampus*) and elephant (*Loxodonta africana*) in five protected park estates in the Zambezi valley, Zimbabwe. *Experimental and Applied Acarology*, 75(4): 409-417.

Mukhebi, A. W., Perry, B. D. and Kruska, R. (1992). Estimated economics of theileriosis control in Africa. *Preventive Veterinary Medicine*, 12(1-2): 73-85.

Musisi, F. L., Jongejan, F., Pegram, R. G. and Munyama, G. (1984). Isolation and transmission of *Theileria mutans* (Chisamba) in Zambia. *Research in Veterinary Science*, 36(1): 129-131.

Nakaghi, A. C. H., Machado, R. Z., Coasta, M. T., André, M. R. and Baldani, C. D. (2008). Canine ehrlichiosis: clinical, hematological, serological and molecular aspects. *Ciência Rural*, 38(3): 766-770.

Nava, S., Beati, L., Venzal, J. M., Labruna, M. B., Szabó, M. P., Petney, T. and Mangold, A. J. (2018). *Rhipicephalus sanguineus* (Latreille, 1806): neotype designation, morphological re-description of all parasitic stages and molecular characterization. *Ticks and tick-borne Diseases*, 9(6): 1573-1585.

Nicholson, W. L., Sonenshine, D. E., Noden, B. H. and Brown, R. N. (2019). Ticks (Ixodida). In: *Medical and Veterinary Entomology* (603-672). Academic Press.

Norval, R. A. I., Perry, B. D., and Young, A. S. (1992). *The epidemiology of theileriosis in Africa*. ILRI (aka ILCA and ILRAD).

Norval, R. A. I., Yunker, C. E., Gibson, J. D. and Deem, S. L. D. (1988). Field sampling of unfed nymphs of *Amblyomma hebraeum*. *Experimental and Applied Acarology*, 4(2): 173-177.

Nutall, P. (2014). Tick-borne viruses. *Biology of ticks*, 2: 180-210.

Ochanda, H., Young, A. S., Wells, C., Medley, G. F. and Perry, B. D. (1996). Comparison of the transmission of *Theileria parva* between different instars of *Rhipicephalus appendiculatus*. *Parasitology*, 113(3) 243-253.

Ojeda Chi, M. M., Rodriguez Vivas, R. I., EsteveGasent, M. D., Pérez de León, A. A., Modarelli, J. J. and Villegas Perez, S. L. (2019). Ticks infesting dogs in rural communities of Yucatan, Mexico and molecular diagnosis of rickettsial infection. *Transboundary and Emerging Diseases*, 66(1): 102-110.

Otranto, D., Dantas-Torres, F., Weigl, S., Latrofa, M. S., Stanneck, D., Decaprariis, D. and Baneth, G. (2011). Diagnosis of *Hepatozoon canis* in young dogs by cytology and PCR. *Parasites & Vectors*, 4(1): 1-6.

Parola, P. and Raoult, D. (2001). Ticks and tickborne bacterial diseases in humans: an emerging infectious threat. *Clinical infectious diseases*, 32(6): 897-928.

Parola, P., Paddock, C. D. and Raoult, D. (2005). Tick-borne rickettsioses around the world: emerging diseases challenging old concepts. *Clinical Microbiology Reviews*, 18(4): 719-756.

Penzhorn, B. L., Vorster, I., Harrison-White, R. F. and Oosthuizen, M. C. (2017). Black-backed jackals (*Canis mesomelas*) are natural hosts of *Babesia rossi*, the virulent causative agent of canine babesiosis in sub-Saharan Africa. *Parasites & Vectors*, 10(1): 1-6.

Pérez de León, A. P., Vannier, E., Almazán, C. and Krause, P. J. (2014). Tick-borne protozoa. In *Biology of Ticks*, Vol 2 (147-179). Oxford University Press New York.

Perez, M., Bodor, M., Zhang, C., Xiong, Q. & Rikihisa, Y. (2006). Human infection with *Ehrlichia canis* accompanied by clinical signs in Venezuela. *Annals of the New York Academy of Sciences,* 1078(1): 110-117.

Pound, J. M., George, J. E., Kammlah, D. M., Lohmeyer, K. H. and Davey, R. B. (2010). Evidence for role of white-tailed deer (Artiodactyla: Cervidae) in epizootiology of cattle ticks and southern cattle ticks (Acari: Ixodidae) in reinfestations along the Texas/Mexico border in south Texas: a review and update. *Journal of Economic Entomology*, 103(2): 211-218.

Quiroz-Castañeda, R. E., Amaro-Estrada, I. and Rodríguez-Camarillo, S. D. (2016). *Anaplasma marginale*: diversity, virulence, and vaccine landscape through a genomics approach. *BioMed Research International*, 2016.

Roberts, F. H. S. Australian Ticks. In: *Melbourne,* Vic: CSIRO; 1970. 267. http://hdl.handle.net/102.100.100/318329?index=1

Rodrigues, D. S. and Leite, R. C. (2013). Economic impact of *Rhipicephalus (Boophilus) microplus*: estimate of decreased milk production on a dairy farm. *Arquivo Brasileiro de Medicina Veterinária e Zootecnia,* 65(5): 1570-1572.

Rodríguez, S. D., Ortiz, M. A. G., Ocampo, R. J. and y Murguía, C. A. V. (2009). Molecular epidemiology of bovine anaplasmosis with a particular focus in Mexico. *Infection, Genetics and Evolution*, 9(6): 1092-1101.

Rodríguez, S. D. (2007). La anaplasmosis bovina en México. *Simposio Internacional Garrapatas, Babesiosis y Anaplasmosis* [Bovine anaplasmosis in Mexico. *International Symposium on Ticks, Babesiosis and Anaplasmosis*]. Universidad Autónoma de Tamaulipas, Mexico, pg 105-118.

Rodríguez-Vivas, R. I., Apanaskevich, D. A., Ojeda-Chi, M. M., Trinidad-Martínez, I., Reyes-Novelo, E., Esteve-Gassent, M. D. and de León, A. P. (2016). Ticks collected from humans, domestic animals, and wildlife in Yucatan, Mexico. *Veterinary Parasitology*, 215: 106-113.

Rodríguez-Vivas, R. I., Grisi, L., de León, A. A. P., Villela, H. S., de Jesús Torres-ACoasta, J. F., Sánchez, H. F. and Carrasco, D. G. (2017). Potential economic impact assessment for cattle parasites in Mexico. Review. *Revista Mexicana de Ciencias Pecuarias*, 8(1): 61-74.

Rodríguez-Vivas, R. I., Ojeda-Chi, M. M., Rosado-Aguilar, J. A., Trinidad-Martínez, I. C., Torres-ACoasta, J. F. J., Ticante-Perez, V. and Vázquez-Gómez, G. (2013). Red deer (*Cervus elaphus*) as a host for the cattle tick *Rhipicephalus microplus* (Acari: Ixodidae) in Yucatan, Mexico. *Experimental and Applied Acarology*, 60(4): 543-552.

Romiti, F., Magliano, A., Antognetti, V., Manna, G., Cersini, A., Scicluna, M. T. and De Liberato, C. (2020). Investigation of Ixodid ticks as vectors of *Babesia caballi* and *Theileria equi* (Protozoa: Apicomplexa) in central Italy. *Journal of Vector Ecology*, 45(1): 25-31.

Roy, B. C., Estrada-Peña, A., Krücken, J., Rehman, A. and Nijhof, A. M. (2018). Morphological and phylogenetic analyses of *Rhipicephalus microplus* ticks from Bangladesh, Pakistan and Myanmar. *Ticks and tick-borne Diseases*, 9(5): 1069-1079.

Sánchez Vizcaíno, J. M., Dixon, L. and Heath, L. (2019). *OIE, Manual of Diagnostic Tests and Vaccines for Terrestrial Animals*.

Silatsa, B. A., Kuiate, J. R., Njiokou, F., Simo, G., Feussom, J. M. K., Tunrayo, A. and Opiyo, S. O. (2019). A countrywide molecular survey leads to a seminal identification of the invasive cattle tick *Rhipicephalus (Boophilus) microplus* in Cameroon, a decade after it was reported in Cote d'Ivoire. *Ticks and tick-borne Diseases,* 10(3): 585-593.

Smith, T. and Kilborne, F. L. (1893). *Investigations into the nature, causation, and prevention of Texas or southern cattle fever* (*No. 1*). US Department of Agriculture, Bureau of Animal Industry.

Solano-Gallego, L. and Baneth, G. (2011). Babesiosis in dogs and cats-expanding parasitological and clinical spectra. *Veterinary Parasitology*, 181(1): 48-60.

Solano-Gallego, L., Sainz, Á., Roura, X., Estrada-Peña, A. and Miró, G. (2016). A review of canine babesiosis: the European perspective. *Parasites & Vectors*, 9(1): 336.

Sonenshine, D. E. and Roe, R. M. (2013). Ticks, people, and animals. *Biology of Ticks* Volume 1, 1.

Stich, R. W., Schaefer, J. J., Bremer, W. G., Needham, G. R. and Jittapalapong, S. (2008). Host surveys, ixodid tick biology and transmission scenarios as related to the tick-borne pathogen, *Ehrlichia canis. Veterinary Parasitology*, 158(4): 256-273.

Strauss, J. H., Strauss, E. G. (2008). Minus-strand RNA viruses. *Viruses and human disease*, 137.

Strickland, R. K. (1976). *Ticks of veterinary importance* (No. 485). US Department of Agriculture, Animal and Plant Health Inspection Service.

Szabó, M. P. J., Pinter, A., and Labruna, M. B. (2013). Ecology, biology and distribution of spotted-fever tick vectors in Brazil. *Frontiers in Cellular and Infection Microbiology*, 3: 27.

Tinoco-Gracia, L., Lomelí, M. R., Hori-Oshima, S., Stephenson, N. & Foley, J. (2018). Molecular Confirmation of Rocky Mountain Spotted Fever Epidemic Agent in Mexicali, Mexico. *Emerging Infectious Diseases*, 24(9): 1723–1725.

Tuppurainen, E. S. M., Lubinga, J. C., Stoltsz, W. H., Troskie, M., Carpenter, S. T., Coetzer, J. A. & Oura, C. A. L. (2013). Mechanical transmission of lumpy skin disease virus by *Rhipicephalus appendiculatus* male ticks. *Epidemiology and Infection*, 141(2): 425-430.

Tuppurainen, E. S., Stoltsz, W. H., Troskie, M., Wallace, D. B., Oura, C. A. L., Mellor, P. S. and Venter, E. H. (2011). A potential role for ixodid (hard) tick vectors in the transmission of lumpy skin disease virus in cattle. *Transboundary and emerging diseases,* 58(2): 93-104.

Uilenberg, G. (1981). Theilerial species of domestic livestock. In *Advances in the Control of Theileriosis* (4-37). Springer, Dordrecht.

Uilenberg, G., Perie, N. M., Lawrence, J. A., De Vos, A. J., Paling, R. W. and Spanjer, A. A. M. (1982). Causal agents of bovine theileriosis in southern Africa. *Tropical Animal Health and Production*, 14(3): 127-140.

Ulloa García, A., Bermudez, S. E., Rosario-Cruz, R., García Rejon, J., Baak Baak, C., López López. Ricketsiosis en México: Revisión y estado actual de la ricketsiosis en el país [Ricketsiosis in Mexico: Review and current status of ricketsiosis in the country]. En: *Algunas enfermedades infecciosas en Mexico: Morbilidad y mortalidad.* (2019) Universidad Autónoma de Chiapas, pp 19-40.

Vega, C. A., Millán, J. V. F., Ramírez, E. E. R., Aragón, J. A. R. and Alarcón, G. J. C. (1999). Insuficiente inmunidad cruzada en bovinos por *Babesia bigemina* y/o *Babesia bovis* derivadas del cultivo in vitro [Insufficient cross immunity in cattle due to Babesia bigemina and / or Babesia bovis derived from in vitro culture]. *Revista Mexicana de Ciencias Pecuarias*, 37(1).

Vial, H. J. and Gorenflot, A. (2006). Chemotherapy against babesiosis. *Veterinary Parasitology*, 138(1-2), 147-160.

Volpato, A., Agostini, M., Tonin, A. A. and Da Silva, A. S. (2013). High infestation by *Rhipicephalus (Boophilus) microplus* in a mare: a rare case report. *Comparative Clinical Pathology*, 22(6): 1263-1265.

Walker, A. R., Bouattour, A., Camicas, J. L., Estrada-Pena, A., Horak, I. G., Latif, A. A. and Preston, P. M. (2003). *Ticks of domestic animals in Africa: a guide to identification of species. Bioscience Reports*. Scotland: The University of Edinburgh.

Wikel, S. K. (2014). Tick-host interactions. *Biology of ticks*, 2: 88-128.

Wilson, A. J., Parker, R. and Trueman, K. F. (1980). Experimental immunization of calves against *Anaplasma marginale* infection: observations on the use of living *A. centrale* and *A. marginale*. *Veterinary Parasitology*, 7(4): 305-311.

Wilson D. (2011). *Clinical Veterinary Advisor: The Horse*. 2011 Elsevier Inc.

In: The Entomological Guide to *Rhipicephalus*
Editors: Sachin Kumar, Raquel Cossio Bayugar et al.
ISBN: 978-1-53619-619-1

Chapter 16

DEVELOPMENT OF VACCINES AGAINST *RHIPICEPHALUS MICROPLUS* TICKS

Juan Mosqueda**, Roberto I. Guerrero-Solorio and Maria Martina E. Pérez-Soria

Immunology and Vaccines Laboratory, Natural Sciences College, Autonomous University of Queretaro. Queretaro, Mexico

ABSTRACT

Vaccines are the preparation of microorganisms (live, attenuated, or inactivated) or their components, such as proteins, carbohydrates, or nucleic acids, to be administered into an organism to induce immunity. The goal of vaccinations is to protect an individual against future pathogen invasion. The *Rhipicephalus microplus* tick is considered one of the most harmful parasites to cattle in tropical and sub-tropical regions, infecting approximately 80% of livestock worldwide. There is an urgent need to develop control strategies that consider the safety of the environment and animal and human health. Vaccination against *R. microplus* is developed from tick antigens and is regarded as an economical and safe alternative to acaricides, is environmentally friendly, and safe for administration. Monoantigenic vaccines are commercially available, containing the *R. microplus* Bm86 antigen. However, these vaccines' efficacy has been highly variable due to allelic variations in the *Bm86* gene, supporting the need to continue vaccination research.

Several antigens have been evaluated as vaccine candidates, including Bm86, subolesin, Ferritin-2, aquaporins, serpins, vitellin, and VDAC. This chapter will describe the key aspects of the immune response required from vaccine antigens and revise the findings reported from the results of tick antigen vaccines. We will address the importance of using reverse vaccinology and immuno-bioinformatics to identify new *R. microplus* antigens isolated from the intestine, salivary gland, and ovary to develop a multiantigenic vaccine against *R. microplus*.

Keywords: vaccination, reverse vaccinology, multiantigenic vaccine, antibody, *Rhipicephalus microplus*

---

* Corresponding Author's Email: joel.mosqueda@uaq.mx.

1. Introduction

Vaccines are the preparation of microorganisms (live, attenuated, or inactivated) or their components, such as proteins, carbohydrates, or nucleic acids, to be administered into an organism to induce immunity. The goal of a vaccine is to protect an individual against future pathogen invasion. Since the first scientific demonstration by Edward Jenner, who identified that protection against diseases could be acquired by introducing material from an attenuated pathogen into the body (Jenner, 1798), the development of vaccines has had a positive impact on human and animal health. Vaccination has greatly reduced the burden of infectious diseases and is considered a fundamental human right (Plotkin & Plotkin, 2012).

For a vaccine to work, it must first induce an innate immune response, followed by an adaptive (humoral, cellular, or both) immune response (Vetter et al., 2018). Vaccines have several advantages against antibiotic treatment sprays to fight diseases since vaccines do not generate resistance, are cost-effective, require few applications, and are ecologically friendly (Donald, 1994). Therefore, scientific efforts are directed towards vaccine development, including those against ticks. Ticks are a major disease transmitting vector, carrying bacteria, viruses, and protozoans (de la Fuente et al., 2008b; Jonsson, 2006; Stuchin et al., 2016). Anti-tick vaccines are a relatively new concept in the era of vaccine development, with the fever cattle tick *Rhipicephalus microplus* vaccination being the most successful (de la Fuente et al., 2007a; Willadsen, 1990). Despite all the efforts invested in discovering and evaluating vaccine candidates against *R. microplus*, there is only one antigen, Bm86 (*Boophilus microplus* 86), which is commercialized for vaccine use (de la Fuente et al., 1999; Willadsen et al., 1995). Bm86 vaccines have a strong efficacy in reducing tick numbers in vaccinated cattle and therefore reduced pathogen transmission (de la Fuente et al., 1998; McKenna et al., 1998). However, Bm86 is a variable antigen, and vaccines based on this antigen have limited efficacy against ticks, which have allelic gene variations, compared to no variation (Garcia-Garcia et al., 1999; Kaewmongkol et al., 2015; Sossai et al., 2005). Therefore, an alternative vaccine against *R. microplus* is urgently needed. This chapter will describe the characteristics to be considered for vaccine development, including the new approaches for tick antigen identification. We will describe antigens that have been successfully discovered as potential vaccine targets, and discuss the prospects for tick vaccine development.

2. Vaccines against Ticks: An Immunological Perspective

The rationale behind all vaccines is to pre-expose an organism to components of a pathogen to build an immune response. Therefore, immediately after vaccination, the immune system of cattle will mount an immune response comprised of two initial phases: an innate immune response followed by and an adaptive immune response. Vaccine components that are recognized as foreign material and initiate the innate immune response (i.e., Bm86) are referred to as pathogen-associated-molecular patterns (PAMP). All PAMPs, have a conserved structure common to pathogens and are recognized by receptors on and in the host cells called pattern-recognition receptors (PRR) (Janeway, 1989). As soon as PRR recognizes PAMPs, a local inflammatory reaction develops to attract specialized cells and molecules to eliminate the foreign material.

The innate immune response is necessary for a vaccine to work; it activates the initial steps of recognizing and processing the foreign material and triggers the adaptive immune response (Iwasaki & Medzhitov, 2010; Robinson et al, 2009). Therefore, it is logical to expect that an initial inflammatory reaction at the application site is generated if the vaccine works. Ideally, the vaccine components are ingested via phagocytosis and endocytosis by antigen-presenting cells (APC), such as macrophages or dendritic cells (DC) (Kapsenberg, 2003). The dermis is rich in DCs called Langerhans cells, which can ingest and process the vaccine components and migrate to the lymph nodes. Langerhans cells express the membrane proteins called class II, major histocompatibility complex (MHC-II) molecules, which they use to present the foreign antigen to other immune cells (Marrack et al., 1989). Mature T helper lymphocytes (Th) recognize the specific peptide through their T-cell receptor and activate, leading to the proliferation of Th cells, which initiates the start of the adaptive response (Igietseme et al., 2004). When B cells recognize specific cytokines produced by activated Th cells, they proliferate and transform into antibody-secreting plasma cells. Antibodies can bind to small parts of antigens called B-cell epitopes (Igietseme et al., 2004). A humoral immune response is based on activated plasma cells that secrete millions of antibodies into the plasma, a component of blood (Shapiro-Shelef & Calame, 2005).

For a tick vaccine to be successful, antibodies in the plasma generated through an immune response from vaccination are bound to ticks via their specific B-cell epitopes, which will then impede tick function. Identifying tick antigens that can generate a specific humoral immune response is not easy. Not all antigens contain B or T-cell epitopes and not all cytokines induce antibody secretion by plasma cells. Moreover, since protein-derived antigens do not contain many PAMPs, these have to be added to the vaccination solution in the form of adjuvants. Not all adjuvants induce a potent innate immune response needed to initiate a humoral response (Hoebe et al., 2004; Pashine et al., 2005). Finally, not all tick antigens are exposed to antibodies during tick feeding, and antigenic variation must be considered before including an antigen in a vaccine. Deep knowledge in immunology and tick physiology and anatomy is needed to develop successful tick-specific vaccines.

3. Vaccine Candidate Antigens for *R. microplus*

In 1979, it was reported for the first time the successful use of tick-derived antigens as a vaccine against ticks. The discovery showed that specific tissues from the tick could elicit an immune response that impeded the tick's reproductive performance by reducing blood feeding and oviposition (Allen & Humphreys, 1979). The first antigen identified from *R. microplus* was a glycoprotein in the midgut cells of about 89 KDa called Bm86 (Bm was from the last scientific name *Boophilus microplus*). Vaccination with cellular fractions containing this antigen reduced tick feeding and oviposition by impeding blood circulation by midgut cells (Willadsen et al., 1988). Since then, several antigens have been described and evaluated for their ability to reduce one or several parameters of the biological fitness of ticks: feeding, oviposition, larvae hatching, and survival. Below is a review of the most characterized antigens identified so far.

3.1. Bm86

Bm86 is a glycoprotein which function has not been completely elucidated (Gough & Kemp, 1993) and was first isolated by Willadsen et al., in 1989 (Willadsen et al., 1989). This protein contains several epidermal growth factor (EGF)-like domains that may be involved in blood coagulation and cell growth (Rand et al., 1989). The nucleotide sequence of the cDNA contains a 1982 base-pair open reading frame. It predicts that Bm86 contains 650 amino acids, including a 19 amino acid long signal sequence and a 23 amino acid long hydrophobic region adjacent to the carboxyl terminus. The deduced Bm86 protein sequence's main feature is the repeated pattern of six cysteine residues, suggesting several epidermal growths factor-like domains, with a molecular weight of 90 KDa (Rand et al., 1989; Richardson et al., 1993). They are predominantly located in the apical membrane of intestinal epithelial cells (Gough & Kemp, 1993). Bm86 vaccines were commercially developed in Cuba (Gavac) and Australia (TickGARD) and were evaluated in field and laboratory testing, which showed an average protection level of 80% (Garcia-Garcia et al., 1998). The Bm96 vaccine caused digestive cell lysis within the intestines of ticks, which reduced their ability to feed on vaccinated animals' blood. In general, this vaccine reduces the number and weight of female tics and in the oviposition (Rand et al., 1989).

To test *P. pastoris* derived Bm86 antigen as a vaccine to control *B. annulatus* infestations (Gavac, Heber Biotec SA, Havana), cattle were vaccinated with the Bm86 antigen and artificially infected with *B. annulatus* larvae. The results demonstrated high protective efficacy of 99.9% compared to unvaccinated cattle (Fragoso et al., 1998). In another trial, adult female *R. annulatus* and *R. microplus* ticks were administered to cattle either vaccinated with Bm86 or unvaccinated to determine the vaccine's efficacy. The efficacy of Bm86 was higher on *R. annulatus* (99.6%) than on *R. microplus* (85.2%), and tick weight was lower in *R. annulatus* (141 ± 59mg) than in *R. microplus* (253 ± 9mg) (Canales et al., 2009). However, its efficacy varies according to the existing strains in a different geographical area, which may be due to allelic variations in the *Bm86* gene (de la Fuente et al., 2007a; de la Fuente & Kocan, 2006). It is considered that a 3.4% difference in the Bm86 amino acid sequence between the vaccinal and the field strains is sufficient to cause an ineffective vaccine immune response (Garcia-Garcia et al., 1999).

3.2. Bm95

Bm95 was isolated from intestinal cells infected with the *R. microplus* strain A from Argentina (Garcia-Garcia et al., 2000). In 1999 the Bm95 antigen sequence was determined and found identical to the Bm86, except for 21 amino acids (Peconick et al., 2008). Bm95 is a glycoprotein with an unknown function (Kumar et al., 2009). The Bm95 protein has a molecular weight of approximately 75 - 80 KDa and contains 569 amino acids (Garcia-Garcia et al., 2000). It is predominantly located in the apical membrane of the intestinal epithelial cells (Gough & Kemp, 1993). In tests carried out in cattle, it was effective against different strains of ticks (Andreotti, 2006), including strains not susceptible to Bm86, presenting an efficacy of 89% from the Camcord strain and 58% from the A strain (Garcia-Garcia et al., 2000). Vaccination with Bm86 reduced tick infestations from *B. annulatus* and *B. microplus* from 71% and 40%, respectively. Further, weight (8% and 15%), oviposition (22% and 5%), and egg fertility (25% and 50%), respectively, were all reduced (Canales et al., 2009). In calves immunized with Bm95, 70.76% reduction in the number of infected adult females, a 25.35% reduction in the average weight of the tick egg masses, an 8.99% reduction in the average weight, a 14.00% reduction in the fertility of the female tick engorgements collected from vaccinated calves with a total 81.27% efficacy in the recombinant Bm95 tick antigen vaccination (Kumar et al., 2009). All these data suggest that Bm95, should be studied further in order to determine if it is worth including it as a candidate for future *R. microplus* vaccines, maybe by using it combined with BM86 to investigate if there is an additive effect in the efficacy.

3.3. Subolesin

The antigen Subolesin (Latin: suboles, progeny) was first identified in *Ixodes scapularis* (Almazan et al., 2003) and is conserved among different species of *Ixodes* ticks (de la Fuente & Kocan, 2006). During evolution, Subolesin maintained the way it interacts and regulates pathogenesis and remained unchanged in conducting biological processes such as digestion, reproduction, and development (Artigas-Jeronimo et al., 2018; De La Fuente et al., 2000). In *Rhipicephalus* spp., Subolesin contains 161 aa amino acids, but the sequence varies in *Ornithodoros*, *Ixodes* and *Haemaphysalis* spp. from 173 to 184 aa. (Artigas-Jeronimo et al., 2018). The expression of Subolesin in unfed and fed ticks is similar and is highly expressed in the salivary glands and midgut tissues (Sultana et al., 2015).

Immunization of white-tailed deer with recombinant subolesin resulted in significant protection (83%) from tick infestations (Carreon et al., 2012). Recent results have shown that in soft ticks such as *Ornithodoros erraticus* and *O. moubata*, RNAi-mediated subolesin gene knockdown inhibits tick oviposition and fertility by more than 90%; however, vaccination with recombinant subolesin resulted in remarkably low protective efficacies (only 5-24.5% reduction in oviposition). Due to inconsistencies in results, four synthetic peptides (OE1, OE2, OM1, and OM2) were designed from the *Ornithodoros* subolesin sequences' disordered regions coupled to keyhole limpet hemocyanin (KLH). All these KLH-peptide conjugates induced the synthesis of antibodies that recognized linear epitopes of B cells located in the unstructured loops of the subolesin protein and provided up to 70.1% and 83.1% vaccine efficacy in *O. erraticus* and *O. moubata*, respectively (Manzano-Roman et al., 2015).

3.4. Ferritin 2

Iron homeostasis is very important in many organisms; iron is essential in biology, it is required for numerous cellular processes, however it is a potentially toxic element. Ticks feed of blood and therefore metabolize large amounts of iron during feeding. Ferritins (FER) are iron storage proteins that play a fundamental role in iron metabolism (Hajdusek et al., 2010). Ticks possess intracellular ferritin 1 (FER1), a common heavy chain ferritin (Kopacek et al., 2003). Ferritin 2 (FER2) was identified and characterized by Hajdusek et al., which is secreted into the hemolymph and has no functional orthologs in vertebrates (Hajdusek et al.,

2009). FER2 is required during iron repletion to distribute and move the iron from intestinal cells to hemolymph. Although it is not clear how iron is delivered to peripheral tissues, it seems that FER2 plays a role in the supply of iron to peripheral tissues to maintain their normal functions and is expressed during all stages of tick development (Hajdusek et al., 2009).

FER2 from *R. microplus* is a protein of 189 amino acids, of which the first 16 constitute the signal peptide to allow protein secretion. It has a molecular weight of 19.8 kDa (Hajdusek et al., 2010). FER2 is expressed in the intestine and is secreted into the hemolymph of ticks, where it acts as an iron transporter to supply iron to the ovaries and salivary glands (Hajdusek et al., 2009). Vaccination trials with FER2 against *R. microplus* and *R. annulatus* infestations achieved a reduction in the percentage of infestation by 30% and 42%, respectively and had an overall efficacy of 64% and 72%, respectively. The FER2 vaccination was slightly lower than the Bm86 immunization, which had a 60% and 100% efficacy response, against *R. microplus* and *R. annulatus*, respectively. FER2 vaccination also caused weight loss in *R. microplus* and *R. annulatus* females by 12% and 4% and reduced oviposition by 14% and 1%, respectively. Overall there was a reduction of *R. microplus* and *R. annulatus* egg hatching by 40% and 51%, respectively. Within the same study, the effectiveness of FER2 vaccination on *I. ricinus* in rabbits was evaluated, which reached a 95% effectiveness against infestation (Hajdusek et al., 2010).

3.5. Aquaporins

Aquaporins (AQPs) or transmembrane water channels are proteins, discovered in 1988, which facilitate water passage through an osmotic gradient. These channels are crucial in maintaining homeostasis and cryoprotection in ticks (Finn & Cerdà, 2015; Perez Di Giorgio et al., 2014). After ingesting blood, ticks are subjected to great osmoregulatory stress; for example, females grow 100 times their normal size by the end of ingestion (Campbell et al., 2010). The salivary glands play a leading role in this process, returning approximately 75% of the water in blood acquired during feeding to the host (Kaufman & Phillips, 1973).

Many AQPs were described as impervious to small solutes and ions in different tissues, however, some recent studies have recognized this as incorrect (Isibashi et al., 1997). AQPs can be divided into water-selective or aquaglyceroporins, the latter permeable to water, glycerol, urea, and other non-electrolyte elements. All AQPs have a similar structure, and contain six transmembrane domains, intracellular C and H terminal ends, and a pair of conserved regions (asparagine-proline-alanine, NPA), which fold towards themselves to form a pore. Aquaglyceroporins are characterized by having an aspartate after the second NPA region (NPARD) (Bowman & Sauer, 2004). Due to the major role AQPs play in water homeostasis, they have been investigated for their vaccination potential. Guerrero et al., (2014) used transcriptome studies of *R. microplus*, several gene coding regions were discovered to encode proteins with similarity to aquaporins, and then, they characterized an AQP in *R. microplus* and named it RmAQP1. It has an open reading frame that codes for a 216 amino acid peptide. This protein had a high expression in the tick synganglion and a lower expression in the male's midgut and ovaries of the female. Vaccination whit RmAQP1 in Holstein calves reduced the number of replete ticks by 29% compared to the control group, and oviposition and hatching were lower with a total efficacy of 75% during the first test and 68% in the second one (Guerrero et al., 2014).

Campbell et al., (2010) silenced the gene of AQP in *I. ricinus* (IrAQP1) by RNA interference (RNAi), using a preparation of double-stranded RNA (AQP1-dsRNA), showing decreased female tick weight when the gene for this protein was silenced. However, they showed an aquaporin expression profile different from the work of Guerrero et al., of 2014 since this aquaporin was not detected in the adult male or the female synganglion. This pattern is consistent with those of aquaporins from *R. sanguineus* (RsAQP1) (Ball et al., 2009) and a Aquporin-like protein of *Dermacentor variabilis* (Holmes et al., 2008). Hussein et al., (2015) identified another AQP in *R. microplus* (RmAQP2) with a different transcription pattern; in non-fed larvae, replete nymphs, and the salivary glands and intestines of semi-replete females. The expression of RmAQP2 was suppressed by RNAi, and its effect was tested in cows that had either an acute phase of *Babesia bovis* infection or free *Babesia* infection and was compared against a control group (Hussein et al., 2015). The adult ticks' average weight was higher in the acute-stage infection than in the control group, and there was no effect on oviposition or egg weight. Still, fertility only reached 50% compared to the control group (90.2%).

An important finding is that no larva survived in the group vaccinated during the acute-infection stage than the control group. This could be due to the deleterious effect that infection of *B. bovis* parasites exerts on the tick (Hussein et al., 2015). In the second group (the free *Babesia* infection group), the average weight of the replete female ticks was higher, there was no change in oviposition, and a decrease in the weight of the eggs was observed compared to the control group. The hatching percentage was also lower, 81.2% for the silenced group than 97.7% for the control group, but in both groups, the larvae had a 100% survival rate (Hussein et al., 2015).

Contreras and de la Fuente (Contreras & de la Fuente, 2017) used an *I. ricinus* AQP (IrAQP, 35 kDa) and a recombinant AQP, formed from regions conserved between the different AQPs of different tick species (CoAQP, 15 kDa) as antigens to test their efficacy of infection control. They vaccinated rabbits infested with *I. ricinus* larvae with these antigens. Mortality increased 6.25 and 5.17 times, molting was reduced by 13 and 73%, and the total effectiveness of 32 and 80%, respectively, was observed compared to the control group (Contreras & de la Fuente, 2017). The effect of the AQP-based vaccines presented by the authors on *I. ricinus* larvae infestation and molting became this antigen as a possible candidate protective antigen for the control of different tick species.

3.6. Voltage-Dependent Anion Channel (VDAC)

VDAC is a protein with a molecular weight between 30 and 35 kDa that polymerizes to form channels in the mitochondria's outer membrane in many species such as *Drosophila melanogaster*, *Saccharomyces cereviciae*, *Mus musculus*, *Triticum aestivium*, *Lotus japonicus*, and *Homo sapiens*. VDAC channels allow the flow of small molecules within the mitochondrial intermembrane space and participate in many complex interactions that regulate cell metabolism and apoptosis (Roman et al., 2006; Sorgato & Moran, 1993; Young et al., 2007). Additionally, VDAC is located on transmembrane and is involved in interactions between host and pathogen. It is expressed in the intestinal cell membrane of different species such as *R. microplus*, *R. annulatus*, *R. sanguineus*, *A. variegatum,* and *Ixodes scapularis*.

In 2012 the group of Rodriguez-Hernandez et al., identified a VDAC-like protein expressed in the intestines of *R. microplus*, naming it BmVDAC (Rodriguez-Hernandez et al., 2012). They also observed a relationship between BmVDAC and the proteins expressed during the sexual stages of *B. bigemina*. This became a promising target for vaccine development to control ticks (Corona-Guerrero et al., 2017; Ribeiro et al., 2011; Rodriguez-Hernandez et al., 2012; Rodriguez-Hernandez et al., 2015). In 2015, Rodriguez-Hernandez et al., further observed overexpression of BmVDAC in intestinal cells of *R. microplus* ticks infected with *B. bigemina*, suggesting that BmVDAC could be an essential target in vaccination strategies against the invasion of apicomplexan parasites into intestinal cells from ticks (Rodriguez-Hernandez et al., 2015).

Recently, Ortega-Sanchez et al., (2020) used an antigen against BmVDAC and obtained effectiveness of 82% against *R. microplus* infestations compared to the control group, although the efficacy decreased when *R. micropus* was infected with *B. bigemina* (34%). Similarly, a reduction in egg fertility (34%) in the group not infected with *B. bigemina* compared to no apicomplexan parasites was observed. In general, cattle who were vaccinated with BmVDAC and infected with *Babesia*-free ticks had fewer full ticks, lower tick weight, and lower egg numbers. More studies are necessary to assess the usefulness of BMVDAC as a transmission-blocking vaccine.

3.7. Vitellogenin

Wigglesworth made the first report of this protein in blood-sucking arthropods in 1943 (Wigglesworth, 1943). Vitellogenin in ticks is synthesized both in the mid-intestine and body fat (Boctor & Kamel, 1976; Coons et al., 1982). Once synthesized, vitellogenin is exported into the hemolymph and accumulates in the oocytes as vitellin (Sappington & Raikhel, 1998). Studies carried out regarding this protein suggest that it is synthesized by replete ticks in the fatty body, consisting of cell branches attached to the tracheal system and internal organs (Taheri et al., 2014). Vitellin is the main component of tick eggs and the primary source of amino acids and energy for developing embryos and larvae that cannot feed (Logullo et al., 2002).

Vitellin also acts as a carrier hormone associated with embryogenesis (Bownes et al., 1988). The gene contains an open reading frame of 5400 base pairs encoding a large protein of 1800 aa (Seixas et al., 2012). Silencing the expression of the vitellogenin receptor in the ovary by RNA interference results in reduced egg production (i.e., reduced egg weight), a lower rate of embryonic development, reduced egg hatching, and reduced transovarial transmission of *B. bovis* (Hussein et al., 2019). In results published by Tellam et al., a reduction in oviposition of *R. microplus* was observed when fed on sheep vaccinated with vitellin (Tellam et al., 2002). Similar results were obtained in *Ornithodoros moubata* ticks when feeding on the blood of rabbits immunized with vitellin, which is why it is considered a potential candidate for developing an anti-tick vaccine (Taheri et al., 2014).

3.8. Serpin

Serine proteinase inhibitors (Serpins) were first recognized in human plasma in 1894 by Fermi and Pernossi (Schulze et al., 1994). The serpin superfamily consists of more than 60 members, which can be further divided into subfamilies based on their amino acid sequence (Potempa et al., 1994). The term serpin was given in 1985 by Carrel and Travis for the entire protein superfamily (Carrell & Travis, 1985). The primary function of serpin in tick saliva is to regulate proteolytic enzymes under physiological and pathological conditions (Kim et al., 2015). Among the physiological processes, serpins regulate coagulation, fibrinolysis, complement activation, angiogenesis, apoptosis, inflammation, neoplasia, viral pathogenesis, vasoconstriction, and modulate host defense mechanisms (Blisnick et al., 2017; Janciauskiene, 2001).

This gene named *Haemaphysalis longicornis serpin-1* (HLS1) was used to clone a 1350bp cDNA which encodes a 378 polypeptide with high sequence similarity to several known serpins. When *Haemaphysalis longicornis* ticks were allowed to feed on rabbits vaccinated with recombinat HLS1 (rHLS1) expressed in *Escherichia coli*, there was a 43.9% and 11.2% mortality in nymphs and adult ticks, respectively, compared with those ticks fed on unvaccinated rabbits (Sugino et al., 2003). Two *Rhipicephalus appendiculatus* serine protease inhibitors (RAS-1 and -2) were used as an anti-tick vaccine in a cocktail. *R. appendiculatus* ticks were fed on cattle vaccinated with rRAS1 and rRAS2. The results showed a 61.4% reduction in nymph engorgement and a 28% and 43% mortality increase in adult female and male ticks, respectively (Imamura et al., 2006). *Rhipicephalus appendiculatus* serpin-3 (RAS-3), *R. appendiculatus* serpin-4 (RAS-4) and a 36-kDa immuno-dominant protein of *R. appendiculatus* (RIM36) were used as anti-tick vaccine to control ixodid ticks. Bovines vaccinated with rRAS3, rRAS4 and rRIM36, displayed 39.5% mortality of *R. appendiculatus* female ticks in the infected group compared with only 12.8% mortality with the control group (Imamura et al., 2008).

Amblyomma americanum serpin 19 is an anti-coagulant protein, inhibiting the activity of five of the eight serine protease blood clotting factors. Serpin 19 has been identified and characterized by its reactive central loop that is 100% conserved in *Ixodid* ticks. In a subsequent study, biochemical characterization revealed that the ubiquitously transcribed *A. americanum* serpin 19 is an anticoagulant protein that inhibits five of the eight serine proteases (Kim et al., 2015). Serpin 27 (AAS27) isolated from *A. americanum* is an immunogenic saliva protein that is injected into the host within the first day of tick feeding and is an anti-inflammatory protein that may act to block plasmin and trypsin functions (Tirloni et al., 2019). There are currently more than twenty serpins that have not been characterized and thus remain unevaluated as tick vaccine antigens.

3.9. Glutathione S-Transferase

Glutathione S-transferase (GST) is a group of ubiquitous enzymes that catalyze the conjugation between glutathione and the first step in forming mercapturic acids (Barnes et al., 1959). It is also part of the cellular detoxification pathway of endogenous and xenobiotic compounds, including those from pesticides and acaricides (insects and ticks) (da Silva Vaz et al., 2004; Freitas et al., 2007; Hayes & Pulford, 1995; Huang et al., 1998; Ranson et al., 1997; Syvanen et al., 1994; Syvanen et al., 1996).

GST in *Rhipicephalus microplus* has a mass of 25.8 kDa, and the cDNA contains an open reading frame that codes for a 223 amino acid protein with a theoretical isoelectric point of 8.53 (He et al., 1999). GST has been used as an antigen in some vaccines directed against some parasites, obtaining some encouraging results (Scott & McManus, 2000; Veerapathran et al., 2009; Vibanco-Perez et al., 1999; Zhan et al., 2005). In an experiment carried out by Bifano et al., (2014), interference RNA was used to decrease protein up-regulation during *A. marginale* infection, however when testing the capacity to infect the vertebrate host, there were no significant changes. On the other hand, Kocan et al., (2009) used the same technology in *Dermacentor variabilis* and found a significant decrease in the dense forms of *A. marginale* in the intestine and salivary gland after feeding, but found a higher concentration of the bacteria in the fatty body, possibly due to changes in bacterial development during the tick's development cycle.

Almazan et al., (2010) silenced different genes, including GST, in *R. microplus* where they found the tick's ability to remain attached to the bovine host decreased by 60%, with no effect on mortality or oviposition. Finally, work done by Parizi et al., (2011) used a GST of recombinant *Haemophysalis longicornis* to immunize cattle and challenged them with *R. microplus*. They obtained fewer full females, lower tick weight, and decreased oviposition and egg fertility, with a total efficacy of 57%. Therefore, more studies are needed on this antigen to consider it as a successful vaccine candidate.

3.10. Voraxin

Studies in *D. variabilitis* first described a substance produced in the male gonads transmitted to the adult female during feeding, allowing the female tick to move from a slow phase to a fast-feeding phase until it was full (Pappas & Oliver, 1972). This protein is also known as the male factor (MF) or engorgement factor (EF), which was later given the name, voraxin (Latin vorax: voracious, gluttonous) (Weiss & Kaufman, 2004). Voraxin stimulates the salivary gland degeneration, development of the ovaries, vitellogenesis, oviposition, and promotes replete feeding (Weiss & Kaufman, 2004). Voraxin is a polypeptide with a molecular weight of 12.5kDa. Its production is initiated in the male accessory gland and stimulates the final phase of tick sperm maturation after the spermatophore has been transferred to the female (Shepherd et al., 1982). The amino acid sequence analysis of voraxin-α in *R. appendiculatus* indicates a high similarity with voraxin-α from *A. hebraeum* and *Dermacentor variabilis*. Therefore, voraxin-α is conserved in different tick genera (Yamada et al., 2009).

When Rabbits were immunized with voraxin and infected with *Amblyomma hebraeum* ticks it was observed that 74% of those ticks did not engorge in the first 14 days. These ticks had reduced saliva production, which affected nutrition-ingestion and had reduced oocyte development (Weiss & Kaufman, 2004). In another study, rabbits immunized with recombinant (r)-voraxin-α and voraxin-β were highly resistant to the tick infestation. Initial vaccination of rabbits with r-voraxin-α elicited humoral immunity and conferred protective immunity against female *R. appendiculatus*, resulting in the reduced fed weight with 42.7%

and 38.6% of ticks detaching from vaccinated rabbits did not lay eggs (Yamada et al., 2009)(Yamada et al., 2009). Voraxin has not been studied in *R. microplus* ticks thus far.

3.11. Chitinase

Chitinase is an enzyme induced by ecdysteroids, which are arthropod steroid hormones, to degrade chitin at the time of molting (You & Fujisaki, 2009). Chitin is an insoluble, structural polysaccharide important in the arthropod exoskeleton (Neville et al., 1976) and is involved in the exoskeleton's shedding process. The *chitinase gene* (*cht1*) contains a 2790 bp open reading frame that encodes 930 amino acid residues with a molecular weight of 104 kDa (You et al., 2003). Native chitinase is highly expressed in tick epidermis and midgut (You et al., 2003). The cuticle protects the tick from the environment and from invading pathogens. Ticks fed on mice immunized with chitinase had longer feeding and molting periods but had decreased molting rates. Recombinant CHT1 (rCHT1) expressed in *Escherichia coli* was used to immunize mice (You & Fujisaki, 2009). Ticks showed longer feeding periods in larval ticks, and the difference in egg weight and shedding periods were longer than controls. Nymphs showed a lower molting rate (76.7%) than 96.7% in the control group (You & Fujisaki, 2009).

Amblyomma americanum (Aam) constitutively expresses long (L) and short (S) acid chitinases (Ach). The full-length AamAch-L and AamAch-S cDNAs are 1959 and 1718 bp in length and contain open reading frames of 1332 and 1104 bp that encode proteins 443 and 367 amino acid residues in length. AamAch-L is predicted to be extracellular, and AamAch-S is predicted to be intracellular and expressed in multiple organs. The function of AamAch-L and AamAch-S is possibly associated with the formation and maintenance of the A. americanum tick cement cone's stability. Silencing the AamAch-L and AamAch-S mRNA's dual RNA interference caused the ticks to adhere lightly to the host's skin, making the ticks \ easily detached from the host, possibly due to malformations of the cementum cone that the tick secretes to anchor itself into the host's skin. Chitinase is also important for the successful feeding of ticks (Kim et al., 2014). To date, there are no studies in *R. microplus* chitinase as a vaccine antigen.

3.11.1. Cytochrome Oxidase Subunit III

Some studies have shown that infections with *A. marginale* alter some genes' transcripts in ticks (de la Fuente et al., 2007b; de la Fuente et al., 2008a; Ramabu et al., 2010; Zivkovic et al., 2009; Zivkovic et al., 2010). Cytochrome oxidase subunit III (COXIII)has not been evaluated yet in a vaccine trial, so its potential has not been elucidated. COXIII is the main subunit of cytochrome oxidase (complex IV) in mitochondria. In humans, COXIII has a molecular weight of 30 kDa and is composed of 261 amino acids (Zong et al., 2013). It is found in the mitochondrial membrane and contains different transmembrane domains (The UniProt Consortium, 2016). It is one of five complexes of the mitochondrial metabolic pathway called oxidation phosphorylation, which includes producing ATP and reactive oxygen species (ROS) (Goshima & Vale, 2003). In work carried out by Bifano and his collaborators (2014), it was discovered that increased COXIII transcription was present in the salivary gland due to infection by *A. marginale*. A test was carried out using interference RNA.When analyzing the transmission of *A. marginale*, in uninfected bovines, this pathogen's transmission was interrupted, resulting in negative nested PCR tests and bacterial presence in blood smears and serological tests (Bifano et al., 2014).

Genetic silencing of COXIII affects the proper transmission of *A. marginale* because it interferes with the release of the bacteria and affects pathogen viability. Since COXIII is involved in ATP production and other essential metabolic functions, the disruption of COXIII leads to reduced ATP production. *A. marginale* leaves cells by fusing with the cell membrane, requiring ATP, therefore the cells outflow by *A. marginale* becomes partially or entirely impeded (Bifano et al., 2014; Blouin & Kocan, 1998). Reduced viability of *A. marginale* is observed when COXIII is silenced, leading to high production of ROS causing cell damage and reduced cell proliferation (Bifano et al., 2014; Drose & Brandt, 2012). The interactions between ticks and pathogens they transmit like *A. marginale* are still unclear.

3.12. Kunitz

Hematophagous arthropods are pool feeders and have developed an important diversity of pharmacological compounds in their saliva like anti-haemostatic, anti-inflammatory and immunomodulatory molecules that contribute to prevent physiological responses. They also require strong vasodilatory substances to increase blood flow perfusion to superficial regions of the skin (Fontaine et al., 2011). These anticoagulants maintain blood in its liquid form within the feeding lesion and the midgut of the tick. This allows the parasites to obtain and digest their blood meal over extended periods (van de Locht et al., 1996). Some studies have shown a specific lineage expansion of several families of secretory proteins predicted for both hard and soft ticks (Mans et al., 2002), such as lipocalins, which are secretory proteins, peptides containing the Kunitz domain, and metalloproteases.

The Kunitz domain-containing family of proteins represents between 5% and 20% of all predicted secretory proteins (Ribeiro et al., 2006). Macedo-Ribeiro et al., (2008) isolated the Kunitz protein from *R. microplus*, which they called boophilin. Mature boophilin is composed of two canonical Kunitz-type domains and inhibits the major procoagulant enzyme, thrombin. In contrast to previously characterized natural thrombin inhibitors, boophilin significantly interferes with the proteolytic activity of other serine proteinases such as trypsin and plasmin (Macedo-Ribeiro et al., 2008). Several Kunitz proteins derived from soft ticks inhibit the serine proteins, which are part of the blood coagulation cascade and platelet aggregation. Ornithodorin, and Savignin were isolated from the blood sucking soft tick *Ornithodoros moubata*, are a potent and highly selective bi-Kunitz thrombin inhibitors (van de Locht et al., 1996). Studies in *R. microplus* reported a Kunitz domain-containing protein family, includes proven tissue factor pathway inhibitor-like (TFPI-like) anti-haemostatics such as ixolaris and penthalaris that play a crucial role during tick feeding. and a protein family of 42 homologs (Louw et al., 2013). Two TFPI homologs identified in *I. scapularis* are: ixolaris, a 140 amino acids-long protein containing ten cysteines and two kunitz-like domains, and penthalaris, a 308-amino acid protein (35 kDa, isoelectric point 8.58) with 12 cysteine bridges, and five tandem kunitz domains, which have been shown to be potent anticoagulants (Francischetti et al., 2004; Francischetti et al., 2002). Immunization against *A. ceylanicum* using kunitz-type inhibitors confer partial protection against growth retardation associated with hookworm, supporting their possible use as a parasitic vaccine candidate (Chu et al., 2004).

3.13. Metalloproteases

Metalloproteases (MPs) are essential in various biological processes of hematophagous organisms, inhibiting coagulation, degrading extracellular matrix proteins, and inhibiting host tissue repair due to their anti-angiogenic activity (Ali et al., 2015; Kessenbrock et al., 2010). MPs have been identified in several tick species and are considered crucial for the maintenance of blood intake. The increase in the transcription of MPs in the salivary gland is correlated with increased blood-feeding of the ticks (Adamson et al., 2013; Barnard et al., 2012; Decrem et al., 2008a; Decrem et al., 2008b; Francischetti et al., 2003; Harnnoi et al., 2007).

Like other hematophagous organisms, ticks produce several molecules that modulate their host's defense to maintain their food source. These molecules interfere with the host homeostasis by affecting host coagulation and the innate and adaptive immune responses. MPs are considered essential for this process (Decrem et al., 2008a; Francischetti, 2010; Imamura et al., 2009). Studies of MPs in ticks are scarce, and their role in parasitism is still not very clear. However, many of the MPs from ticks belong to the Metzicine group (Barnard et al., 2012; Decrem et al., 2008a; Decrem et al., 2008b; Harnnoi et al., 2007). Metzicines participate in the development and remodeling of tissue, maintenance, nonspecific protein degradation of ingested proteins, as well as highly specific cleavage events to activate or inactivate themselves, proenzymes or bioactive peptides (Chang & Werb, 2001; Gomis-Ruth, 2009). Metzicins are further subdivided into families: astacins, serralisins, matrix metalloproteases, snapalysins, leishmanolysins, eucolysins, pappalisins, and reprolysins (Gomis-Ruth, 2009).

The *R. microplus* Metalloprotease 4 (BrRm-MP4) is a reprolysin, was characterized in work carried out by Ali et al., to understand the role MPs have in tick feeding and their physiology. BrRm-MP4 has an estimated molecular weight of 62 kDa. It contains 555 amino acids: the first 18 make up the signal peptide, amino acids 19 to 170 belong to the propeptide, amino acids 171 to 390 make up the reprolycin domain with the zinc-binding site and the methionine turn, and amino acids 391 to 555 is the cysteine-rich domain at the c terminal end. Reprolysin has an isoelectric point of 6.43 (Ali et al., 2014). BrRm-MP4 was selected for a vaccine trial (Ali et al., 2015) because it is a polypeptide with high antigenicity determined by *in silico* analysis. It has a comparable transcriptional pattern in most tissues compared to other MPs. It is also transcribed in males, larvae, and ovaries, suggesting an important role during tick development and reproduction (Ali et al., 2015; Ali et al., 2014).

In this vaccination trial, the number of replete females, the collected females' weight, the oviposition percentage, and the hatching percentage were lower than the control group, a general effectiveness result of 60% was obtained (Ali et al., 2015). In other work that used MPs as antigens, increased mortality in adults and nymphs with *Haemophysalis longicornis* is observed (Imamura et al., 2009). The use of interfering RNA against MPs was found to interfere with blood consumption termination and alters the fibrinolytic and fibrinogenolytic activities of tick saliva in *I. ricinus*. In vaccination trials involving MPs, a decreased intake of blood, weight gain, and oviposition rate in the same ticks were observed (Decrem et al., 2008a; Decrem et al., 2008b). Metalloproteases have potential as antigens in anti-tick vaccines, their efficacy maybe improved by multi-antigenic formulations with other proteins that confer some degree of immunity against ticks.

REFERENCES

Adamson, S. W., Browning, R. E., Chao, C. C., Bateman, R. C., Jr., Ching, W. M., & Karim, S. (2013). Molecular characterization of tick salivary gland glutaminyl cyclase. *Insect Biochem Mol Biol,* 43(9), 781-793. doi: 10.1016/j.ibmb.2013.05.011.

Ali, A., Parizi, L. F., Guizzo, M. G., Tirloni, L., Seixas, A., Vaz Ida, S., Jr., & Termignoni, C. (2015). Immunoprotective potential of a *Rhipicephalus* (Boophilus) microplus metalloprotease. *Vet Parasitol,* 207(1-2), 107-114. doi: 10.1016/j.vetpar.2014.11.007.

Ali, A., Tirloni, L., Isezaki, M., Seixas, A., Konnai, S., Ohashi, K., da Silva Vaz Junior, I., & Termignoni, C. (2014). Reprolysin metalloproteases from Ixodes persulcatus, *Rhipicephalus sanguineus* and *Rhipicephalus microplus* ticks. *Exp Appl Acarol,* 63(4), 559-578. doi: 10.1007/s10493-014-9796-9.

Allen, J. R., & Humphreys, S. J. (1979). Immunisation of guinea pigs and cattle against ticks. *Nature,* 280(5722), 491-493. doi: 10.1038/280491a0.

Almazan, C., Kocan, K. M., Bergman, D. K., Garcia-Garcia, J. C., Blouin, E. F., & de la Fuente, J. (2003). Identification of protective antigens for the control of Ixodes scapularis infestations using cDNA expression library immunization. *Vaccine,* 21(13-14), 1492-1501. doi: 10.1016/s0264-410x(02)00683-7.

Almazan, C., Lagunes, R., Villar, M., Canales, M., Rosario-Cruz, R., Jongejan, F., & de la Fuente, J. (2010). Identification and characterization of *Rhipicephalus* (*Boophilus*) microplus candidate protective antigens for the control of cattle tick infestations. *Parasitol Res,* 106(2), 471-479. doi: 10.1007/s00436-009-1689-1.

Andreotti, R. (2006). Performance of two Bm86 antigen vaccin formulation against tick using crossbreed bovines in stall test. *Rev Bras Parasitol Vet,* 15(3), 97-100.

Artigas-Jeronimo, S., Villar, M., Cabezas-Cruz, A., Valdes, J. J., Estrada-Pena, A., Alberdi, P., & de la Fuente, J. (2018). Functional Evolution of Subolesin/Akirin. *Front Physiol,* 9, 1612. doi: 10.3389/fphys.2018.01612.

Ball, A., Campbell, E. M., Jacob, J., Hoppler, S., & Bowman, A. S. (2009). Identification, functional characterization and expression patterns of a water-specific aquaporin in the brown dog tick, *Rhipicephalus sanguineus*. *Insect Biochem Mol Biol,* 39(2), 105-112. doi: 10.1016/j.ibmb.2008.10.006.

Barnard, A. C., Nijhof, A. M., Gaspar, A. R., Neitz, A. W., Jongejan, F., & Maritz-Olivier, C. (2012). Expression profiling, gene silencing and transcriptional networking of metzincin metalloproteases in the cattle tick, *Rhipicephalus* (*Boophilus*) *microplus*. *Vet Parasitol,* 186(3-4), 403-414. doi: 10.1016/j.vetpar.2011.11.026.

Barnes, M. M., James, S. P., & Wood, P. B. (1959). The formation of mercapturic acids. 1. Formation of mercapturic acid and the levels of glutathione in tissues *Biochem J,* 71(4).

Bifano, T. D., Ueti, M. W., Esteves, E., Reif, K. E., Braz, G. R., Scoles, G. A., Bastos, R. G., White, S. N., & Daffre, S. (2014). Knockdown of the *Rhipicephalus microplus* cytochrome c oxidase subunit III gene is associated with a failure of Anaplasma marginale transmission. *PLoS One, 9*(5), e98614. doi: 10.1371/journal.pone.0098614.

Blisnick, A. A., Foulon, T., & Bonnet, S. I. (2017). Serine Protease Inhibitors in Ticks: An Overview of Their Role in Tick Biology and Tick-Borne Pathogen Transmission. *Front Cell Infect Microbiol,* 7, 199. doi: 10.3389/fcimb.2017.00199.

Blouin, E. F., & Kocan, K. M. (1998). Morphology and development of Anaplasma marginale (Rickettsiales: Anaplasmataceae) in cultured Ixodes scapularis (Acari: Ixodidae) cells. *J Med Entomol,* 35(5), 788-797. doi: 10.1093/jmedent/35.5.788.

Boctor, F. N., & Kamel, M. Y. (1976). Purification and characterization of two lipovitellins from eggs of the tick, Dermacentor andersoni. *Insect Biochem Mol Biol,* 6, 233–240.

Boldt, H. B., Overgaard, M. T., Laursen, L. S., Weyer, K., Sottrup-Jensen, L., & Oxvig, C. (2001). Mutational analysis of the proteolytic domain of pregnancy-associated plasma protein-A (PAPP-A): classification as a metzincin. *Biochem J,* 358, 359–367.

Bowman, A. S., & Sauer, J. R. (2004). Tick salivary glands: function, physiology and future. *Parasitology,* 129 Suppl, S67-81. doi: 10.1017/s0031182004006468.

Bownes, M., Shirras, A., Blair, M., Collins, J., & Coulson, A. (1988). Evidence that insect embryogenesis is regulated by ecdysteroids released from yolk proteins. *Proc Natl Acad Sci U S A,* 85(5), 1554-1557. doi: https://doi.org/10.1073/pnas.85.5.1554.

Campbell, E. M., Burdin, M., Hoppler, S., & Bowman, A. S. (2010). Role of an aquaporin in the sheep tick Ixodes ricinus: assessment as a potential control target. *Int J Parasitol,* 40(1), 15-23. doi: 10.1016/j.ijpara.2009.06.010.

Canales, M., Almazan, C., Naranjo, V., Jongejan, F., & de la Fuente, J. (2009). Vaccination with recombinant Boophilus annulatus Bm86 ortholog protein, Ba86, protects cattle against B. annulatus and B. microplus infestations. *BMC Biotechnol,* 9, 29. doi: 10.1186/1472-6750-9-29.

Carrell, R., & Travis, J. (1985). α1-Antitrypsin and the serpins: variation and countervariation. *Trends Biochem Sci,* 10(1), 20-24. doi: 10.1016/0968-0004(85)90011-8.

Carreon, D., de la Lastra, J. M., Almazan, C., Canales, M., Ruiz-Fons, F., Boadella, M., Moreno-Cid, J. A., Villar, M., Gortazar, C., Reglero, M., Villarreal, R., & de la Fuente, J. (2012). Vaccination with BM86, subolesin and akirin protective antigens for the control of tick infestations in white tailed deer and red deer. *Vaccine,* 30(2), 273-279. doi: 10.1016/j.vaccine.2011.10.099.

Chang, C., & Werb, Z. (2001). The many faces of metalloproteases: cell growth, invasion, angiogenesis and metastasis. *Trends Cell Biol,* 11, s37-s43.

Chu, D., Bungiro, R. D., Ibanez, M., Harrison, L. M., Campodonico, E., Jones, B. F., Mieszczanek, J., Kuzmic, P., & Cappello, M. (2004). Molecular Characterization of Ancylostoma ceylanicum Kunitz-Type Serine Protease Inhibitor: Evidence for a Role in Hookworm-Associated Growth Delay. *Infection and Immunity,* 72(4), 2214–2221. doi: 10.1128/IAI.72.4.2214-2221.2004.

Contreras, M., & de la Fuente, J. (2017). Control of infestations by Ixodes ricinus tick larvae in rabbits vaccinated with aquaporin recombinant antigens. *Vaccine,* 35(9), 1323-1328. doi: 10.1016/j.vaccine.2017.01.052.

Coons, L. B., Tarnowski, B., & Ourth, D. D. (1982). *Rhipicephalus sanguinius*: localization of vitellogenin synthesis by immunological methods and electron microscopy. *Exp Parasitol,* 54(3), 331-339. doi: 10.1016/0014-4894(82)90042-x.

Corona-Guerrero, I., Camacho-Nuez, M., Carbajal-Gámez, B. I., & Mosqueda, J. (2017). Caracterización molecular del gen del canal de aniones dependiente de voltaje (VDAC) en *Rhipicephalus sanguineus* Latreile, 1806 (Ixodida: Ixodidae) y *Rhipicephalus* annulatus Say (Ixodida: Ixodidae) [Molecular characterization of the voltage-gated anion channel (VDAC) gene in *Rhipicephalus sanguineus* Latreile, 1806 (Ixodida: Ixodidae) and *Rhipicephalus* annulatus Say (Ixodida: Ixodidae)]. *Entomología Mexicana,* 4, 633 - 638.

da Silva Vaz, I., Jr., Torino Lermen, T., Michelon, A., Sanchez Ferreira, C. A., Joaquim de Freitas, D. R., Termignoni, C., & Masuda, A. (2004). Effect of acaricides on the activity of a *Boophilus microplus* glutathione S-transferase. *Vet Parasitol,* 119(2-3), 237-245. doi: 10.1016/j.vetpar.2003.11.004.

de la Fuente, J., Almazan, C., Canales, M., Perez de la Lastra, J. M., Kocan, K. M., & Willadsen, P. (2007a). A ten-year review of commercial vaccine performance for control of tick infestations on cattle. *Anim Health Res Rev,* 8(1), 23-28. doi: 10.1017/S1466252307001193.

de la Fuente, J., Blouin, E. F., Manzano-Roman, R., Naranjo, V., Almazan, C., Perez de la Lastra, J. M., Zivkovic, Z., Jongejan, F., & Kocan, K. M. (2007b). Functional genomic studies of tick cells in response to infection with the cattle pathogen, Anaplasma marginale. *Genomics,* 90(6), 712-722. doi: 10.1016/j.ygeno.2007.08.009.

de la Fuente, J., Blouin, E. F., Manzano-Roman, R., Naranjo, V., Almazan, C., Perez de la Lastra, J. M., Zivkovic, Z., Massung, R. F., Jongejan, F., & Kocan, K. M. (2008a). Differential expression of the tick protective antigen subolesin in anaplasma marginale- and A. phagocytophilum-infected host cells. *Ann N Y Acad Sci,* 1149, 27-35. doi: 10.1196/annals.1428.056.

de la Fuente, J., Estrada-Pena, A., Venzal, J. M., Kocan, K. M., & Sonenshine, D. E. (2008b). Overview: Ticks as vectors of pathogens that cause disease in humans and animals. *Front Biosci,* 13, 6938-6946. doi: 10.2741/3200.

de la Fuente, J., & Kocan, K. M. (2006). Strategies for development of vaccines for control of ixodid tick species. *Parasite Immunol,* 28(7), 275-283. doi: 10.1111/j.1365-3024.2006.00828.x.

De La Fuente, J., Rodriguez, M., & Garcia-Garcia, J. C. (2000). Immunological control of ticks through vaccination with Boophilus microplus gut antigens. *Ann N Y Acad Sci,* 916, 617-621. doi: 10.1111/j.1749-6632.2000.tb05347.x.

de la Fuente, J., Rodriguez, M., Montero, C., Redondo, M., Garcia-Garcia, J. C., Mendez, L., Serrano, E., Valdes, M., Enriquez, A., Canales, M., Ramos, E., Boue, O., Machado, H., & Lleonart, R. (1999). Vaccination against ticks (Boophilus spp.): the experience with the Bm86-based vaccine Gavac. *Genet Anal,* 15(3-5), 143-148. doi: 10.1016/s1050-3862(99)00018-2.

de la Fuente, J., Rodriguez, M., Redondo, M., Montero, C., Garcia-Garcia, J. C., Mendez, L., Serrano, E., Valdes, M., Enriquez, A., Canales, M., Ramos, E., Boue, O., Machado, H., Lleonart, R., de Armas, C. A., Rey, S., Rodriguez, J. L., Artiles, M., & Garcia, L. (1998). Field studies and cost-effectiveness analysis of vaccination with Gavac against the cattle tick Boophilus microplus. *Vaccine,* 16(4), 366-373. doi: 10.1016/s0264-410x(97)00208-9.

Decrem, Y., Beaufays, J., Blasioli, V., Lahaye, K., Brossard, M., Vanhamme, L., & Godfroid, E. (2008a). A family of putative metalloproteases in the salivary glands of the tick Ixodes ricinus. *FEBS J,* 275(7), 1485-1499. doi: 10.1111/j.1742-4658.2008.06308.x.

Decrem, Y., Mariller, M., Lahaye, K., Blasioli, V., Beaufays, J., Zouaoui Boudjeltia, K., Vanhaeverbeek, M., Cerutti, M., Brossard, M., Vanhamme, L., & Godfroid, E. (2008b). The impact of gene knock-down and vaccination against salivary metalloproteases on blood feeding and egg laying by Ixodes ricinus. *Int J Parasitol,* 38(5), 549-560. doi: 10. 1016/j.ijpara.2007.09.003.

Donald, A. D. (1994). Parasites, animal production and sustainable development. *Vet Parasitol,* 54(1-3), 27-47.

Drose, S., & Brandt, U. (2012). Molecular mechanisms of superoxide production by the mitochondrial respiratory chain. *Adv Exp Med Biol,* 748, 145–169.

Finn, R. N., & Cerdà, J. (2015). Evolution and functional diversity of aquaporins. *Biol Bull,* 229, 6 – 23.

Fontaine, A., Diouf, I., Bakkali, N., Missé, D., Pagès, F., Fusai, T., Rogier, C., & Almeras, L. (2011). Implication of haematophagous arthropod salivary proteins in host-vector interactions. *Parasites & Vectors,* 4, 187. doi: 10.1186/1756-3305-4-187.

Fragoso, H., Rad, P. H., Ortiz, M., Rodriguez, M., Redondo, M., Herrera, L., & de la Fuente, J. (1998). Protection against Boophilus annulatus infestations in cattle vaccinated with the B. microplus Bm86-containing vaccine Gavac. off. *Vaccine,* 16(20), 1990-1992. doi: 10.1016/s0264-410x(98)00116-9.

Francischetti, I. M. (2010). Platelet aggregation inhibitors from hematophagous animals. *Toxicon,* 56(7), 1130-1144. doi: 10.1016/j.toxicon.2009.12.003.

Francischetti, I. M., Mather, T. N., & Ribeiro, J. M. (2003). Cloning of a salivary gland metalloprotease and characterization of gelatinase and fibrin(ogen)lytic activities in the saliva of the Lyme disease tick vector Ixodes scapularis. *Biochem Biophys Res Commun,* 305(4), 869-875. doi: 10.1016/s0006-291x(03)00857-x.

Francischetti, I. M., Mather, T. N., & Ribeiro, J. M. (2004). Penthalaris, a novel recombinant five-Kunitz tissue factor pathway inhibitor (TFPI) from the salivary gland of the tick vector of Lyme disease, Ixodes scapularis. *Thromb Haemost,* 91(5), 886-898. doi: 10. 1160/TH03-11-0715.

Francischetti, I. M., Valenzuela, J. G., Andersen, J. F., Mather, T. N., & Ribeiro, J. M. (2002). Ixolaris, a novel recombinant tissue factor pathway inhibitor (TFPI) from the salivary gland of the tick, Ixodes scapularis: identification of factor X and factor Xa as scaffolds for the inhibition of factor VIIa/tissue factor complex. *Blood,* 99(10), 3602-3612. doi: 10.1182/blood-2001-12-0237.

Freitas, D. R., Rosa, R. M., Moraes, J., Campos, E., Logullo, C., Da Silva Vaz, I., Jr., & Masuda, A. (2007). Relationship between glutathione S-transferase, catalase, oxygen consumption, lipid peroxidation and oxidative stress in eggs and larvae of Boophilus microplus (Acarina: Ixodidae). *Comp Biochem Physiol A Mol Integr Physiol,* 146(4), 688-694. doi: 10.1016/j.cbpa.2006.04.032.

Garcia-Garcia, J. C., Gonzalez, I. L., Gonzalez, D. M., Valdes, M., Mendez, L., Lamberti, J., D'Agostino, B., Citroni, D., Fragoso, H., Ortiz, M., Rodriguez, M., & de la Fuente, J. (1999). Sequence variations in the Boophilus microplus Bm86 locus and implications for immunoprotection in cattle vaccinated with this antigen. *Exp Appl Acarol,* 23(11), 883-895. doi: 10.1023/a:1006270615158.

Garcia-Garcia, J. C., Montero, C., Redondo, M., Vargas, M., Canales, M., Boue, O., Rodriguez, M., Joglar, M., Machado, H., Gonzalez, I. L., Valdes, M., Mendez, L., & de la Fuente, J. (2000). Control of ticks resistant to immunization with Bm86 in cattle vaccinated with the recombinant antigen Bm95 isolated from the cattle tick, Boophilus microplus. *Vaccine,* 18(21), 2275-2287. doi: 10.1016/s0264-410x(99)00548-4.

Garcia-Garcia, J. C., Soto, A., Nigro, F., Mazza, M., Joglar, M., Hechevarria, M., Lamberti, J., & de la Fuente, J. (1998). Adjuvant and immunostimulating properties of the recombinant Bm86 protein expressed in Pichia pastoris. *Vaccine,* 16(9-10), 1053-1055. doi: 10.1016/s0264-410x(97)00266-1.

Gomis-Ruth, F. X. (2009). Catalytic domain architecture of metzincin metalloproteases. *J Biol Chem,* 284, 15353–15357.

Goshima, G., & Vale, R. D. (2003). The roles of microtubule-based motor proteins in mitosis: comprehensive RNAi analysis in the Drosophila S2 cell line. *J Cell Biol,* 162, 1003–1016.

Gough, J. M., & Kemp, D. H. (1993). Localization of a low abundance membrane protein (Bm86) on the gut cells of the cattle tick Boophilus microplus by immunogold labeling. *J Parasitol,* 79(6), 900-907.

Guerrero, F. D., Andreotti, R., Bendele, K. G., Cunha, R. C., Miller, R. J., Yeater, K., & Perez de Leon, A. A. (2014). *Rhipicephalus* (*Boophilus*) *microplus* aquaporin as an effective vaccine antigen to protect against cattle tick infestations. *Parasit Vectors,* 7, 475. doi: 10.1186/s13071-014-0475-9.

Hajdusek, O., Almazan, C., Loosova, G., Villar, M., Canales, M., Grubhoffer, L., Kopacek, P., & de la Fuente, J. (2010). Characterization of ferritin 2 for the control of tick infestations. *Vaccine,* 28(17), 2993-2998. doi: 10.1016/j.vaccine.2010.02.008.

Hajdusek, O., Sojka, D., Kopacek, P., Buresova, V., Franta, Z., Sauman, I., Winzerling, J., & Grubhoffer, L. (2009). Knockdown of proteins involved in iron metabolism limits tick reproduction and development. *Proc Natl Acad Sci USA,* 106(4), 1033-1038. doi: 10. 1073/pnas.0807961106.

Harnnoi, T., Sakaguchi, T., Nishikawa, Y., Xuan, X., & Fujisaki, K. (2007). Molecular characterization and comparative study of 6 salivary gland metalloproteases from the hard tick, Haemaphysalis longicornis. *Comp Biochem Physiol B Biochem Mol Biol,* 147(1), 93-101. doi: 10.1016/j.cbpb.2006.12.008.

Hayes, J. D., & Pulford, D. J. (1995). The glutathione S-transferase supergene family: regulation of GST and the contribution of the isoenzymes to cancer chemoprotection and drug resistance. *Crit Rev Biochem Mol Biol,* 30, 445-600.

He, H., Chen, A. C., Davey, R. B., Ivie, G. W., & George, J. E. (1999). Characterization and molecular cloning of a glutathione S-transferase gene from the tick, Boophilus microplus (Acari: Ixodidae). *Insect Biochem Mol Biol,* 29(8), 737-743. doi: 10.1016/s0965-1748(99)00056-9.

Hoebe, K., Janssen, E., & Beutler, B. (2004). The interface between innate and adaptive immunity. *Nat Immunol,* 5, 971–974. doi: 10.1038/ni1004-971.

Huang, H. S., Hu, N. T., Yao, Y. E., Wu, C. Y., Chiang, S. W., & Sun, C. N. (1998). Molecular cloning and heterologous expression of a glutathione S-transferase involved in insecticide resistance from the diamondback moth, Plutella xylostella. *Insect Biochem. Mol. Biol.,* 28, 651–658.

Hussein, H. E., Johnson, W. C., Taus, N. S., Suarez, C. E., Scoles, G. A., & Ueti, M. W. (2019). Silencing expression of the *Rhipicephalus microplus* vitellogenin receptor gene blocks Babesia bovis transmission and interferes with oocyte maturation. *Parasit Vectors,* 12(1), 7. doi: 10.1186/s13071-018-3270-1.

Hussein, H. E., Scoles, G. A., Ueti, M. W., Suarez, C. E., Adham, F. K., Guerrero, F. D., & Bastos, R. G. (2015). Targeted silencing of the Aquaporin 2 gene of *Rhipicephalus*

(*Boophilus*) *microplus* reduces tick fitness. *Parasit Vectors,* 8, 618. doi: 10.1186/s13071-015-1226-2.

Igietseme, J. U., Eko, F. O., He, Q., & Black, M. C. (2004). Antibody regulation of T-cell immunity: implications for vaccine strategies against intracellular pathogens. *Expert Review of Vaccines,* 3(1), 23-34. doi: 10.1586/14760584.3.1.23.

Imamura, S., da Silva Vaz, I., Jr., Konnai, S., Yamada, S., Nakajima, C., Onuma, M., & Ohashi, K. (2009). Effect of vaccination with a recombinant metalloprotease from Haemaphysalis longicornis. *Exp Appl Acarol,* 48(4), 345-358. doi: 10.1007/s10493-009-9245-3.

Imamura, S., Konnai, S., Vaz Ida, S., Yamada, S., Nakajima, C., Ito, Y., Tajima, T., Yasuda, J., Simuunza, M., Onuma, M., & Ohashi, K. (2008). Effects of anti-tick cocktail vaccine against *Rhipicephalus* appendiculatus. *Jpn J Vet Res,* 56(2), 85-98.

Imamura, S., Namangala, B., Tajima, T., Tembo, M. E., Yasuda, J., Ohashi, K., & Onuma, M. (2006). Two serine protease inhibitors (serpins) that induce a bovine protective immune response against *Rhipicephalus* appendiculatus ticks. *Vaccine,* 24(13), 2230-2237. doi: 10.1016/j.vaccine.2005.10.055.

Iwasaki, A., & Medzhitov, R. (2010). Regulation of adaptive immunity by the innate immune system. *Science,* 327(5963), 291-295. doi: 10.1126/science.1183021.

Janciauskiene, S. (2001). Conformational properties of serine proteinase inhibitors (serpins) confer multiple pathophysiological roles. *Biochimica et Biophysica Acta (BBA) - Molecular Basis of Disease,* 1535(3), 221–235. doi: 10.1016/S0925-4439(01)00025-4.

Janeway, C. A. (1989). Approaching the asymptote? Evolution and revolution in immunology. 1989; 54 Pt 1:1-13. doi: 10.1101/sqb.1989.054.01.003. PMID: 2700931. *Cold Spring Harb Symp Quant Biol,* 54(1), 1-13. doi: 10.1101/sqb.1989.054.01.003.

Jenner, E. (1798). An Inquiry into the Causes and Effects of the Variole Vaccine, or Cow-Pox, 1798. In E. Jenner (Ed.), *Vaccination against Smallpox* (Lit2Go ed.). United States of America.

Joh, J. H., Kim, B. G., Kong, W. S., Yoo, Y. B., Kim, N. K., Park, H. R., Cho, B. G., & Lee, C. S. (2004). Cloning and developmental expression of a metzincin family metalloprotease cDNA from oyster mushroom Pleurotus ostreatus. *FEMS Microbiol Lett,* 239, 57–62.

Jonsson, N. N. (2006). The productivity effects of cattle tick (Boophilus microplus) infestation on cattle, with particular reference to Bos indicus cattle and their crosses. *Vet Parasitol,* 137(1-2), 1-10. doi: 10.1016/j.vetpar.2006.01.010.

Kaewmongkol, S., Kaewmongkol, G., Inthong, N., Lakkitjaroen, N., Sirinarumitr, T., Berry, C. M., Jonsson, N. N., Stich, R. W., & Jittapalapong, S. (2015). Variation among Bm86 sequences in *Rhipicephalus* (*Boophilus*) *microplus* ticks collected from cattle across Thailand. *Exp Appl Acarol,* 66(2), 247-256. doi: 10.1007/s10493-015-9897-0.

Kapsenberg, M. L. (2003). Dendritic-cell control of pathogen-driven T-cell polarization. *Nat Rev Immunol,* 3(12), 984-993. doi: 10.1038/nri1246.

Kaufman, W. R., & Phillips, J. E. (1973). Ion and water balance in the ixodid tick Dermacentor andersoni. I Routes of ion and water excretion. *J Exp Biol,* 58, 523 – 536.

Kessenbrock, K., Plaks, V., & Werb, Z. (2010). Matrix metalloproteinases: regulators of the tumor microenvironment. *Cell*(141), 52-67.

Kim, T. K., Curran, J., & Mulenga, A. (2014). Dual silencing of long and short Amblyomma americanum acidic chitinase forms weakens the tick cement cone stability. *J Exp Biol,* 217(Pt 19), 3493-3503. doi: 10.1242/jeb.107979.

Kim, T. K., Tirloni, L., Radulovic, Z., Lewis, L., Bakshi, M., Hill, C., da Silva Vaz, I., Jr., Logullo, C., Termignoni, C., & Mulenga, A. (2015). Conserved Amblyomma americanum tick Serpin19, an inhibitor of blood clotting factors Xa and XIa, trypsin and plasmin, has anti-haemostatic functions. *Int J Parasitol,* 45(9-10), 613-627. doi: 10.1016/ j.ijpara.2015.03.009.

King, L. S., & Agre, p. (1996). Pathophysiology of the aquaporin water channel. *Annu Rev Physiol,* 58, 619-648.

Knepper, M. A. (1994). The aquaporin family of molecular water channels. *Proc Natl Acad Sci U S A,* 91, 6255-6258.

Kocan, K. M., Zivkovic, Z., Blouin, E. F., Naranjo, V., Almazan, C., Mitra, R., & de la Fuente, J. (2009). Silencing of genes involved in Anaplasma marginale-tick interactions affects the pathogen developmental cycle in Dermacentor variabilis. *BMC Dev Biol,* 9, 42. doi: 10.1186/1471-213X-9-42.

Kopacek, P., Zdychova, J., Yoshiga, T., Weise, C., Rudenko, N., & Law, J. H. (2003). Molecular cloning, expression and isolation of ferritins from two tick species--Ornithodoros moubata and Ixodes ricinus. *Insect Biochem Mol Biol,* 33(1), 103-113. doi: 10.1016/s0965-1748(02)00181-9.

Kumar, A., Garg, R., Yadav, C. L., Vatsya, S., Kumar, R. R., Sugumar, P., Chandran, D., Mangamoorib, L. N., & Bedarkar, S. N. (2009). Immune responses against recombinant tick antigen, Bm95, for the control of *Rhipicephalus* (*Boophilus*) *microplus* ticks in cattle. *Vet Parasitol,* 165(1-2), 119-124. doi: 10.1016/j.vetpar.2009.06.030.

Logullo, C., Moraes, J., Dansa-Petretski, M., Vaz, I. S., Masuda, A., Sorgine, M. H., Braz, G. R., Masuda, H., & Oliveira, P. L. (2002). Binding and storage of heme by vitellin from the cattle tick, *Boophilus microplus*. *Insect Biochem Mol Biol,* 32(12), 1805-1811. doi: 10.1016/s0965-1748(02)00162-5.

Louw, E., van der Merwe, N. A., Neitz, A. W., & Maritz-Olivier, C. (2013). Evolution of the tissue factor pathway inhibitor-like Kunitz domain-containing protein family in *Rhipicephalus microplus*. *Int J Parasitol,* 43(1), 81-94. doi: 10.1016/j.ijpara.2012.11.006.

Macedo-Ribeiro, S., Almeida, C., Calisto, B. M., Friedrich, T., Mentele, R., Sturzebecher, J., Fuentes-Prior, P., & Pereira, P. J. (2008). Isolation, cloning and structural characterisation of boophilin, a multifunctional Kunitz-type proteinase inhibitor from the cattle tick. *PLoS One,* 3(2), e1624. doi: 10.1371/journal.pone.0001624.

Mans, B. J., Louw, A. I., & Neitz, A. W. (2002). Evolution of hematophagy in ticks: common origins for blood coagulation and platelet aggregation inhibitors from soft ticks of the genus Ornithodoros. *Mol Biol Evol,* 19(10), 1695-1705. doi: 10.1093/oxfordjournals. molbev.a003992.

Manzano-Roman, R., Diaz-Martin, V., Oleaga, A., & Perez-Sanchez, R. (2015). Identification of protective linear B-cell epitopes on the subolesin/akirin orthologues of Ornithodoros spp. soft ticks. *Vaccine,* 33(8), 1046-1055. doi: 10.1016/j.vaccine.2015.01.015.

Marrack, P., McCormack, J., & Kappler, J. (1989). Presentation of antigen, foreign major histocompatibility complex proteins and self by thymus cortical epithelium. *Nature,* 338(6215), 503-505. doi: 10.1038/338503a0.

McKenna, R. V., Riding, G. A., Jarmey, J. M., Pearson, R. D., & Willadsen, P. (1998). Vaccination of cattle against the Boophilus microplus using a mucin-like membrane glycoprotein. *Parasite Immunol,* 20(7), 325-336. doi: 10.1046/j.1365-3024.1998.00149.x.

N.I.A.I.D. *Vaccines*. Retrieved from https://www.niaid.nih.gov/research/vaccines.

Neville, A. C., Parry, D. A. D., & Woodhead-Galloway, J. (1976). The chitin crystallite in arthropod cuticle. *J Cell Sci,* 21, 73 – 82.

Ortega-Sanchez, R., Camacho-Nuez, M., Castaneda-Ortiz, E. J., Martinez-Benitez, M. B., Hernandez-Silva, D. J., Aguilar-Tipacamu, G., & Mosqueda, J. (2020). Vaccine efficacy of recombinant BmVDAC on *Rhipicephalus microplus* fed on Babesia bigemina-infected and uninfected cattle. *Vaccine*. doi: 10.1016/j.vaccine.2019.12.040.

Pappas, P. J., & Oliver, J. H., Jr. (1972). Reproduction in ticks (Acari: Ixodoidea). 2. Analysis of the stimulus for rapid and complete feeding of female Dermacentor variabilis (Say). *J Med Entomol,* 9(1), 47-50. doi: 10.1093/jmedent/9.1.47.

Parizi, L. F., Utiumi, K. U., Imamura, S., Onuma, M., Ohashi, K., Masuda, A., & da Silva Vaz, I., Jr. (2011). Cross immunity with Haemaphysalis longicornis glutathione S-transferase reduces an experimental *Rhipicephalus* (*Boophilus*) *microplus* infestation. *Exp Parasitol,* 127(1), 113-118. doi: 10.1016/j.exppara.2010.07.001.

Pashine, A., Valiante, N. M., & Ulmer, J. B. (2005). Targeting the innate immune response with improved vaccine adjuvants. *Nat Med,* 11(4 Suppl), S63-68. doi: 10.1038/nm1210.

Peconick, A. P., Sossai, S., Girao, F. A., Rodrigues, M. Q., Souza, E. S. C. H., Guzman, Q. F., Patarroyo, V. A., Vargas, M. I., & Patarroyo, J. H. (2008). Synthetic vaccine (SBm7462) against the cattle tick *Rhipicephalus* (*Boophilus*) *microplus*: preservation of immunogenic determinants in different strains from South America. *Exp Parasitol,* 119(1), 37-43. doi: 10.1016/j.exppara.2007.12.007.

Perez Di Giorgio, J., Soto, G., Alleva, K., Jozefkowicz, C., Amodeo, G., & Muschietti, J. P. (2014). Prediction of aquaporin function by integrating evolutionary and functional analyses. *J Membr Biol,* 247, 107 – 125.

Plotkin, S. L., & Plotkin, S. A. (2012). A short history of vaccination. In S. A. Plotkin & W. A. Orenstein (Eds.), *Vaccines* (4th ed., pp. 1-15). Philadelphia: WB Saunders.

Potempa, J., Korzus, E., & Travis, J. (1994). The serpin superfamily of proteinase inhibitors: Structure, function, and regulation. *J Biol Chem,* 269 15957–15960.

Ramabu, S. S., Ueti, M. W., Brayton, K. A., Baszler, T. V., & Palmer, G. H. (2010). Identification of Anaplasma marginale proteins specifically upregulated during colonization of the tick vector. *Infect Immun,* 78(7), 3047-3052. doi: 10.1128/IAI.00300-10.

Rand, K. N., Moore, T., Sriskantha, A., Spring, K., Tellam, R., Willadsen, P., & Cobon, G. S. (1989). Cloning and expression of a protective antigen from the cattle tick Boophilus microplus. *Proc Natl Acad Sci USA,* 86(24), 9657-9661. doi: 10.1073/pnas.86.24.9657.

Ranson, H., Prapanthadara, L. A., & Hemingway, J. (1997). Cloning and characterization of two glutathione S-transferases from a DDT resistant strain of Anopheles gambiae. *Biochem J,* 324, 97–102.

Ribeiro, J. M., Alarcon-Chaidez, F., Francischetti, I. M., Mans, B. J., Mather, T. N., Valenzuela, J. G., & Wikel, S. K. (2006). An annotated catalog of salivary gland transcripts from Ixodes scapularis ticks. *Insect Biochem Mol Biol,* 36(2), 111-129. doi: 10.1016/j.ibmb.2005.11.005.

Ribeiro, J. M., Anderson, J. M., Manoukis, N. C., Meng, Z., & Francischetti, I. M. (2011). A further insight into the sialome of the tropical bont tick, Amblyomma variegatum. *BMC Genomics,* 12, 136. doi: 10.1186/1471-2164-12-136.

Richardson, M. A., Smith, D. R., Kemp, D. H., & Tellam, R. L. (1993). Native and baculovirus-expressed forms of the immuno-protective protein BM86 from Boophilus microplus are anchored to the cell membrane by a glycosyl-phosphatidyl inositol linkage. *Insect Mol Biol,* 1(3), 139-147. doi: 10.1111/j.1365-2583.1993.tb00115.x.

Robinson, M. J., Osorio, F., Rosas, M., Freitas, R. P., Schweighoffer, E., Gross, O., Verbeek, J. S., Ruland, J., Tybulewicz, V., Brown, G. D., Moita, L. F. T., P. R., & C., R. e. S. (2009). Dectin-2 is a Syk-coupled pattern recognition receptor crucial for Th17 responses to fungal infection. *J Exp Med,* 206(9), 2037-2051. doi: 10.1084/jem.20082818.

Rodriguez-Hernandez, E., Mosqueda, J., Alvarez-Sanchez, M. E., Neri, A. F., Mendoza-Hernandez, G., & Camacho-Nuez, M. (2012). The identification of a VDAC-like protein involved in the interaction of Babesia bigemina sexual stages with *Rhipicephalus microplus* midgut cells. *Vet Parasitol,* 187(3-4), 538-541. doi: 10.1016/j.vetpar.2012.01. 028.

Rodriguez-Hernandez, E., Mosqueda, J., Leon-Avila, G., Castaneda-Ortiz, E. J., Alvarez-Sanchez, M. E., Camacho, A. D., Ramos, A., & Camacho-Nuez, M. (2015). BmVDAC upregulation in the midgut of *Rhipicephalus microplus*, during infection with Babesia bigemina. *Vet Parasitol,* 212(3-4), 368-374. doi: 10.1016/j.vetpar.2015.06.016.

Roman, I., Figys, J., & Zizi, M. (2006). Hunting interactomes of a membrane protein: obtaining the largest set of VDAC - interacting protein epitopes. *Mol Cell Proteomics,* 5, 1667 - 1680.

Sappington, T. W., & Raikhel, A. S. (1998). Molecular characteristics of insect vitellogenins and vitellogenin receptors. *Insect Biochem. Mol. Biol.,* 28(5-6), 277-300.

Schulze, A. J., Huber, R., Bode, W., & Engh, R. A. (1994). Structural aspects of serpin inhibition. *FEBS Lett,* 344(2-3), 117-124. doi: 10.1016/0014-5793(94)00369-6.

Scott, J. C., & McManus, D. P. (2000). Molecular cloning and enzymatic expression of the 28-kDa glutathione S-transferase of Schistosoma japonicum: evidence for sequence variation but lack of consistent vaccine efficacy in the murine host. *Parasitology International,* 49, 289–300.

Seixas, A., Oliveira, P., Termignoni, C., Logullo, C., Masuda, A., & da Silva Vaz, I., Jr. (2012). *Rhipicephalus* (*Boophilus*) *microplus* embryo proteins as target for tick vaccine. *Vet Immunol Immunopathol,* 148(1-2), 149-156. doi: 10.1016/j.vetimm.2011.05.011.

Shapiro-Shelef, M., & Calame, K. (2005). Regulation of plasma-cell development. *Nat Rev Immunol,* 5(3), 230-242. doi: 10.1038/nri1572.

Shepherd, J., Jr, J. H. O., & Hall, J. D. (1982). A polypeptide from male accessory glands which triggers maturation of tick spermatozoa. *International Journal of Invertebrate Reproduction,* 5(3), 129-137. doi: 10.1080/01651269.1982.10553462.

Sorgato, M. C., & Moran, O. (1993). Channels in mitochondrial membranes: knowns, uknowns, and prospects for the future. *Crit Rev Biochem Mol Biol,* 28., 127 - 171.

Sossai, S., Peconick, A. P., Sales-Junior, P. A., Marcelino, F. C., Vargas, M. I., Neves, E. S., & Patarroyo, J. H. (2005). Polymorphism of the bm86 gene in South American strains of the cattle tick Boophilus microplus. *Exp Appl Acarol,* 37(3-4), 199-214. doi: 10.1007/ s10493-005-3262-7.

Stuchin, M., Machalaba, C., & Karesh, W. (2016). Vector-borne diseases: Animals and patterns. In *Forum on Microbial Threats; Board on Global Health; Health and Medicine Division; National Academies of Sciences, Engineering, and Medicine. Global Health Impacts of Vector-Borne Diseases: Workshop Summary*. Washington, DC, USA: National Academies Press.

Sugino, M., Imamura, S., Mulenga, A., Nakajima, M., Tsuda, A., Ohashi, K., & Onuma, M. (2003). A serine proteinase inhibitor (serpin) from ixodid tick Haemaphysalis longicornis; cloning and preliminary assessment of its suitability as a candidate for a tick vaccine. *Vaccine,* 21(21-22), 2844-2851. doi: 10.1016/s0264-410x(03)00167-1.

Sultana, H., Patel, U., Sonenshine, D. E., & Neelakanta, G. (2015). Identification and comparative analysis of subolesin/akirin ortholog from Ornithodoros turicata ticks. *Parasit Vectors,* 8, 132. doi: 10.1186/s13071-015-0749-x.

Syvanen, M., Zhou, Z. H., Wang, J. Y., G. t., Genet., g. f. f. t. h. M. d. M. G., & 245. (1994). Glutathione transferase gene family from the housefly Musca domestica. *Mol Gen Genet,* 245, 25-31.

Syvanen, M., Zhou, Z. H., Wharton, J., Goldsbury, C., & Clark, A. (1996). Heterogeneity of the glutathione transferase genes encoding enzymes responsible for insecticide degradation in the housefly. *J Mol Evol,* 43, 236–240.

Taheri, M., Nabian, S., Ranjbar, M., Mazaheri Nezhad, R., Gerami Sadeghian, A., & Sazmand, A. (2014). Study of vitellogenin in Boophilus annulatus tick larvae and its immunological aspects. *Trop Biomed,* 31(3), 398-405.

Tellam, R. L., Kemp, D., Riding, G., Briscoe, S., Smith, D., Sharp, P., Irving, D., & Willadsen, P. (2002). Reduced oviposition of Boophilus microplus feeding on sheep vaccinated with vitellin. *Vet Parasitol,* 103(1-2), 141-156. doi: 10.1016/s0304-4017(01) 00573-8.

The UniProt Consortium. (2016). UniProt: the universal protein knowledgebase. *Nucleic Acids Research,* 45(D1), D158-D169. doi: 10.1093/nar/gkw1099.

Tirloni, L., Kim, T. K., Berger, M., Termignoni, C., da Silva Vaz, I., Jr., & Mulenga, A. (2019). Amblyomma americanum serpin 27 (AAS27) is a tick salivary anti-inflammatory protein secreted into the host during feeding. *PLoS Negl Trop Dis,* 13(8), e0007660. doi: 10.1371/journal.pntd.0007660.

van de Locht, A., Stubbs, M. T., Bode, W., Friedrich, T., Bollschweiler, C., Hoffken, W., & Huber, R. (1996). The ornithodorin-thrombin crystal structure, a key to the TAP enigma? *EMBO J,* 15(22), 6011-6017.

Veerapathran, A., Dakshinamoorthy, G., Gnanasekar, M., Reddy, M. V. R., & Kalyanasundaram, R. (2009). Evaluation of Wuchereria bancrofti GST as a Vaccine Candidate for Lymphatic Filariasis *PLoS Negl Trop Dis,* 3(6), e457.

Vetter, V., Denizer, G., Friedland, L. R., Krishnan, J., & Shapiro, M. (2018). Understanding modern-day vaccines: what you need to know. *Ann Med,* 50(2), 110-120. doi: 10.1080/ 07853890.2017.1407035.

Vibanco-Perez, N., Jimenez, L., Merchant, M. T., & Landa, A. (1999). Characterization of glutathione S-transferase of Taenia solium. *Journal of Parasitology,* 85, 448–453.

Weiss, B. L., & Kaufman, W. R. (2004). Two feeding-induced proteins from the male gonad trigger engorgement of the female tick Amblyomma hebraeum. *Proc Natl Acad Sci U S A,* 101(16), 5874-5879. doi: 10.1073/pnas.0307529101.

Wigglesworth, V. B. (1943). The fate of haemoglobin in Rhodnius prolixus (Hemiptera) and other blood-sucking arthropods. *Proc R Soc Lond [Biol],* 131, 313-339.

Willadsen, P. (1990). Perspectives for subunit vaccines for the control of ticks. *Parassitologia,* 32(1), 195-200.

Willadsen, P., Bird, P., Cobon, G. S., & Hungerford, J. (1995). Commercialisation of a recombinant vaccine against Boophilus microplus. *Parasitology,* 110 Suppl, S43-50. doi: 10.1017/s0031182000001487.

Willadsen, P., McKenna, R. V., & Riding, G. A. (1988). Isolation from the cattle tick, Boophilus microplus, of antigenic material capable of eliciting a protective immunological response in the bovine host. *Int J Parasitol,* 18(2), 183-189. doi: 10.1016/0020-7519(88)90059-8.

Willadsen, P., Riding, G. A., McKenna, R. V., Kemp, D. H., Tellam, R. L., Nielsen, J. N., Lahnstein, J., Cobon, G. S., & Gough, J. M. (1989). Immunologic control of a parasitic arthropod. Identification of a protective antigen from Boophilus microplus. *J Immunol,* 143(4), 1346-1351.

Yamada, S., Konnai, S., Imamura, S., Ito, T., Onuma, M., & Ohashi, K. (2009). Cloning and characterization of *Rhipicephalus* appendiculatus voraxinalpha and its effect as anti-tick vaccine. *Vaccine,* 27(43), 5989-5997. doi: 10.1016/j.vaccine.2009.07.072.

You, M., & Fujisaki, K. (2009). Vaccination effects of recombinant chitinase protein from the hard tick Haemaphysalis longicornis (Acari: Ixodidae). *J Vet Med Sci,* 71(6), 709-712. doi: 10.1292/jvms.71.709.

You, M., Xuan, X., Tsuji, N., Kamio, T., Taylor, D., Suzuki, N., & Fujisaki, K. (2003). Identification and molecular characterization of a chitinase from the hard tick Haemaphysalis longicornis. *J Biol Chem,* 278(10), 8556-8563. doi: 10.1074/jbc.M20683 1200.

Young, M., Bay, D., Hausner, G., & Court, D. A. (2007). The evolutionary history of mitochondrial porins. *BMC Evol Biol,* 7, 1 - 21.

Zhan, B., Liu, S., Perally, S., Xue, J., Fujiwara, R., Brophy, P., Xiao, S. H., Liu, Y. Y., Feng, J. J., Williamson, A., Wang, Y., Bueno, L. L., Mendez, S., Goud, G., Bethony, J. M., Hawdon, J. M., Loukas, A., Jones, K., & Hotez, P. J. (2005). Biochemical characterization and vaccine potential of a heme-binding glutathione transferase from the adult hookworm Ancylostoma caninum. *Infection and Immunity,* 73, 6903–6911.

Zivkovic, Z., Blouin, E. F., Manzano-Roman, R., Almazan, C., Naranjo, V., Massung, R. F., Jongejan, F., Kocan, K. M., & de la Fuente, J. (2009). Anaplasma phagocytophilum and Anaplasma marginale elicit different gene expression responses in cultured tick cells. *Comp Funct Genomics*, 705034. doi: 10.1155/2009/705034.

Zivkovic, Z., Esteves, E., Almazan, C., Daffre, S., Nijhof, A. M., Kocan, K. M., Jongejan, F., & de la Fuente, J. (2010). Differential expression of genes in salivary glands of male *Rhipicephalus* (*Boophilus*) *microplus* in response to infection with Anaplasma marginale. *BMC Genomics,* 11, 186. doi:10.1186/1471-2164-11-186.

Zong, N. C., Li, H., Li, H., Lam, M. P. Y., Jimenez, R. C., Kim, C. S., Deng, N., Kim, A. K., Choi, J. H., Zelaya, I., Liem, D., Meyer, D., Odeberg, J., Fang, C., Lu, H., Xu, T., Weiss, J., Duan, H., Uhlen, M., YatesIII, J. R., Apweiler, R., Ge, J., Hermjakob, H., & Pin, P. (2013). Integration of Cardiac Proteome Biology and Medicine by a Specialized Knowledgebase. *Circulation Research,* 113(9), 1043-1053.

INDEX

A

B

C

D

E

F

G

H

I

J

K

L

M

N

O

P

Q

R

S

T

U

V

W

Z